Collins

FURTHER MATHS

Practice Book
for the AQA Level 2 Certificate

Trevor Senior

Introduction

Welcome to Collins Further Maths Practice Book. This book will help you to progress with ease through more challenging maths work, with detailed examples and plenty of practice in the key areas needed to succeed in maths at a higher level.

This book is suitable to support the AQA Level 2 Certificate in Further Mathematics.

KEY POINTS AND WORKED EXAMPLES

Remember the important points of each topic with key words and succinct topic explanations. Gain a greater understanding of each area by studying the detailed worked examples before each exercise.

GRADED EXERCISES

Consolidate your knowledge of topics and know exactly what level you are working at, with hundreds of graded practice questions. Stretch and challenge yourself with questions ranging from Grades C to A* with Distinction.

HINTS AND TIPS

Find valuable hints and tips boxes throughout the book that give you handy methods to try when answering questions.

EXAM PREPARATION

Ensure you are ready for your exams by tackling the exam-style questions at the end of each chapter.

ANSWERS

All answers to the practice exercises and exam-style questions in this book can be found at the back of the book.

Contents

1 Number recall ... 4
1.1 Number: Recall and extension ... 4
1.2 Manipulation of surds ... 7
2 Algebra recall ... 12
2.1 Recall of basic algebra ... 12
2.2 Expanding brackets and collecting like terms ... 16
2.3 Factorising ... 24
3 Geometry recall 1 ... 32
3.1 Perimeter of compound shapes ... 32
3.2 Area of basic shapes ... 33
3.3 Circumference and area of a circle ... 35
3.4 Volume of a cube, cuboid, prism and pyramid ... 37
3.5 Volume of a cone and a sphere ... 42
4 Geometry recall 2 ... 46
4.1 Special triangles and quadrilaterals ... 46
4.2 Angles in polygons ... 49
4.3 Circle theorems ... 53
4.4 Cyclic quadrilaterals ... 56
4.5 Tangents and chords ... 58
4.6 Alternate segment theorem ... 61
5 Functions ... 64
5.1 Function notation ... 64
5.2 Domain and range ... 66
5.3 Sketching graphs of linear and quadratic functions ... 68
5.4 The significant points of a quadratic graph ... 73
6 Matrices ... 80
6.1 Introduction to matrices ... 80
6.2 The zero matrix and the identity matrix ... 83
6.3 Transformations ... 85
6.4 Combinations of transformations ... 88
7 Algebra ... 90
7.1 Manipulation of rational expressions ... 90
7.2 Use and manipulation of formulae and expressions ... 93
7.3 The factor theorem ... 95
8 Sequences ... 98
8.1 Number sequences ... 98
8.2 The *n*th term of a sequence ... 99
8.3 The *n*th term of a linear sequence ... 100
8.4 The *n*th term of a quadratic sequence ... 102
8.5 The limiting value of a sequence as $n \to \infty$ ... 104
9 Pythagoras' theorem and trigonometry ... 106
9.1 Pythagoras' theorem ... 106
9.2 Trigonometry in right-angled triangles ... 113
9.3 The sine rule and the cosine rule ... 126
10 Solving equations ... 134
10.1 Solving linear equations ... 134
10.2 Setting up equations ... 137
10.3 Solving quadratic equations by factorisation or the quadratic formula ... 140
10.4 Solving quadratic equations by completing the square ... 146
11 Simultaneous equations ... 150
11.1 Simultaneous linear equations ... 150
11.2 Linear and non-linear simultaneous equations ... 155
11.3 Using graphs to solve simultaneous linear equations ... 157
11.4 Using graphs to solve simultaneous equations, one linear and one non-linear ... 159
11.5 Solving equations by the method of intersection ... 161
12 Inequalities ... 168
12.1 Linear inequalities ... 168
12.2 Quadratic inequalities ... 171
12.3 Graphical inequalities ... 173
13 Coordinate geometry ... 180
13.1 Cartesian grids and straight-line graphs ... 180
13.2 The equation of a straight line in the forms $y = mx + c$ and $y - y_1 = m(x - x_1)$ ... 187
13.3 Parallel and perpendicular lines ... 194
13.4 Applications of coordinate geometry ... 198
14 The equation of a circle ... 202
14.1 The equation of a circle centred on the origin (0, 0) ... 202
14.2 The equation of a circle centred on any point (a, b) ... 204
15 Indices ... 208
15.1 Using indices ... 208
15.2 Rules of indices ... 210
15.3 Negative indices ... 211
15.4 Fractional indices ... 213
15.5 Solving equations with indices ... 217
16 Calculus ... 220
16.1 The gradient of a curve ... 220
16.2 More complex curves ... 223
16.3 Stationary points and curve sketching ... 225
16.4 The equation of a tangent and normal at any point on a curve ... 230
17 Ratios of angles and their graphs ... 234
17.1 Trigonometric ratios of angles between 90° and 360° ... 234
17.2 The circular function graphs ... 239
17.3 Special right-angled triangles ... 242
17.4 Trigonometrical expressions and equations ... 244
18 Proof ... 248
18.1 Algebraic proof ... 248
18.2 Geometric proof ... 251
Answers ... 255

1 Number recall

1.1 Number: Recall and extension

THIS SECTION WILL SHOW YOU HOW TO ...

✓ understand and use the correct hierarchy of operations (BIDMAS / BODMAS)
✓ understand and use decimals, fractions and percentages
✓ understand and use ratio and proportion
✓ understand and use numbers in index form and standard form
✓ understand rounding and give answers to an appropriate degree of accuracy

KEY WORDS

✓ ratio
✓ proportion
✓ standard form

Most of the calculations in this chapter will require the combination of **at least two** basic number skills. For example, questions involving fractions will also test order of operations, a conversion of units may have several steps and a ratio problem may be combined with a percentage problem.

Questions may be answered with or without a calculator. If you use a non-calculator method, it is good practice to use one to check answers.

EXAMPLE 1

Work out $5\frac{1}{2}+3\frac{1}{4}\times1\frac{2}{3}$

$3\frac{1}{4}\times1\frac{2}{3}=\frac{13}{4}\times\frac{5}{3}=\frac{65}{12}=5\frac{5}{12}$

$5\frac{1}{2}+5\frac{5}{12}=10\frac{11}{12}\Rightarrow5\frac{1}{2}+3\frac{1}{4}\times1\frac{2}{3}=10\frac{11}{12}$

HINTS AND TIPS

Remember from BIDMAS / BODMAS that multiplication is carried out before addition.

EXAMPLE 2

Work out $\frac{6.5+6.7}{8.9-2.3}$

Put in brackets. $\frac{(6.5+6.7)}{(8.9-2.3)}$

Working out the brackets first gives $\frac{13.2}{6.6}$ or $\frac{132}{66}=2$

HINTS AND TIPS

It is sensible to insert brackets before carrying out the calculation.

EXAMPLE 3

For every 3 boys in a class, there are 5 girls. There are 6 more girls than boys. How many students are there in the class altogether?

The ratio of boys to girls is 3 : 5, giving 8 'parts'.
So the difference is $5 - 3 = 2$ parts, which is equivalent to 6 students.

1 part is equivalent to 3 students.
There are 9 boys and 15 girls in the class, giving 24 students altogether.

EXAMPLE 4

I increase the amount I save each week by the same amount.
In week 1, I save £20. In week 7, I save £65.

How much have I saved in the 7 weeks?

In the 7-week period there are 6 weeks when the amount increases.
The increase is £65 – £20 = £45

£45 ÷ 6 = £7.50
So Week 1 = £20 Week 2 = £27.50 Week 3 = £35, and so on.
The total saved in the seven weeks
= £20 + £27.50 + £35 + £42.50 + £50 + £57.50 + £65 = £342.50

EXAMPLE 5

Calculate $4^3 - 8^2 \div 2^{10}$.
Give your answer as a mixed number.

$$8^2 \div 2^{10} = (2^3)^2 \div 2^{10} = 2^6 \div 2^{10} = 2^{-4} = \frac{1}{2^4} = \frac{1}{16}$$

$$4^3 - \frac{1}{16} = 63\frac{15}{16}$$

HINTS AND TIPS

Remember from BIDMAS / BODMAS that division is carried out before subtraction.

EXERCISE 1A

GRADE C

1. Work out each of these. Give your answers as mixed numbers.

a) $4\frac{1}{2} + 2\frac{1}{3}$ **b)** $3\frac{3}{5} + 4\frac{2}{3}$ **c)** $5\frac{2}{7} - 1\frac{5}{6}$

d) $7\frac{1}{4} - 2\frac{3}{8}$ **e)** $2\frac{3}{8} + 1\frac{2}{5} - \frac{2}{3}$ **f)** $5\frac{4}{9} + 4\frac{3}{4} - 7\frac{1}{2}$

HINTS AND TIPS

Use a calculator to check your answers.

GRADE B

2. Work out each of these. Give your answers as mixed numbers.

a) $4\frac{1}{2} \times 2\frac{1}{3}$ **b)** $5\frac{2}{7} \div 1\frac{5}{6}$ **c)** $7\frac{1}{4} \div 2\frac{3}{8}$

d) $2\frac{3}{8} + 1\frac{2}{5} \times \frac{2}{3}$ **e)** $5\frac{4}{9} + 4\frac{3}{4} \div 7\frac{1}{2}$ **f)** $2\frac{2}{3} - 1\frac{1}{4} \div 2\frac{1}{2}$

g) $\left(3\frac{3}{4} + 3\frac{1}{5}\right) \div \left(\frac{2}{3} - \frac{1}{2}\right)$ **h)** $\dfrac{\frac{1}{2}+\frac{1}{6}}{\frac{1}{2}-\frac{1}{6}}$ **i)** $\dfrac{\left(\frac{1}{3}\right)^2}{1-\frac{1}{4}}$

HINTS AND TIPS

Use a calculator to check your answers.

3. The ratio of $A : B = 5 : 4$ and the ratio of $B : C = 3 : 7$.

Work out the ratio of $A : C$. Give your answer in its simplest form.

4. z is 10% more than y. y is 20% more than x.

By what percentage is z more than x?

5. At a show, two-fifths of the people are men and the rest are women. 30% of the men are over 60 years old. 40% of the women are over 60 years old.

What proportion of all the people at the show are over 60?

6. Work out the number of minutes in January.

Give your answer in standard form.

7. In a school, there are equal numbers of boys and girls. At the end of the school year, 20% of the boys leave and 25% of the girls leave. There are now 450 girls in the school.

Work out the number of boys now in the school.

8. How many times is 2.56×10^{27} bigger than 5×10^{5}?

Give your answer in standard form.

9. A fuel tank is three-quarters full . Then 10% of the fuel is taken out. There are now 54 litres in the tank.

How much will the tank hold when full?

10. The price of a loaf is increased by 14%. This week a shop sells 12% fewer loaves than last week.

Does the shop take more or less money from selling loaves this week? You **must** show your working.

11. These three numbers are written in standard form.

2.3×10^{7} 8.1×10^{5} 5.7×10^{6}

a) Work out the smallest answer when two of these numbers are multiplied together.

b) Work out the largest answer when two of these numbers are multiplied together.
Give your answers in standard form.

12. At the start of the year, a train is half full. Each week the number of people catching this train increases by 1%.

How many weeks will it be before the train is full?

1.2 Manipulation of surds

THIS SECTION WILL SHOW YOU HOW TO ...

✓ simplify expressions by manipulating surds
✓ expand brackets which contain surds
✓ rationalise the denominator, including denominators in the form $a\sqrt{b} + c\sqrt{d}$ where a, b, c and d are integers

KEY WORDS

✓ surd ✓ exact value

A **surd** is a root of a rational number, for example $\sqrt{5}$, $\sqrt{6}$.

Surds are also called **exact values**.

Here are some rules of surds.

$\sqrt{a} \times \sqrt{a} = a$ $\sqrt{a} \times \sqrt{b} = \sqrt{ab}$ $\sqrt{a} \div \sqrt{b} = \sqrt{\frac{a}{b}}$

$a\sqrt{b} \times c\sqrt{d} = ac\sqrt{bd}$ $a\sqrt{b} \div c\sqrt{d} = \frac{a}{c}\sqrt{\frac{b}{d}}$

EXAMPLE 6

Simplify $\sqrt{6} \times \sqrt{2}$

$\sqrt{6} \times \sqrt{2} = \sqrt{12} = \sqrt{4 \times 3} = \sqrt{4} \times \sqrt{3} = 2\sqrt{3}$

EXERCISE 1B

GRADE A

1. Simplify each of the following. Leave your answers in surd form if necessary.

a) $\sqrt{2} \times \sqrt{3}$ **b)** $\sqrt{5} \times \sqrt{3}$ **c)** $\sqrt{2} \times \sqrt{2}$ **d)** $\sqrt{2} \times \sqrt{8}$
e) $\sqrt{2} \times \sqrt{7}$ **f)** $\sqrt{2} \times \sqrt{18}$ **g)** $\sqrt{6} \times \sqrt{6}$ **h)** $\sqrt{5} \times \sqrt{6}$

2. Simplify each of the following. Leave your answers in surd form if necessary.

a) $\sqrt{12} \div \sqrt{3}$ **b)** $\sqrt{15} \div \sqrt{3}$ **c)** $\sqrt{12} \div \sqrt{2}$ **d)** $\sqrt{24} \div \sqrt{8}$
e) $\sqrt{28} \div \sqrt{7}$ **f)** $\sqrt{48} \div \sqrt{8}$ **g)** $\sqrt{6} \div \sqrt{6}$ **h)** $\sqrt{54} \div \sqrt{6}$

3. Simplify each of the following. Leave your answers in surd form if necessary.

a) $\sqrt{2} \times \sqrt{3} \times \sqrt{2}$ **b)** $\sqrt{2} \times \sqrt{2} \times \sqrt{8}$ **c)** $\sqrt{5} \times \sqrt{8} \times \sqrt{8}$
d) $\sqrt{6} \times \sqrt{2} \times \sqrt{48}$ **e)** $\sqrt{2} \times \sqrt{7} \times \sqrt{2}$ **f)** $\sqrt{6} \times \sqrt{6} \times \sqrt{3}$

4. Simplify each of the following. Leave your answers in surd form.

a) $\sqrt{2} \times \sqrt{3} \div \sqrt{2}$ **b)** $\sqrt{5} \times \sqrt{3} \div \sqrt{15}$ **c)** $\sqrt{32} \times \sqrt{2} \div \sqrt{8}$
d) $\sqrt{5} \times \sqrt{8} \div \sqrt{8}$ **e)** $\sqrt{3} \times \sqrt{3} \div \sqrt{3}$ **f)** $\sqrt{8} \times \sqrt{12} \div \sqrt{48}$
g) $\sqrt{2} \times \sqrt{7} \div \sqrt{2}$ **h)** $\sqrt{2} \times \sqrt{18} \div \sqrt{3}$ **i)** $\sqrt{5} \times \sqrt{6} \div \sqrt{30}$

5. Simplify each of these expressions.

a) $\sqrt{a}\times\sqrt{a}$ **b)** $\sqrt{a}\div\sqrt{a}$ **c)** $\sqrt{a}\times\sqrt{a}\div\sqrt{a}$

6. Simplify each of the following surds into the form $a\sqrt{b}$.

a) $\sqrt{18}$ **b)** $\sqrt{24}$ **c)** $\sqrt{12}$ **d)** $\sqrt{50}$ **e)** $\sqrt{8}$ **f)** $\sqrt{27}$
g) $\sqrt{32}$ **h)** $\sqrt{200}$ **i)** $\sqrt{1000}$ **j)** $\sqrt{250}$ **k)** $\sqrt{98}$ **l)** $\sqrt{243}$

7. Simplify each of these.

a) $2\sqrt{18}\times3\sqrt{2}$ **b)** $4\sqrt{24}\times2\sqrt{5}$ **c)** $3\sqrt{12}\times3\sqrt{3}$ **d)** $2\sqrt{8}\times2\sqrt{8}$
e) $2\sqrt{27}\times4\sqrt{8}$ **f)** $2\sqrt{48}\times3\sqrt{8}$ **g)** $2\sqrt{45}\times3\sqrt{3}$ **h)** $2\sqrt{63}\times2\sqrt{7}$

8. Simplify each of these.

a) $4\sqrt{2}\times5\sqrt{3}$ **b)** $4\sqrt{2}\times3\sqrt{2}$ **c)** $2\sqrt{2}\times2\sqrt{8}$ **d)** $3\sqrt{3}\times2\sqrt{3}$
e) $5\sqrt{7}\times2\sqrt{3}$ **f)** $2\sqrt{2}\times3\sqrt{7}$ **g)** $2\sqrt{2}\times3\sqrt{18}$ **h)** $4\sqrt{5}\times3\sqrt{6}$

9. Simplify each of these.

a) $6\sqrt{12}\div2\sqrt{3}$ **b)** $3\sqrt{15}\div\sqrt{3}$ **c)** $6\sqrt{12}\div\sqrt{2}$ **d)** $4\sqrt{24}\div2\sqrt{8}$
e) $9\sqrt{28}\div3\sqrt{7}$ **f)** $12\sqrt{56}\div6\sqrt{8}$ **g)** $25\sqrt{6}\div5\sqrt{6}$ **h)** $32\sqrt{54}\div4\sqrt{6}$

GRADE A*

10. Simplify each of these.

a) $4\sqrt{2}\times\sqrt{3}\div2\sqrt{2}$ **b)** $4\sqrt{5}\times\sqrt{3}\div\sqrt{15}$ **c)** $2\sqrt{32}\times3\sqrt{2}\div2\sqrt{8}$
d) $6\sqrt{2}\times2\sqrt{8}\div3\sqrt{8}$ **e)** $3\sqrt{8}\times3\sqrt{12}\div3\sqrt{48}$ **f)** $4\sqrt{7}\times2\sqrt{3}\div8\sqrt{3}$
g) $8\sqrt{2}\times2\sqrt{18}\div4\sqrt{3}$ **h)** $5\sqrt{6}\times5\sqrt{6}\div5\sqrt{3}$ **i)** $2\sqrt{5}\times3\sqrt{6}\div\sqrt{30}$

11. Simplify each of these expressions.

a) $a\sqrt{b}\times c\sqrt{b}$ **b)** $a\sqrt{b}\div c\sqrt{b}$ **c)** $a\sqrt{b}\times c\sqrt{b}\div a\sqrt{b}$

12. Find the value of a that makes each of these surds true.

a) $\sqrt{5}\times\sqrt{a}=10$ **b)** $\sqrt{6}\times\sqrt{a}=12$ **c)** $\sqrt{10}\times2\sqrt{a}=20$
d) $2\sqrt{6}\times3\sqrt{a}=72$ **e)** $2\sqrt{a}\times\sqrt{a}=6$ **f)** $3\sqrt{a}\times3\sqrt{a}=54$

13. Simplify the following.

a) $\left(\frac{\sqrt{3}}{2}\right)^2$ **b)** $\left(\frac{5}{\sqrt{3}}\right)^2$ **c)** $\left(\frac{\sqrt{5}}{4}\right)^2$ **d)** $\left(\frac{6}{\sqrt{3}}\right)^2$ **e)** $\left(\frac{\sqrt{8}}{2}\right)^2$

14. Decide whether each statement is true or false. Show your working.

a) $\sqrt{(a+b)}=\sqrt{a}+\sqrt{b}$ **b)** $\sqrt{(a-b)}=\sqrt{a}-\sqrt{b}$

15. Write down a product of two different surds which has an integer answer.

16. **a)** Write $\sqrt{8}+\sqrt{18}$ in the form $a\sqrt{2}$.
b) Write $\sqrt{48}-\sqrt{12}$ in the form $b\sqrt{3}$.
c) Write $\sqrt{50}+\sqrt{72}+\sqrt{18}$ in the form $a\sqrt{2}$.
d) Write $\sqrt{12}+\sqrt{75}-\sqrt{3}$ in the form $b\sqrt{c}$.
e) Write $\sqrt{20}+\sqrt{180}-\sqrt{125}$ in the form $d\sqrt{5}$.
f) Write $4\sqrt{200}-\sqrt{162}-8\sqrt{2}$ in the form $e\sqrt{2}$.

Rationalising a denominator

To rationalise a denominator of the form $a\sqrt{b} + c\sqrt{d}$, multiply the numerator and denominator by $a\sqrt{b} - c\sqrt{d}$.

EXAMPLE 7

Rationalise the denominator and simplify $\frac{7}{3-4\sqrt{2}}$.

Multiply the numerator and denominator by $3 + 4\sqrt{2}$ to give:

$$\frac{7}{3-4\sqrt{2}} = \frac{7}{(3-4\sqrt{2})} \times \frac{(3+4\sqrt{2})}{(3+4\sqrt{2})} = \frac{7(3+4\sqrt{2})}{9-32} = -\frac{7}{23}(3+4\sqrt{2})$$

EXAMPLE 8

$\frac{21}{5+x\sqrt{2}}$ simplifies to $15 - y\sqrt{2}$ where $x > 0$.

Work out the values of x and y.

METHOD 1

Multiplying the denominator by $5 - x\sqrt{2}$ gives:

$$\frac{21}{5+x\sqrt{2}} = \frac{21}{(5+x\sqrt{2})} \times \frac{(5-x\sqrt{2})}{(5-x\sqrt{2})} = \frac{21(5-x\sqrt{2})}{25-2x^2} \Rightarrow \frac{21 \times 5}{25-2x^2} = 15$$

$7 = 25 - 2x^2 \Rightarrow x = 3$

So $\frac{21(5-x\sqrt{2})}{25-2x^2} = 15 - 9\sqrt{2}$ giving $y = 9$.

METHOD 2

Alternatively if $\frac{21}{5+x\sqrt{2}} \equiv 15 - y\sqrt{2}$

$21 \equiv (15 - y\sqrt{2})(5 + x\sqrt{2}) \Rightarrow 21 \equiv 75 + 15x\sqrt{2} - 5y\sqrt{2} - 2xy$

Equating surds gives $15x - 5y = 0 \Rightarrow y = 3x$.

Equating rational parts gives $21 = 75 - 2xy$.

$21 = 75 - 6x^2 \Rightarrow 6x^2 = 54 \Rightarrow x^2 = 9 \Rightarrow x = 3$

$y = 3x \Rightarrow y = 9$

EXERCISE 1C

GRADE A*

1. Show that:

a) $(2 + \sqrt{3})(1 + \sqrt{3}) = 5 + 3\sqrt{3}$

b) $(1 + \sqrt{2})(2 + \sqrt{3}) = 2 + 2\sqrt{2} + \sqrt{3} + \sqrt{6}$

c) $(4 - \sqrt{3})(4 + \sqrt{3}) = 13$

2. Expand and simplify where possible.

a) $\sqrt{3}(2-\sqrt{3})$ **b)** $\sqrt{2}(3-4\sqrt{2})$ **c)** $\sqrt{5}(2\sqrt{5}+4)$

d) $3\sqrt{7}(4-2\sqrt{7})$ **e)** $3\sqrt{2}(5-2\sqrt{8})$ **f)** $\sqrt{3}(\sqrt{27}-1)$

3. Expand and simplify where possible.

a) $(1+\sqrt{3})(3-\sqrt{3})$ **b)** $(2+\sqrt{5})(3-\sqrt{5})$ **c)** $(1-\sqrt{2})(3+2\sqrt{2})$

d) $(3-2\sqrt{7})(4+3\sqrt{7})$ **e)** $(2-3\sqrt{5})(2+3\sqrt{5})$ **f)** $(\sqrt{3}+\sqrt{2})(\sqrt{3}+\sqrt{8})$

g) $(2+\sqrt{5})^2$ **h)** $(1-\sqrt{2})^2$ **i)** $(3+\sqrt{2})^2$

4. Work out the missing lengths in each of these triangles, giving the answer in as simple a form as possible.

a)

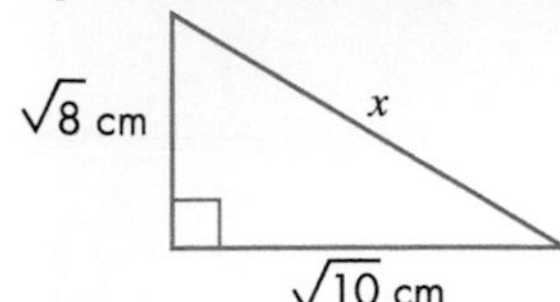

b)

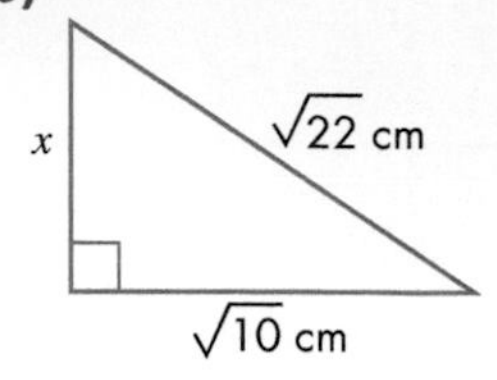

c)

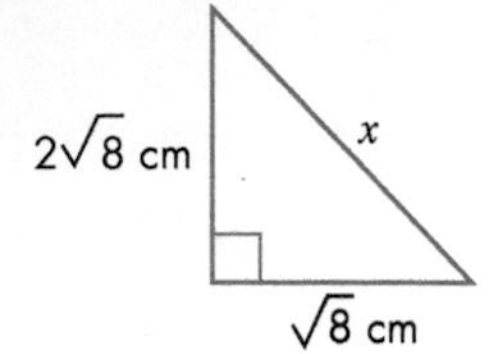

5. Calculate the area of each of these rectangles, simplifying your answers where possible. (The area of a rectangle with length l and width w is $A = l \times w$.)

a)

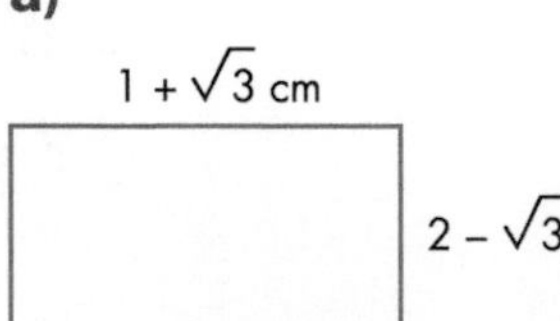

b)

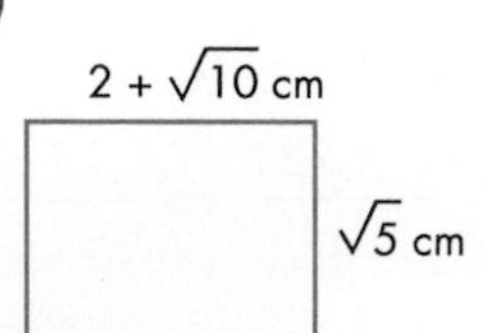

c)

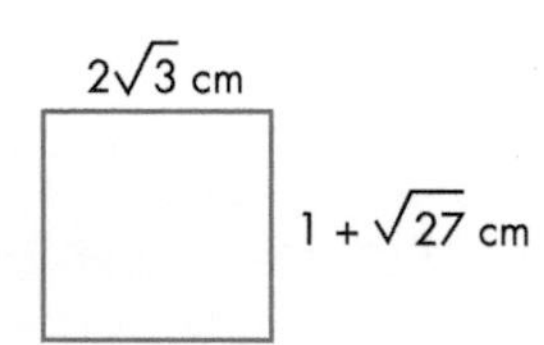

6. Rationalise the denominators of these expressions.

a) $\frac{1}{\sqrt{3}}$ **b)** $\frac{1}{\sqrt{2}}$ **c)** $\frac{1}{\sqrt{5}}$ **d)** $\frac{1}{2\sqrt{3}}$ **e)** $\frac{3}{\sqrt{3}}$ **f)** $\frac{5}{\sqrt{2}}$

g) $\frac{3\sqrt{2}}{\sqrt{8}}$ **h)** $\frac{5\sqrt{3}}{\sqrt{6}}$ **i)** $\frac{\sqrt{7}}{\sqrt{3}}$ **j)** $\frac{1+\sqrt{2}}{\sqrt{2}}$ **k)** $\frac{2-\sqrt{3}}{\sqrt{3}}$ **l)** $\frac{5+2\sqrt{3}}{\sqrt{3}}$

7. a) Expand and simplify the following.

i) $(2+\sqrt{3})(2-\sqrt{3})$ **ii)** $(1-\sqrt{5})(1+\sqrt{5})$ **iii)** $(\sqrt{3}-1)(\sqrt{3}+1)$

iv) $(3\sqrt{2}+1)(3\sqrt{2}-1)$ **v)** $(2-4\sqrt{3})(2+4\sqrt{3})$

b) What happens in the answers to part **a**? Why?

8. An engineer uses a formula to work out the number of metres of cable he needs to complete a job. His calculator displays the answer as $10\sqrt{70}$. The button for converting this to a decimal is not working.

He has 80 metres of cable. Without using a calculator, decide whether he has enough cable. Show clearly how you decide.

9. Write $(3+\sqrt{2})^2-(1-\sqrt{8})^2$ in the form $a+b\sqrt{c}$ where a, b and c are integers.

10. $x^2-y^2 \equiv (x+y)(x-y)$ is an identity which means it is true for any values of x and y whether they are numeric or algebraic.

Show that it is true for $x = 1+\sqrt{2}$ and $y = 1-\sqrt{8}$

11. Rationalise the denominator and simplify in each part.

a) $\frac{1}{3\sqrt{2}+1}$ **b)** $\frac{5}{3-\sqrt{3}}$ **c)** $\frac{\sqrt{2}}{7+2\sqrt{2}}$ **d)** $\frac{3+\sqrt{2}}{3-\sqrt{2}}$

e) $\frac{3\sqrt{5}+1}{\sqrt{5}+2}$ **f)** $\frac{2+\sqrt{3}}{5-\sqrt{3}}$ **g)** $\frac{\sqrt{2}+\sqrt{3}}{\sqrt{2}-\sqrt{3}}$ **h)** $\frac{5\sqrt{3}+\sqrt{2}}{2\sqrt{3}-\sqrt{2}}$

GRADE A**

12. Work out the values of x and y such that:

a) $(\sqrt{3}+5)(\sqrt{3}+4)=x+y\sqrt{3}$ **b)** $(4\sqrt{2})^3+(6\sqrt{3})^3=x\sqrt{2}+y\sqrt{3}$

c) $\frac{2}{\sqrt{5}+1}=x\sqrt{5}+y$ **d)** $\frac{3}{\sqrt{5}+x}=3\sqrt{5}+y.$

Exam-style questions

Do not use a calculator for questions 1, 2 or 5.

1. $2\frac{1}{2}+1\frac{2}{3}=3\frac{3}{4}\times k$

Write k as a mixed number. Give your answer as simply as possible. GRADE B

2. Light travels at 3.00×10^8 m/s. GRADE B

a) How many kilometres does light travel in one hour? Give your answer in standard form.

b) How long does light take to travel one metre? Give your answer in standard form correct to 3 significant figures.

3. The number of boys in a sports club increased by 35%.

The number of girls in the sports club decreased by 10%.

There are now equal numbers of boys and girl in the club.

What was the initial ratio of boys : girls? GRADE A

4. Over a year the price of a litre of petrol has increased from 122.9p to 136.8p.

The price of a litre of diesel has increased in the same proportion and is now 155.3p.

Find the price of diesel a year ago. GRADE A

5. $x=2-\sqrt{5}$ and $y=4-\sqrt{5}$

Write in the form $p+q\sqrt{5}$ where p and q are rational numbers:

a) xy **b)** $\frac{x}{y}$ GRADE A*

2 Algebra recall

2.1 Recall of basic algebra

THIS SECTION WILL SHOW YOU HOW TO ...

✓ understand and use commutative, associative and distributive laws
✓ understand and use the correct hierarchy of operations (BIDMAS / BODMAS)
✓ recall and apply knowledge of the basic processes of algebra, extending to more complex expressions, equations, formulae and identities

KEY WORDS

✓ brackets
✓ coefficient
✓ constant term
✓ equation
✓ expand
✓ expand and simplify
✓ expression
✓ factor
✓ factorise
✓ formula
✓ identity
✓ simplify
✓ substitute
✓ terms
✓ variable

Here are some words used in algebra that you need to know.

Variable: This is what the letters used to represent numbers are called. They can take on any value so they 'vary'.

Coefficient: This is the number in front of a letter, so in $2x$ the coefficient of x is 2.

Expression: This is any combination of letters and numbers. For example, $2x + 4y$ and $\frac{p-6}{5}$ are expressions.

Equations: These contain an equals sign and at least one variable. For example, $2x + 9 = 9$ and $2x + 3y = 15$ are equations. This is called solving the equation.

Formula: You may already have seen many formulae (the plural of formula). These are like equations in that they contain an equals sign, but there is more than one variable and there are rules for working out quantities such as area or the cost of taxi fares.

For example, $V = x^3$, $A = \frac{1}{2}bh$ and $C = 3 + 4m$ are formulae.

Identity: This looks like a formula, but the important fact about an identity is that it is true for all values, whether numerical or algebraic. For example, $5n \equiv 2n + 3n$ and $(x + 1)^2 \equiv x^2 + 2x + 1$ are identities. Note that the special sign $\equiv$ is used in an identity.

Terms: These are the separate parts of expressions, equations, formulae and identities. In $3x + 2y - 7$, there are three terms, $3x$, $+2y$ and -7. Expressions inside **brackets** are treated as a single term in calculations, although they can be multiplied out in expansions.

Constant term: This is any single number in an expression or equation, so in $x^2 + 3x + 7$, the constant term is 7.

EXAMPLE 1

Expand **a)** $3(2x + 7)$ **b)** $2x(3x - 4y)$

a) $3 \times 2x + 3 \times 7 = 6x + 21$ **b)** $2x \times 3x - 2x \times 4y = 6x^2 - 8xy$

EXAMPLE 2

Expand and **simplify** $6(2x + 5) - 3(3x - 1)$

$6(2x + 5) - 3(3x - 1) = 12x + 30 - 9x + 3 = 12x - 9x + 30 + 3 = 3x + 33$

EXAMPLE 3

Factorise the following expressions.

a) $4my + 12mx$ **b)** $5kp - 10k^2p + 15kp^2$

a) The common **factor** is $4m$, so $4my + 12mx = 4m(y + 3x)$

b) The common factor is $5kp$, so $5kp - 10k^2p + 15kp^2 = 5kp(1 - 2k + 3p)$

EXAMPLE 4

State whether each of the following is an expression (E), equation (Q), formula (F) or identity (I).

A: $x^2 - 5x$ **B:** $c = \sqrt{(a^2 + b^2)}$ **C:** $2x - 3 = 1$ **D:** $4n - 3n = n$

A is an expression (E) with two terms.

B is a formula (F). This is the formula for finding the hypotenuse of a right-angled triangle with short sides of lengths a and b.

C is an equation (Q) which can be solved to give $x = 2$.

D is an identity (I).

EXERCISE 2A

GRADE C

1. a) Which of the following expressions are equivalent?

$3m \times 8n$ $2m \times 12n$ $4n \times 6m$ $m \times 24n$

b) The expressions $\frac{x}{2}$ and x^2 are the same for only one positive value of x. What is the value?

2. Expand these expressions.

a) $5(3 - m)$ **b)** $3(2x + 7)$ **c)** $x(x + 2)$ **d)** $2m(5 - m)$ **e)** $3n(m - p)$

3. Factorise the following expressions.

a) $18 - 3m$ **b)** $x^2 + 5x$ **c)** $10m - m^2$ **d)** $15s^2 + 3$ **e)** $3n - pn$

4. Find the missing terms to make these equations true.

a) $8x + 12y - \square - \square = 5x + 4y$ **b)** $3a - 5b - \square + \square = a - b$

5. ABCDEF is an L-shape.

$AB = DE = x$ $AF = 4x - 1$ and $EF = 3x + 1$

a) Explain why the length $BC = 3x - 1$.

b) Find the perimeter of the shape in terms of x.

c) If $x = 6$ cm what is the perimeter of the shape?

6. Darren wrote the following:

$2(3x - 5) = 5x - 3$

Explain the **two** mistakes that Darren has made.

HINTS AND TIPS

It is not enough to give the right answer. You must try to explain why Darren wrote 5 for 2×3 instead of 6.

7. The expansion $3(x + 4) = 3x + 12$ can be shown by the diagram.

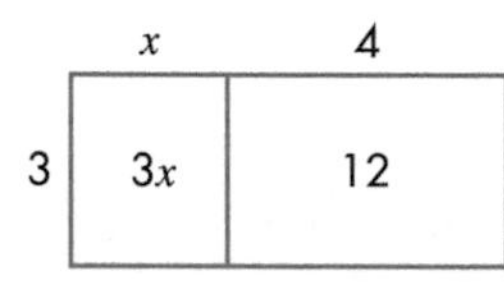

a) What expansion is shown in this diagram?

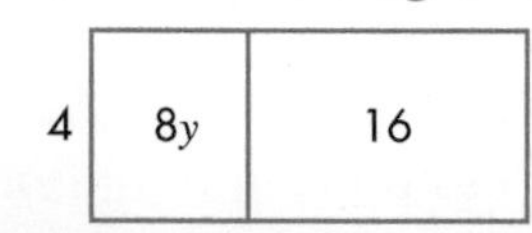

b) Write down an expansion that is shown on this diagram.

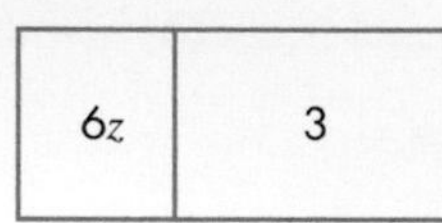

8. A square and a rectangle have the same perimeter. The rectangle has one side that is three times as long as the other. The square has a side of 8 cm.

What are the dimensions of the rectangle?

9. Find the value of each expression when $x = 1.4$, $y = 2.5$ and $z = 0.8$.

a) $\frac{3x+4}{2}$ **b)** $\frac{x+2y}{z}$ **c)** $\frac{y}{z}+x$

10. The formula for the area, A, of a square with side x is $A = x^2$.

The formula for the area, T, of a triangle with base b and height h is $T = \frac{1}{2}bh$.

Find **different** values of x, b and h so that $A = T$.

11. x and y are different prime numbers.

Choose values for x and y so that the formula $5x + 2y$:

a) evaluates to an even number

b) evaluates to an odd number.

12. Kaz knows that x, y and z have the values 3, 6 and 9 but he does not know which variable has which value.

a) What is the maximum value that $x + 3y - 5z$ could have?
b) What is the minimum value that $2x - y + 3z$ could have?

13. A car costs £90 per day to rent. Some friends rent the car for five days.

a) Which of the following formulae would represent the cost per day if there are n people in the car and they share the cost equally?

$\frac{450}{5n}$ $\quad\quad$ $\frac{450}{n+5}$ $\quad\quad$ $\frac{450}{n}$

b) Three friends rent the car. When they get the bill they find that there is a special discount for a five-day rental. They each find it cost them £20 less than they expected.

How much does a five-day rental cost?

14. A rectangle with sides 6 and $3x + 5$ has a smaller rectangle with sides 2 and $x - 2$ cut from it.

Work out the remaining area.

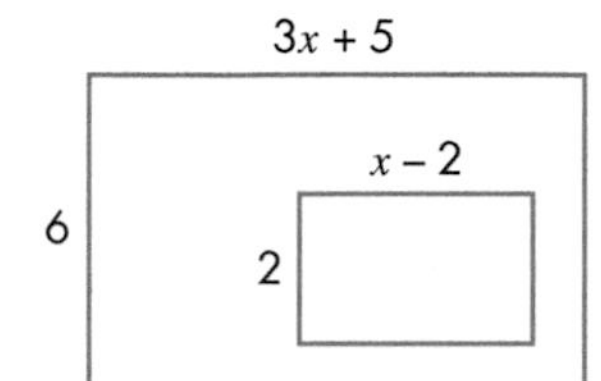

15. Five friends have a meal together. They each have a main course costing £7.25 and a dessert costing £2.75.

Colin says that the bill will be $5 \times £7.25 + 5 \times £2.75$.

Kim says that she has an easier way to work out the bill as $5 \times £(7.25 + 2.75)$.

a) Explain why Colin's and Kim's methods both give the correct answer.
b) Explain why Kim's method is better.
c) What is the total bill?

16. Explain why $7x - 9y$ cannot be factorised.

17. $3(x + 1) = 3x + 3$ is an identity.

a) Substitute $x = 5$ in both sides to show that it is true for a numerical value.
b) Substitute $x = n + 2$ and expand and simplify both sides to show that it is true for an algebraic expression.

GRADE B

18. Expand these expressions.

a) $4p^2(3p - q)$ **b)** $5t^2(2t^2 + 7)$ **c)** $5x(2x + 7y)$
d) $2m^2(5 - m^3)$ **e)** $8s^3(s + 3t)$ **f)** $6nm^2(m - n)$

19. Expand and simplify the following expressions.

a) $5(4x + 1) + 3(x + 2)$ **b)** $4(y - 2) + 5(y + 3)$ **c)** $2(3x - 2) - 4(x + 1)$
d) $5(2x + 3) + 6(2x - 1)$ **e)** $6x(2x - 3) + 2x(x + 4)$ **f)** $3(4x^2 - 3) + x^2(5 + 2x)$

20. Factorise the following expressions.

a) $9p^2 + 6pt$ **b)** $12mp - 8m^2$ **c)** $16a^2b + 4ab$
d) $4a^2 - 6a + 2$ **e)** $20xy^2 + 10x^2y + 5xy$ **f)** $8mt^2 - 4m^2t$

2.2 Expanding brackets and collecting like terms

THIS SECTION WILL SHOW YOU HOW TO ...

✓ expand and simplify brackets
✓ simplify algebraic expressions by collecting like terms
✓ expand two linear brackets to obtain a quadratic expression

KEY TERMS

✓ expand and simplify
✓ expression
✓ like terms
✓ quadratic expression

When you **expand and simplify** a product of two brackets you should collect **like terms**. Algebraic **expressions** should always be simplified as much as possible.

EXAMPLE 5

$3(4 + m) + 2(5 + 2m) = 12 + 3m + 10 + 4m = 22 + 7m$

EXAMPLE 6

$3t(5t + 4) - 2t(3t - 5) = 15t^2 + 12t - 6t^2 + 10t = 9t^2 + 22t$

EXERCISE 2B

GRADE C

1. Simplify these expressions.

a) $3d + 2d + 4de$ **b)** $3t - t$ **c)** $6y^2 - 2y^2$ **d)** $7a^2d - 4a^2d$

2. Find the missing terms to make these equations true.

a) $4x + 5y + \ldots - \ldots = 6x + 3y$ **b)** $3a - 6b - \ldots + \ldots = 2a + b$

3. ABCDEF is an L-shape.

AB = DE = x

AF = $3x - 1$

EF = $2x + 1$

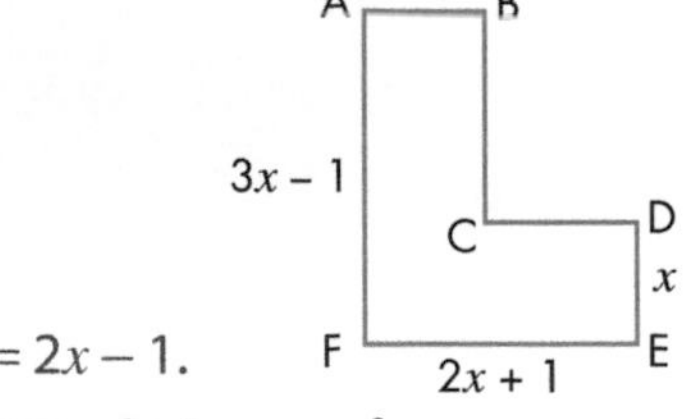

a) Explain why the length BC = $2x - 1$.

b) Find the perimeter of the shape in terms of x.

c) If $x = 2.5$ cm what is the perimeter of the shape?

HINTS AND TIPS

Make sure your explanation uses expressions. Do not try to explain in words alone.

4. Expand and simplify.

a) $3(4 + t) + 2(5 + t)$ **b)** $5(1 + 3g) + 3(3 - 4g)$

5. Expand and simplify.

a) $4(3 + 2h) - 2(5 + 3h)$ **b)** $4(4e + 3) - 2(5e - 4)$

6. Expand and simplify.

a) $m(4 + p) + p(3 + m)$ **b)** $k(3 + 2h) + h(4 + 3k)$

c) $4r(3 + 4p) + 3p(8 - r)$ **d)** $5k(3m + 4) - 2m(3 - 2k)$

GRADE B

7. Expand and simplify.

a) $t(3t + 4) + 3t(3 + 2t)$ **b)** $2y(3 + 4y) + y(5y - 1)$

c) $4e(3e - 5) - 2e(e - 7)$ **d)** $3k(2k + p) - 2k(3p - 4k)$

HINTS AND TIPS

Be careful with minus signs. For example, $-2(5e - 4) = -10e + 8$

8. Expand and simplify.

a) $4a(2b + 3c) + 3b(3a + 2c)$ **b)** $3y(4w + 2t) + 2w(3y - 4t)$

c) $5m(2n - 3p) - 2n(3p - 2m)$ **d)** $2r(3r + r^2) - 3r^2(4 - 2r)$

9. A two-carriage train has f first-class seats and $2s$ standard-class seats.

A three-carriage train has $2f$ first-class seats and $3s$ standard-class seats.

On a weekday, 5 two-carriage trains and 2 three-carriage trains travel from Hull to Liverpool.

a) Write down an expression for the total number of first-class and standard-class seats available during the day.

b) On average in any day, half of the first-class seats are used at a cost of £60. On average in any day, three-quarters of the standard-class seats are used at a cost of £40.
How much money does the rail company earn in an average day on this route? Give your answer in terms of f and s.

c) $f = 15$ and $s = 80$. It costs the rail company £30 000 per day to operate this route. How much profit do they make on an average day?

10. Fill in whole-number values so that this expansion is true.

$3(\ldots x + \ldots y) + 2(\ldots x + \ldots y) = 11x + 17y$

HINTS AND TIPS

There is more than one answer. You don't have to give them all.

11. A rectangle with sides 5 and $3x + 2$ has a smaller rectangle with sides 3 and $2x - 1$ cut from it.

Work out the remaining area.

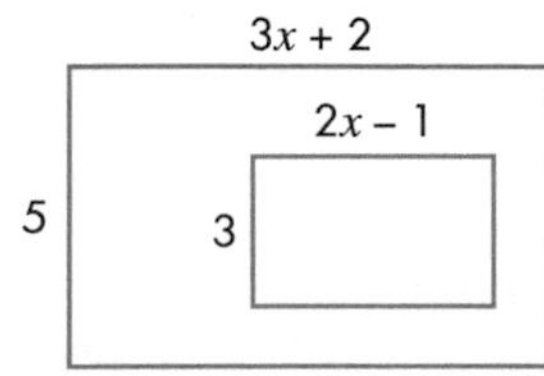

HINTS AND TIPS

Write out the expression for the difference between the two rectangles and then work it out.

Quadratic expansion

A **quadratic expression** is one in which the highest power of the variables is 2. For example,

y^2 $3t^2 + 5t$ $5m^2 + 3m + 8$

An expression such as $(3y + 2)(4y - 5)$ can be expanded to give a **quadratic expression.**

Multiplying out such pairs of brackets is called quadratic expansion.

The rule for expanding expressions such as $(t + 5)(3t - 4)$ is similar to that for expanding single brackets: multiply everything in one set of brackets by everything in the other set of brackets.

There are several methods for doing this. Examples 7 to 9 show the three main methods: expansion, FOIL and the box method.

EXAMPLE 7

In the expansion method, split the terms in the first set of brackets, so that each of them multiplies both terms in the second set of brackets, then simplify the outcome.

Expand $(x + 3)(x + 4)$

$$\begin{aligned}(x + 3)(x + 4) &= x(x + 4) + 3(x + 4)\\ &= x^2 + 4x + 3x + 12\\ &= x^2 + 7x + 12\end{aligned}$$

EXAMPLE 8

FOIL stands for First, Outer, Inner and Last. This is the order of multiplying the terms from each set of brackets.

Expand $(t + 5)(t - 2)$

First terms give: $t \times t = t^2$

Outer terms give: $t \times -2 = -2t$

Inner terms give: $5 \times t = 5t$

Last terms give: $+5 \times -2 = -10$

$$\begin{aligned}(t + 5)(t - 2) &= t^2 - 2t + 5t - 10\\ &= t^2 + 3t - 10\end{aligned}$$

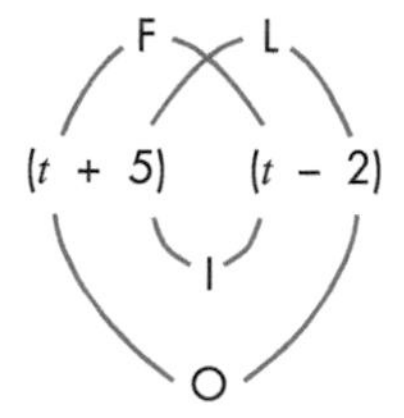

EXAMPLE 9

The box method is similar to that used to do long multiplication.

Expand $(k - 3)(k - 2)$

$$\begin{aligned}(k - 3)(k - 2) &= k^2 - 2k - 3k + 6\\ &= k^2 - 5k + 6\end{aligned}$$

×	k	-3
k	k^2	$-3k$
-2	$-2k$	$+6$

Warning: Be careful with the signs. This is the main place where mistakes are made in questions involving the expansion of brackets.

EXERCISE 2C

GRADE B

Expand the expressions in questions **1–10**.

1. $(x+3)(x+2)$ **2.** $(w+1)(w+3)$ **3.** $(a+4)(a+1)$

4. $(x+4)(x-2)$ **5.** $(w+3)(w-1)$ **6.** $(f+2)(f-3)$

7. $(y+4)(y-3)$ **8.** $(x-3)(x+4)$ **9.** $(p-2)(p+1)$

10. $(a-1)(a+3)$

The expansions of the expressions in questions **11–16** follow a pattern. Work out the first few and try to spot the pattern that will allow you immediately to write down the answers to the rest.

11. $(x+3)(x-3)$ **12.** $(t+5)(t-5)$ **13.** $(t+2)(t-2)$

14. $(y+8)(y-8)$ **15.** $(5+x)(5-x)$ **16.** $(x-6)(x+6)$

HINTS AND TIPS

A common error is to get minus signs wrong. $-2x - 3x = -5x$ and $-2 \times -3 = +6$

17. This rectangle is made up of four parts with areas of x^2, $2x$, $3x$ and 6 square units.

Work out expressions for the sides of the rectangle, in terms of x.

x^2	$2x$
$3x$	6

18. This square has an area of x^2 square units. It is split into four rectangles.

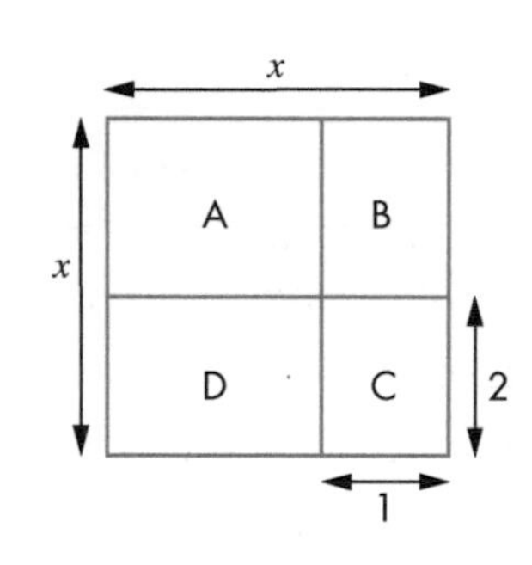

a) Fill in the table below to show the dimensions and area of each rectangle.

Rectangle	Width	Height	Area
A	$x-1$	$x-2$	$(x-1)(x-2)$
B			
C			
D			

b) Add together the areas of rectangles B, C and D. Expand any brackets and collect terms together.

c) Use the results to explain why $(x-1)(x-2) = x^2 - 3x + 2$.

19. a) Expand $(x-3)(x+3)$

b) Use the result in **a** to write down the answers to these. (Do not use a calculator or do a long multiplication.)

i) 97×103 **ii)** 197×203

Quadratic expansion with non-unit coefficients

All the algebraic terms in x^2 in Exercise 2C have a **coefficient** of 1 or -1. The next two examples show what to do if you have to expand brackets containing terms in x^2 with coefficients that are not 1 or -1.

EXAMPLE 10

Expand $(2t + 3)(3t + 1)$

$(2t + 3)(3t + 1) = 6t^2 + 2t + 9t + 3$

$= 6t^2 + 11t + 3$

×	$2t$	$+3$
$3t$	$6t^2$	$+9t$
$+1$	$+2t$	$+3$

EXAMPLE 11

Expand $(4x - 1)(3x - 5)$

$(4x - 1)(3x - 5) = 4x(3x - 5) - (3x - 5)$

$= 12x^2 - 20x - 3x + 5$

$= 12x^2 - 23x + 5$

HINTS AND TIPS

$(3x - 5)$ is the same as $1(3x - 5)$.

EXERCISE 2D

GRADE A

Expand the expressions in questions **1–10**.

1. $(2x + 3)(3x + 1)$ **2.** $(4t + 3)(2t - 1)$ **3.** $(4k + 3)(3k - 5)$

4. $(2a - 3)(3a + 1)$ **5.** $(3g - 2)(5g - 2)$ **6.** $(4d - 1)(3d + 2)$

7. $(5 + 2p)(3 + 4p)$ **8.** $(6 + 5t)(1 - 2t)$ **9.** $(2 + 3f)(2f - 3)$

10. $(4 - 2t)(3t + 1)$

11. Expand:

a) $(x + 1)(x + 1)$ **b)** $(x - 1)(x - 1)$ **c)** $(x + 1)(x - 1)$

d) Use the results in parts **a, b** and **c** to show that $(p + q)^2 \equiv p^2 + 2pq + q^2$ is an identity.

HINTS AND TIPS

Take $p = x + 1$ and $q = x - 1$

12. a) Without expanding the brackets, match each expression on the left with an expression on the right. One is done for you.

$(3x - 2)(2x + 1)$	$4x^2 - 4x + 1$
$(2x - 1)(2x - 1)$	$6x^2 - x - 2$
$(6x - 3)(x + 1)$	$6x^2 + 7x + 2$
$(4x + 1)(x - 1)$	$6x^2 + 3x - 3$
$(3x + 2)(2x + 1)$	$4x^2 - 3x - 1$

b) Taking any expression on the left, explain how you can match it with an expression on the right without expanding the brackets.

EXERCISE 2E

GRADE A

Try to spot the pattern in each of the expressions in questions **1–10** so that you can immediately write down the expansion.

1. $(2x + 1)(2x - 1)$ **2.** $(5y + 3)(5y - 3)$ **3.** $(4m + 3)(4m - 3)$

4. $(4h - 1)(4h + 1)$ **5.** $(2 + 3x)(2 - 3x)$ **6.** $(6 - 5y)(6 + 5y)$

7. $(a + b)(a - b)$ **8.** $(2m - 3p)(2m + 3p)$ **9.** $(ab + cd)(ab - cd)$

10. $(a^2 + b^2)(a^2 - b^2)$

11. Imagine a square of side a units with a square of side b units cut from one corner.

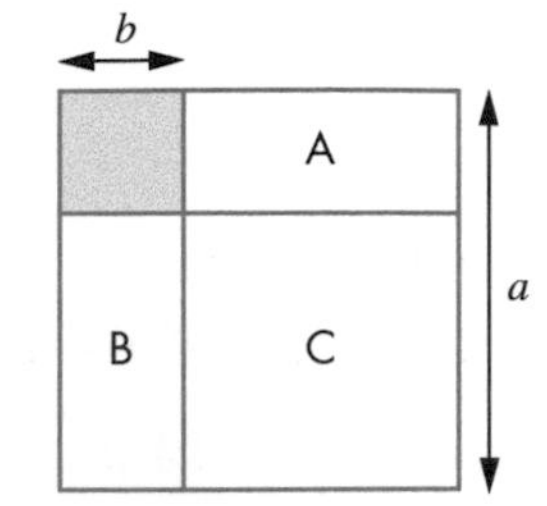

a) What is the area remaining after the small square is cut away?

b) The remaining area is cut into rectangles, A, B and C, and rearranged as shown.

Write down the dimensions and area of the rectangle formed by A, B and C.

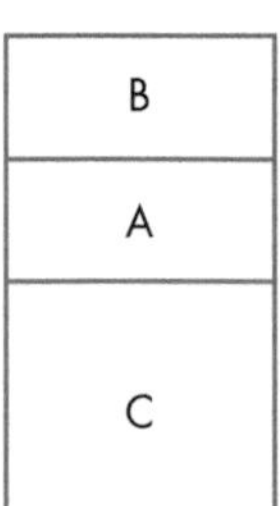

c) Explain why $a^2 - b^2 = (a + b)(a - b)$.

12. Explain why the areas of the shaded regions are the same.

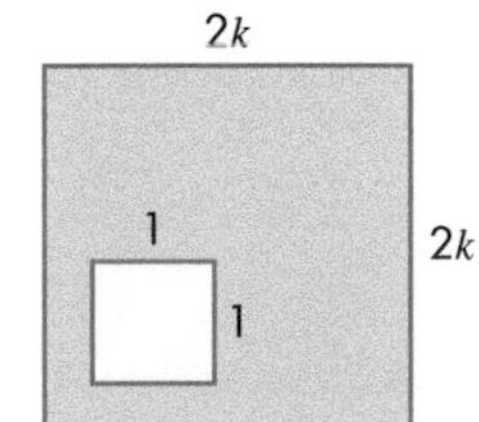

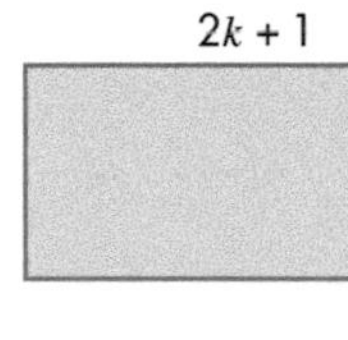

Expanding squares

Whenever you see a linear bracketed term squared, you must write the brackets down twice and then use whichever method you prefer to expand.

EXAMPLE 12

Expand $(x + 3)^2$

$(x + 3)^2 = (x + 3)(x + 3) = x(x + 3) + 3(x + 3) = x^2 + 3x + 3x + 9$

$= x^2 + 6x + 9$

EXAMPLE 13

Expand $(3x - 2)^2$

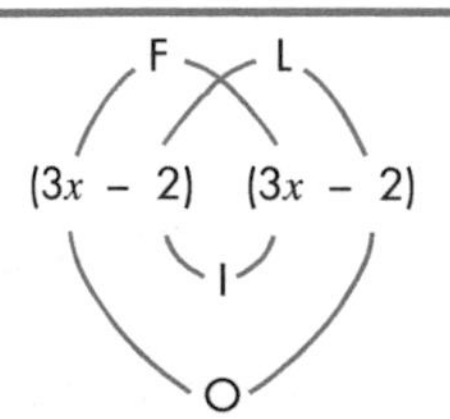

$(3x - 2)^2 = (3x - 2)(3x - 2)$

$= 9x^2 - 6x - 6x + 4$

$= 9x^2 - 12x + 4$

EXERCISE 2F

GRADE A

Expand the squares in questions **1–12** and simplify.

1. $(x + 5)^2$ **2.** $(6 + t)^2$ **3.** $(m - 3)^2$

4. $(7 - k)^2$ **5.** $(3x + 1)^2$ **6.** $(2 + 5y)^2$

7. $(3x - 2)^2$ **8.** $(x + y)^2$ **9.** $(m - n)^2$

10. $(m - 3n)^2$ **11.** $(x - 5)^2 - 25$ **12.** $(x - 2)^2 - 4$

HINTS AND TIPS

Remember *always* write down the brackets twice. Do not try to take any short cuts.

13. A teacher asks her class to expand $(3x + 1)^2$.

Bernice's answer is $9x^2 + 1$.

Pete's answer is $3x^2 + 6x + 1$.

a) Explain the mistakes that Bernice has made.

b) Explain the mistakes that Pete has made.

c) Work out the correct answer.

14. Use the diagram to show algebraically and diagrammatically that:

$(2x - 1)^2 = 4x^2 - 4x + 1$

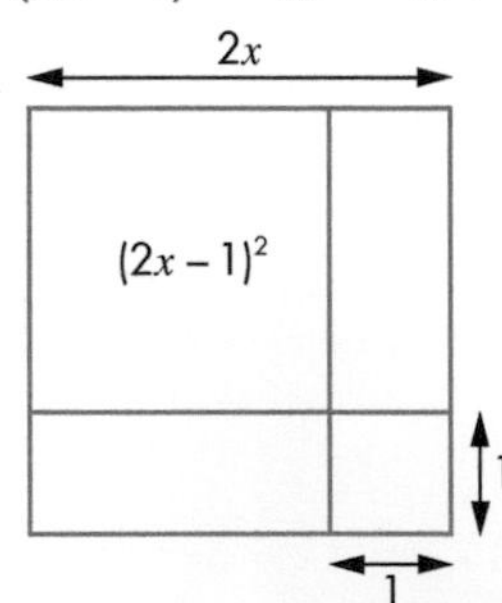

When you multiply out two terms in brackets, always multiply every term in the first expression by every term in the second expression.

EXAMPLE 14

Expand and simplify $(x^4 + 2x^3 - 4x^2 + 5x - 1)(x + 1)$

$$\begin{aligned}(x^4 + 2x^3 - 4x^2 + 5x - 1)(x + 1) &= x^5 + 2x^4 - 4x^3 + 5x^2 - x \\ &\quad + x^4 + 2x^3 - 4x^2 + 5x - 1 \\ &= x^5 + 3x^4 - 2x^3 + x^2 + 4x - 1\end{aligned}$$

EXAMPLE 15

Expand and simplify $(3x + 1)^3$

$$\begin{aligned}(3x + 1)^3 &= (3x + 1)(3x + 1)(3x + 1) \\ &= (9x^2 + 6x + 1)(3x + 1) \\ &= 27x^3 + 9x^2 + 18x^2 + 6x + 3x + 1 \\ &= 27x^3 + 27x^2 + 9x + 1\end{aligned}$$

EXAMPLE 16

Expand and simplify $(x^{\frac{1}{2}} - 2x^{-\frac{1}{2}})(x^{\frac{1}{2}} + x^{-\frac{1}{2}})$

$$(x^{\frac{1}{2}} - 2x^{-\frac{1}{2}})(x^{\frac{1}{2}} + x^{-\frac{1}{2}}) = x + 1 - 2 - 2x^{-1}$$

$$= x - 1 - \frac{2}{x}$$

EXERCISES 2G

GRADE A

1. Expand and simplify:

a) $(x^2 + 3x - 1)(x + 1)$

b) $(2x^2 + 5x + 1)(4x + 1)$

c) $(x^3 - 6x^2 + 2x - 3)(2x + 3)$

d) $x^2(x^4 + x^3 - 6x^2 + 2x - 3) + x(2x^2 + 3x - 1)$

GRADE A*

2. Expand and simplify:

a) $(x + 1)^3$

b) $(2x - 1)^3$

c) $(3x + 2)^3$

d) $(4x - 3)^3$

GRADE A**

3. Expand and simplify:

a) $2x^{\frac{1}{2}}(x^{\frac{1}{2}} - x^{-\frac{1}{2}})$

b) $x^{\frac{1}{3}}(x^{\frac{2}{3}} - x^{-\frac{1}{3}})$

c) $(3x^{\frac{1}{2}} - 2x^{-\frac{1}{2}})(2x^{\frac{1}{2}} - 3x^{-\frac{1}{2}})$

d) $(4x^{\frac{1}{2}} - x^{-\frac{1}{3}})(4x^{\frac{1}{2}} + x^{-\frac{1}{3}})$

2.3 Factorising

THIS SECTION WILL SHOW YOU HOW TO …

✓ factorise an algebraic expression
✓ factorise a quadratic expression into two linear brackets

KEY WORDS

✓ brackets
✓ coefficient
✓ common factor
✓ difference of two squares
✓ factorisation
✓ quadratic expression

Factorisation is the opposite of expansion. It puts an expression back into the **brackets** it may have come from.

To factorise an expression, you have to look for the **common factors** in *every* term of the expression.

EXAMPLE 17

Factorise each expression.

a) $6my + 4py$ **b)** $10a^2b - 15ab^2$

a) First look at the numbers.

These have a common factor of 2. m and p do not occur in both terms but y does, and is a common factor, so the factorisation is:

$6my + 4py = 2y(3m + 2p)$

b) 5 is a common factor of 10 and 15 and ab is a common factor of ab and ab^2.

$10a^{2b} - 15ab^2 = 5ab(2a - 3b)$

EXERCISE 2H

GRADE C

1. Factorise each expression.

a) $8m + 12k$ **b)** $mn + 3m$ **c)** $5g^2 + 3g$

d) $3y^2 + 2y$ **e)** $4t^2 - 3t$ **f)** $3m^2 - 3mp$

g) $6p^2 + 9pt$ **h)** $8pt + 6mp$ **i)** $4a^2 + 6a + 8$

j) $6ab + 9bc + 3bd$ **k)** $5t^2 + 4t + at$ **l)** $6mt^2 - 3mt + 9m^2t$

m) $8ab^2 + 2ab - 4a^2b$ **n)** $10pt^2 + 15pt + 5p^2t$

2. Factorise the following expressions where possible. List those that do not factorise.

a) $5m + 2mp$ **b)** $t^2 - 7t$ **c)** $8pt + 5ab$

d) $a^2 + b$ **e)** $4a^2 - 5ab$ **f)** $5ab - 3b^2c$

3. Three students are asked to factorise the expression $12m - 8$. These are their answers.

Aidan	Bernice	Craig
$2(6m - 4)$	$4(3m - 2)$	$4m(3 - \frac{2}{m})$

All the answers are accurately factorised, but only one is the normally accepted answer.

a) Which student gave the correct answer?

b) Explain why the other two students' answers are not acceptable as correct answers.

4. Explain why $5m + 6p$ cannot be factorised.

5. Alvin has correctly factorised the top and bottom of an algebraic fraction and cancelled out the terms to give a final answer of $2x$. Unfortunately some of his work has had coffee spilt on it. What was the original fraction?

$$\frac{4x\ldots}{2\ldots} = \frac{4\ldots}{2(x - 3)} = 2x$$

Quadratic factorisation

Factorisation involves changing a **quadratic expression** into the product of two linear expressions, written as two terms in **brackets** (if possible). Start with the factorisation of quadratic expressions of the type:

$x^2 + ax + b$

where a and b are integers.

Sometimes it is easy to put a quadratic expression back into its brackets, other times it seems hard. However, there are some simple rules that will help you to factorise.

- The expression inside each set of brackets will start with an x, and the signs in the quadratic expression show which signs to put after the xs.
- When the second sign in the expression is a plus, the signs in both sets of brackets are the same as the first sign.

  $x^2 + ax + b = (x + ?)(x + ?)$ Since everything is positive.

  $x^2 - ax + b = (x - ?)(x - ?)$ Since –ve × –ve = +ve
- When the second sign is a minus, the signs in the brackets are different.

  $x^2 + ax - b = (x + ?)(x - ?)$ Since +ve × –ve = –ve

  $x^2 - ax - b = (x + ?)(x - ?)$

- Next, look at the last number, b, in the expression. When multiplied together, the two numbers in the brackets must give b.
- Finally, look at the **coefficient** of x, a. The sum of the two numbers in the brackets will give a.

EXAMPLE 18

Factorise $x^2 - x - 6$.

Because of the signs you know the brackets must be $(x + ?)(x - ?)$.

Two numbers that have a product of -6 and a sum of -1 are -3 and $+2$.

So, $x^2 - x - 6 = (x + 2)(x - 3)$

EXAMPLE 19

Factorise $x^2 - 9x + 20$.

Because of the signs you know the brackets must be $(x - ?)(x - ?)$.

Two numbers that have a product of $+20$ and a sum of -9 are -4 and -5.

So, $x^2 - 9x + 20 = (x - 4)(x - 5)$

EXERCISE 2I

GRADE B

Factorise the expressions in questions **1–24**.

1. $x^2 + 5x + 6$ **2.** $m^2 + 7m + 10$ **3.** $p^2 + 14p + 24$

4. $w^2 + 11w + 18$ **5.** $a^2 + 8a + 12$ **6.** $b^2 + 20b + 96$

7. $t^2 - 5t + 6$ **8.** $d^2 - 5d + 4$ **9.** $x^2 - 15x + 36$

10. $t^2 - 13t + 36$ **11.** $y^2 - 16y + 48$ **12.** $j^2 - 14j + 48$

13. $y^2 + 5y - 6$ **14.** $m^2 - 4m - 12$ **15.** $n^2 - 3n - 18$

16. $m^2 - 7m - 44$ **17.** $t^2 - t - 90$ **18.** $h^2 - h - 72$

19. $t^2 - 2t - 63$ **20.** $y^2 + 20y + 100$ **21.** $m^2 - 18m + 81$

22. $x^2 - 24x + 144$ **23.** $d^2 - d - 12$ **24.** $q^2 - q - 56$

25. This rectangle is made up of four parts. Two of the parts have areas of x^2 and 6 square units.

The sides of the rectangle are of the form $x + a$ and $x + b$.

There are two possible answers for a and b.

Work out both answers and copy and complete the areas in the other parts of the rectangle.

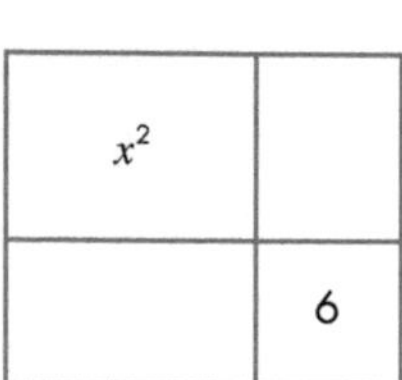

26. a) Expand $(x+a)(x+b)$

b) If $x^2+7x+12=(x+p)(x+q)$, use your answer to part **a** to write down the values of: **i)** $p+q$ **ii)** pq

c) Explain how you can tell that $x^2+12x+7$ will not factorise.

Difference of two squares

In an earlier exercise you multiplied out, for example, $(a+b)(a-b)$ and obtained a^2-b^2. This type of quadratic expression, with only two terms, both of which are perfect squares separated by a minus sign, is called the **difference of two squares**. You should have found that all the expansions in Exercise 2E involved the differences of two squares.

The exercise illustrates a system of factorisation that will always work for the difference of two squares such as these.

x^2-9 x^2-25 x^2-4 x^2-100

There are three conditions that must be met if the difference of two squares works.

- There must be two terms.
- They must separated by a negative sign.
- Each term must be a perfect square, say x^2 and n^2.

When these three conditions are met, the factorisation is:

$x^2-n^2=(x+n)(x-n)$

EXAMPLE 20

Factorise x^2-36.

- Recognise the difference of two squares x^2 and 6^2.
- So it factorises to $(x+6)(x-6)$.

Expanding the brackets shows that they do come from the original expression.

EXAMPLE 21

Factorise $9x^2-169$.

- Recognise the difference of two squares $(3x)^2$ and 13^2.
- So it factorises to $(3x+13)(3x-13)$.

EXERCISE 2J

GRADE A

Each of the expressions in questions **1–9** is the difference of two squares. Factorise them.

HINTS AND TIPS

Learn how to spot the difference of two squares.

1. $x^2 - 9$ **2.** $t^2 - 25$ **3.** $m^2 - 16$

4. $9 - x^2$ **5.** $49 - t^2$ **6.** $k^2 - 100$

7. $4 - y^2$ **8.** $x^2 - 64$ **9.** $t^2 - 81$

10. a) Expand and simplify: $(x + 2)^2 - (x + 1)^2$

b) Factorise: $a^2 - b^2$

c) In your answer for part **b**, replace a with $(x + 2)$ and b with $(x + 1)$. Expand and simplify the answer.

d) What can you say about the answers to parts **a** and **c**?

e) Simplify: $(x + 1)^2 - (x - 1)^2$

Each of the expressions in questions **11–19** is the difference of two squares. Factorise them.

11. $x^2 - y^2$ **12.** $x^2 - 4y^2$ **13.** $x^2 - 9y^2$

14. $9x^2 - 1$ **15.** $16x^2 - 9$ **16.** $25x^2 - 64$

17. $4x^2 - 9y^2$ **18.** $9t^2 - 4w^2$ **19.** $16y^2 - 25x^2$

Factorising $ax^2 + bx + c$

You can adapt the method for factorising $x^2 + ax + b$ to take into account the factors of the coefficient of x^2 when it is not 1.

EXAMPLE 22

Factorise $3x^2 + 8x + 4$

- First, note that both signs are positive. So the signs in the brackets must be $(?x + ?)(?x + ?)$.
- As 3 has only 3×1 as factors, the brackets must be $(3x + ?)(x + ?)$.
- Next, note that the factors of 4 are 4×1 and 2×2.
- Now find which pair of factors of 4 combine with 3 and 1 to give 8.

③	4	②
①	1	②

You can see that the combination 3×2 and 1×2 gives 8.

- So, the complete factorisation becomes $(3x + 2)(x + 2)$.

EXAMPLE 23

Factorise $6x^2 - 7x - 10$

- First, note that both signs are negative. So the signs in the brackets must be $(?x + ?)(?x - ?)$.
- As 6 has 6×1 and 3×2 as factors, the brackets could be $(6x \pm ?)(x \pm ?)$ or $(3x \pm ?)(2x \pm ?)$.
- Next, note that the factors of 10 are 5×2 and 1×10.
- Now find which pair of factors of 10 combine with the factors of 6 to give -7.

3	(6)	|	±1	(±2)
2	(1)	|	±10	(±5)

You can see that the combination 6×-2 and 1×5 gives -7.

- So, the complete factorisation becomes $(6x + 5)(x - 2)$.

EXERCISE 2K

GRADE A

Factorise the expressions in questions **1–12**.

1. $2x^2 + 5x + 2$ **2.** $7x^2 + 8x + 1$ **3.** $4x^2 + 3x - 7$

4. $24t^2 + 19t + 2$ **5.** $15t^2 + 2t - 1$ **6.** $16x^2 - 8x + 1$

7. $6y^2 + 33y - 63$ **8.** $4y^2 + 8y - 96$ **9.** $8x^2 + 10x - 3$

10. $6t^2 + 13t + 5$ **11.** $3x^2 - 16x - 12$ **12.** $7x^2 - 37x + 10$

13. This rectangle is made up of four parts, with areas of $12x^2$, $3x$, $8x$ and 2 square units.

Work out expressions for the sides of the rectangle, in terms of x.

$12x^2$	$3x$
$8x$	2

14. Three students are asked to factorise the expression $6x^2 + 30x + 36$. These are their answers.

Adam	**Bertie**	**Cara**
$(6x + 12)(x + 3)$	$(3x + 6)(2x + 6)$	$(2x + 4)(3x + 9)$

All the answers are correctly factorised.

a) Explain why one quadratic expression can have three different factorisations.

b) Which of the following is the most complete factorisation?
$2(3x + 6)(x + 3)$ $6(x + 2)(x + 3)$ $3(x + 2)(2x + 6)$
Explain your choice.

EXAMPLE 24

Factorise $3x^2 + 8xy + 4y^2$.

This is similar to Example 22. Look at the pattern with the y terms.

$3x^2 + 8x + 4 = (3x + 2)(x + 2)$

So $3x^2 + 8xy + 4y^2 = (3x + 2y)(x + 2y)$

This can be checked by expanding $(3x + 2y)(x + 2y)$.

EXAMPLE 25

Factorise $6x^2 - 7xy - 10y^2$.

This is similar to Example 23. Look at the pattern with the y terms.

$6x^2 - 7x - 10 = (6x + 5)(x - 2)$

So $6x^2 - 7xy - 10y^2 = (6x + 5y)(x - 2y)$

This can be checked by expanding $(6x + 5y)(x - 2y)$.

EXAMPLE 26

Factorise fully $(3x + 5)^2 - (2x - 3)^2$.

This is a difference of two squares $(3x + 5)^2$ and $(2x - 3)^2$.

$(3x + 5)^2 - (2x - 3)^2 = (3x + 5 + 2x - 3)(3x + 5 - (2x - 3))$

$= (5x + 2)(x + 8)$

EXAMPLE 27

Factorise fully $x^4 - 81y^4$.

This is a difference of two squares $(x^2)^2$ and $(9y^2)^2$.

$x^4 - 81y^4 = (x^2 + 9y^2)(x^2 - 9y^2)$

$= (x^2 + 9y^2)(x + 3y)(x - 3y)$

as $x^2 - 9y^2$ is also a difference of two squares.

EXERCISE 2L

GRADE A

Factorise the expressions.

1. $x^2 + 5xy + 6y^2$ **2.** $x^2 + 10xy + 21y^2$ **3.** $x^2 - 5xy + 4y^2$

4. $x^2 - 6xy - 7y^2$ **5.** $x^2 - xy - 72y^2$ **6.** $2x^2 + 5xy + 2y^2$

7. $3x^2 - 16xy - 12y^2$ **8.** $5x^2 - 13xy - 6y^2$ **9.** $6x^2 + 13xy + 5y^2$

GRADE A*

10. $15x^2 + 2xy - y^2$ **11.** $x^4 - 25y^4$ **12.** $8x^3 - 50x$

13. $16x^4 - 25y^4$ **14.** $(x + 5)^2 - (x - 3)^2$ **15.** $(2x + 1)^2 - (2x - 1)^2$

16. $(3x + 2)^2 - (2x + 3)^2$ **17.** $4x^2 - (2x + 1)^2$ **18.** $(5x + 1)^2 - 9x^2$

Exam-style questions

1. Solve the equation $2(15-4a)=4(3a+7)$. GRADE C

2. The lengths of the sides of this triangle are shown on the diagram.

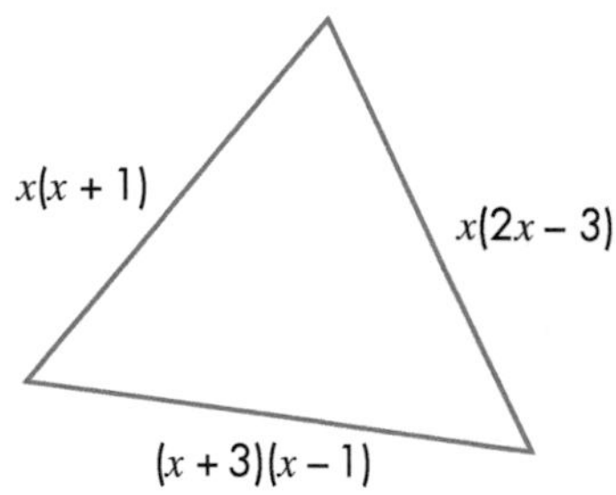

Find an expression for the perimeter of the triangle. Write your answer as simply as possible. GRADE B

3. Simplify as much as possible $\dfrac{(6y)^2+(4y)^2}{6y-4y}$. GRADE B

4. a) Factorise $9x^2-4$.

b) Factorise $6x^2-x-2$.

c) Simplify $\dfrac{6x^2-x-2}{9x^2-4}$. GRADE A

5. Simplify $(2d+3)(d-2)-2(d-2)^2$. GRADE A

6. Expand and simplify. GRADE A

a) $(x-3)(x^2-3x-6)$

b) $(4-x)^3$

c) $x^{\frac{1}{2}}\left(x^{\frac{1}{2}}+x^{-\frac{1}{2}}\right)$ GRADE A*

7. Factorise as much as possible $(5x-3)^2-(3x-5)^2$. GRADE A*

8. Factorise as much as possible $12x^4+30x^3-72x^2$. GRADE A*

3 Geometry recall 1

3.1 Perimeter of compound shapes

THIS SECTION WILL SHOW YOU HOW TO ...

✓ find the perimeter of compound shapes

KEY WORDS

✓ compound shape ✓ perimeter

A **compound shape** is any 2D shape that is made up of other simple shapes such as rectangles and triangles.

EXAMPLE 1

Find the **perimeter** of this compound shape.
The missing lengths of the two unmarked sides are 6 cm and 4 cm.
So, the perimeter = 2 + 6 + 4 + 2 + 6 + 8 = 28 cm

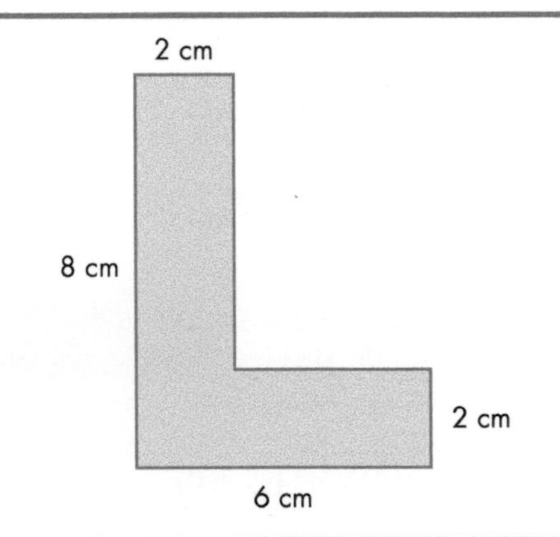

EXERCISE 3A

GRADE C

Find the perimeter of each shape.

1.

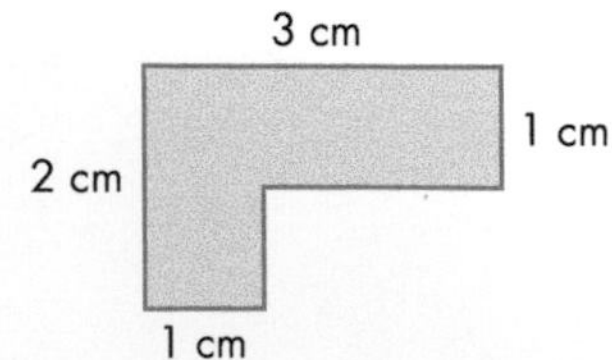

2.

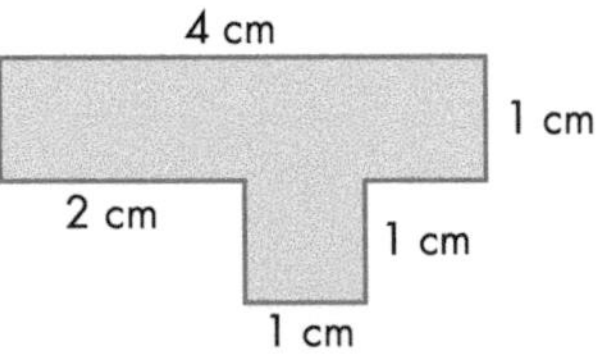

3.

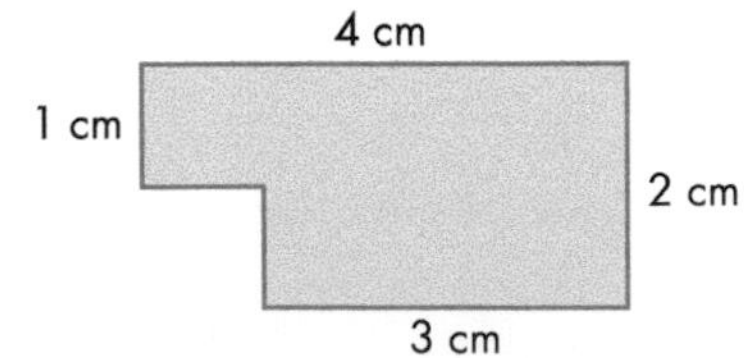

4.

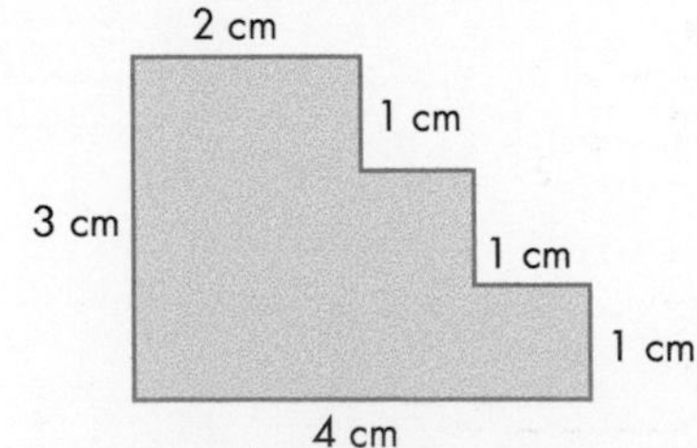

5.

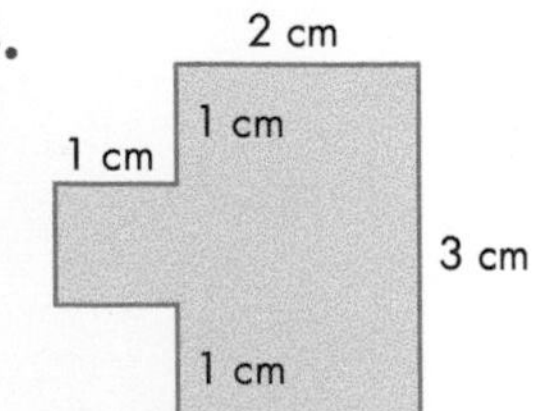

6.

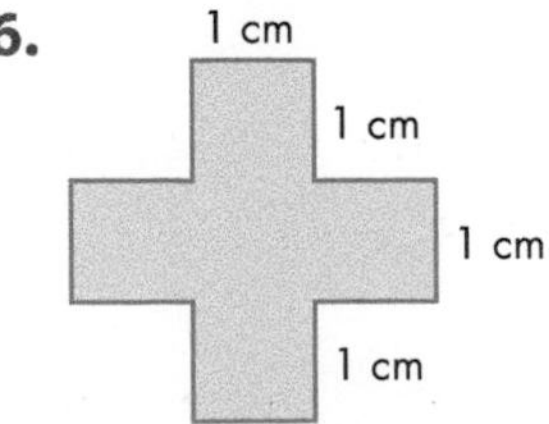

3.2 Area of basic shapes

THIS SECTION WILL SHOW YOU HOW TO ...

✓ use the formula for the area of a rectangle
✓ use the formula $\frac{1}{2}\times$ base $\times$ height for the area of a triangle
✓ use the formula for the area of a trapezium

KEY WORDS

✓ area
✓ triangle
✓ parallelogram
✓ trapezium

- The **area** of a **triangle** = $\frac{1}{2}\times$ base $\times$ perpendicular height.
  $A = \frac{1}{2}bh$
- The area of a **parallelogram** = base $\times$ perpendicular height.
  $A = bh$
- The area of a **trapezium** = $\frac{1}{2}(a + b)h$ where a and b are the lengths of the parallel sides and h is the perpendicular height.

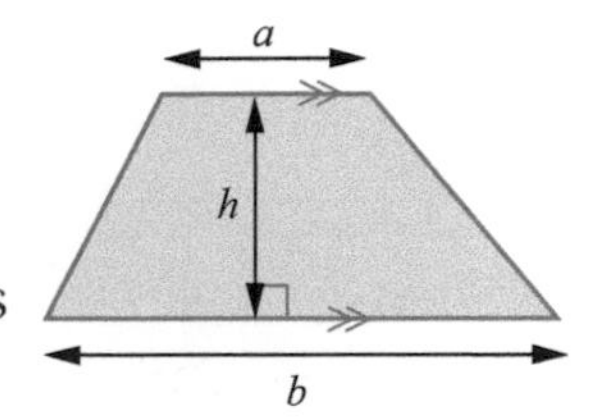

EXAMPLE 2

Calculate the area of the shape shown.

This is a compound shape that can be split into a rectangle (R) and a triangle (T).
Area of the shape

= area of R + area of T = $7\times 2 + \frac{1}{2}\times 2\times 3$
= $14 + 3 = 17$ cm^2

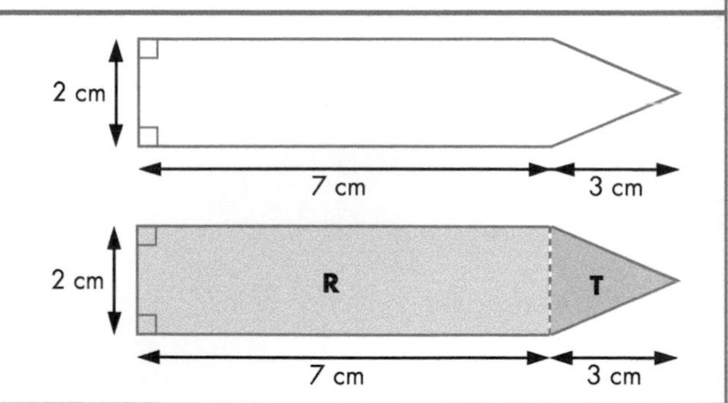

EXAMPLE 3

Find the area of this parallelogram.
Area = 8×6 cm^2
= 48 cm^2

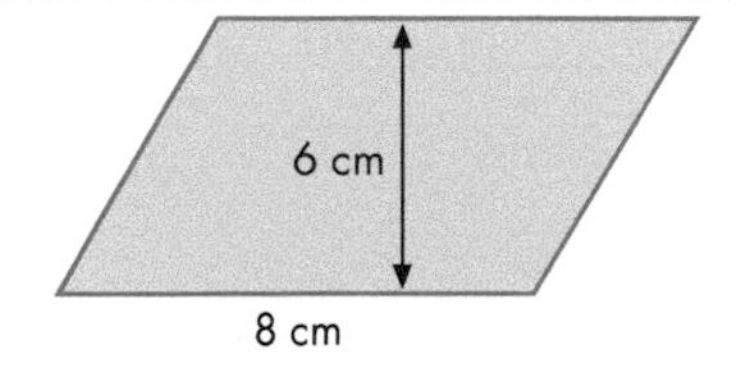

EXAMPLE 4

Find the area of the trapezium ABCD.
$A = \frac{1}{2}(4 + 7)\times 3$ cm^2
= 16.5 cm^2

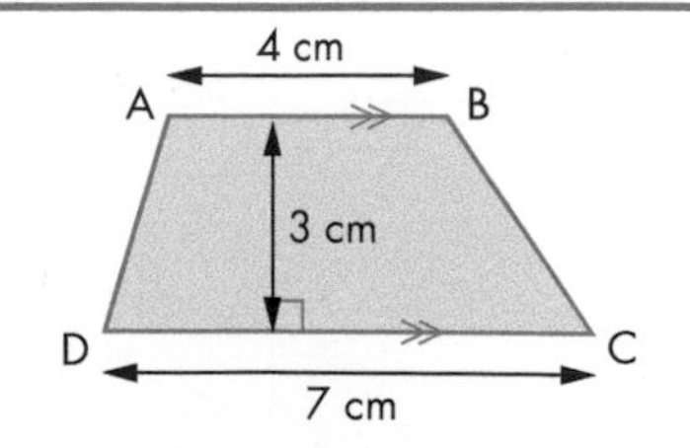

EXERCISE 3B

GRADE C

1. Calculate the area of each shape.

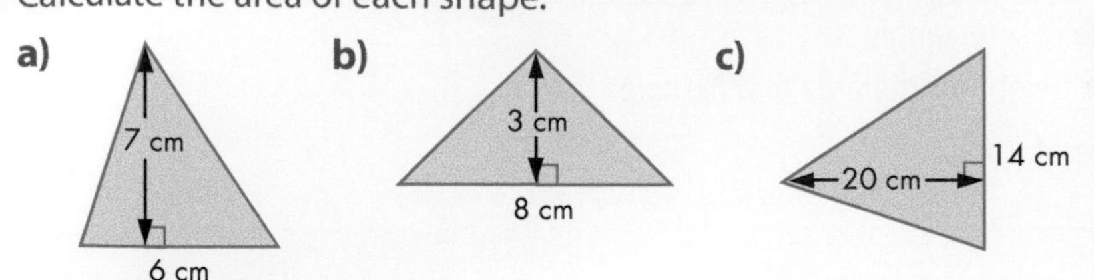

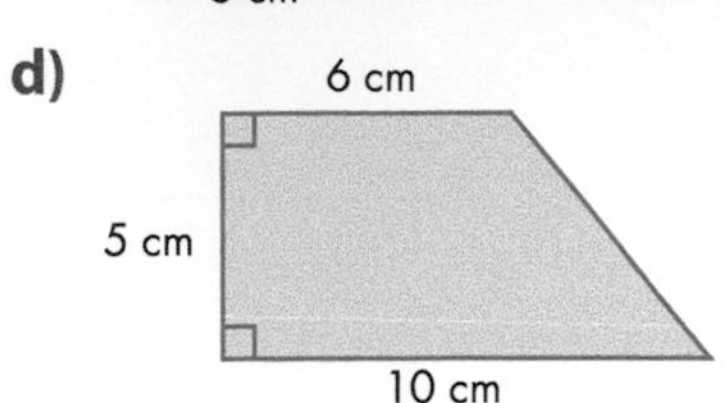

e)
4 m
6 m
4 m
13 m

f)
12 cm
4 cm
10 cm

HINTS AND TIPS

Refer to Example 2 on how to find the area of a compound shape.

2. Find the area of the shaded part of each shape.

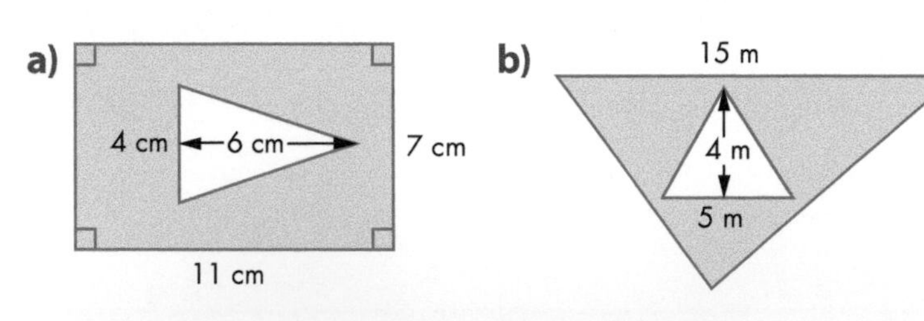

HINTS AND TIPS

Find the area of the outer shape and subtract the area of the inner shape.

3. Calculate the area of each parallelogram below.

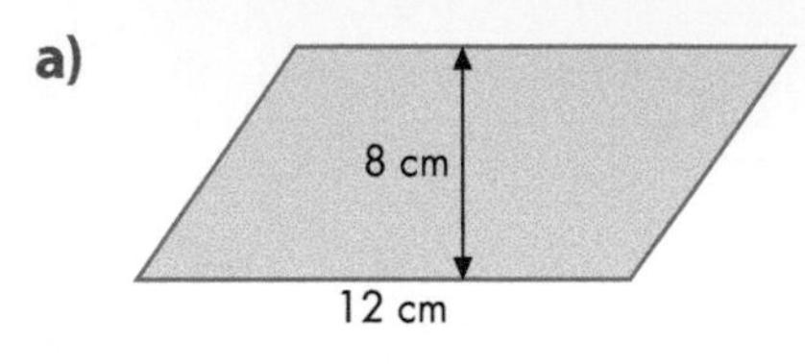

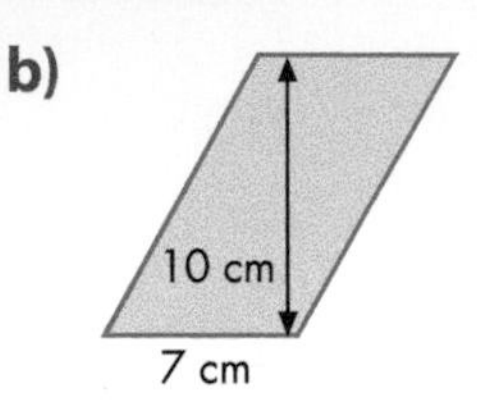

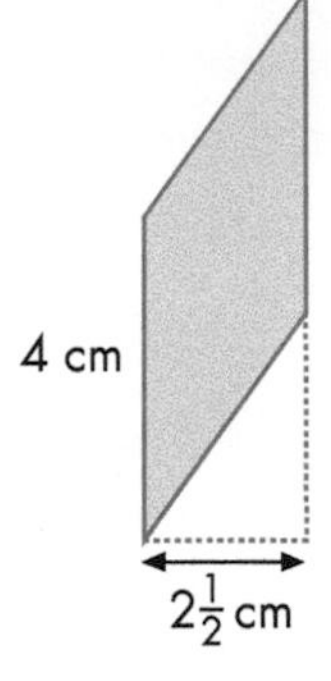

4. Calculate the perimeter and the area of each trapezium.

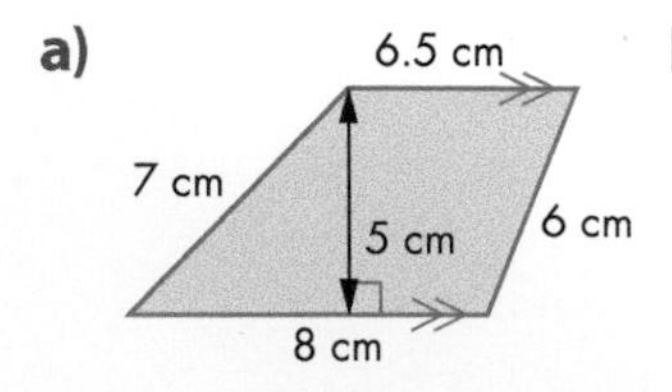

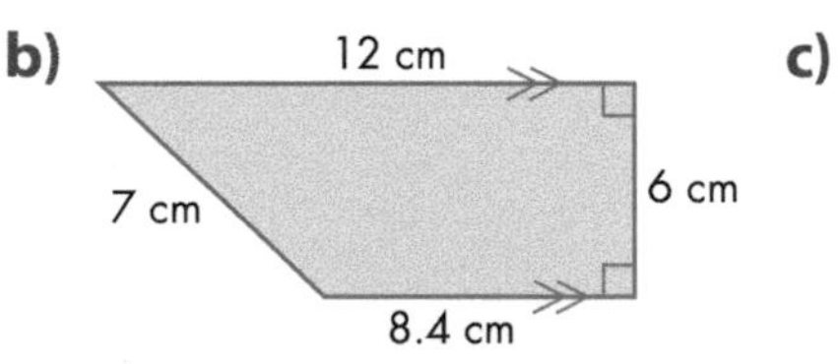

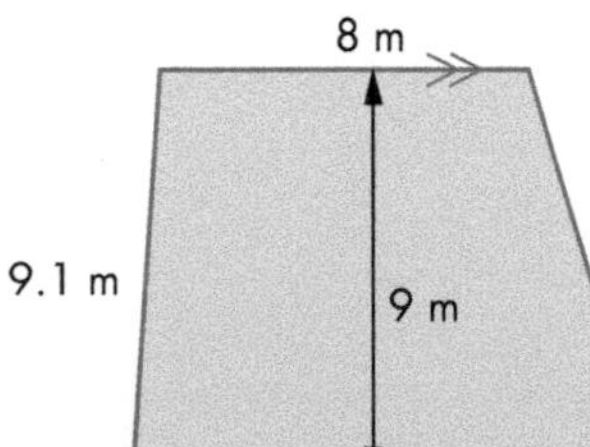

5. Calculate the area of each of these compound shapes.

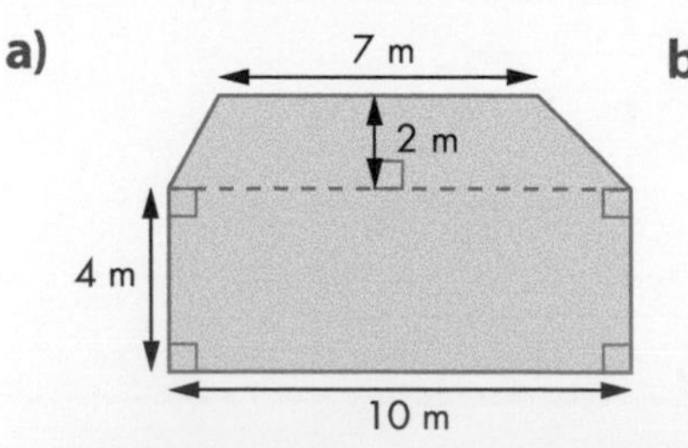

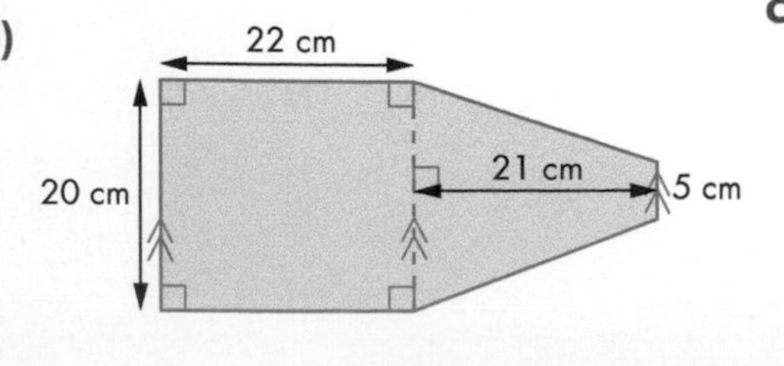

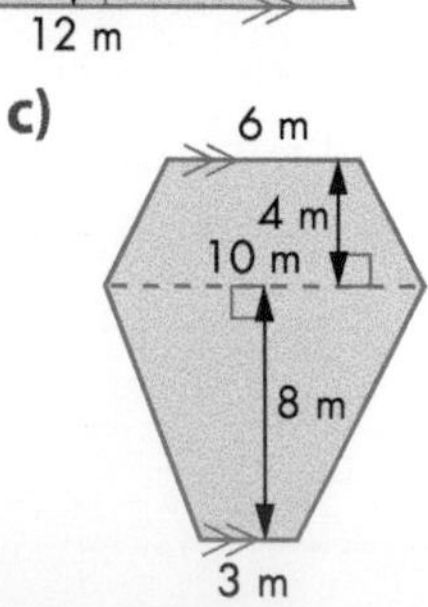

3.3 Circumference and area of a circle

THIS SECTION WILL SHOW YOU HOW TO ...

✓ calculate the circumference and area of a circle

KEY WORDS

✓ circumference ✓ area ✓ π

EXAMPLE 5

Calculate the **circumference** of the circle. Give your answer to 3 significant figures.

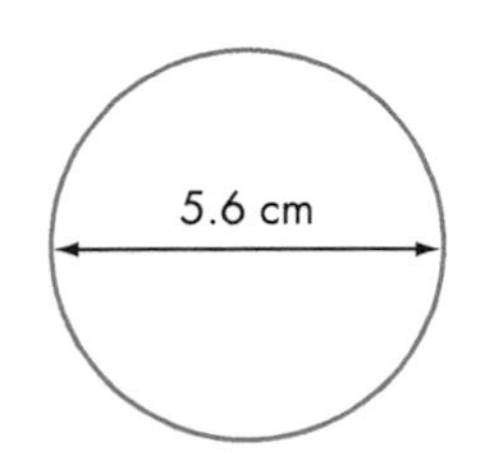

$C = \pi d$

$= \pi \times 5.6$ cm

$= 17.6$ cm (to 3 significant figures)

EXAMPLE 6

Calculate the **area** of the circle. Give your answer in terms of π.

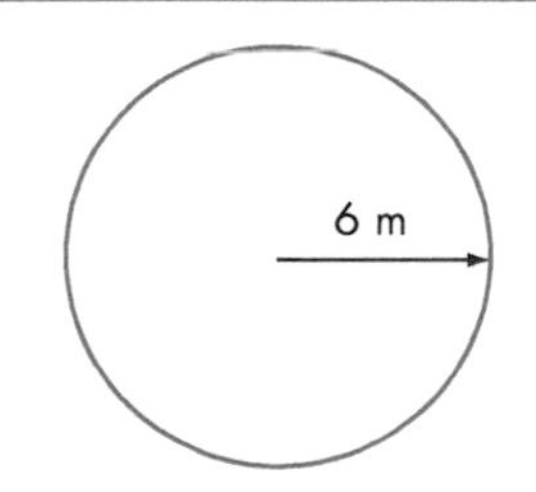

$A = \pi r^2$

$= \pi \times 6^2$ m^2

$= 36\pi$ m^2

EXERCISE 3C

GRADE C

1. Copy and complete the table for each circle. Give your answers to 3 significant figures or one decimal place.

	Radius	Diameter	Circumference	Area
a	4.0 cm			
b	2.6 m			
c		12.0 cm		
d		3.2 m		

2. Find the circumference of each circle. Give your answers in terms of π.

   a) Diameter 5 cm **b)** Radius 4 cm **c)** Radius 9 m **d)** Diameter 12 cm

3. Find the area of each circle. Give your answers in terms of π.

   a) Radius 5 cm **b)** Diameter 12 cm **c)** Radius 10 cm **d)** Diameter 1 m

4. A rope is wrapped eight times around a capstan (a cylindrical post), the diameter of which is 35 cm. How long is the rope?

5. The roller used on a cricket pitch has a radius of 70 cm.

   A cricket pitch has a length of 20 m. How many complete revolutions does the roller make when rolling the pitch?

6. The diameter of each of the following coins is as follows.

   1p: 2.0 cm   2p: 2.6 cm   5p: 1.7 cm   10p: 2.4 cm

   Calculate the area of one face of each coin. Give your answers to 1 decimal place.

7. The distance around the outside of a large pipe is 2.6 m. What is the diameter of the pipe?

8. What is the total perimeter of a semicircle of diameter 15 cm?

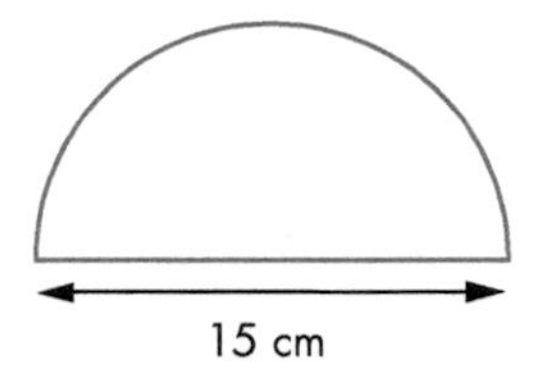

9. A restaurant sells two sizes of pizza. The diameters are 24 cm and 30 cm. The restaurant claims that the larger size is 50% bigger.

   Your friend disagrees and wants to complain to the local trading standards officer. What would you advise? Give a reason for your answer.

10. Calculate the area of each shape, giving your answers in terms of π.

    a)

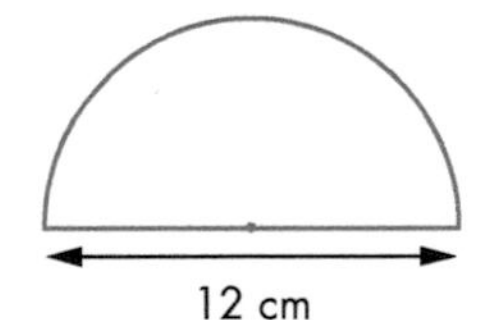

    b)

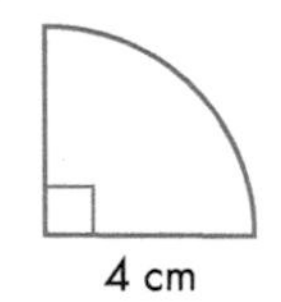

11. Calculate the area of the shaded part of the diagram, giving your answer in terms of π.

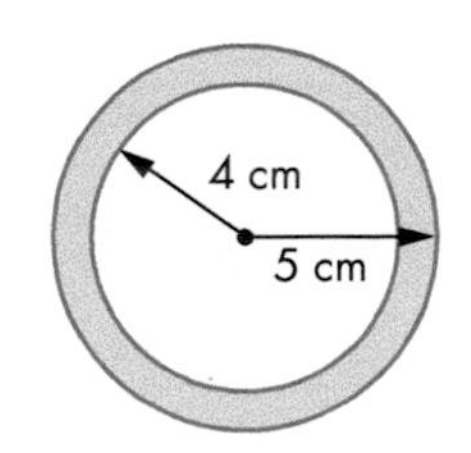

12. This is the plan of a large pond with a gravel path all around it. What area needs to be covered with gravel?

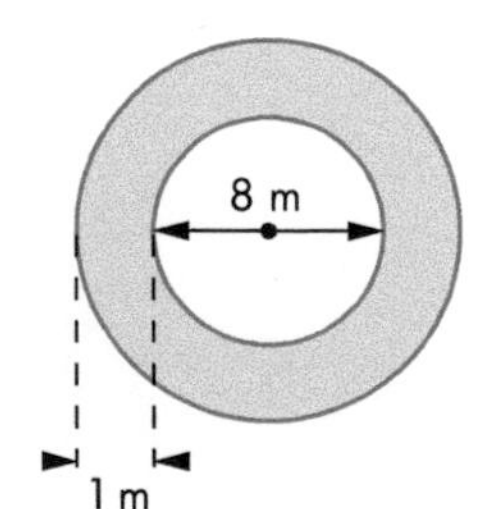

13. A tree in Sequoia National Park in USA is considered to be the largest in the world. It has a circumference at the base of 31.3 m. Would the base of the tree fit inside your classroom?

14. The wheel of a bicycle has a diameter of 70 cm.

    The bicycle travels 100 m.

    How many complete revolutions does the wheel make?

3.4 Volume of a cube, cuboid, prism and pyramid

THIS SECTION WILL SHOW YOU HOW TO ...

- ✓ calculate the volume of a prism
- ✓ calculate the volume and surface area of a cylinder
- ✓ calculate the volume of a pyramid

KEY WORDS

✓ cross-section	✓ cylinder	✓ volume	✓ frustum
✓ prism	✓ surface area	✓ apex	✓ pyramid

A **prism** is a 3D shape that has the same **cross-section** perpendicular to its length.

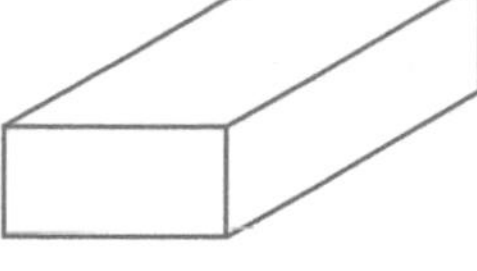
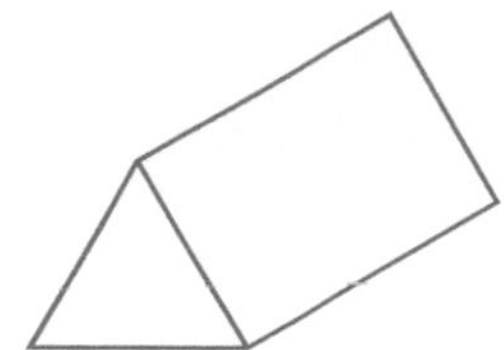
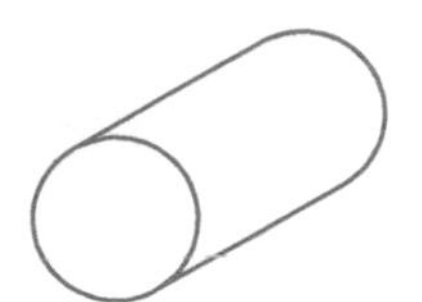

Name:	Cuboid	Triangular prism	Cylinder
Cross-section:	Rectangle	Isosceles triangle	Circle

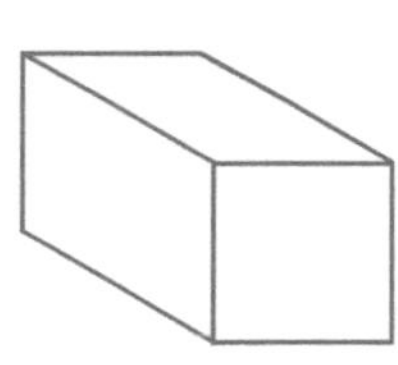
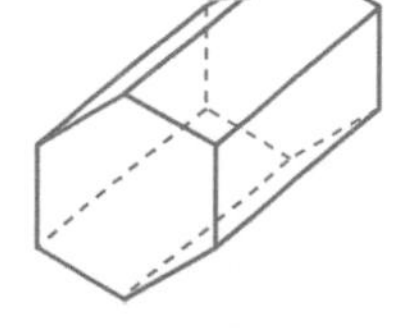

Name:	Cuboid	Hexagonal prism
Cross-section:	Square	Regular hexagon

You find the volume of a prism by multiplying the area of its cross-section by the length of the prism (or height if the prism is stood on end).

That is, volume of prism = area of cross-section × length $V = Al$

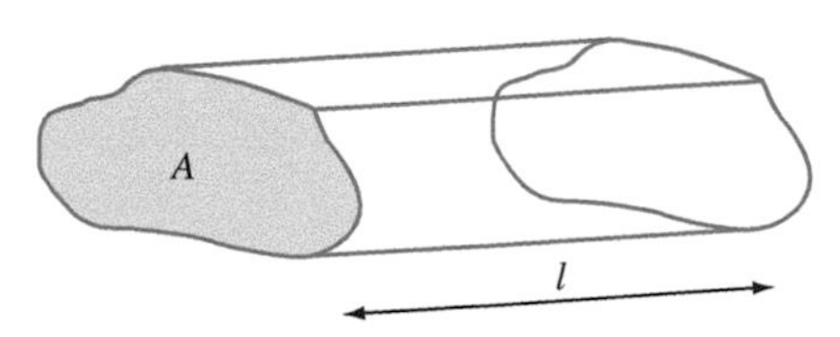

EXAMPLE 7

Find the volume of the triangular prism.

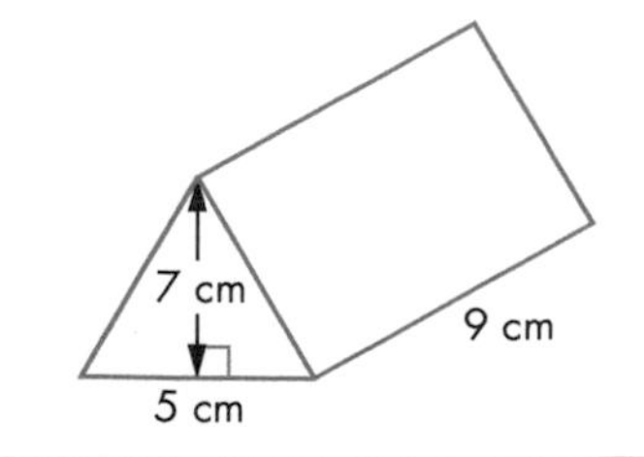

The area of the triangular cross-section $= A = \frac{5 \times 7}{2} = 17.5\text{ cm}^2$

The volume is the area of its cross-section × length =

$Al = 17.5 \times 9 = 157.5\text{ cm}^3$

EXERCISE 3D

GRADE C

1. For each prism shown:

   i) calculate the area of the cross-section **ii)** calculate the volume.

   a)

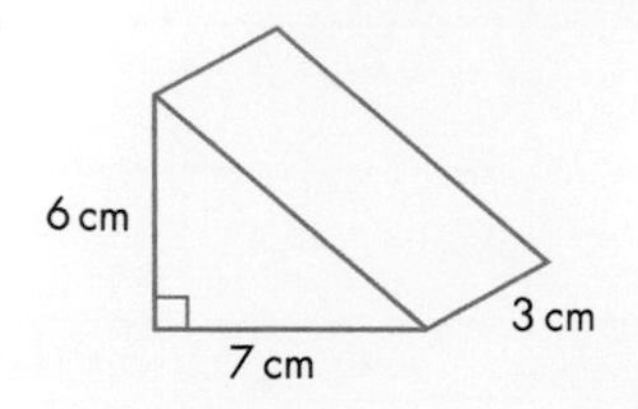

   b)

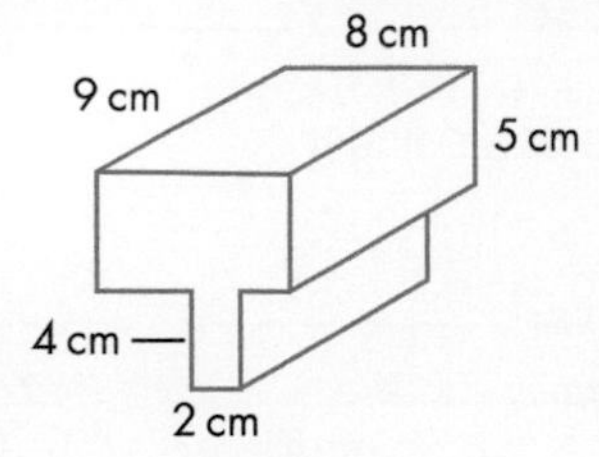

   c)

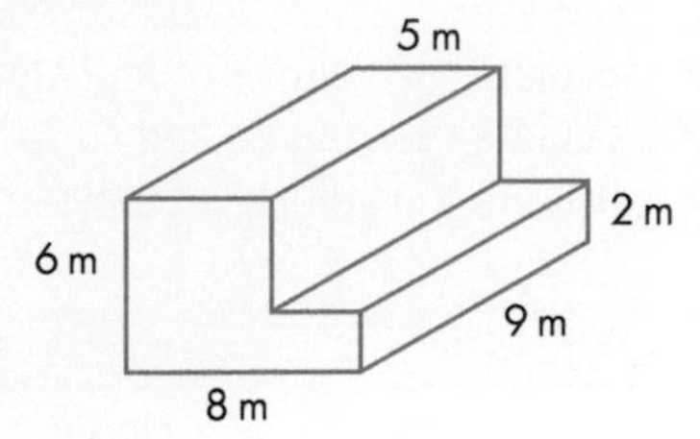

2. A swimming pool is 10 m wide and 25 m long.

   It is 1.2 m deep at one end and 2.2 m deep at the other end. The floor slopes uniformly from one end to the other.

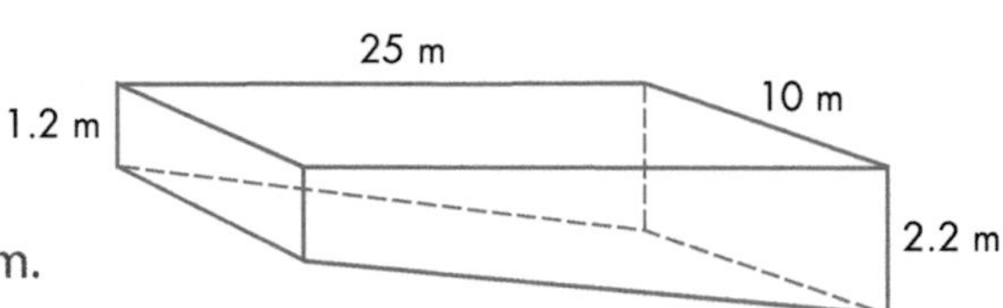

   a) Explain why the shape of the pool is a prism.

   b) The pool is filled with water at a rate of 2 m^3 per minute. How long will it take to fill the pool?

3. Each of these prisms has a uniform cross-section in the shape of a right-angled triangle.

   a) Find the volume of each prism.

   b) Find the total surface area of each prism.

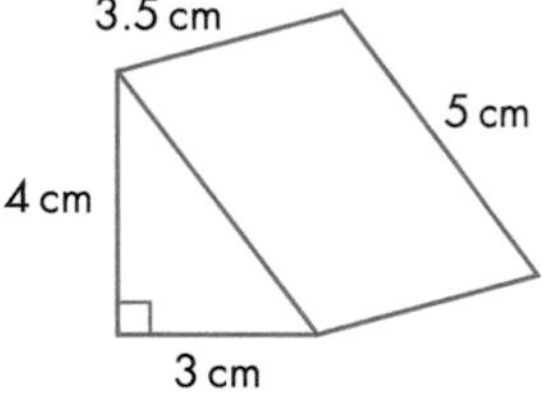

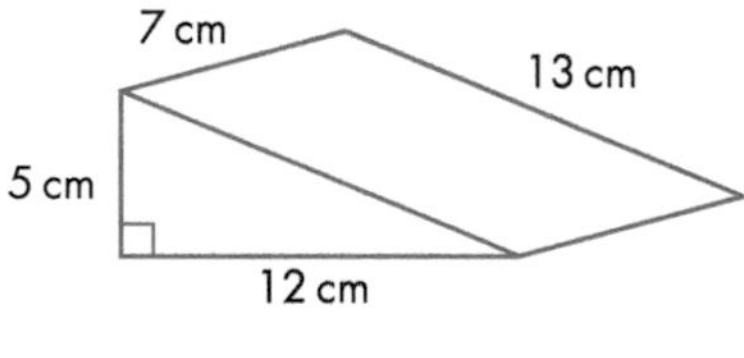

GRADE B

4. The top and bottom of the container shown here are the same size, both consisting of a rectangle, 4 cm by 9 cm, with a semicircle at each end. The depth is 3 cm.

   Find the volume of the container.

5. A horse trough is in the shape of a semicircular prism as shown.

   What volume of water will the trough hold when it is filled to the top? Give your answer in litres.

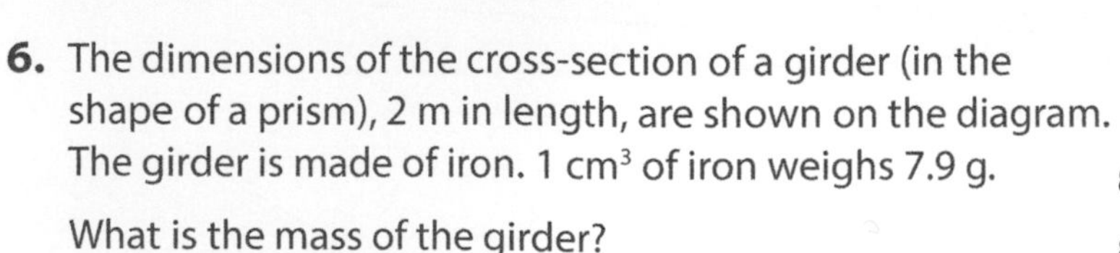

6. The dimensions of the cross-section of a girder (in the shape of a prism), 2 m in length, are shown on the diagram. The girder is made of iron. 1 cm^3 of iron weighs 7.9 g.

   What is the mass of the girder?

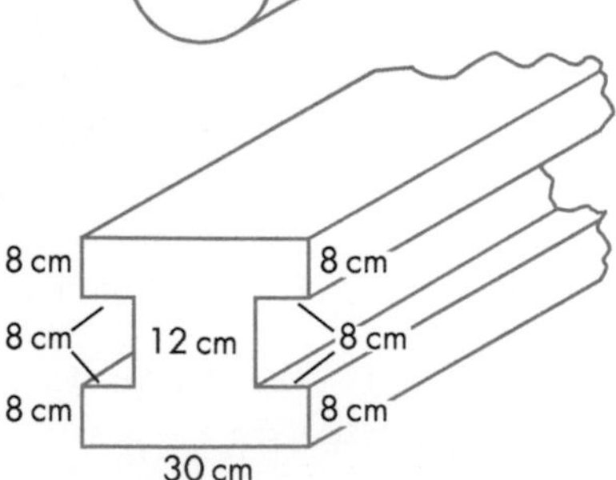

Volume

Since a **cylinder** is an example of a prism, its **volume** is found by multiplying the area of one of its circular ends by the height.

That is, volume = $\pi r^2 h$

where r is the radius of the cylinder and h is its height or length.

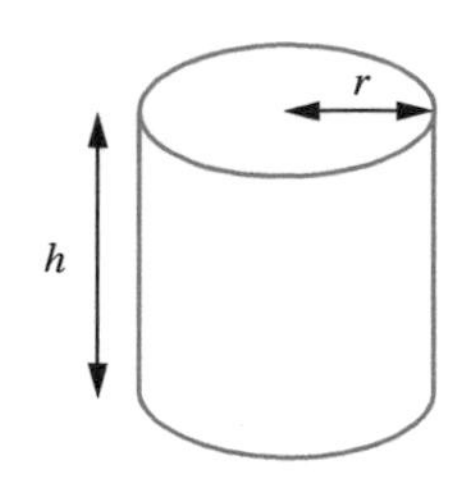

EXAMPLE 8

What is the volume of a cylinder having a radius of 5 cm and a height of 12 cm?

Volume = area of circular base × height

$= \pi r^2 h$

$= \pi \times 5^2 \times 12 \text{ cm}^3$

$= 942 \text{ cm}^3$ (3 significant figures)

Surface area

The total **surface area** of a cylinder is made up of the area of its curved surface plus the area of its two circular ends.

The curved surface, when opened out, is a rectangle with length equal to the circumference of the circular end.

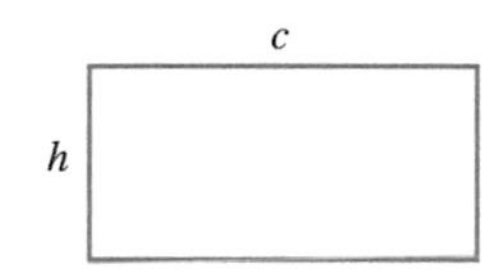

curved surface area = circumference of end × height of cylinder

$= 2\pi rh$ **or** πdh

area of one end = πr^2

Therefore, total surface area = $2\pi rh + 2\pi r^2$ **or** $\pi dh + 2\pi r^2$

EXAMPLE 9

What is the total surface area of a cylinder with a radius of 15 cm and a height of 2.5 m?

First, you must change the dimensions to a *common unit*. Use centimetres in this case.

Total surface area = $\pi dh + 2\pi r^2$

$= \pi \times 30 \times 250 + 2 \times \pi \times 15^2 \text{ cm}^2$

$= 23\,562 + 1414 \text{ cm}^2$

$= 24\,976 \text{ cm}^2$

$= 25\,000 \text{ cm}^2$ (3 significant figures)

EXERCISE 3E

GRADE C

1. For the cylinders below find: **i)** the volume **ii)** the total surface area.

a)

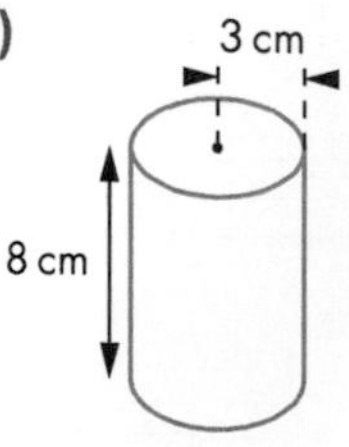

b)

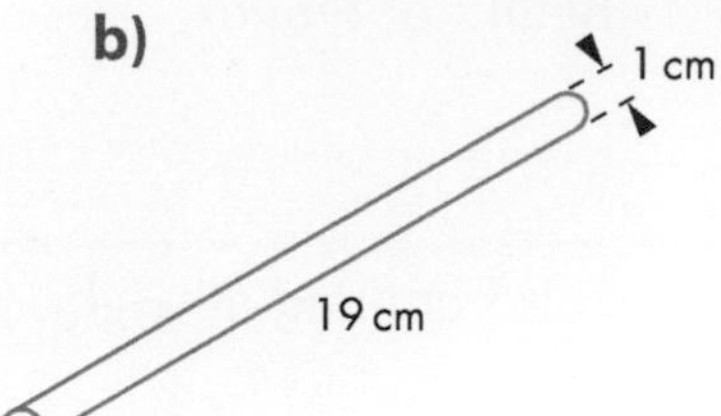

c)

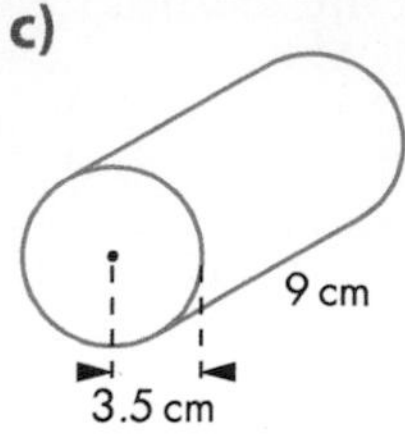

d)

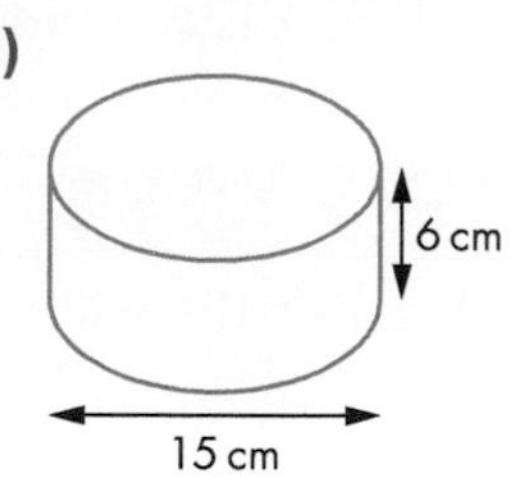

2. For each of these cylinder dimensions find: **i)** the volume **ii)** the curved surface area.

Give your answers in terms of π.

a) Base radius 3 cm and height 8 cm **b)** Base diameter 8 cm and height 7 cm

3. The diameter of a marble, cylindrical column is 60 cm and its height is 4.2 m. The cost of making this column is quoted as £67.50 per cubic metre. What is the estimated total cost of making the column?

4. Find the mass of a solid iron cylinder 55 cm high with a base diameter of 60 cm. 1 cm³ iron has a mass of 7.9 g.

5. A cylindrical container is 65 cm in diameter. Water is poured into the container until it is 1 m deep. How much water is in the container? Give your answer in litres.

A **pyramid** is a 3D shape with a base from which triangular faces rise to a common vertex, called the **apex**. The base can be any polygon, but is usually a triangle, a rectangle or a square.

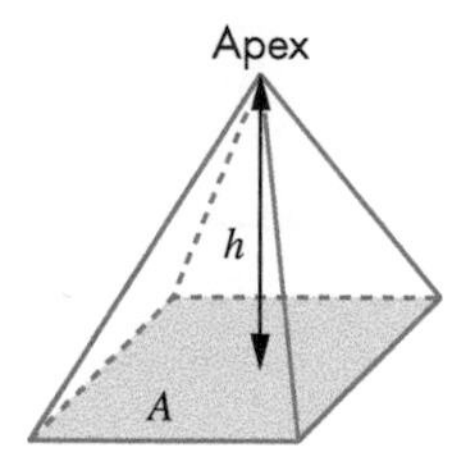

The **volume** of a pyramid is given by:

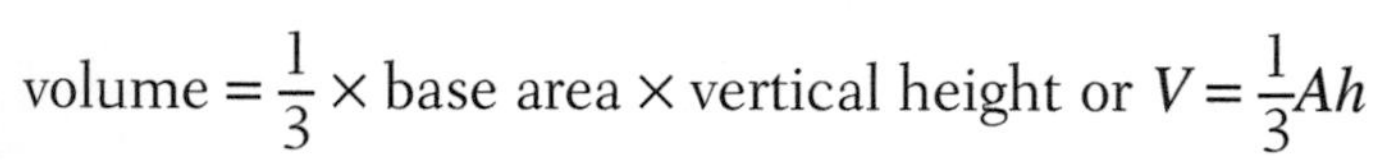

$$\text{volume} = \frac{1}{3} \times \text{base area} \times \text{vertical height or } V = \frac{1}{3}Ah$$

where A is the base area and h is the vertical height.

EXAMPLE 10

Calculate the volume of the pyramid on the right.

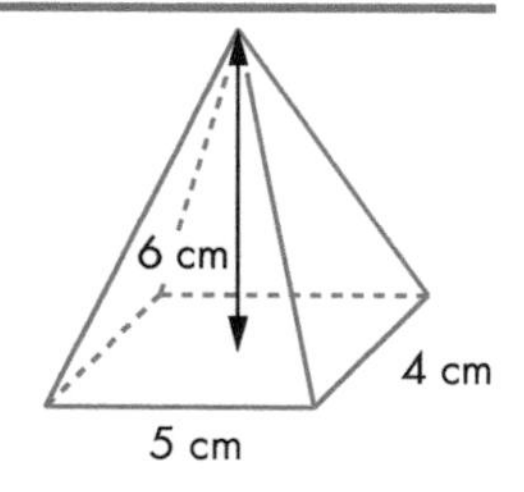

Base area = 5 × 4 = 20 cm²

$$\text{Volume} = \frac{1}{3} \times 20 \times 6 = 40 \text{ cm}^3$$

EXAMPLE 11

A pyramid, with a square base of side 8 cm, has a volume of 320 cm³.
What is the vertical height of the pyramid?

Let h be the vertical height of the pyramid. Then,

$$\text{volume} = \frac{1}{3} \times 64 \times h = 320 \text{ cm}^3 \quad \Rightarrow \frac{64h}{3} = 320 \text{ cm}^3 \quad \Rightarrow h = \frac{960}{64} \text{ cm} \quad \Rightarrow h = 15 \text{ cm}$$

EXERCISE 3F

GRADE A

1. Calculate the volume of each pyramid, all with rectangular bases.

a) 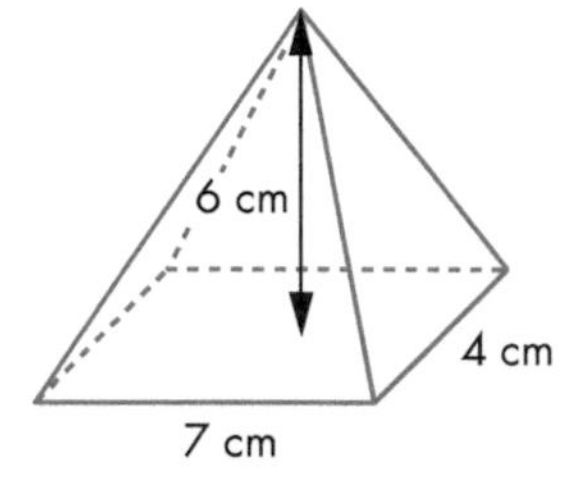

b) 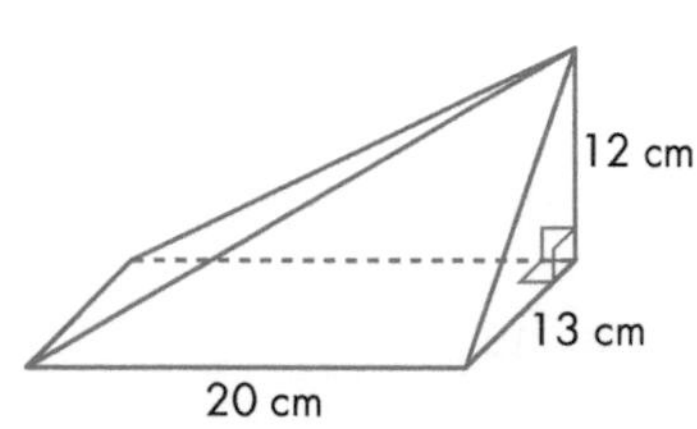

c) 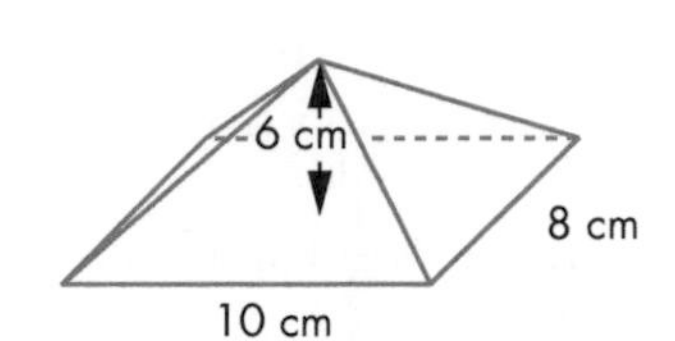

2. Suppose you have six pyramids each with a height that is equal to half the length of a side of the square base.
   a) Explain how they can fit together to make a cube.
   b) How does this show that the formula for the volume of a pyramid is correct?

3. Calculate the volume of each of these shapes.

a) 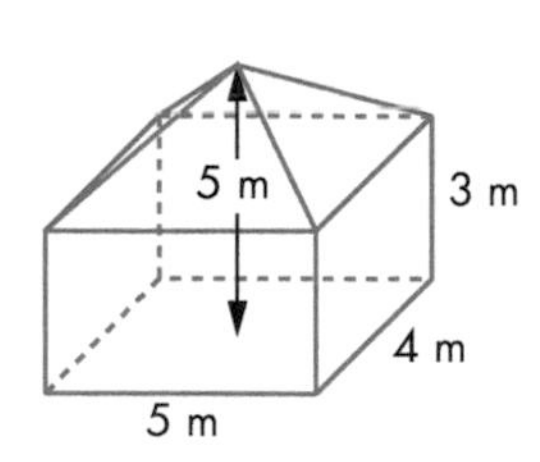

b) 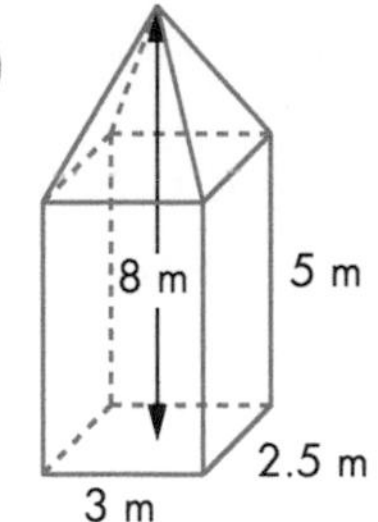

c) 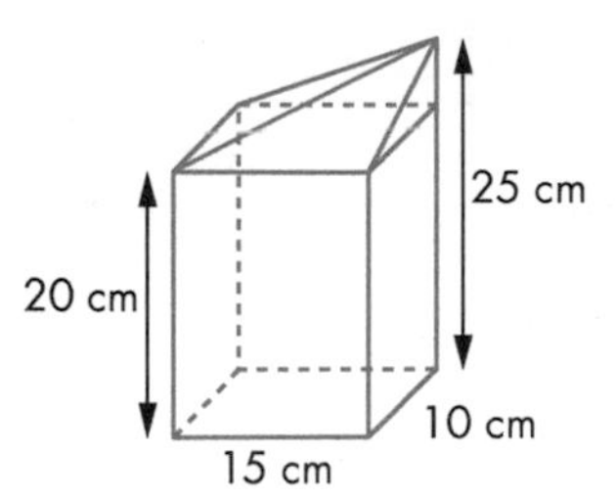

4. A pyramid has a square base of side 6.4 cm. Its volume is 81.3 cm³. Calculate the height of the pyramid.

5. The pyramid in the diagram has its top 5 cm cut off as shown. The shape which is left is called a **frustum**. Calculate the volume of the frustum.

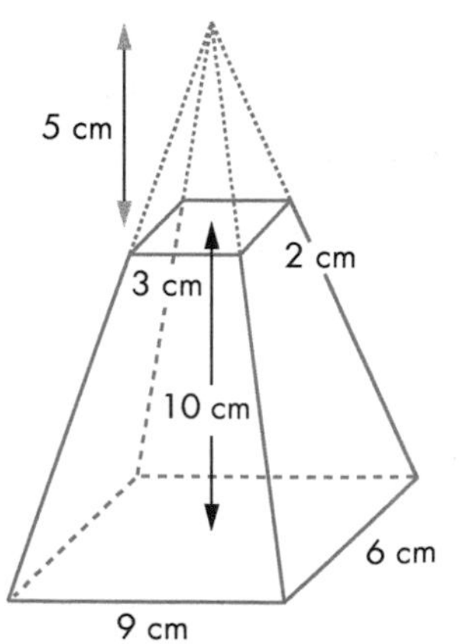

3.5 Volume of a cone and a sphere

THIS SECTION WILL SHOW YOU HOW TO ...

✓ calculate the volume and surface area of a cone
✓ calculate the volume and surface area of a sphere

KEY WORDS

✓ slant height ✓ surface area ✓ volume
✓ sphere ✓ vertical height

Cones

A cone can be treated as a pyramid with a circular base. Therefore, the formula for the **volume** of a cone is the same as that for a pyramid.

So volume $= \frac{1}{3} \times$ base area $\times$ vertical height

$$V = \frac{1}{3}\pi r^2 h$$

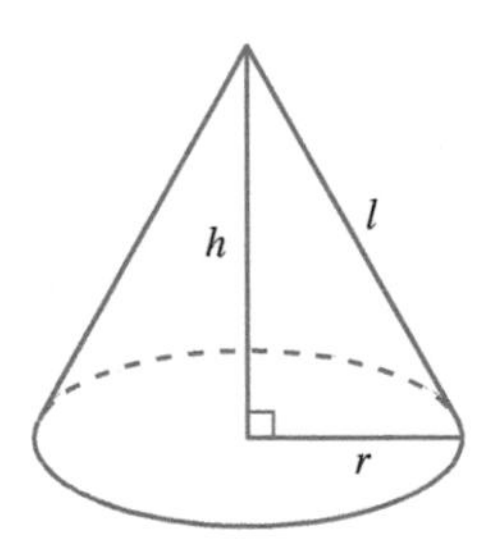

where r is the radius of the base and h is the **vertical height** of the cone.

The curved **surface area** of a cone is given by:

curved surface area $= \pi \times$ radius $\times$ slant height

$$S = \pi r l$$

where l is the **slant height** of the cone.

So the total surface area of a cone is given by the curved surface area plus the area of its circular base.

$A = \pi r l + \pi r^2$

EXAMPLE 12

For the cone in the diagram, calculate:

a) its volume **b)** its total surface area.

Give your answers in terms of π.

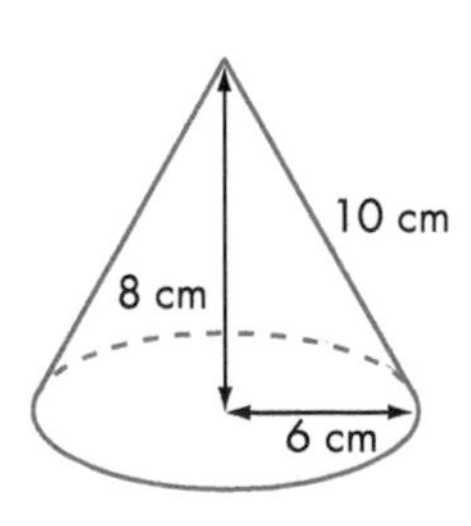

a) The volume is given by $V = \frac{1}{3}\pi r^2 h$

$$= \frac{1}{3} \times \pi \times 36 \times 8 = 96\pi \text{ cm}^3$$

b) The total surface area is given by $A = \pi r l + \pi r^2$

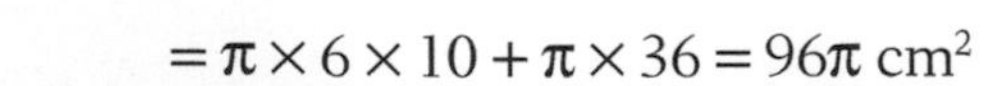

$$= \pi \times 6 \times 10 + \pi \times 36 = 96\pi \text{ cm}^2$$

EXERCISE 3G

GRADE A

1. For each cone, calculate:

i) its volume **ii)** its total surface area.

Give your answers to 3 significant figures.

a)

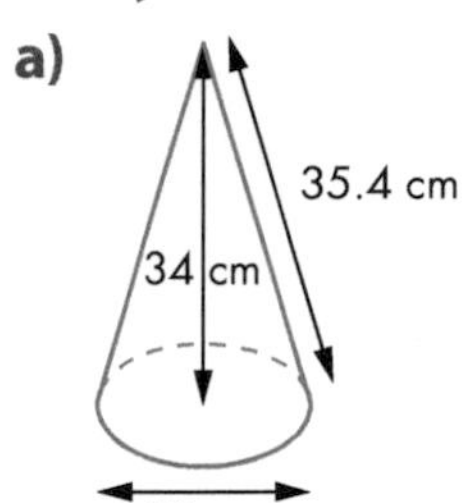

b)

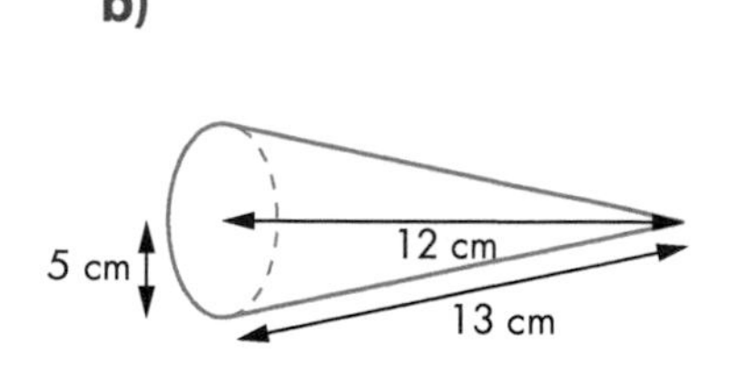

c)

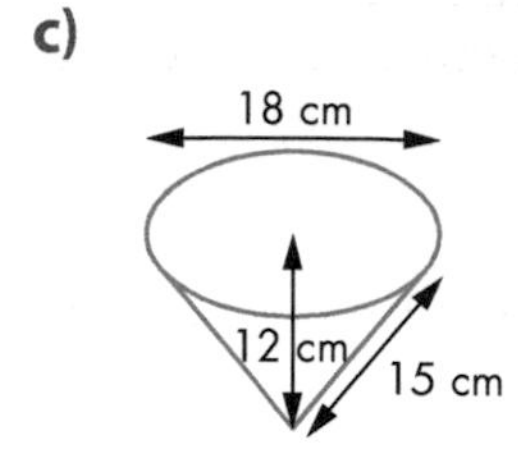

2. Find the total surface area of a cone whose base radius is 3 cm and slant height is 5 cm. Give your answer in terms of π.

GRADE A*

3. Calculate the volume of each of these shapes. Give your answers in terms of π.

a)

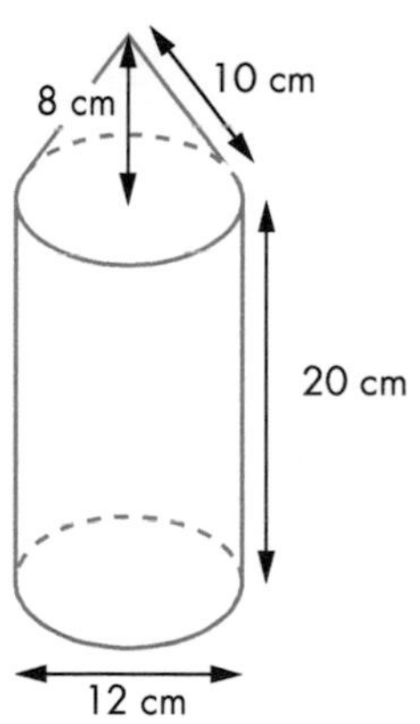

b)

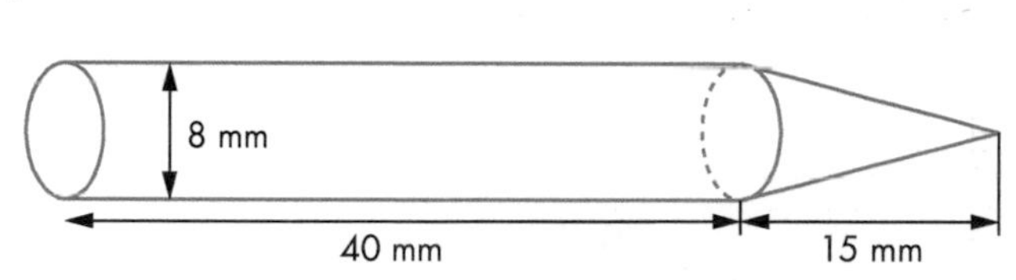

4. You could work with a partner on this question.

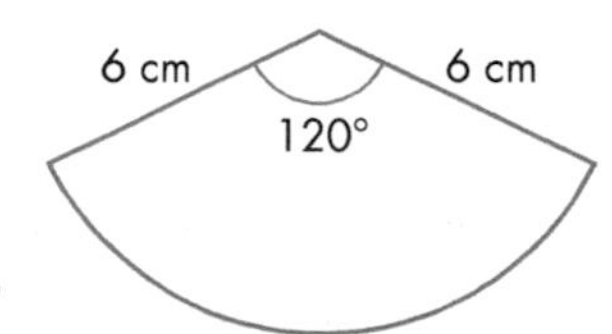

A sector of a circle, as in the diagram, can be made into a cone (without a base) by sticking the two straight edges together.

a) What would be the diameter of the base of the cone in this case?

b) What is the diameter if the angle is changed to 180°?

c) Investigate other angles.

5. A cone has base diameter 6 cm, vertical height 4 cm and slant height 5 cm. Calculate the total surface area, leaving your answer in terms of π.

6. If the slant height of a cone is equal to the base diameter, show that the area of the curved surface is twice the area of the base.

7. A container in the shape of a cone, base radius 10 cm and vertical height 19 cm, is full of water. The water is poured into an empty cylinder of radius 15 cm. How high is the water in the cylinder?

Spheres

The **volume** of a **sphere**, radius r, is given by:

$V = \frac{4}{3}\pi r^3$

Its **surface area** is given by:

$A = 4\pi r^2$

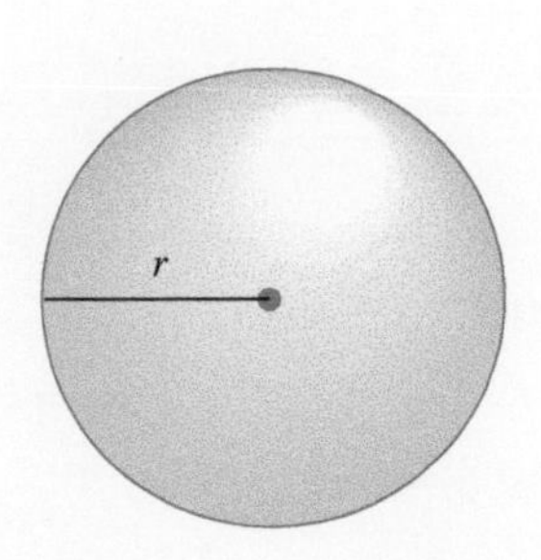

EXAMPLE 13

For a sphere of radius of 8 cm, calculate **a** its volume **b** its surface area.

a) The volume is given by:

$$V = \frac{4}{3}\pi r^3$$

$$= \frac{4}{3} \times \pi \times 8^3 = \frac{2048}{3} \times \pi = 2140 \text{ cm}^3 \quad \text{(3 significant figures)}$$

b) The surface area is given by:

$$A = 4\pi r^2$$

$$= 4 \times \pi \times 8^2 = 256 \times \pi = 804 \text{ cm}^2 \text{ (3 significant figures)}$$

EXERCISE 3H

GRADE A

1. Calculate the volume of each sphere. Give your answers in terms of π.

a) Radius 3 cm **b)** Radius 6 cm **c)** Diameter 20 cm

2. Calculate the surface area of each sphere. Give your answers in terms of π.

a) Radius 3 cm **b)** Radius 5 cm **c)** Diameter 14 cm

3. Calculate the volume and the surface area of a sphere with diameter 50 cm.

4. A sphere fits exactly into an open cubical box of side 25 cm. Calculate:

a) the surface area of the sphere **b)** the volume of the sphere.

5. A metal sphere of radius 15 cm is melted down and recast into a solid cylinder of radius 6 cm. Calculate the height of the cylinder.

6. A sphere has a radius of 5.0 cm. A cone has a base radius of 8.0 cm. The sphere and the cone have the same volume. Calculate the height of the cone.

7. A sphere of diameter 10 cm is carved out of a wooden block in the shape of a cube of side 10 cm. What percentage of the wood is wasted?

8. A manufacturer is making cylindrical packaging for a sphere as shown. The curved surface of the cylinder is made from card.

Show that the area of the card is the same as the surface area of the sphere.

Exam-style questions

Give your answers to a reasonable degree of accuracy.

1. The length of the side of a wooden cube is 10 cm.

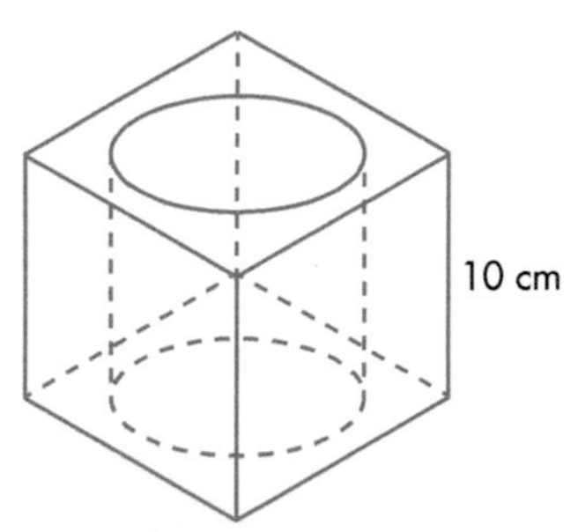

A cylindrical hole is drilled through the cube.
The volume of wood removed is half the original volume of the cube.
Calculate the diameter of the hole. GRADE B

2. The sides of a cuboid are $2a$ cm, $3a$ cm and $4a$ cm. GRADE B

a) Find an expression, in terms of a, for the volume of the cuboid.
b) Find an expression, in terms of a, for the surface area of the cuboid.

3. The radius of a sphere is r cm.

The value of the surface area of the sphere in cm² is greater than the value of the volume in cm³.
What does this imply about the value of r? GRADE A

4. The diagram shows a trapezium.

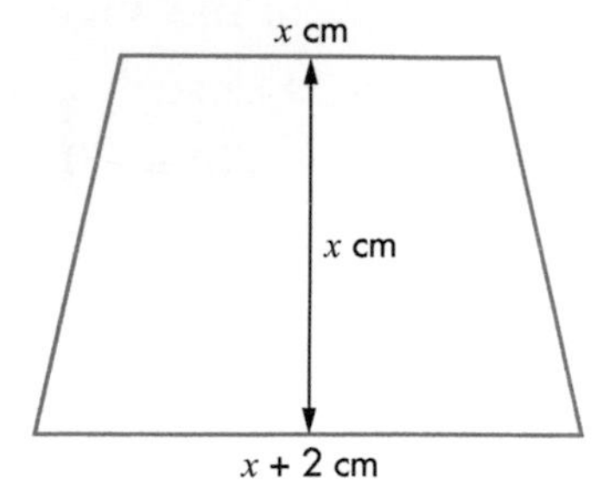

a) Find an expression, in terms of x, for the area of the trapezium.
b) A circle of radius r cm has the same area as the trapezium.
Find an expression for r in terms of x.

5. Calculate the volume of a pyramid having a square base of side 9 cm and a vertical height of 10 cm.

6. A pyramid has the same volume as a cube of side 10.0 cm.
The height of the pyramid is the same as the length of the side of its square base.
Calculate the height of the pyramid. GRADE A*

7. The volume of a piece of steel is 30 cm³.

How many ball bearings with a diameter of 3 mm can be made from this?

Geometry recall 2

4.1 Special triangles and quadrilaterals

THIS SECTION WILL SHOW YOU HOW TO ...

✓ work out the sizes of angles in triangles and quadrilaterals

KEY WORDS

✓ equilateral triangle
✓ isosceles triangle
✓ kite
✓ parallelogram
✓ rhombus
✓ trapezium

Special triangles

An **equilateral triangle** is a triangle with all its sides equal.

Therefore, all three interior angles are 60°.

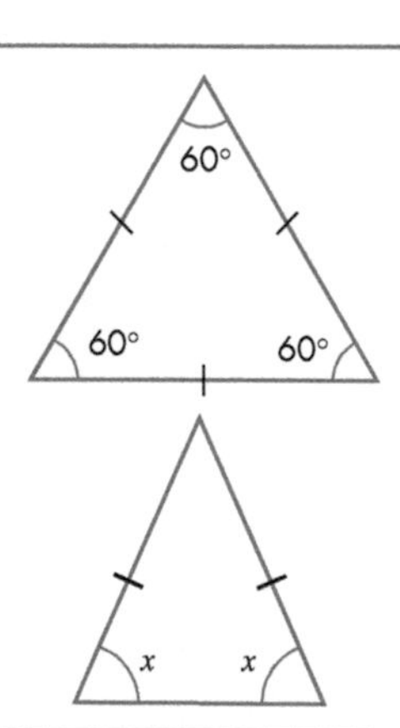

An **isosceles triangle** is a triangle with two equal sides, and therefore with two equal angles.

Notice how to mark the equal sides and equal angles.

EXAMPLE 1

Find the size of the angle marked a in the triangle.

The triangle is isosceles, so both base angles are 70°.

So $a = 180° - (70° + 70°) = 180° - 140° = 40°$

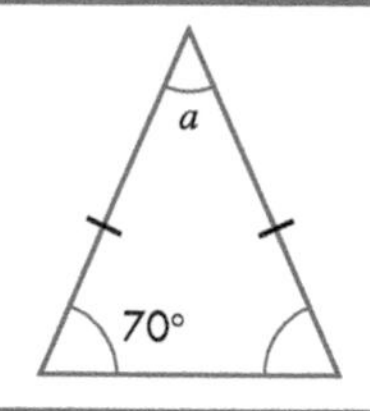

Special quadrilaterals

A **parallelogram** has opposite sides that are parallel.

Its opposite sides are equal. Its diagonals bisect each other. Its opposite angles are equal: that is, ∠A = ∠C and ∠B = ∠D.

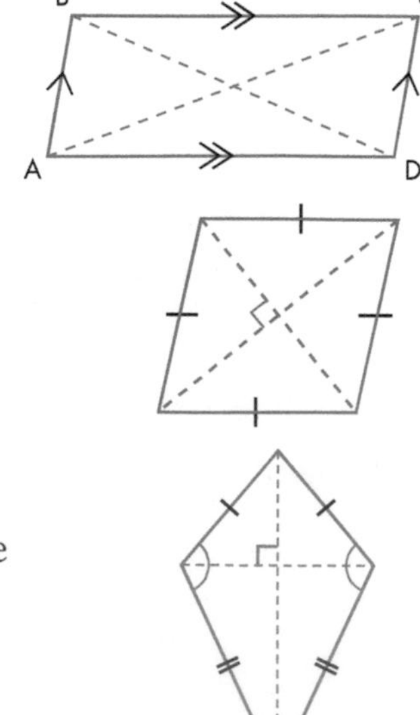

A **rhombus** is a parallelogram with all its sides equal.

Its diagonals bisect each other at right angles. Its diagonals also bisect the angles at the vertices.

A **kite** is a quadrilateral with two pairs of equal adjacent sides. Its longer diagonal bisects its shorter diagonal at right angles. The opposite angles between the sides of different lengths are equal.

A **trapezium** has two parallel sides.

The sum of the interior angles between the parallel sides is 180°: that is, ∠A + ∠D = 180° and ∠B + ∠C = 180°.

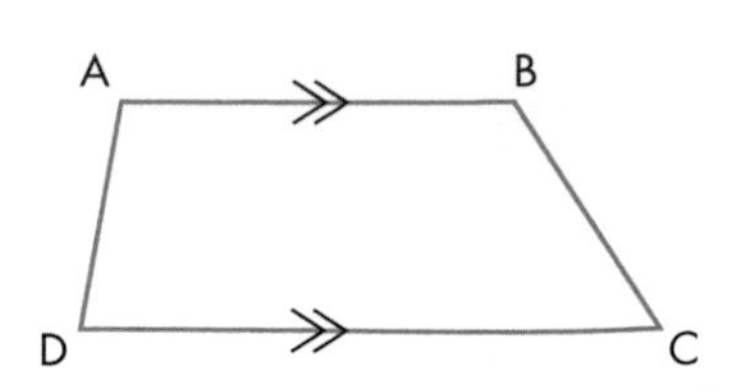

EXAMPLE 2

Find the size of the angles marked x and y in this parallelogram.

$x = 55°$ (opposite angles are equal) and $y = 125°$
$(x + y = 180°)$

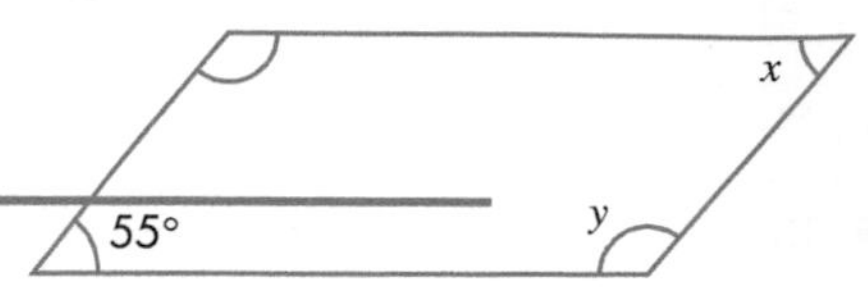

EXERCISE 4A

GRADE C

1. Calculate the sizes of the lettered angles in each shape.

a)

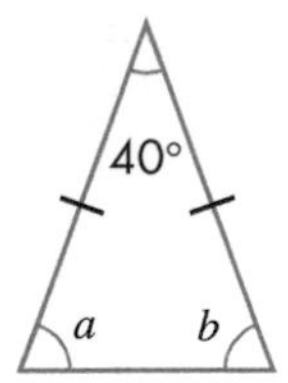

b)

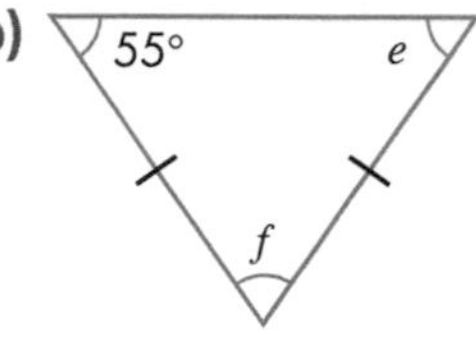

c)

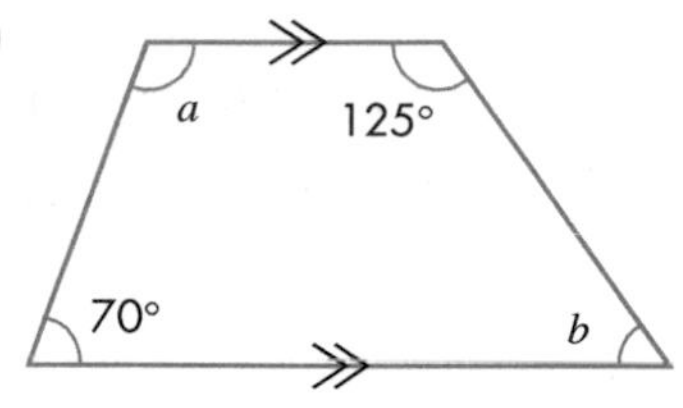

d)

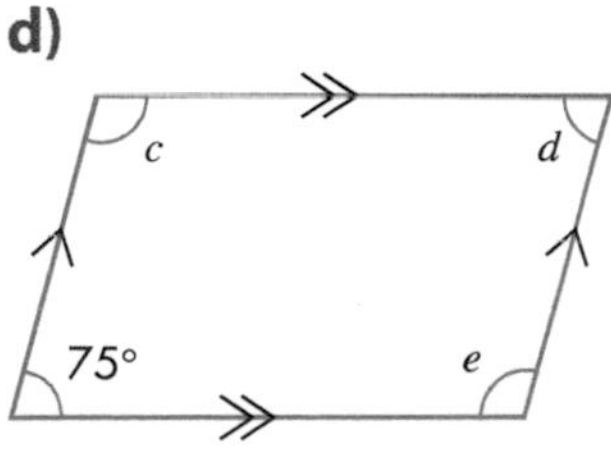

e)

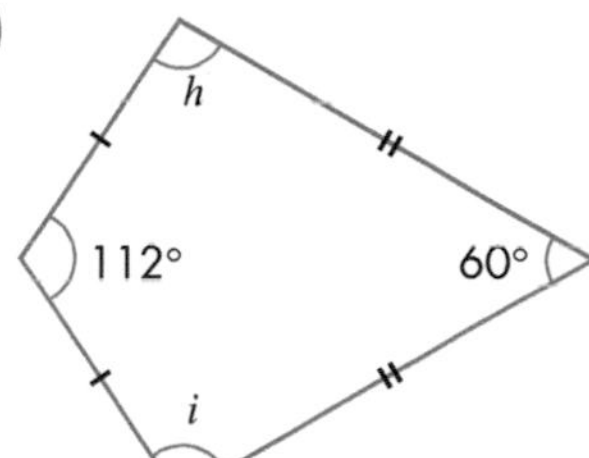

f)

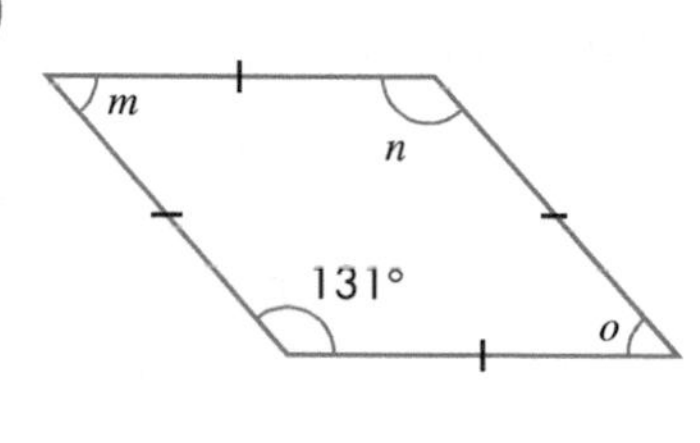

2. The three angles of an isosceles triangle are $2x$, $x - 10$ and $x - 10$. What is the actual size of each angle?

3. Calculate the sizes of the lettered angles in these diagrams.

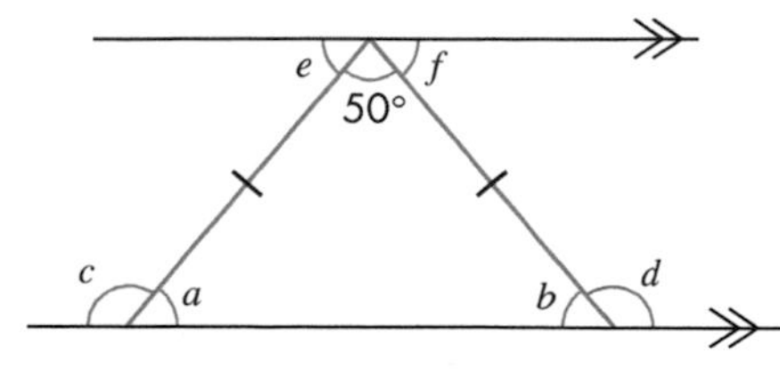

4. Calculate the values of x and y in each of these quadrilaterals.

a)

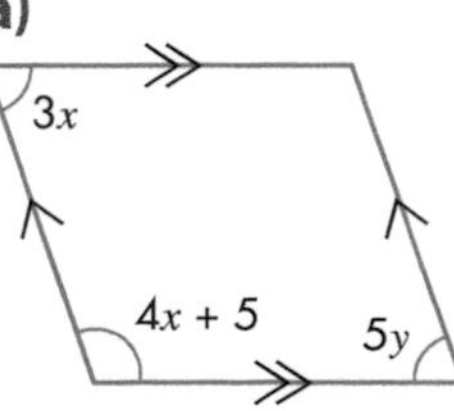

b)

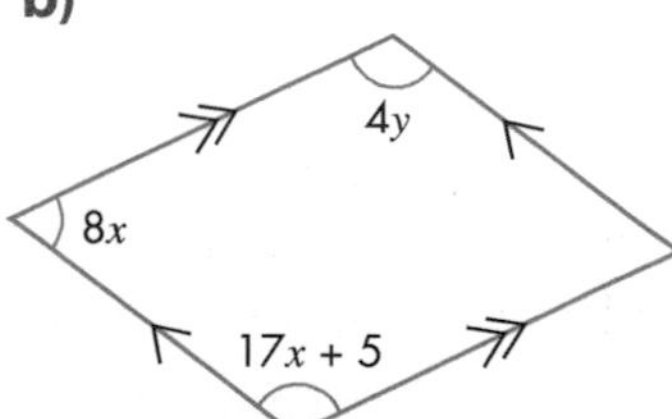

c)

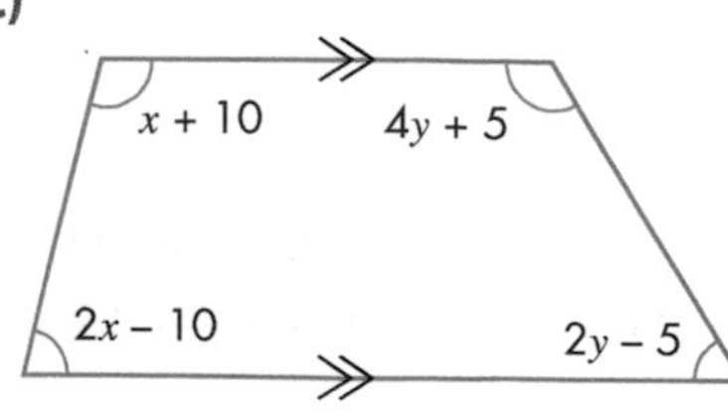

5. Find the value of x in each of these quadrilaterals and hence state what type of quadrilateral it could be.

a) A quadrilateral with angles $x + 10°$, $x + 20°$, $2x + 20°$, $2x + 10°$

b) A quadrilateral with angles $x - 10°$, $2x + 10°$, $x - 10°$, $2x + 10°$

c) A quadrilateral with angles $x - 10°$, $2x$, $5x - 10°$, $5x - 10°$

d) A quadrilateral with angles $4x + 10°$, $5x - 10°$, $3x + 30°$, $2x + 50°$

6. The diagram shows a parallelogram ABCD.

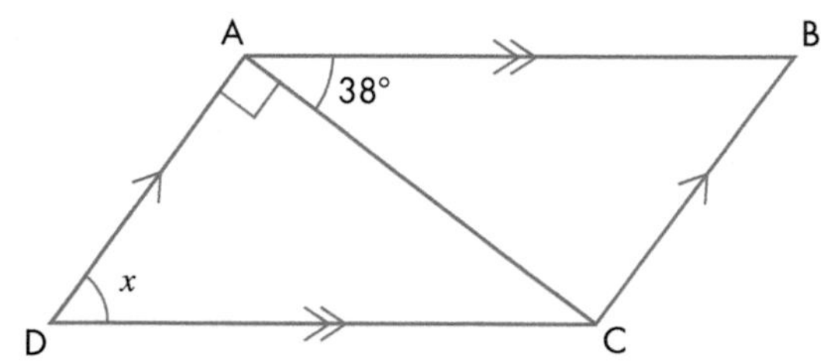

Work out the size of the angle marked x on the diagram.

7. Dani is making a kite and wants angle C to be half of angle A.

Work out the size of angles B and D.

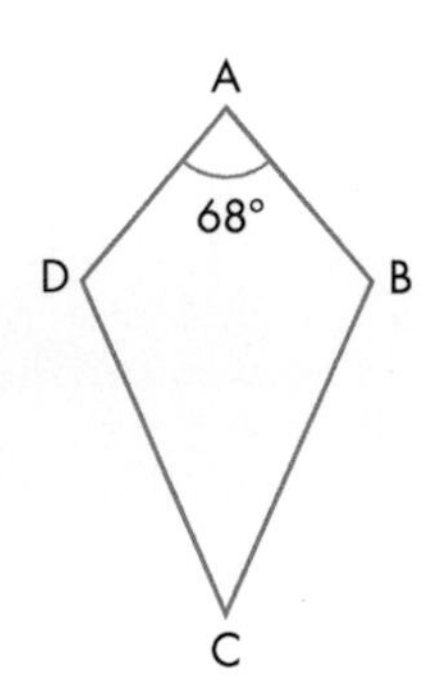

8. This quadrilateral is made from two isosceles triangles.

They are both the same size.

Find the value of y in terms of x.

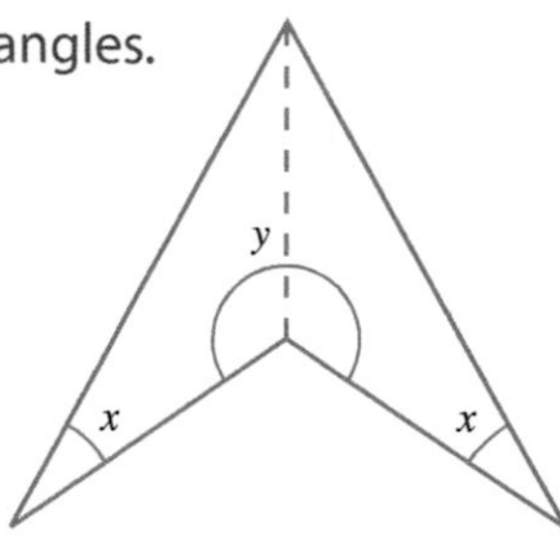

9. The diagram shows a quadrilateral ABCD.

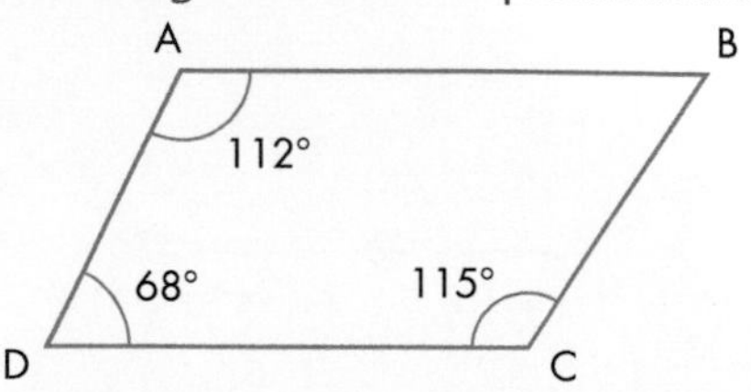

a) Calculate the size of angle B.

b) What special name is given to the quadrilateral ABCD? Explain your answer.

4.2 Angles in polygons

THIS SECTION WILL SHOW YOU HOW TO …

✓ work out the sizes of interior angles and exterior angles in a polygon

KEY WORDS

✓ decagon
✓ exterior angle
✓ heptagon
✓ hexagon
✓ interior angle
✓ nonagon
✓ octagon
✓ pentagon
✓ polygon
✓ regular polygon

A **polygon** has two kinds of angles.

- **Interior angles** are angles made by adjacent sides of the polygon and lying inside the polygon.
- **Exterior angles** are angles on the outside of the polygon between one side and the extended adjacent side. So the interior angle + the exterior angle = 180°.

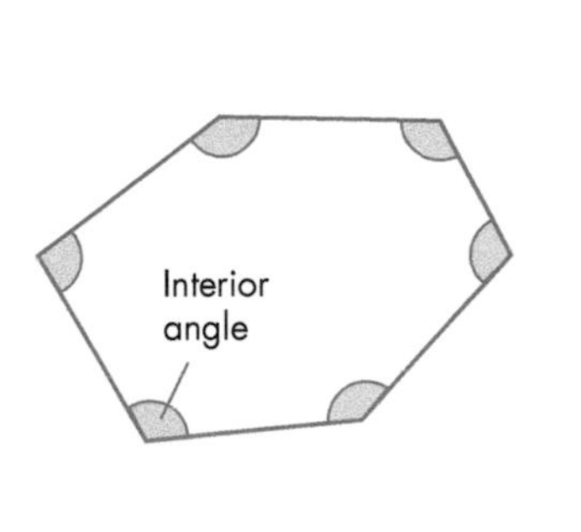

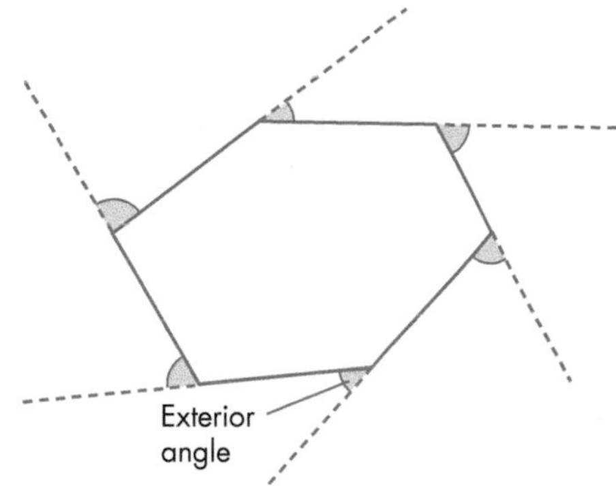

The *exterior* angles of *any* polygon add up to 360°.

Interior angles

You can find the sum of the interior angles of any polygon by splitting it into triangles.

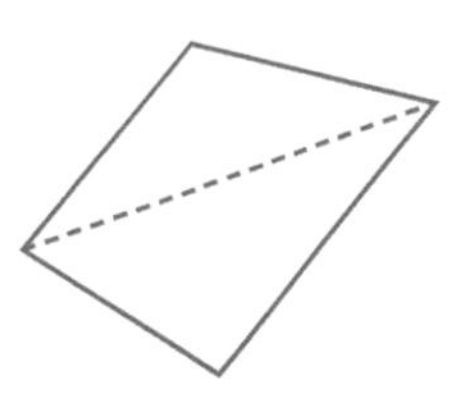

Two triangles

Pentagon

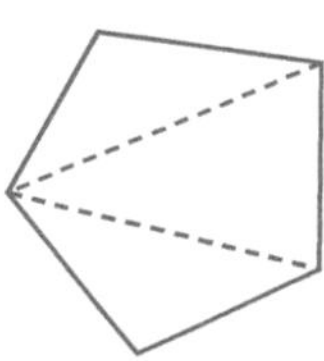

Three triangles

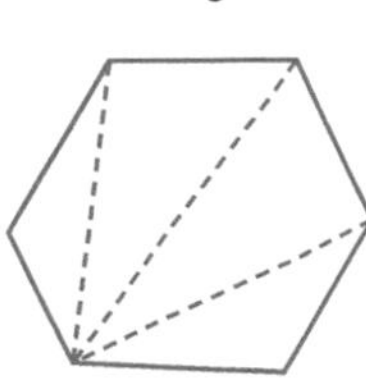

Four triangles

Heptagon

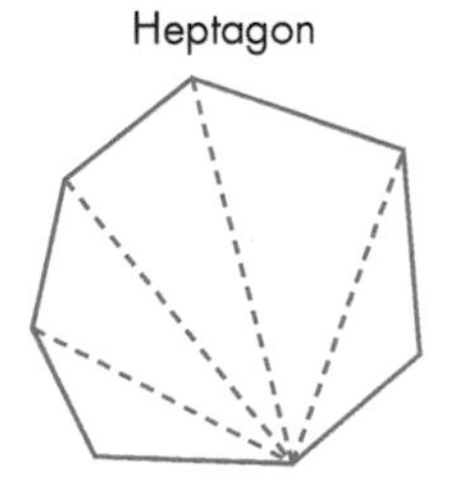

Five triangles

Since you already know that the angles in a triangle add up to 180°, you find the sum of the interior angles in a polygon by multiplying the number of triangles in the polygon by 180°, as shown in the table on the next page.

Shape	Name	Sum of interior angles
4-sided	Quadrilateral	$2 \times 180° = 360°$
5-sided	**Pentagon**	$3 \times 180° = 540°$
6-sided	**Hexagon**	$4 \times 180° = 720°$
7-sided	**Heptagon**	$5 \times 180° = 900°$
8-sided	**Octagon**	$6 \times 180° = 1080°$
9-sided	**Nonagon**	$7 \times 180° = 1260°$
10-sided	**Decagon**	$8 \times 180° = 1440°$

As you can see from the table, for an n-sided polygon, the sum of the interior angles, S, is given by the formula:

$$S = 180(n - 2)°$$

Exterior angles

As you can see from the diagram, the sum of an exterior angle and its adjacent interior angle is 180°.

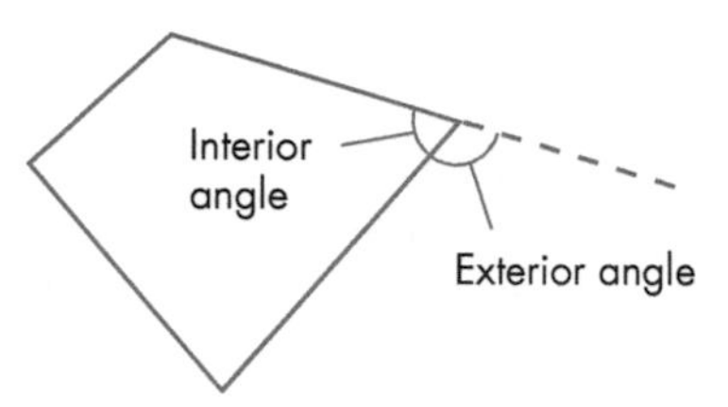

Regular polygons

A polygon is regular if all its interior angles are equal and all its sides have the same length.

This means that all the exterior angles of a **regular polygon** are also equal.

Here are two simple formulae for calculating the interior and the exterior angles of **regular polygons**.

The exterior angle, E, of a regular n-sided polygon is $E = \frac{360°}{n}$

The interior angle, I, of a regular n-sided polygon is $I = 180° - E$

$$= 180° - \frac{360°}{n}$$

This can be summarised in the following table.

Regular polygon	Number of sides	Size of each exterior angle	Size of each interior angle
Square	4	90°	90°
Pentagon	5	72°	108°
Hexagon	6	60°	120°
Heptagon	7	$51\frac{3}{7}°$	$128\frac{4}{7}°$
Octagon	8	45°	135°
Nonagon	9	40°	140°
Decagon	10	36°	144°
n-sided	n	$\frac{360°}{n}$	$180° - \frac{360°}{n}$

EXAMPLE 3

Find the exterior angle, x, and the interior angle, y, for this regular octagon.

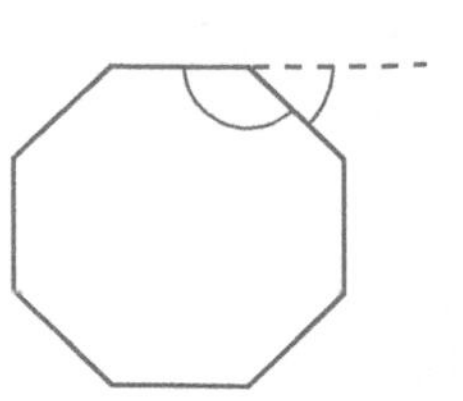

$$x = \frac{360^\circ}{8} = 45^\circ \text{ and } y = 180^\circ - 45^\circ = 135^\circ$$

EXERCISE 4B

GRADE C

1. Calculate the sum of the interior angles of polygons with:
   a) 10 sides **b)** 15 sides **c)** 100 sides **d)** 45 sides.

2. Calculate the size of the interior angle of regular polygons with:
   a) 12 sides **b)** 20 sides **c)** 9 sides **d)** 60 sides.

3. Find the number of sides of polygons with these interior angle sums.
   a) 1260° **b)** 2340° **c)** 18 000° **d)** 8640°

4. Find the number of sides of regular polygons with these exterior angles.
   a) 24° **b)** 10° **c)** 15° **d)** 5°

5. Find the number of sides of regular polygons with these interior angles.
   a) 150° **b)** 140° **c)** 162° **d)** 171°

GRADE B

6. Calculate the size of the unknown angle in each of these polygons.

a)

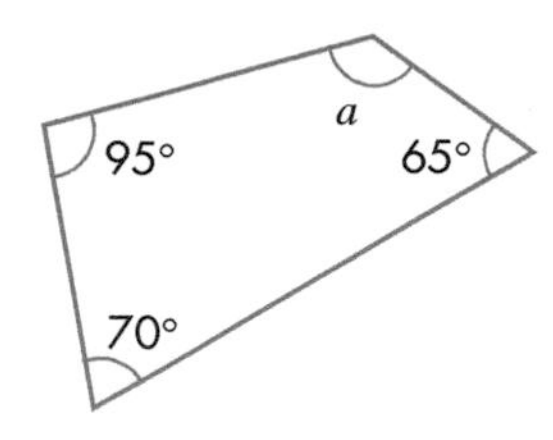

b)

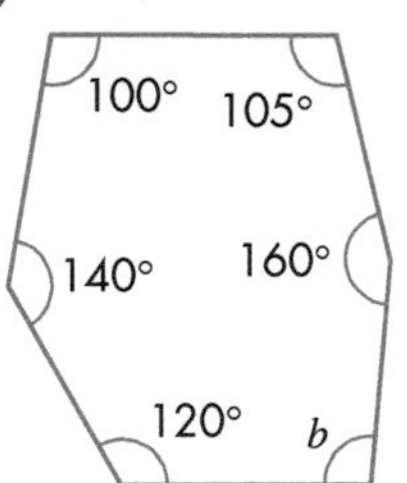

c)

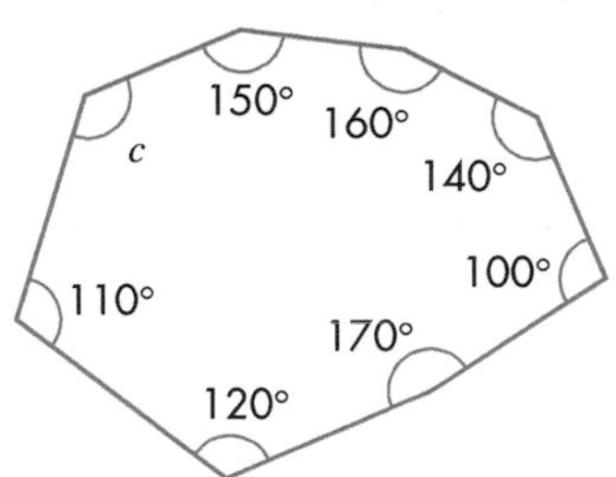

7. Find the value of x in each of these polygons.

a)

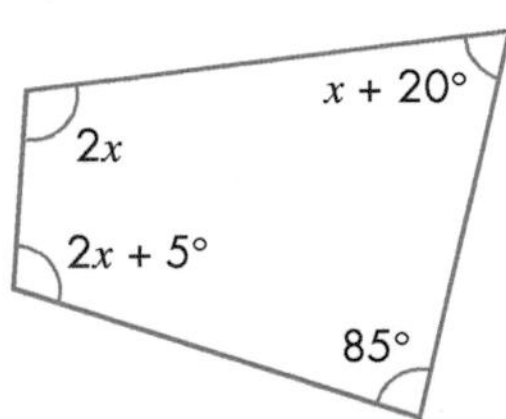

b)

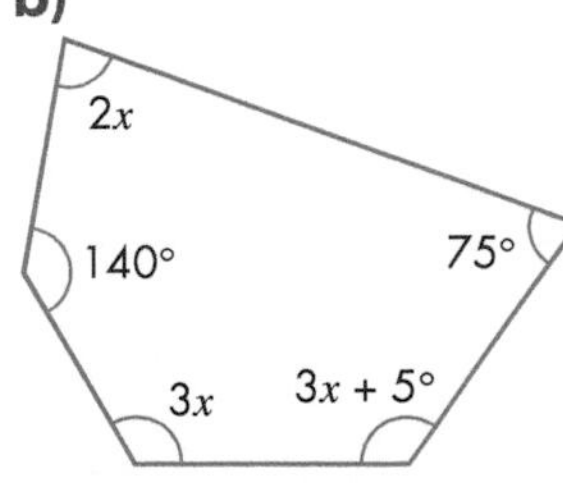

c)

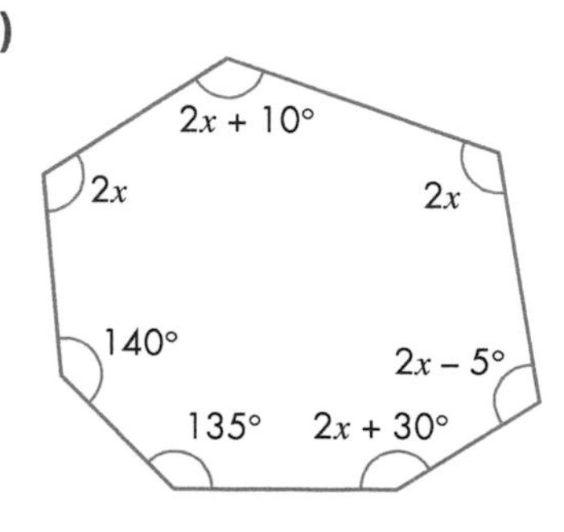

What is the name of the regular polygon in which the interior angles are twice its exterior angles?

9. Wesley measured all the interior angles in a polygon. He added them up to make 991°, but he had missed out one angle.

a) What type of polygon did Wesley measure?

b) What is the size of the missing angle?

10. a) In the triangle ABC, angle A is 42° and angle B is 67°.

i) Calculate the value of angle C.

ii) What is the value of the exterior angle at C?

iii) What connects the exterior angle at C with the sum of the angles at A and B?

b) Prove that any exterior angle of a triangle is equal to the sum of the two opposite interior angles.

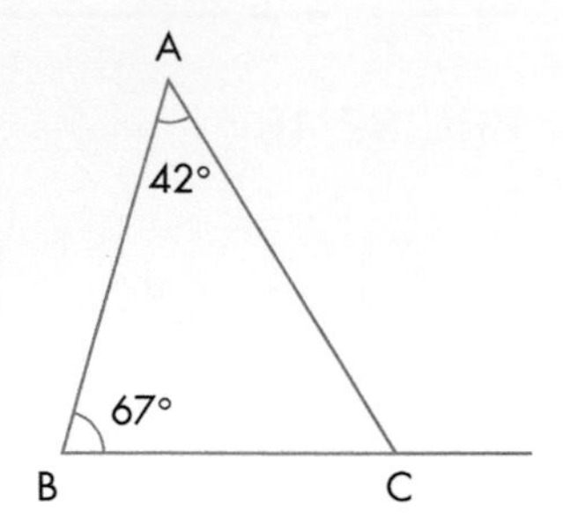

11. Two regular pentagons are placed together.

Work out the value of a.

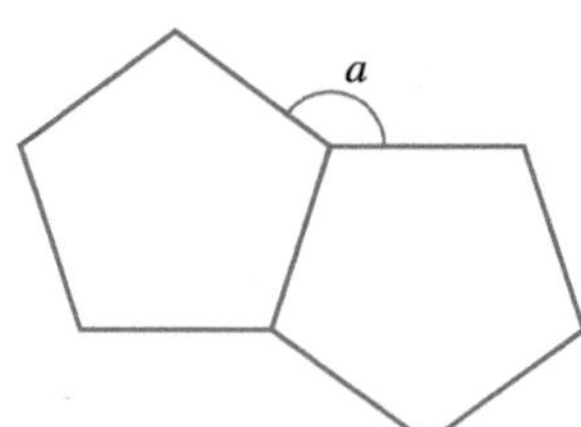

12. A joiner is making tables so that the shape of each one is half a regular octagon, as shown in the diagram. He needs to know the size of each angle on the table top. What are the sizes of the angles?

13. This star shape has 10 sides that are equal in length.

Each reflex interior angle is 200°.

Work out the size of each acute interior angle.

HINTS AND TIPS

Find the sum of the interior angles of a decagon first.

14. The diagram shows part of a regular polygon.

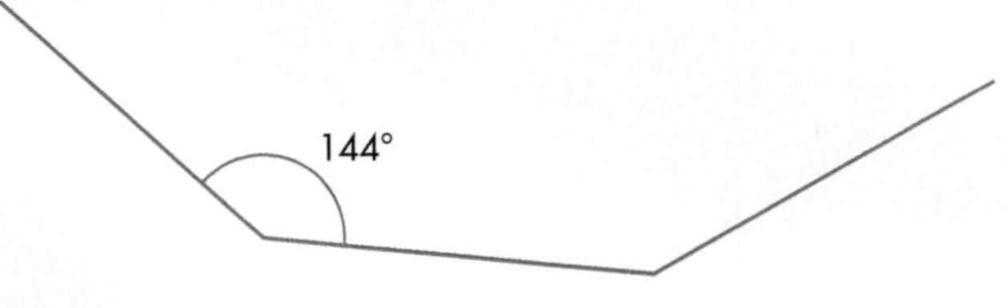

Each interior angle is 144°.

a) What is the size of each exterior angle of the polygon?

b) How many sides does the polygon have?

4.3 Circle theorems

THIS SECTION WILL SHOW YOU HOW TO ...

✓ work out the sizes of angles in circles

KEY WORDS

✓ arc
✓ circle
✓ circumference
✓ diameter
✓ segment
✓ semicircle
✓ subtended

Here are three **circle** theorems you need to know.

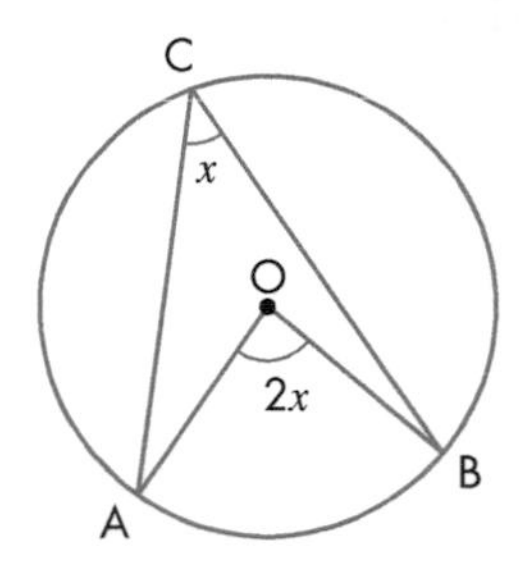

- **Circle theorem 1**

  The angle at the centre of a circle is twice the angle at the **circumference** that is **subtended** by the same **arc**.

  $\angle AOB = 2 \times \angle ACB$

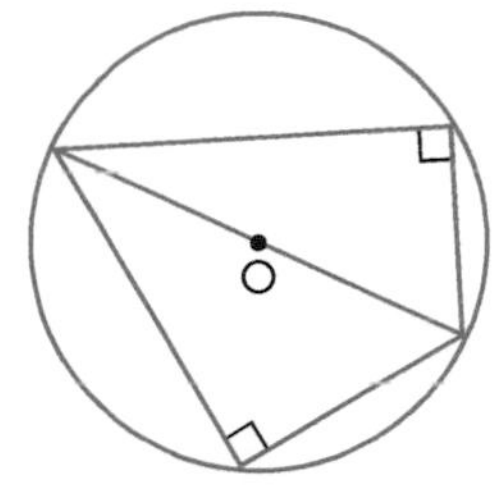

- **Circle theorem 2**

  Every angle at the circumference of a **semicircle** that is subtended by a **diameter** of the semicircle is a right angle.

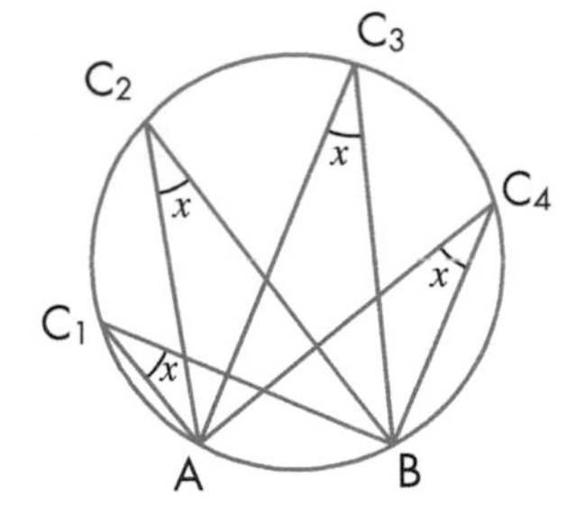

- **Circle theorem 3**

  Angles subtended at the circumference in the same **segment** of a circle are equal.

  Points C_1, C_2, C_3 and C_4 on the circumference are subtended by the same arc AB.

  So $\angle AC_1B = \angle AC_2B = \angle AC_3B = \angle AC_4B$

  Examples 4–6 show how these theorems are applied.

EXAMPLE 4

O is the centre of each circle. Find the sizes of the angles marked *a* and *b* in each circle.

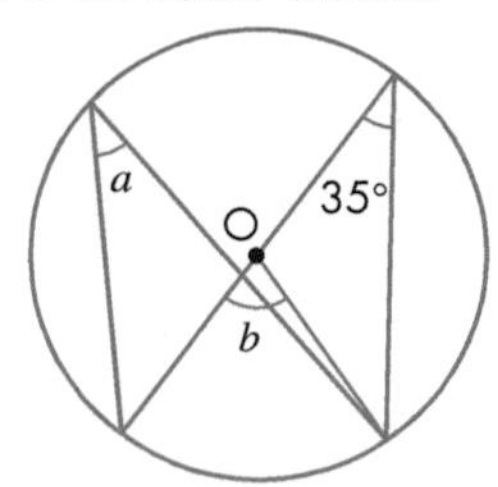

i) $a = 35°$ (angles in same segment)

$b = 2 \times 35°$ (angle at centre = twice angle at circumference)
$= 70°$

d) With OP = OQ, triangle OPQ is isosceles and the sum of the angles in this triangle = 180°.

So $a + (2 \times 25°) = 180° \Rightarrow a = 180° - (2 \times 25°) = 130°$

$b = 130° \div 2$ (angle at centre = twice angle at circumference)

$= 65°$

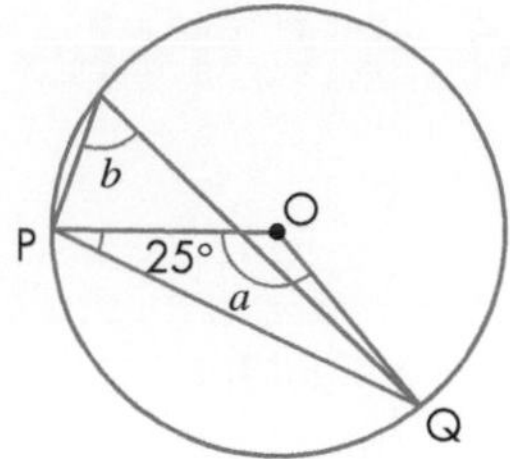

EXAMPLE 5

O is the centre of the circle. PQR is a straight line.

Find the size of the angle labelled a.

$\angle PQT = 180° - 72° = 108°$ (angles on straight line)

The reflex angle $\angle POT = 2 \times 108°$

(angle at centre = twice angle at circumference)

$= 216°$

$a + 216° = 360°$ (sum of angles around a point)

$a = 360° - 216° = 144°$

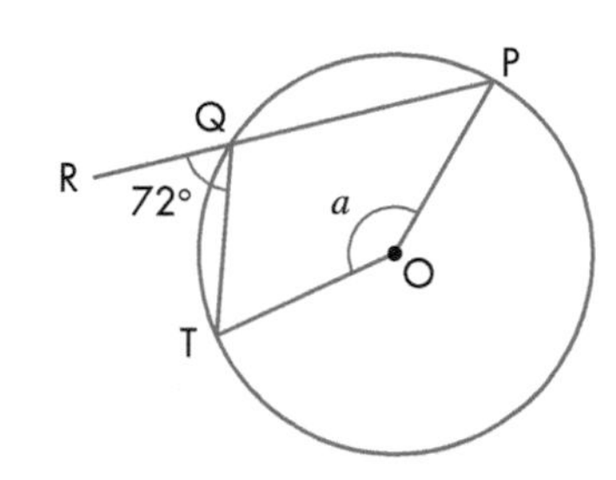

EXAMPLE 6

O is the centre of the circle. POQ is parallel to TR.

Find the sizes of the angles labelled a and b.

$a = 64° \div 2$ (angle at centre = twice angle at circumference)

$a = 32°$

$\angle TQP = a$ (alternate angles)

$= 32°$

$\angle PTQ = 90°$ (angle in a semicircle)

$b + 90° + 32° = 180°$ (sum of angles in ΔPQT)

$b = 180° - 122° = 58°$

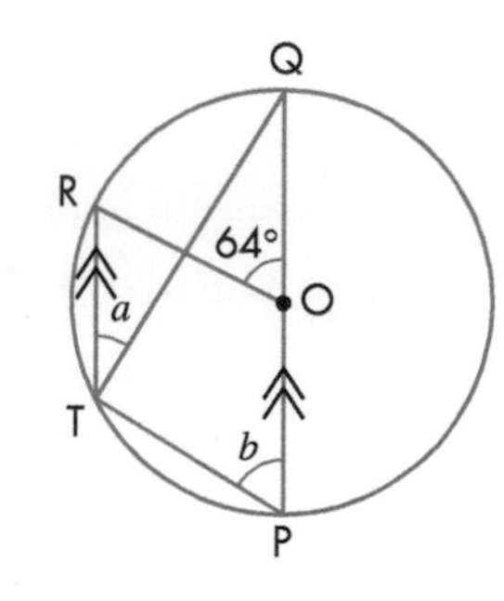

EXERCISE 4C

GRADE B

1. Find the size of the angle marked x in each of these circles with centre O.

a)

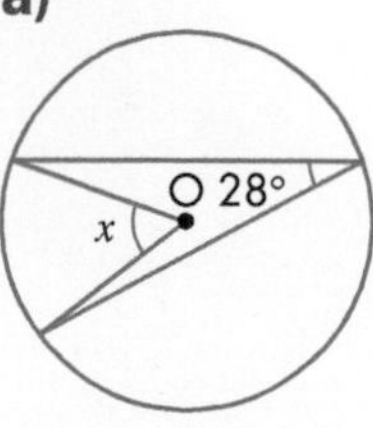

b)

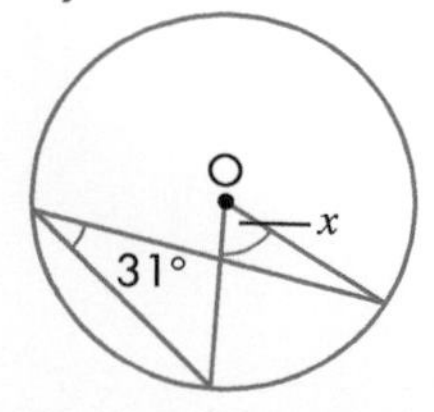

c)

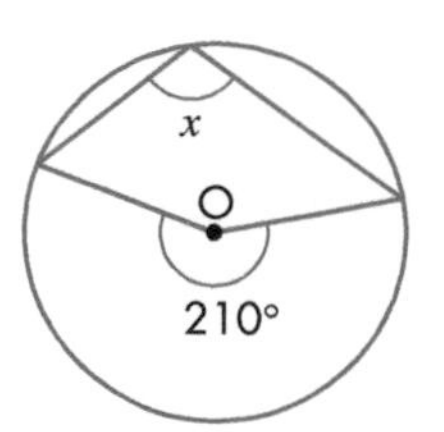

d)

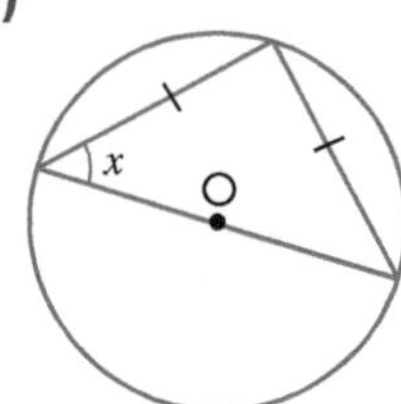

e)

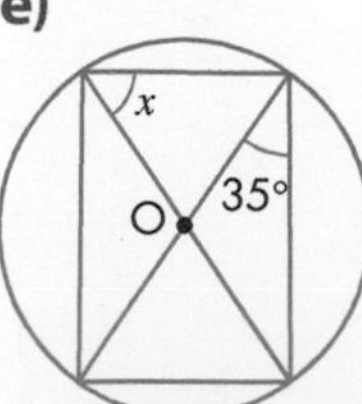

f)

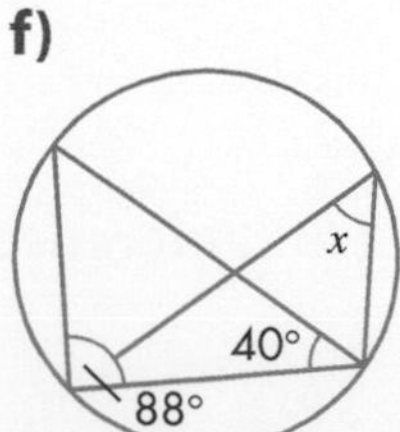

g)

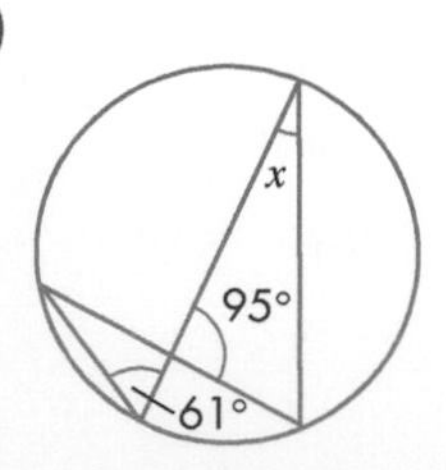

h)

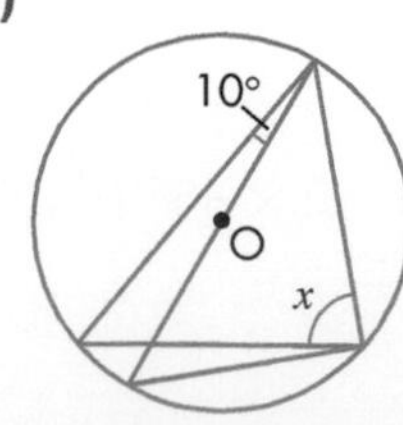

2. In the diagram, O is the centre of the circle. Find the sizes of these angles.

   a) $\angle$ADB
   b) $\angle$DBA
   c) $\angle$CAD

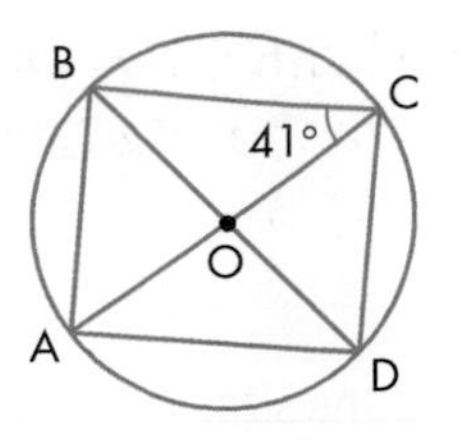

3. In the diagram, O is the centre of the circle. Find the sizes of these angles.

   a) $\angle$EDF
   b) $\angle$DEG
   c) $\angle$EGF

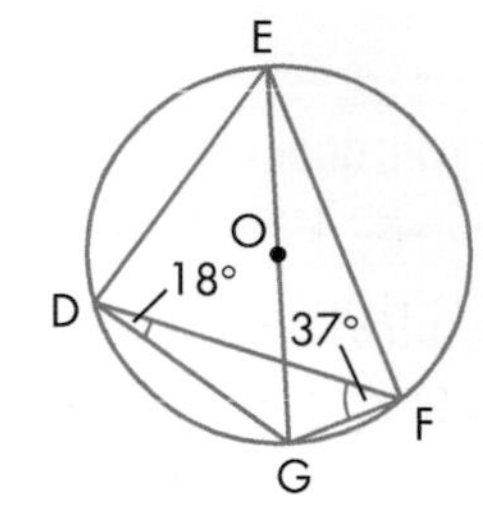

4. In the diagram XY is a diameter of the circle and $\angle$AZX is a.

   Ben says that the value of a is 55°.

   Give reasons to explain why he is wrong.

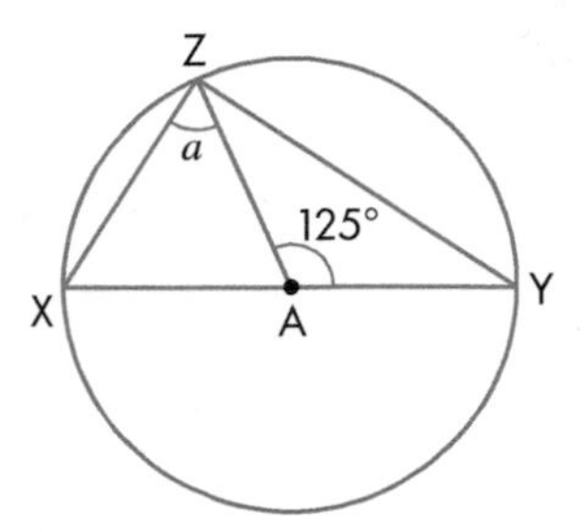

5. In the diagram, O is the centre and AD a diameter of the circle.

   Find the value of x.

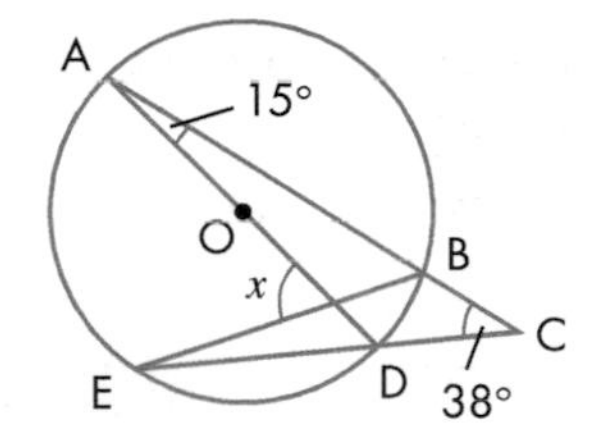

6. In the diagram, O is the centre of the circle and $\angle$CBD is x.

   Show that the reflex $\angle$AOC is $2x$, giving reasons to explain your answer.

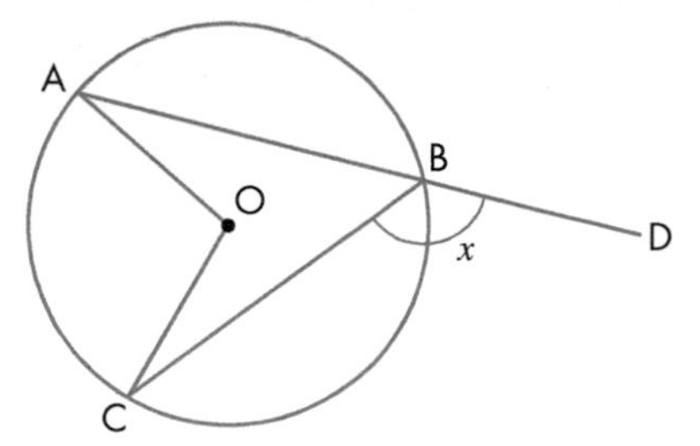

7. A, B, C and D are points on the circumference of a circle with centre O.

   Angle ABO is x and angle CBO is y.

   a) State the value of angle BAO.
   b) State the value of angle AOD.
   c) Prove that the angle subtended by the chord AC at the centre of the circle is twice the angle subtended at the circumference.

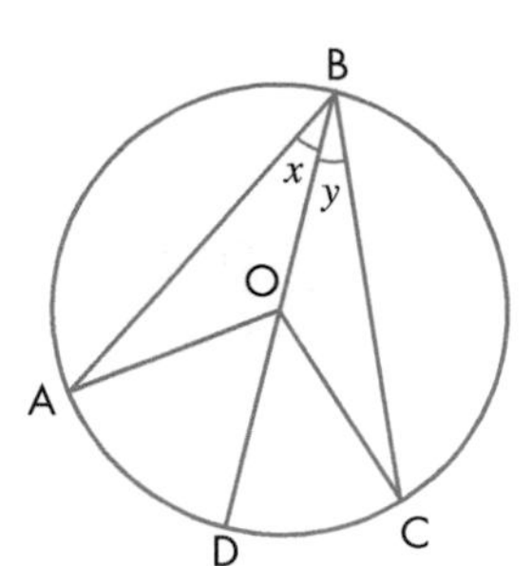

4.4 Cyclic quadrilaterals

THIS SECTION WILL SHOW YOU HOW TO …

✓ find the sizes of angles in cyclic quadrilaterals

KEY WORDS

✓ cyclic quadrilateral

A quadrilateral whose four vertices lie on the circumference of a circle is called a **cyclic quadrilateral**.

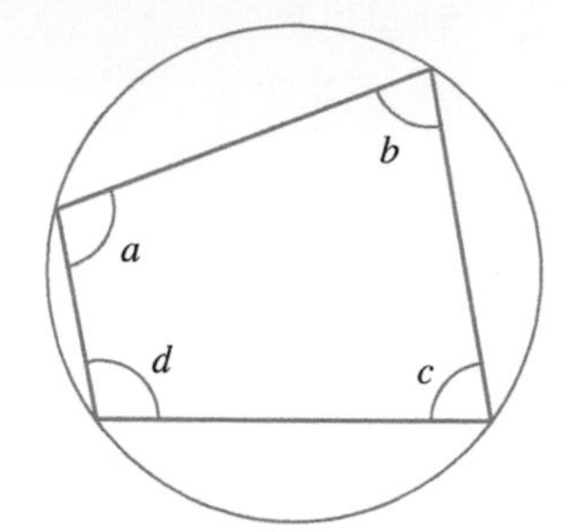

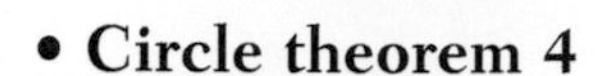

- **Circle theorem 4**

  The sum of the opposite angles of a cyclic quadrilateral is 180°.

  $a + c = 180°$ and $b + d = 180°$

EXAMPLE 7

Find the sizes of the angles marked x and y in the diagram.

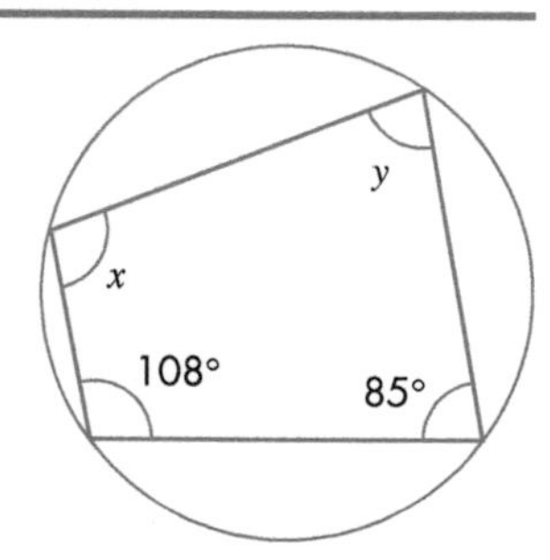

$x + 85° = 180°$ (angles in a cyclic quadrilateral)

So, $x = 95°$

$y + 108° = 180°$ (angles in a cyclic quadrilateral)

So, $y = 72°$

EXERCISE 4D

GRADE B

1. Find the sizes of the lettered angles in each of these circles.

a)

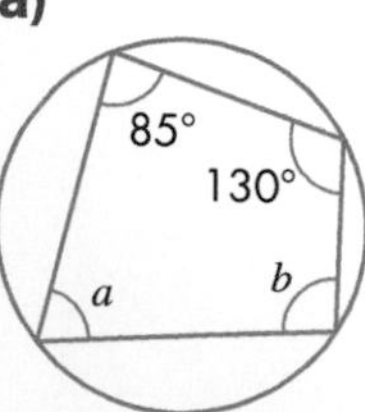

b)

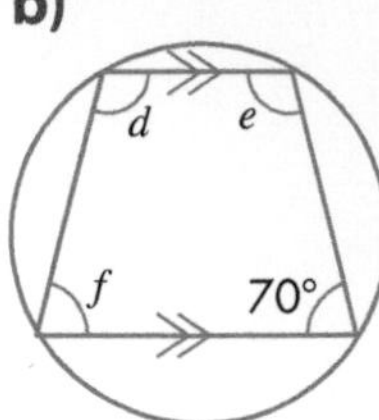

c)

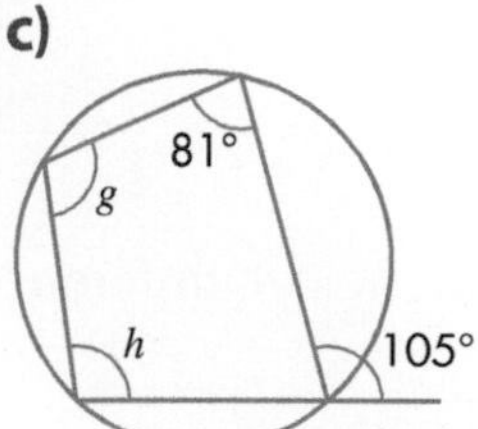

d)

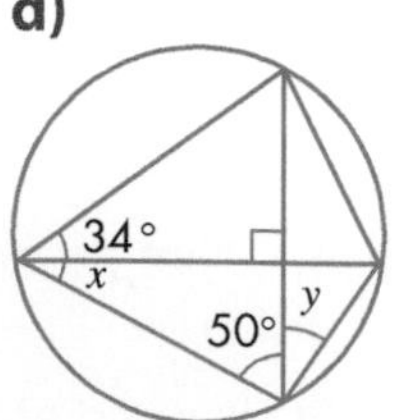

2. Find the values of x and y in each of these circles. Where shown, O marks the centre of the circle.

a)

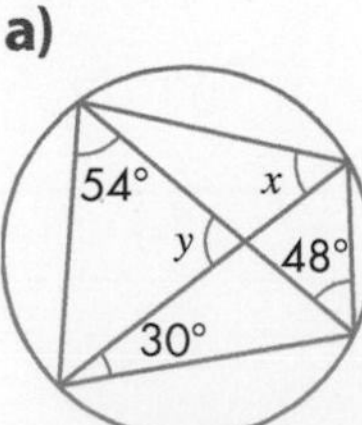

b)

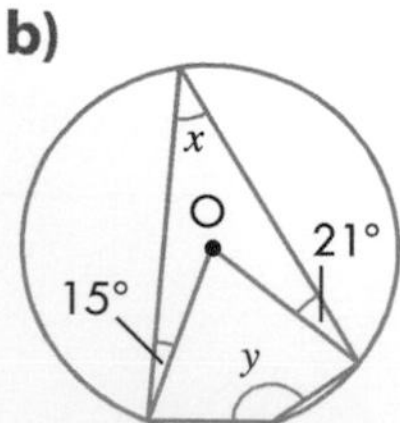

c)

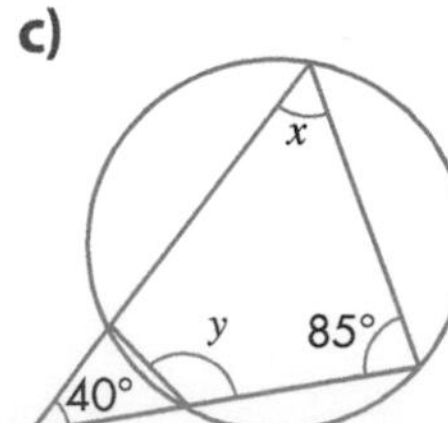

d)

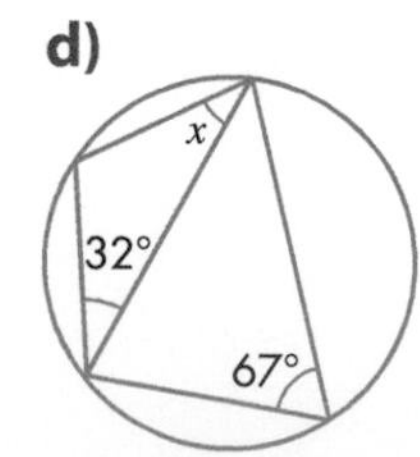

3. Find the values of x and y in each of these circles. Where shown, O marks the centre of the circle.

a)

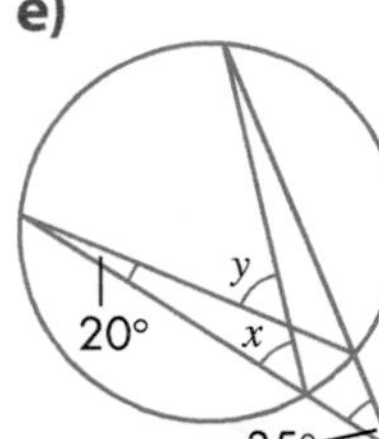

b)

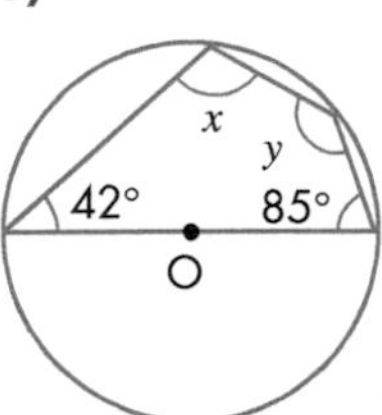

c)

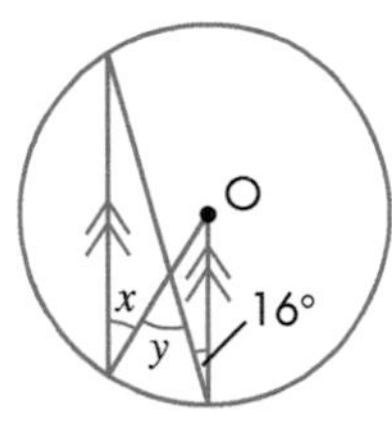

d)

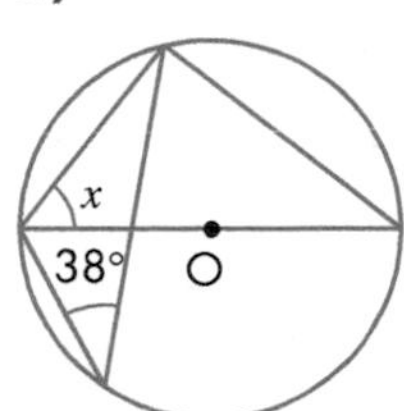

e)

f)

g)

h)

4. In cyclic quadrilateral PQRT, $\angle$ROQ = 38° where O is the centre of the circle. POT is a diameter and parallel to QR. Calculate these angles.

a) $\angle$ROT **b)** $\angle$QRT **c)** $\angle$QPT

5. In the diagram, O is the centre of the circle.

a) Explain why $3x - 30° = 180°$.

b) Work out the size of $\angle$CDO, marked y on the diagram.

c) Give reasons in your working.

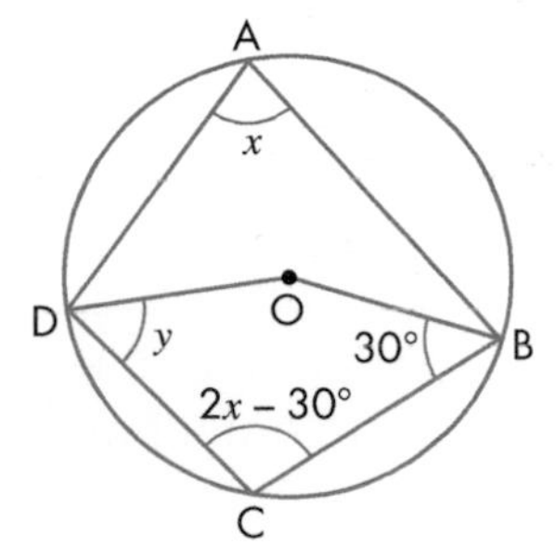

6. ABCD is a cyclic quadrilateral within a circle centre O and $\angle$AOC is $2x$°.

a) Write down the value of $\angle$ABC.

b) Write down the value of the reflex angle AOC.

c) Prove that the sum of a pair of opposite angles of a cyclic quadrilateral is 180°.

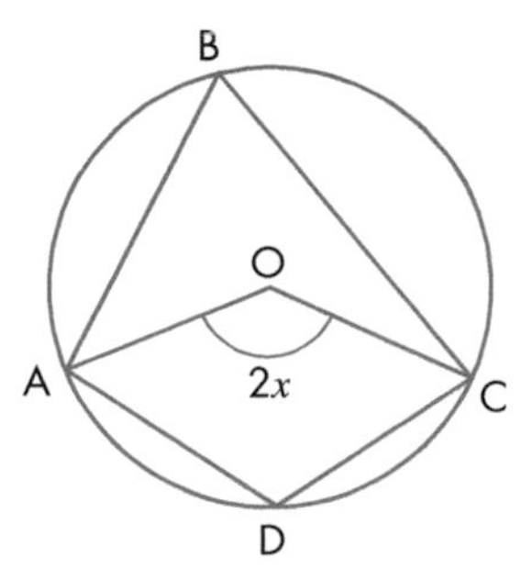

7. In the diagram, ABCE is a parallelogram.

Prove $\angle$AED = $\angle$ADE.

Give reasons in your working.

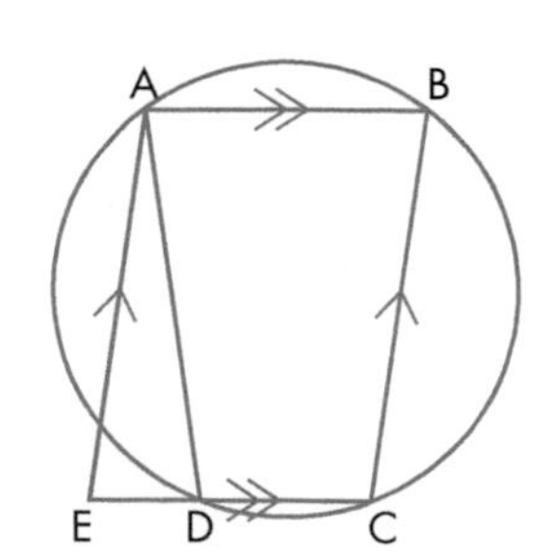

4.5 Tangents and chords

THIS SECTION WILL SHOW YOU HOW TO ...

✓ use tangents and chords to find the sizes of angles in circles

KEY WORDS

✓ chord
✓ point of contact
✓ radius
✓ tangent

A **tangent** is a straight line that touches a circle at one point only. This point is called the **point of contact**. A **chord** is a line that joins two points on the circumference.

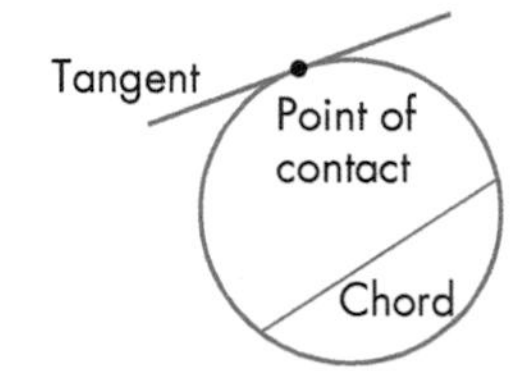

- **Circle theorem 5**

  A tangent to a circle is perpendicular to the **radius** drawn to the point of contact.

  The radius OX is perpendicular to the tangent AB.

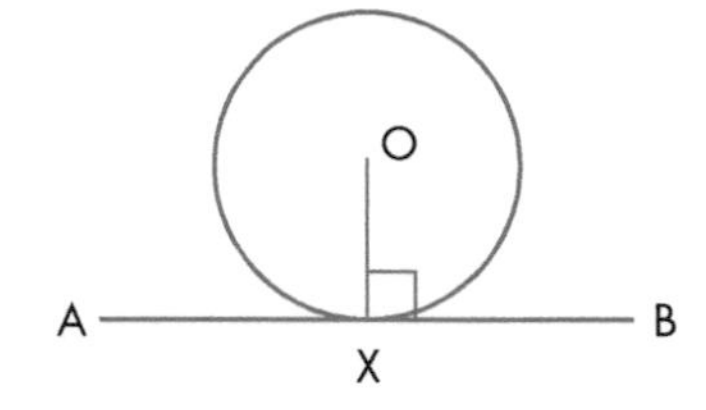

- **Circle theorem 6**

  Tangents to a circle from an external point to the points of contact are equal in length.

  AX = AY

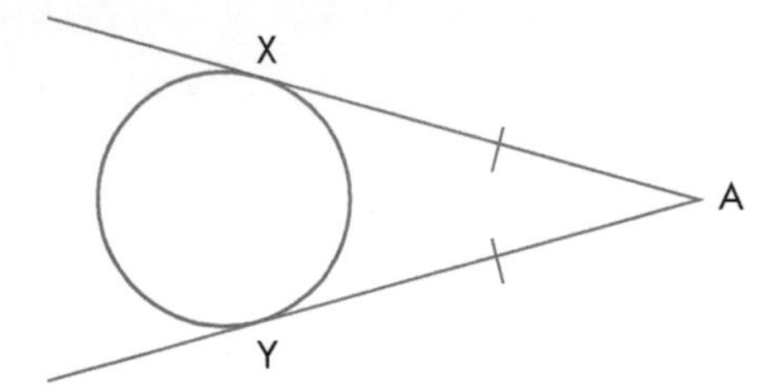

- **Circle theorem 7**

  The line joining an external point to the centre of the circle bisects the angle between the tangents to the circle from that external point.

  $\angle OAX = \angle OAY$

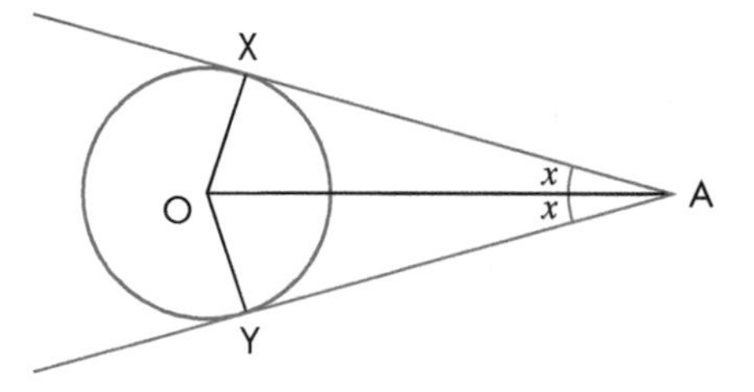

- **Circle theorem 8**

  A radius bisects a chord at 90°.

  If O is the centre of the circle,

  $\angle BMO = 90°$ and BM = CM

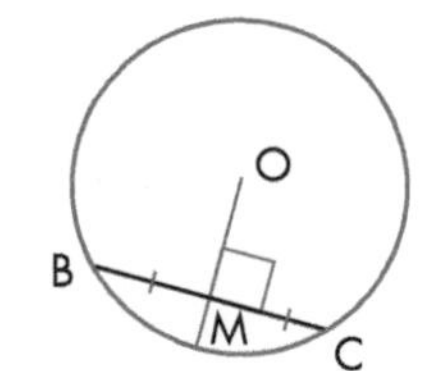

EXAMPLE 8

OA is the radius of the circle and AB is a tangent.

OA = 5 cm and AB = 12 cm

Calculate the length OB.

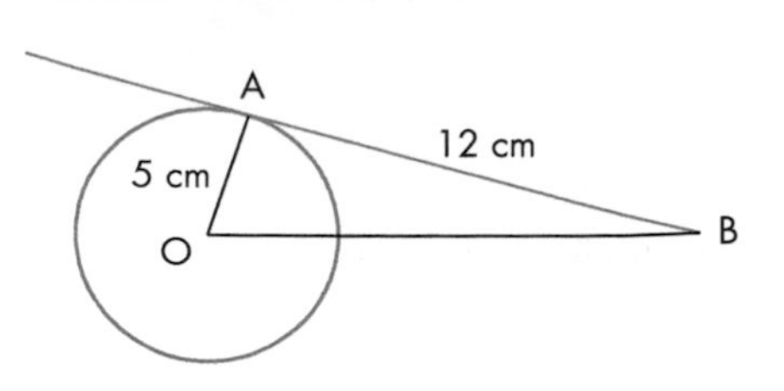

$\angle OAB = 90°$ (radius is perpendicular to the tangent)

By Pythagoras' theorem,

$OB^2 = 5^2 + 12^2 = 169$

So $OB = \sqrt{169} = 13$ cm

EXERCISE 4E

GRADE B

1. In each diagram, TP and TQ are tangents to a circle with centre O.
Find the value of x in each case.

a)

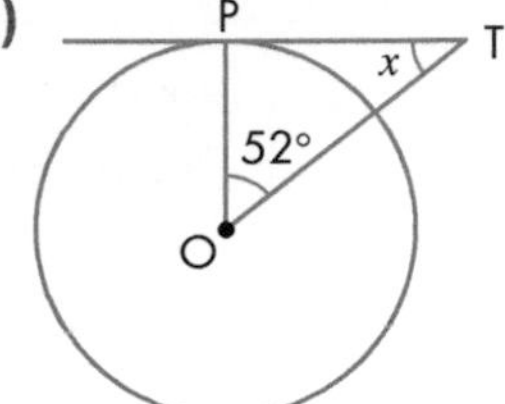

b)

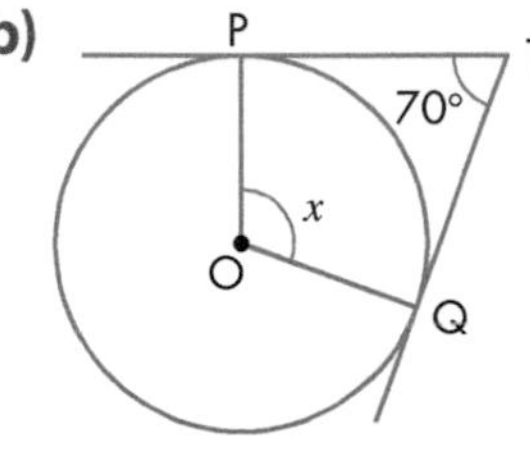

c)

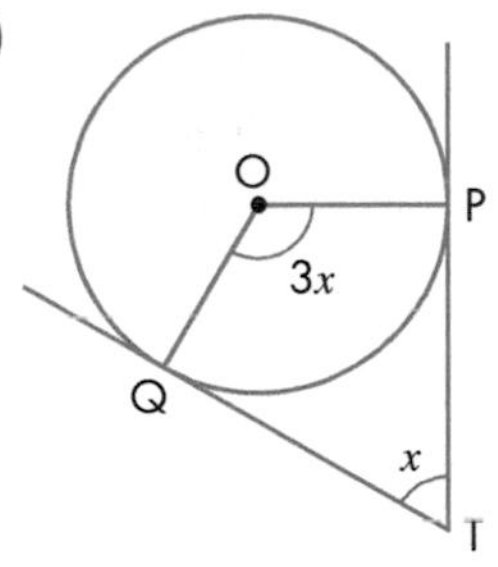

2. Each diagram shows tangents to a circle with centre O. Find the value of y in each case.

a)

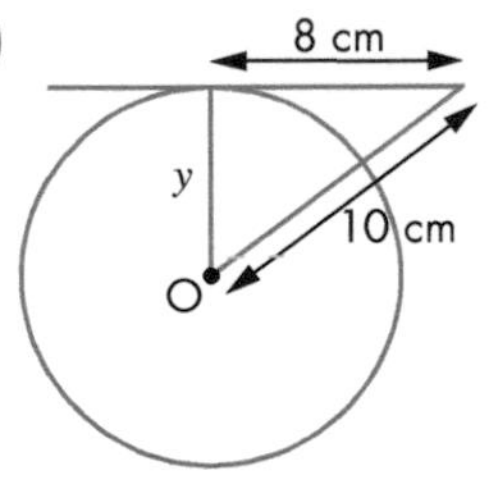

b)

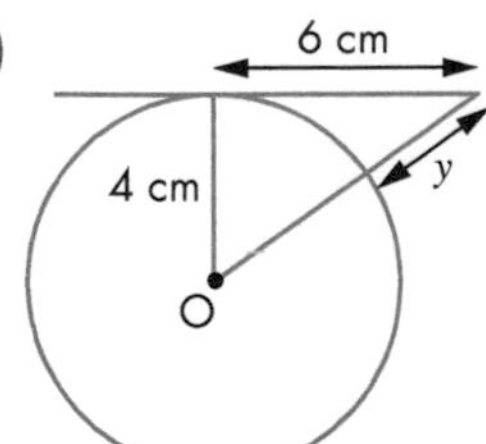

c)

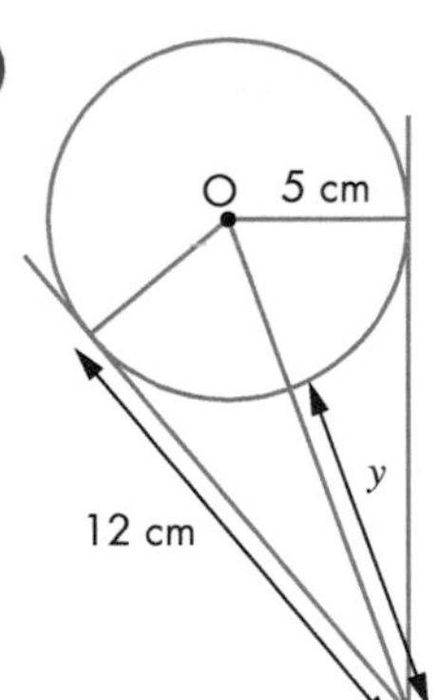

3. Each diagram shows a tangent to a circle with centre O. Find the value of x and y in each case.

a)

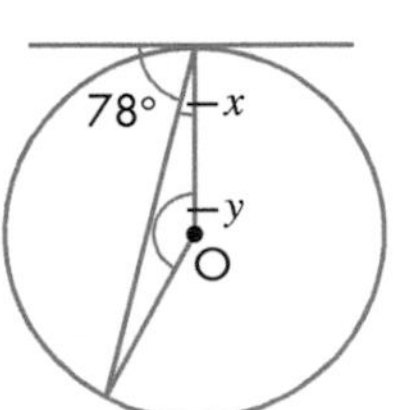

b)

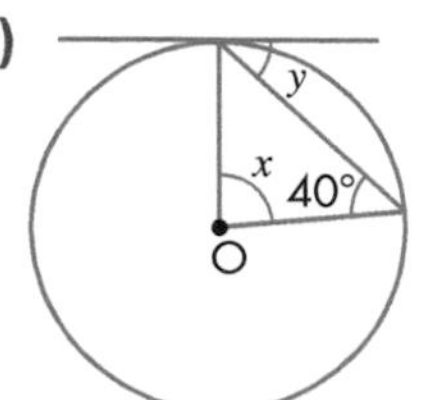

c)

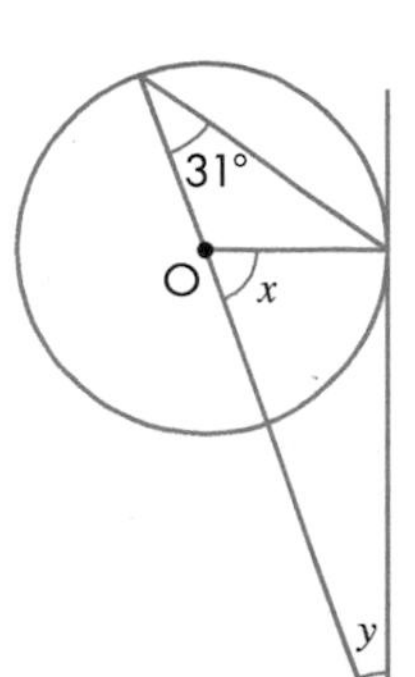

d)

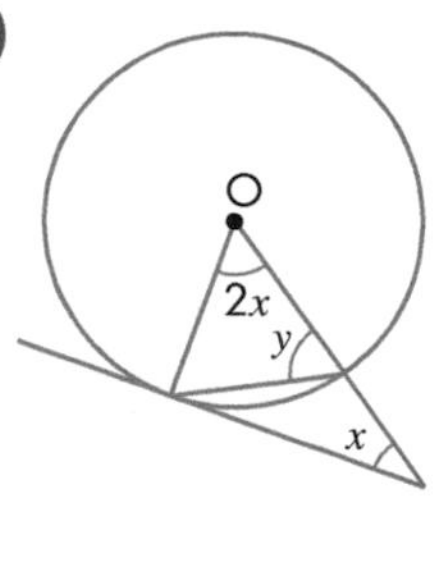

4. In each of the diagrams, TP and TQ are tangents to the circle with centre O. Find the value of x in each case.

a)

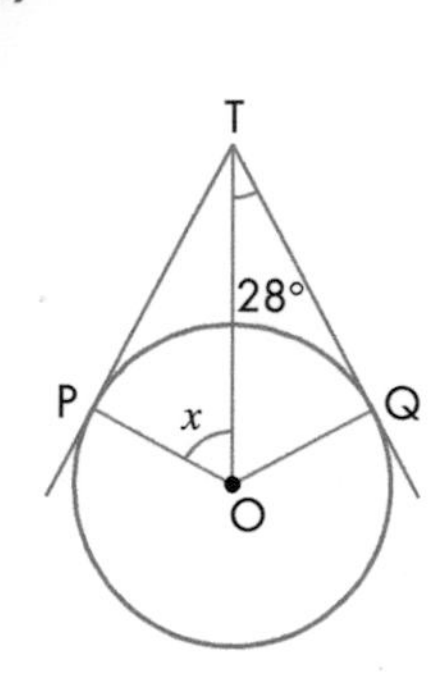

b)

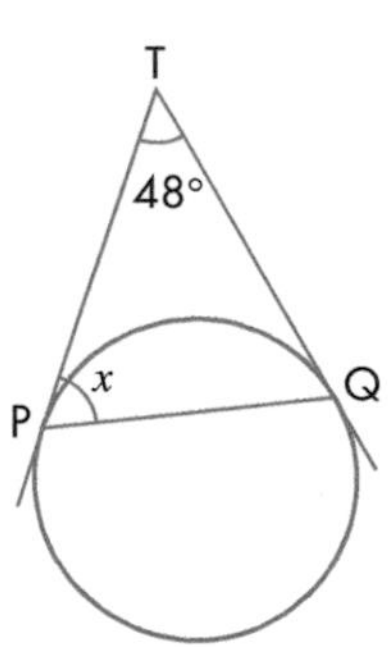

c)

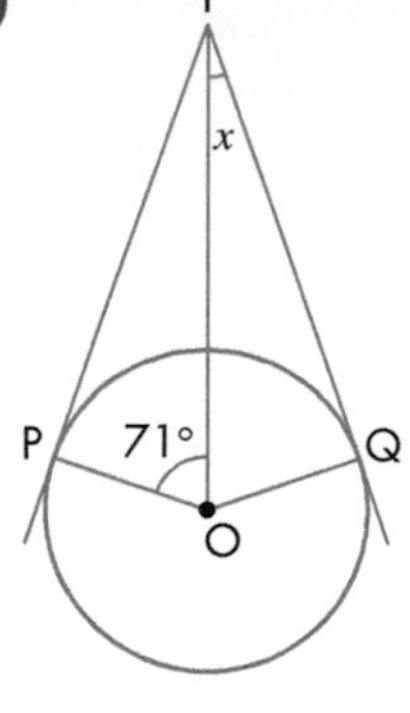

d)

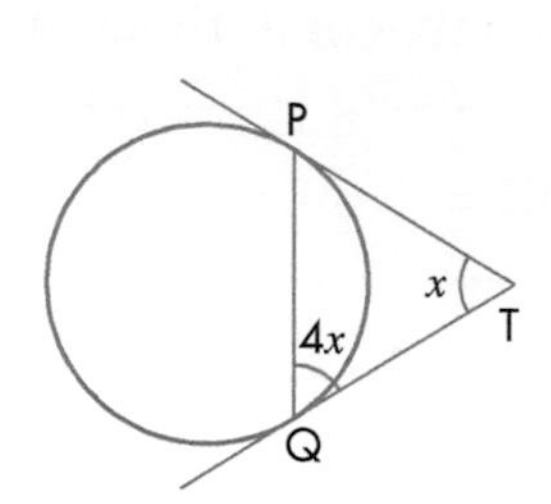

5. Two circles with the same centre have radii of 7 cm and 12 cm respectively. A tangent to the inner circle cuts the outer circle at A and B. Find the length of AB.

HINTS AND TIPS

Remember that you can use Pythagoras' theorem and trigonometry to solve problems.

6. The diagram shows a circle with centre O.

The circle fits exactly inside an equilateral triangle XYZ.

The lengths of the sides of the triangle are 20 cm.

Work out the radius of the circle.

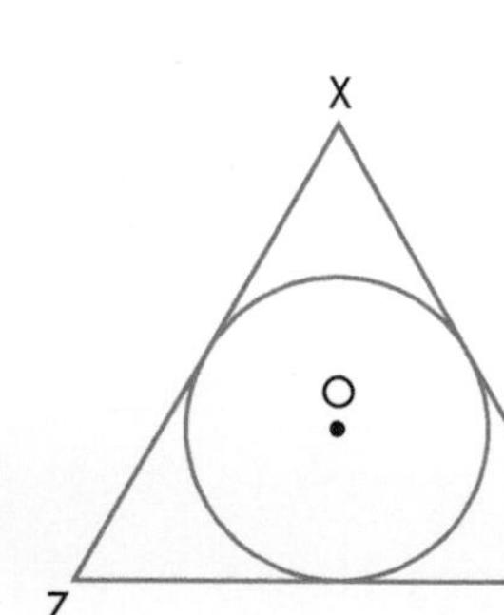

7. In the diagram, O is the centre of the circle and AB is a tangent to the circle at C.

Explain why triangle BCD is isosceles.

Give reasons to justify your answer.

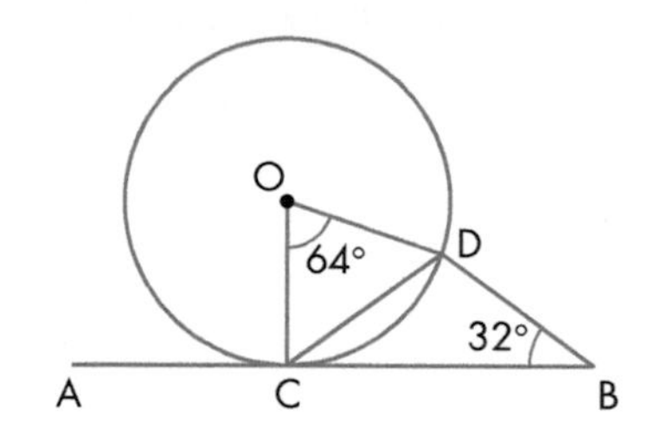

8. AB and CB are tangents from B to the circle with centre O. OA and OC are radii.

a) Prove that angles AOB and COB are equal.

b) Prove that OB bisects the angle ABC.

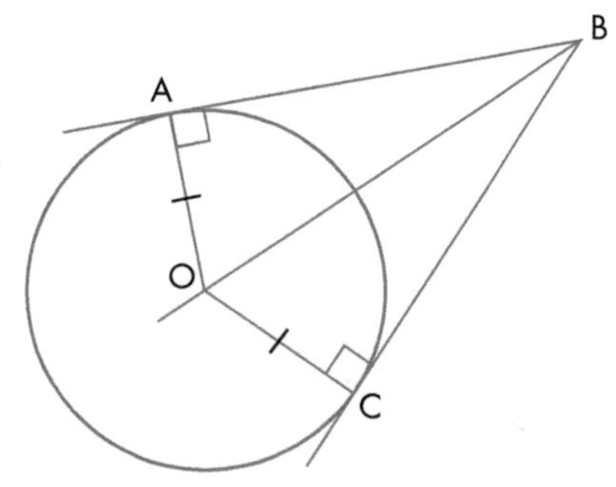

4.6 Alternate segment theorem

THIS SECTION WILL SHOW YOU HOW TO ...

✓ use the alternate segment theorem to find the sizes of angles in circles

KEY WORDS

✓ alternate segment
✓ chord
✓ tangent

PTQ is the **tangent** to a circle at T. The segment containing ∠TBA is known as the **alternate segment** of ∠PTA, because it is on the other side of the **chord** AT from ∠PTA.

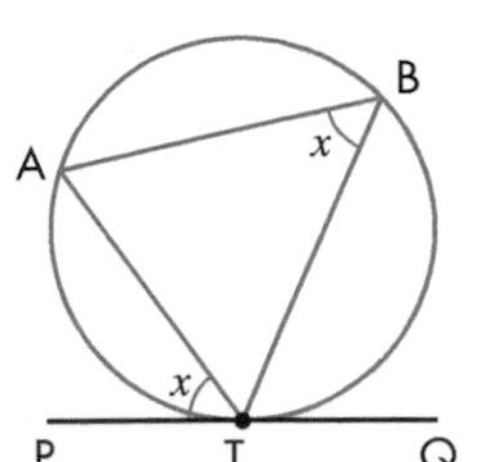

- **Circle theorem 9**

  The angle between a tangent and a chord through the point of contact is equal to the angle in the alternate segment.

  ∠PTA = ∠TBA

EXAMPLE 9

In the diagram, find the size of **a** ∠ATS and **b** ∠TSR.

a) ∠ATS = 80° (angle in alternate segment)

b) ∠TSR = 70° (angle in alternate segment)

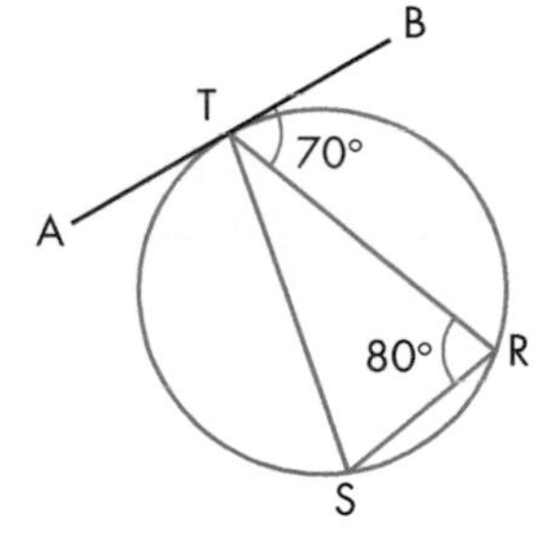

EXERCISE 4F

GRADE A

1. Find the size of each lettered angle.

a)

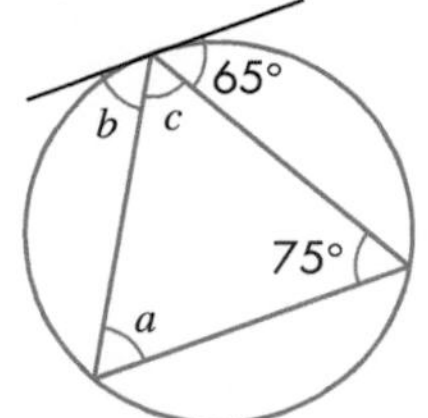

b)

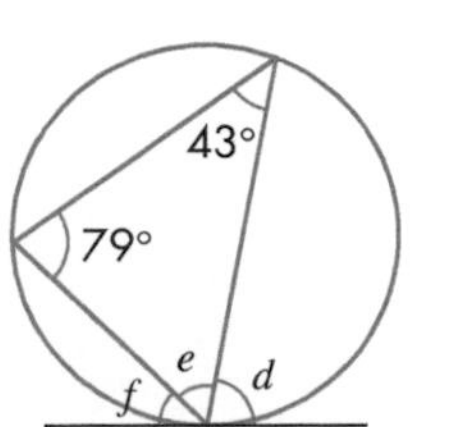

c)

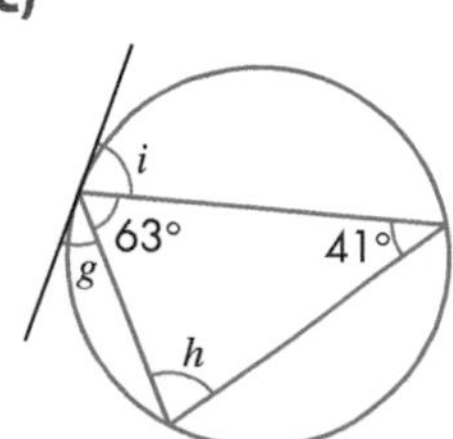

d)

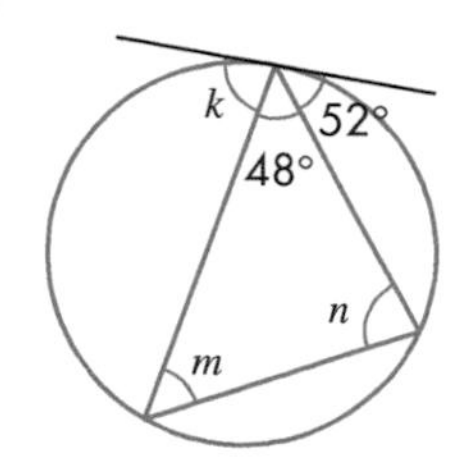

2. In each diagram, find the size of each lettered angle.

a) c, b, d, 75°, a

b)

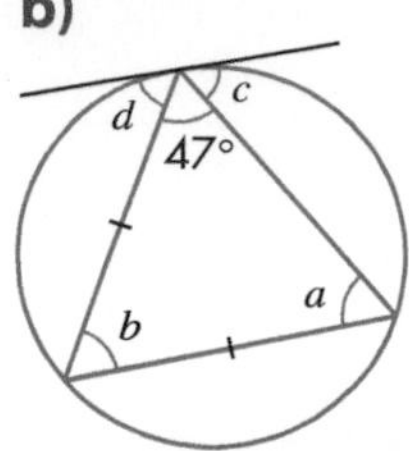

c)

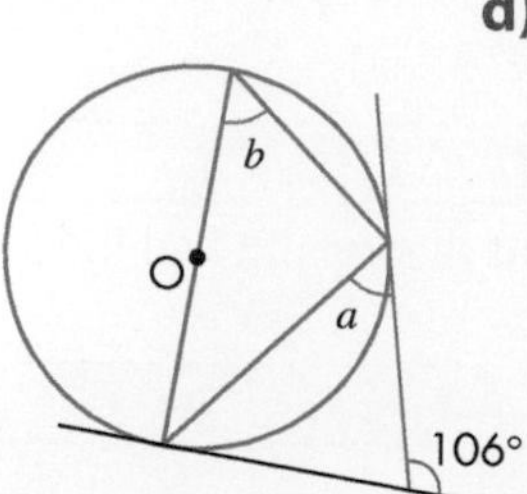

d)

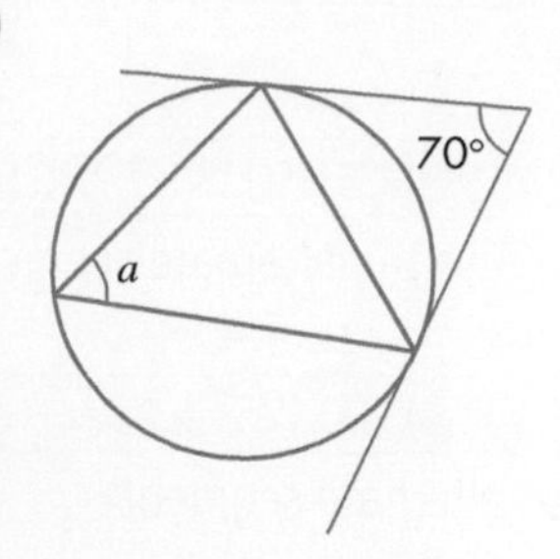

3. In each diagram, find the value of x.

a)

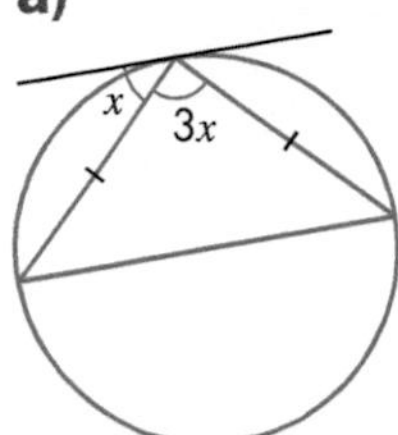

b)

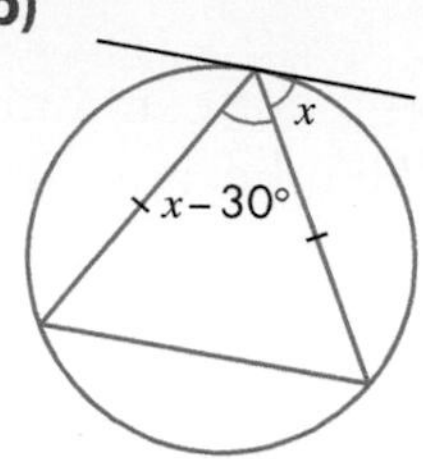

4. ATB is a tangent to each circle with centre O.

Find the size of each lettered angle.

a)

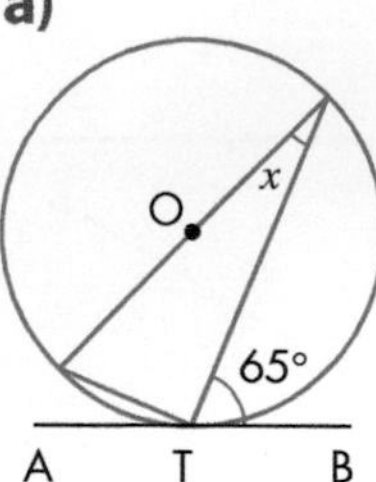

b)

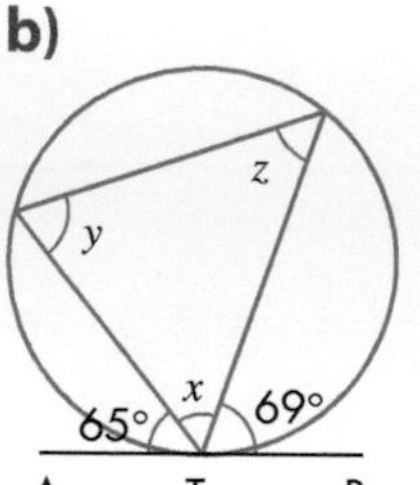

c)

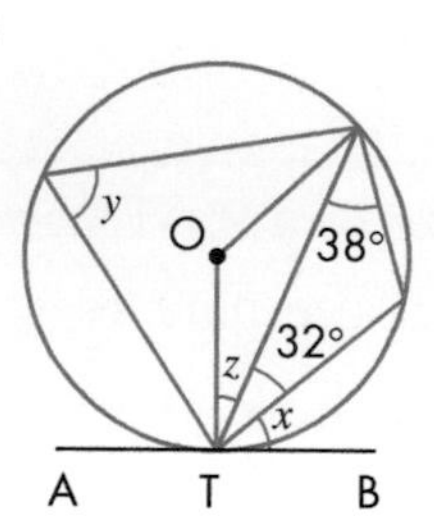

d)

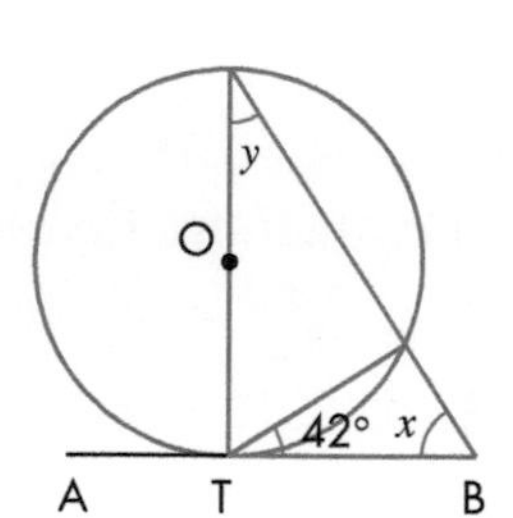

GRADE A*

5. In the diagram, O is the centre of the circle.

XY is a tangent to the circle at A.

BCX is a straight line.

Show that triangle ACX is isosceles.

Give reasons to justify your answer.

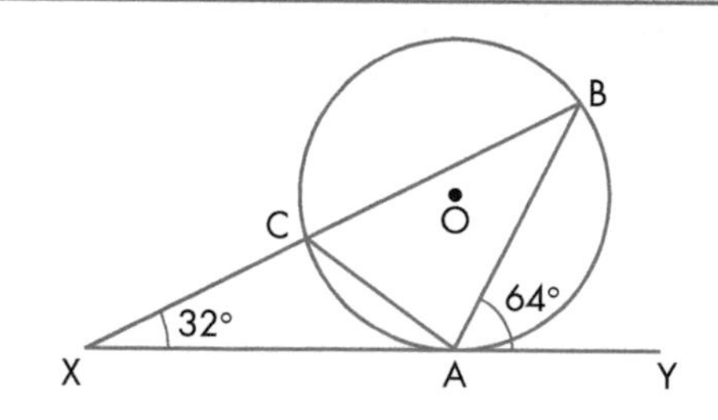

6. AB and AC are tangents to the circle at X and Y.

Work out the size of ∠XYZ.

Give reasons to justify your answer.

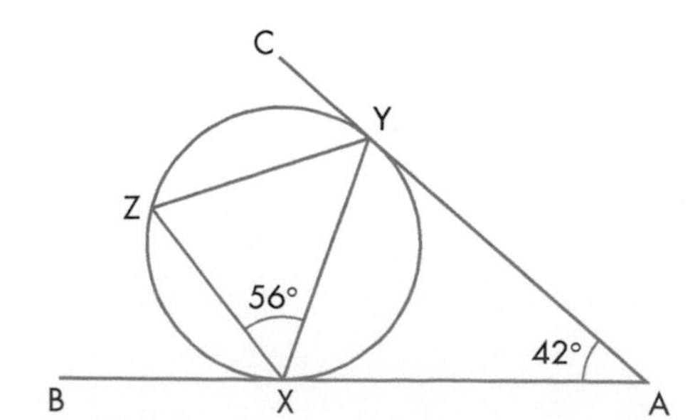

Exam-style questions

1. ABCDE is a regular pentagon.

   EB is a straight line.

   Show that angle DEB is twice the size of angle AEB. GRADE C

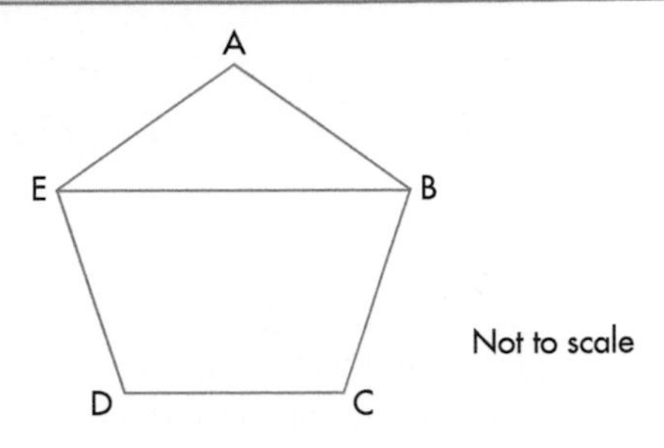

2. ABCD is a quadrilateral. The size of each angle is shown.

   Show that ABCD is a trapezium. GRADE C

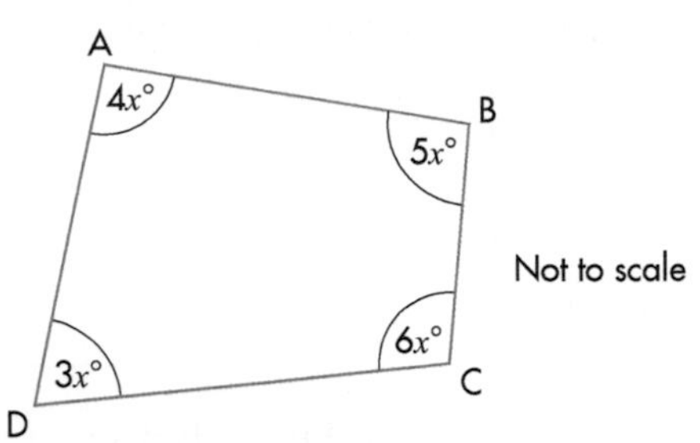

3. Two equilateral triangles overlap to make a hexagon ABCDEF.

   Find the value of $x + y$. GRADE B

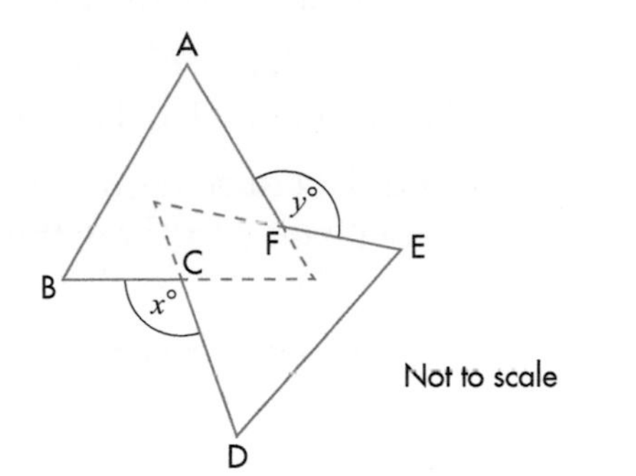

4. B is the centre of the circle.

   AS and AT are tangents to the circle.

   AT = 12 cm and AB = 13 cm.

   Work out the area of the quadrilateral ASBT. GRADE B

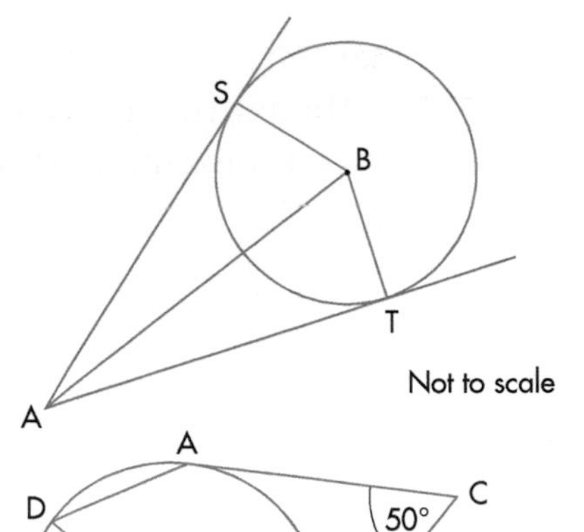

5. AC and BC are tangents to a circle.

   D is on the circumference of the circle.

   Work out the size of angle ADB. GRADE A*

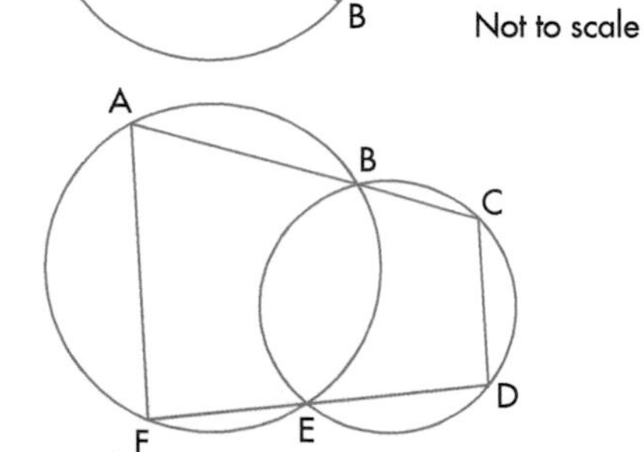

6. Two circles cross at B and E.

   ABC and DEF are straight lines.

   Show that AF and CD are parallel. GRADE A*

5 Functions

5.1 Function notation

THIS SECTION WILL SHOW YOU HOW TO ...

- ✓ understand what functions are
- ✓ understand and use function notation, for example f(x)
- ✓ substitute numbers into a function, knowing that, for example f(2) is the value of the function when $x = 2$
- ✓ solve equations that use function notation

KEY WORDS

- ✓ function

A **function** is a relation between two sets of values.

Equations written in terms of x and y such as $y = 3x - 4$ or $y = 2x^2 + 5x - 3$ show that y is a function of x. The value of y depends on the value of x.

You can use a different notation, writing $f(x) = 3x - 4$ and calling it 'function f'. Then you can show the result of using different values for x, for example:

- 'the value of $f(x)$ when x is 5' can be written as f(5).
  So $f(5) = 3 \times 5 - 4 = 11$
- f(1) means 'the value of $f(x)$ when x is 1'.
  So $f(1) = 3 \times 1 - 4 = -1$ and $f(-1) = -7$
- If there are different functions in the same problem, you can use different letters, for example, $g(x) = 2x^2 + 5x - 3$ or 'function g'.

EXAMPLE 1

$f(x) = 3x - 5$

a) Work out $f(-4)$. **b)** Work out $f(0.6)$. **c)** Solve $f(x) = 0$.

d) Write expressions for $f(2x) + f(x + 1)$ as simply as possible.

a) $f(-4) = 3 \times -4 - 5 = -17$ **b)** $f(0.6) = 3 \times 0.6 - 5 = -3.2$

c) $3x - 2 = 0 \Rightarrow x = \frac{2}{3}$

d) Replacing x by $2x$ gives $f(2x) = 3(2x) - 5 = 6x - 5$

Replacing x by $x + 1$ gives $f(x + 1) = 3(x + 1) - 5 = 3x - 2$

$f(2x) + f(x + 1) = 6x - 5 + 3x - 2 = 9x - 7$

EXAMPLE 2

$f(x) = x^2 + 4x - 5$

Solve $f(2x) = 0$.

$f(2x) = (2x)^2 + 4(2x) - 5 = 4x^2 + 8x - 5$

So $4x^2 + 8x - 5 = 0 \Rightarrow (2x - 1)(2x + 5) = 0 \Rightarrow x = \frac{1}{2}$ or $x = -\frac{5}{2}$

EXERCISE 5A

GRADE C

1. $f(x) = 2x + 6$ Work out:

a) $f(3)$ **b)** $f(10)$ **c)** $f\left(\frac{1}{2}\right)$ **d)** $f(-4)$ **e)** $f(-1.5)$

2. $g(x) = \frac{x^2+1}{2}$ Work out:

a) $g(0)$ **b)** $g(3)$ **c)** $g(10)$ **d)** $g(-2)$ **e)** $g\left(-\frac{1}{2}\right)$

3. $f(x) = x^3 - 2x + 1$ Work out:

a) $f(2)$ **b)** $f(-2)$ **c)** $f(100)$ **d)** $f(0)$ **e)** $f\left(\frac{1}{2}\right)$

4. $g(x) = 2^x$ Work out:

a) $g(2)$ **b)** $g(5)$ **c)** $g(0)$ **d)** $g(-1)$ **e)** $g(-3)$

5. $h(x) = \frac{x+1}{x-1}$ Work out:

a) $h(2)$ **b)** $h(3)$ **c)** $h(-1)$ **d)** $h(0)$ **e)** $h\left(1\frac{1}{2}\right)$

6. $f(x) = 2x + 5$ **a)** If $f(a) = 20$, what is the value of a?

b) If $f(b) = 0$, what is the value of b? **c)** If $f(c) = c$, what is the value of c?

7. $g(x) = \sqrt{x+3}$ **a)** Work out $g(33)$.

b) If $g(a) = 10$, work out a. **c)** If $g(b) = 2.5$, work out b.

8. $f(x) = 2x - 8$ and $g(x) = 10 - x$. **a)** If $f(x) = g(x)$, what is the value of x?

b) Sketch the graphs of $y = f(x)$ and $y = g(x)$. At what point do they cross?

GRADE B

9. $h(x) = \frac{12}{x} + 1$ $k(x) = 2^x - 1$

a) Work out $h(6)$. **b)** Work out $k(-1)$.

c) Solve the equation $h(x) = k(3)$. **d)** Solve the equation $k(x) = h(-12)$.

10. $f(x) = 4x - 1$ Work out:

a) $f(x + 2)$ **b)** $f(3x)$ **c)** $3f(x) + 5$ **d)** $f(x + 1) - f(x)$

11. $f(x) = x^2 - 16$

a) Solve $f(x) = 0$ **b)** Solve $f(2x) = 0$ **c)** Solve $f(3x) = 0$

d) What do you notice about your answers above?

GRADE A

12. $f(x) = x^2 + 5x - 1$

a) Solve $f(x) = 0$ Give your answer to 2 decimal places.

b) Use your answers to part **a** to write down the solution to $f(2x) = 0$

13. $g(x) = 2x^2 - 3x - 20$

a) Solve $g(x) = 0$ **b)** Solve $g(x) = -20$ **c)** Solve $g(x) = 2x^2$

5.2 Domain and range

THIS SECTION WILL SHOW YOU HOW TO ...

✓ define the domain of a function
✓ work out the range of a function
✓ express a domain in a variety of forms
✓ express a range in a variety of forms

KEY WORDS

✓ domain ✓ range ✓ mapping

Consider function $f(x) = \sqrt{x - 3}$.

$f(3) = 0$

$f(4) = 1$

$f(19) = 4$

$f(103) = 10$

This function shows a connection between two sets of numbers. The starting set is called the **domain**.

The resulting set of numbers is called the **range**.

This connection between the two sets is called a **mapping**.

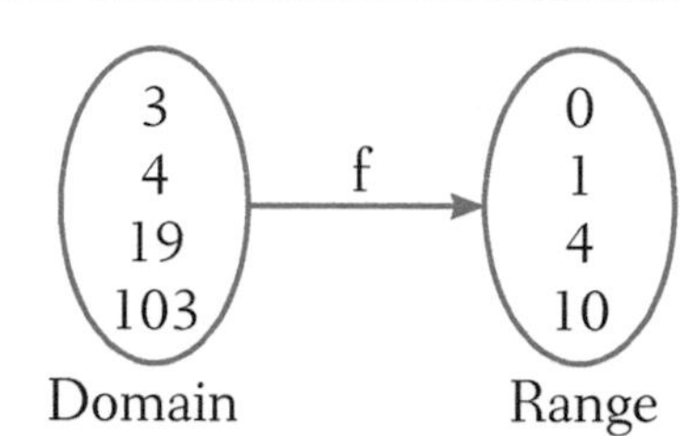

You cannot find the square root of a negative number so numbers less than 3 must be excluded from the domain of f in this case.

EXAMPLE 3

$g(x) = \frac{1}{x}$

a) What number must be excluded from the domain of g?

b) If the domain is $\{x: 1 \leqslant x \leqslant 2\}$, what is the range?

a) You cannot evaluate $\frac{1}{0}$ so 0 must be excluded from the domain of g.

b) $g(1) = \frac{1}{1} = 1$ $\quad$ $g(2) = \frac{1}{2} + \frac{1}{2}$

The range will be all the numbers from $\frac{1}{2}$ to 1.

You could write this as $\left\{y: \frac{1}{2} \leqslant y \leqslant 1\right\}$

EXAMPLE 4

$f(x) = x^2$ has domain $x > 5$.

State the range of $f(x)$.

$f(x) > 5^2$, so the range of $f(x)$ is $f(x) > 25$.

EXAMPLE 5

$f(x) = 3x - 2$ and $f(x) > 4$.

Work out the domain of x.

$3x - 2 > 4$

$3x > 6$

So the domain is $x > 2$.

EXERCISE 5B

GRADE A

1. What values of x must be excluded from the domains of these functions?

a) $f(x) = \sqrt{x}$ **b)** $g: = \frac{1}{x+1}$ **c)** $h: = \sqrt{x+1}$

d) $j: = \frac{1}{2x+1}$ **e)** $k(x) = \frac{1}{x^2 - 3x + 2}$

2. $f(x) = x^2 + 1$. Work out the ranges for each domain.

a) $\{3, 4, 5\}$ **b)** $\{-2, -1, 0, 1, 2\}$ **c)** $\{x: 1 \leqslant x \leqslant 2\}$

d) $\{x: x \geqslant 10\}$ **e)** $\{x: x \leqslant -10\}$

3. Suppose the domain is $\{1, 2, 3, 4\}$. Work out the range for each function.

a) $f(x) = (x - 2)^2$ **b)** $g(x) = \frac{1}{x}$ **c)** $h(x) = 2x + 3$

d) $f(x) = 6 - x$ **e)** $g(x) = (x - 1)(x - 4)$

4. $f(x) = x^2$. Explain why -2 could be in the domain but cannot be in the range.

5. The domain of a function f is $\{1, 2, 3, 4\}$ and the range is $\{2, 3, 4, 5\}$.
Say whether each of these is a possible description of:

a) $f(x) = x + 1$ **b)** $f(x) = 2x$ **c)** $f(x) = 6 - x$

6. $f(x) = 2x - 9$ and $f(x) > 0$. State the smallest whole number value of x.

7. $f(x) = 3x + 1$ and $x > 5$

a) State the range of $f(x)$. **b)** Work out the domain and range of $f(4x)$.

8. $f(x) = x^2 + 3$ and $x > 0$

a) State the range of $f(x)$. **b)** Work out the domain and range of $f(2x)$.

GRADE A*

9. $f(x)$ is a quadratic graph of the form $y = x^2 + c$.
The range of $f(x)$ is $f(x) > 7$.

a) Write down the value of c. **b)** Work out $f(2)$.

10. The function $f(x)$ is defined as $f(x) = 11 - 2x$ for $a < x < b$.
The range of $f(x)$ is $-5 < f(x) < 7$. Work out the values of a and b.

11. The graph of $y = f(x)$ is shown on the right for values of x from -4 to 2. Write down the range of $f(x)$.

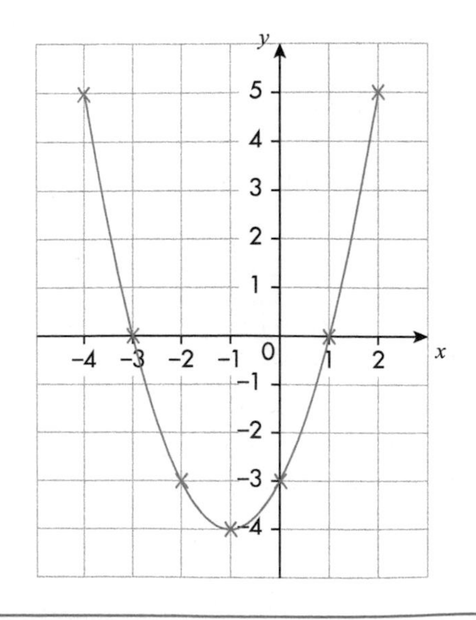

5.3 Sketching graphs of linear and quadratic functions

THIS SECTION WILL SHOW YOU HOW TO ...

- ✓ draw or sketch graphs of linear and quadratic functions with up to three domains
- ✓ label points of intersection of graphs with the axes
- ✓ understand that graphs should only be drawn within the given domain
- ✓ identify any symmetries on a quadratic graph and from this determine the coordinates of the turning point.

KEY WORDS

- ✓ quadratic
- ✓ parabola

Quadratic graphs

A **quadratic** graph can be drawn from a quadratic equation of the form $y = ax^2 + bx + c$, where a, b and c are constants, $a \neq 0$. All of the following are quadratic equations and each would produce a quadratic graph.

$y = x^2$ $\quad$ $y = x^2 + 5$ $\quad$ $y = x^2 - 3x$

$y = x^2 + 5x + 6$ $\quad$ $y = 3x^2 - 5x + 4$

EXAMPLE 6

Draw the graph of $y = x^2 + 5x + 6$ for $-5 \leqslant x \leqslant 3$.

Make a table, as shown below. Work out the values in each row (x^2, $5x$, 6) separately, adding them together to obtain the values of y.

x	–5	–4	–3	–2	–1	0	1	2	3
y^2	25	16	9	4	1	0	1	4	9
$+5x$	–25	–20	–15	–10	–5	0	5	10	15
$+6$	6	6	6	6	6	6	6	6	6
y	6	2	0	0	2	6	12	20	30

Now plot the points from the table.

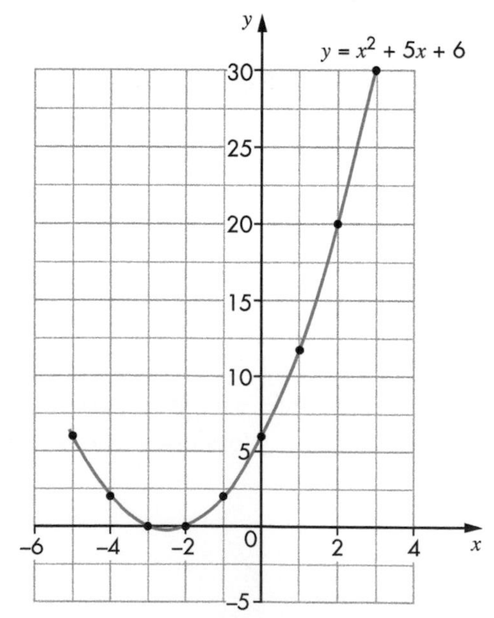

Note that in an examination paper you may be given only the first and last rows, with some values filled in. For example:

x	−5	−4	−3	−2	−1	0	1	2	3
y	6		0		2				30

In this case, you would either construct your own table, or work out the remaining y-values with a calculator.

EXAMPLE 7

a) Complete the table for $y = 3x^2 - 5x + 4$ for $-1 \leqslant x \leqslant 3$, then draw the graph.

x	−1	−0.5	0	0.5	1	1.5	2	2.5	3
y	12		0	2.25	2			10.25	16

b) Use your graph to find the value of y when $x = 2.2$.

c) Use your graph to find the values of x that give a y-value of 9.

a) The table gives only some values. So you either set up your own table with $3x^2$, $-5x$ and $+4$, or calculate each y-value. For example, on the majority of scientific calculators, the value for −0.5 will be worked out as:

3 × ((−) 0 . 5) x^2 − 5 × (−) 0 . 5 + 4 =

Check that you get an answer of 7.25.

If you want to make sure that you are doing the correct arithmetic with your calculator, try some values for x for which you know the answer. For example, try $x = 0.5$, and see whether your answer is 2.25.

The complete table should be:

x	−1	−0.5	0	0.5
y	12	7.25	4	2.25

1	1.5	2	2.5	3
2	3.25	6	10.25	16

The graph is shown on the right.

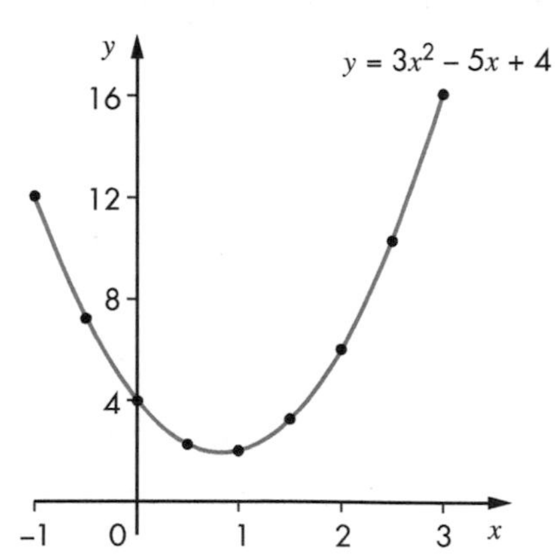

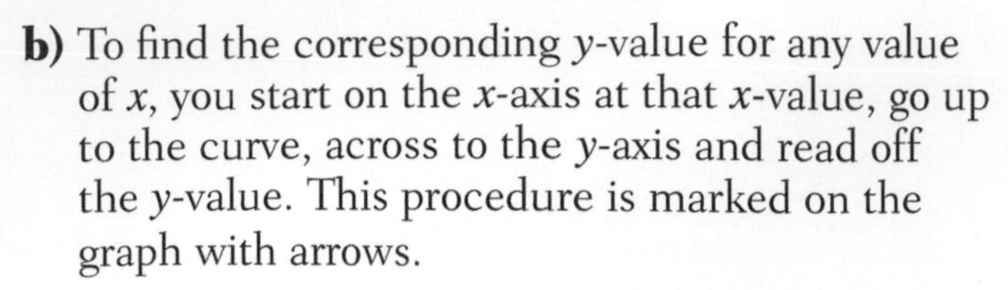

b) To find the corresponding y-value for any value of x, you start on the x-axis at that x-value, go up to the curve, across to the y-axis and read off the y-value. This procedure is marked on the graph with arrows.

Always show these arrows because even if you make a mistake and misread the scales, you may still get a mark.

When $x = 2.2$, $y = 7.5$.

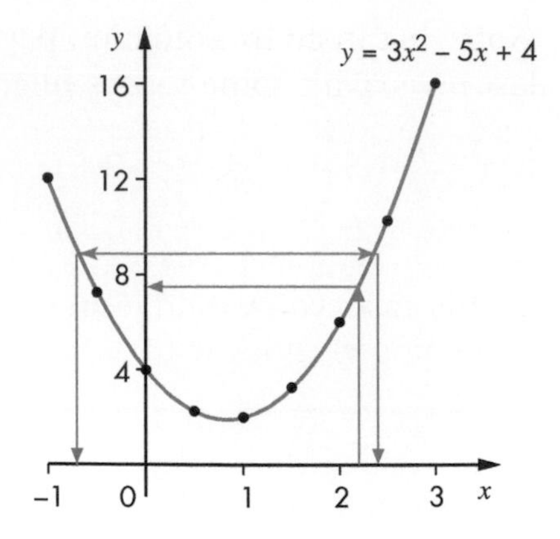

c) This time start at 9 on the y-axis and read off the two x-values that correspond to a y-value of 9. Again, this procedure is marked on the graph with arrows.

When $y = 9$, $x = -0.7$ or $x = 2.4$.

A quadratic curve drawn correctly will always give a smooth curve, called a **parabola**.

Drawing accurate graphs

Although it is difficult to draw accurate curves, examiners work to a tolerance of only 1 mm. Here are some of the more common ways in which marks are lost in an examination.

- When the points are too far apart, a curve tends to 'wobble'.
- Drawing curves in small sections leads to 'feathering'.
- The place where a curve should turn smoothly is drawn 'flat'.
- A line is drawn through a point that, clearly, has been incorrectly plotted.

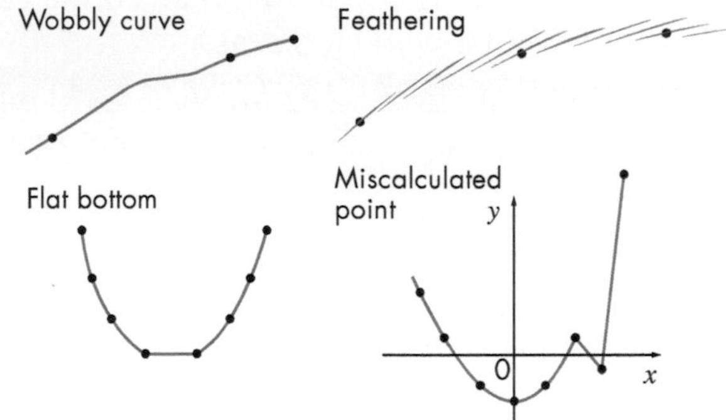

Here are some tips that will make it easier for you to draw smooth, curved lines.

- If you are *right-handed*, turn your paper or exercise book round so that you draw from left to right. Your hand is steadier this way than when you are trying to draw from right to left or away from your body. If you are *left-handed*, you should find drawing from right to left the more accurate way.
- Move your pencil over the points as a practice run without drawing the curve.
- Do one continuous curve and only stop at a plotted point.
- Use a *sharp* pencil and do not press too heavily, so that you may easily rub out mistakes.

Normally, in an examination, grids are provided with the axes clearly marked, so the examiner can place a transparent master over a graph and see immediately whether any lines are badly drawn or points are misplotted. Remember: a tolerance of 1 mm is all that you are allowed.

You do not need to work out all values in a table. You need only to work out the y-value. The other rows in the table are just working lines to break down the calculation. Learn how to calculate y-values with a calculator as there is no credit given for setting up tables in examinations.

EXERCISE 5C

GRADE C

In this exercise, suitable ranges are suggested for the axes. You can use any type of graph paper.

1. a) Copy and complete the table or use a calculator to work out values for the graph of $y = 3x^2$ for values of x from –3 to 3.

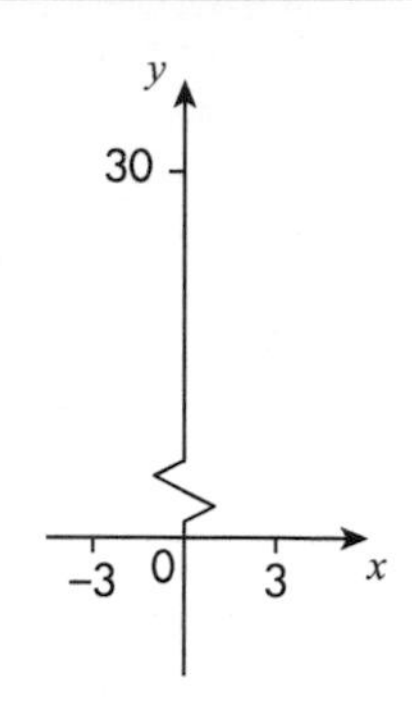

x	–3	–2	–1	0	1	2	3
y	27		3			12	

b) Use your graph to find the value of y when $x = –1.5$.

c) Use your graph to find the values of x that give a y-value of 10.

2. a) Copy and complete the table or use a calculator to work out values for the graph of $y = x^2 + 2$ for values of x from –5 to 5.

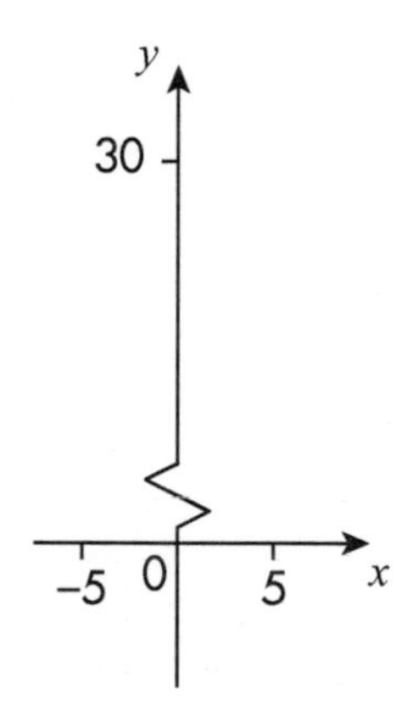

x	–5	–4	–3	–2	–1	0	1	2	3	4	5
$y = x^2 + 2$	27		11					6			

b) Use your graph to find the value of y when $x = –2.5$.

c) Use your graph to find the values of x that give a y-value of 14.

3. a) Copy and complete the table or use a calculator to work out values for the graph of $y = x^2 – 2x – 8$ for values of x from –5 to 5.

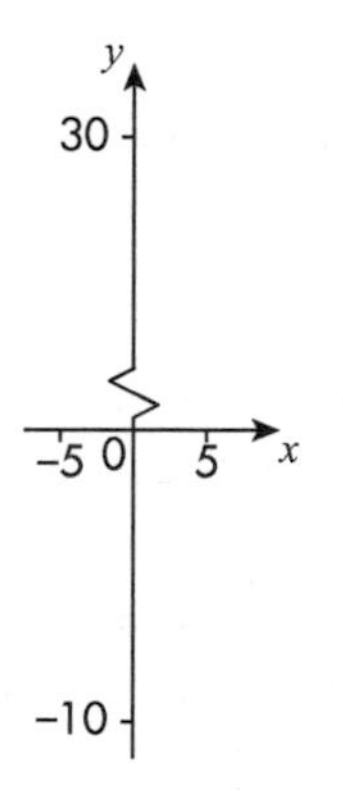

x	–5	–4	–3	–2	–1	0	1	2	3	4	5
x^2	25		9					4			
$–2x$	10							–4			
$–8$	–8							–8			
y	27							–8			

b) Use your graph to find the value of y when $x = 0.5$.

c) Use your graph to find the values of x that give a y-value of –3.

4. a) Copy and complete the table or use a calculator to work out the values for the graph of $y = x^2 + 2x – 1$ for values of x from –3 to 3.

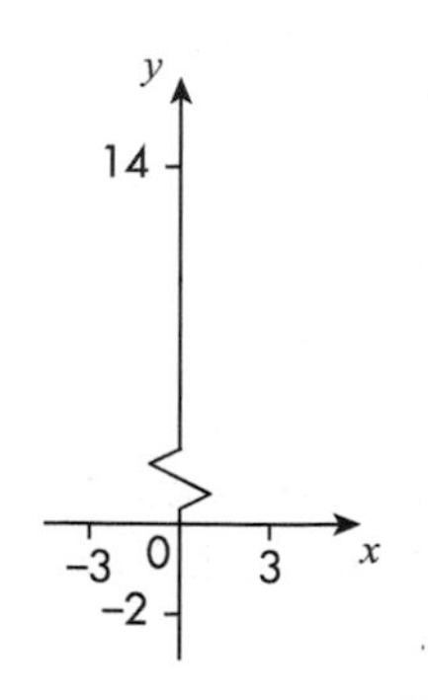

x	–3	–2	–1	0	1	2	3
x^2	9				1	4	
$+2x$	–6		–2			4	
$–1$	–1	–1				–1	
y	2					7	

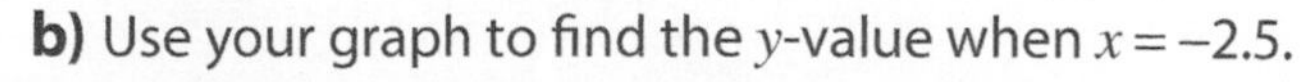
b) Use your graph to find the y-value when $x = –2.5$.

c) Use your graph to find the values of x that give a y-value of 1.

d) On the same axes, draw the graph of $y = \frac{x}{2} + 2$.

e) Where do the graphs $y = x^2 + 2x - 1$ and $y = \frac{x}{2} + 2$ cross?

5. a) Copy and complete the table or use a calculator to work out values for the graph of $y = x^2 - x + 6$ for values of x from −3 to 3.

x	−3	−2	−1	0	1	2	3
x^2	9				1	4	
$-x$	3					−2	
+6	6					6	
y	18					8	

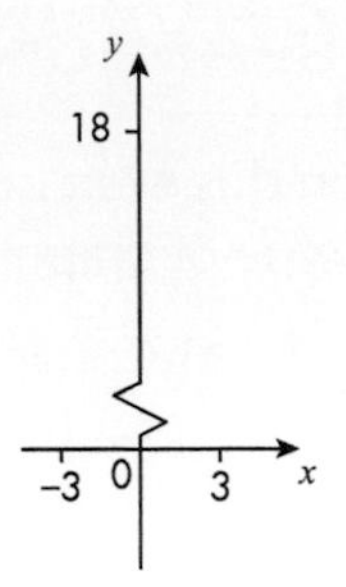

b) Use your graph to find the y-value when $x = 2.5$.

c) Use your graph to find the values of x that give a y-value of 8.

d) Copy and complete the table or use a calculator to draw the graph of $y = x^2 + 5$ on the same axes.

x	−3	−2	−1	0	1	2	3
y	14		6				14

e) Where do the graphs $y = x^2 - x + 6$ and $y = x^2 + 5$ cross?

6. a) Copy and complete the table or use a calculator to work out values for the graph of $y = x^2 + 2x + 1$ for values of x from −3 to 3.

x	−3	−2	−1	0	1	2	3
x^2	9				1	4	
+2x	−6					4	
+1	1					1	
y	4						

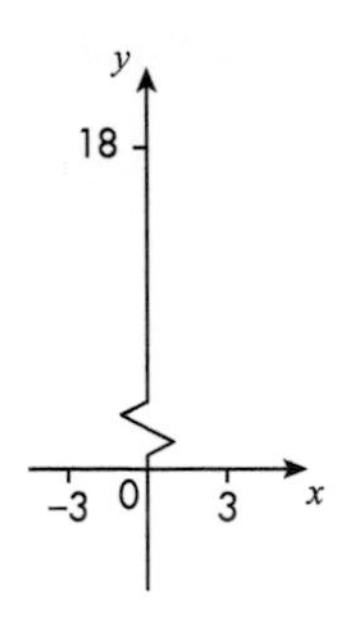

b) Use your graph to find the y-value when $x = 1.7$.

c) Use your graph to find the values of x that give a y-value of 2.

d) On the same axes, draw the graph of $y = 2x + 2$.

e) Where do the graphs $y = x^2 + 2x + 1$ and $y = 2x + 2$ cross?

7. a) Copy and complete the table or use a calculator to work out values for the graph of $y = 2x^2 - 5x - 3$ for values of x from −2 to 4.

x	−2	−1.5	−1	−0.5	0	0.5	1	1.5	2	2.5	3	3.5	4
y	15	9			−3	−5				−3			9

b) Where does the graph cross the x-axis?

5.4 The significant points of a quadratic graph

THIS SECTION WILL SHOW YOU HOW TO ...

✓ recognise and calculate the significant points of a quadratic graph

KEY WORDS

✓ intercept
✓ maximum
✓ minimum
✓ roots
✓ vertex

A quadratic graph has four points that are of interest to a mathematician. These are the points A, B, C and D on the diagram. The x-values at A and B are called the **roots**, and are where the graph crosses the x-axis. C is the point where the graph crosses the y-axis (the **intercept**) and D is the **vertex**, which is the lowest or highest point of the graph.

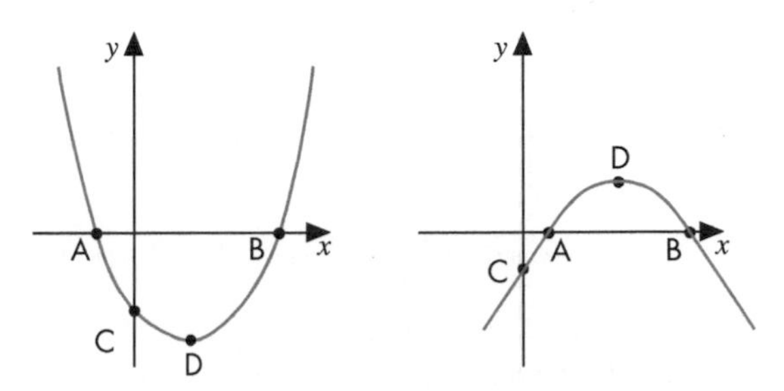

The roots

If you look at your answer to question **7** in Exercise 5C, you will see that the graph crosses the x-axis at $x = -0.5$ and $x = 3$. Since the x-axis is the line $y = 0$, the y-value at any point on the x-axis is zero. So, you have found the solution to the equation:

$0 = 2x^2 - 5x - 3$ that is $2x^2 - 5x - 3 = 0$

Equations of this type are known as *quadratic equations*.

You can solve quadratic equations by finding the values of x that make them true. Such values are called the roots of the equation. On the graph, these occur where the curve cuts the x-axis. So the roots of the quadratic equation $2x^2 - 5x - 3 = 0$ are -0.5 and 3.

Now check these values.

For $x = 3.0$ $2(3)^2 - 5(3) - 3 = 18 - 15 - 3 = 0$

For $x = 0.5$ $2(-0.5)^2 - 5(-0.5) - 3 = 0.5 + 2.5 - 3 = 0$

You can find the roots of a quadratic equation by drawing its graph and finding where the graph crosses the x-axis.

EXAMPLE 8

a) Draw the graph of $y = x^2 - 3x - 4$ for $-2 \leqslant x \leqslant 5$.

b) Use your graph to find the roots of the equation $x^2 - 3x - 4 = 0$.

a) Set up a table.

x	−2	−1	0	1	2	3	4	5
x	4	1	0	1	4	9	16	25
$-3x$	6	3	0	−3	−6	−9	−12	−15
−4	−4	−4	−4	−4	−4	−4	−4	−4
y	6	0	−4	−6	−6	−4	0	6

Draw the graph.

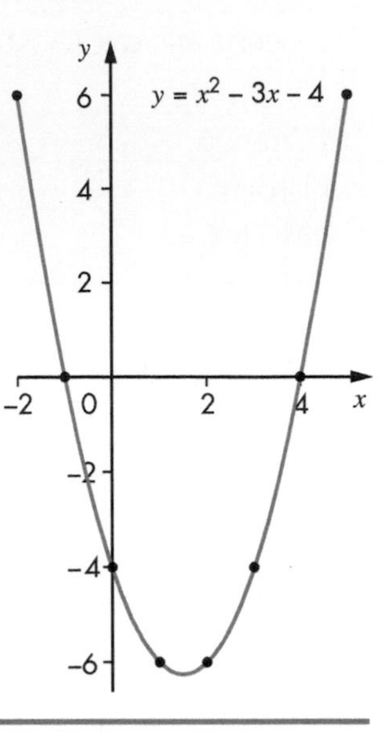

b) The points where the graph crosses the x-axis are −1 and 4.

So, the roots of $x^2 - 3x - 4 = 0$ are $x = -1$ and $x = 4$.

Note that sometimes the quadratic graph may not cross the x-axis. In this case there are no roots.

The y-intercept

If you look at all the quadratic graphs you have drawn so far you will see a connection between the equation and the point where the graph crosses the y-axis. Very simply, the constant term of the equation $y = ax^2 + bx + c$ (that is, the value c) is where the graph crosses the y-axis. The intercept is at $(0, c)$.

The vertex

The lowest (or highest) point of a quadratic graph is called the *vertex*.

If it is the highest point, it is called the **maximum**.

If it is the lowest point, it is called the **minimum**.

It is difficult to find a general rule for this point, but the x-coordinate is always half-way between the roots. The easiest way to find the y-value is to substitute the x-value into the original equation.

EXAMPLE 9

Sketch the graph of $y = x^2 - 5x + 6$ for $0 < x < 5$.

Setting up a table of values gives:

x	0	1	2	3	4	5
y	6	2	0	0	2	6

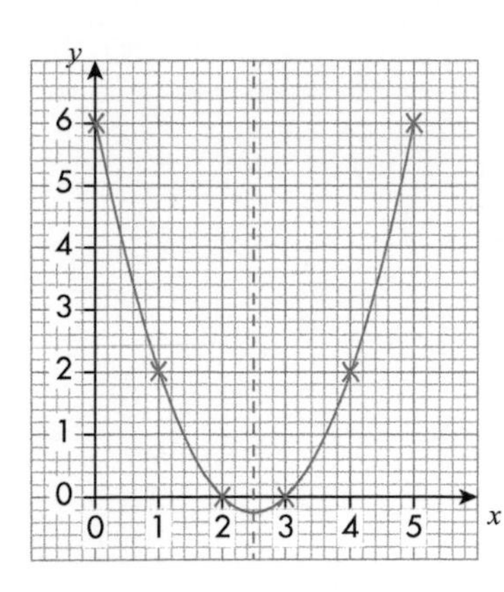

The points of intersection with the axes are (0, 6), (2, 0) and (3, 0).

The equation of the line of symmetry is $x = 2.5$.

The turning point of the graph is at (2.5, –0.25).

EXERCISE 5D

GRADE B

1. a) Copy and complete the table to draw the graph of $y = x^2 - 4$ for $-4 \leqslant x \leqslant 4$.

x	–4	–3	–2	–1	0	1	2	3	4
y	12			–3				5	

b) Use your graph to find the roots of $x^2 - 4 = 0$.

2. a) Look at the equation of the graph you drew in question **1**. Is there a connection between the number in the equation and its roots?

b) Before you draw the graphs in parts **c)** and **d)**, try to predict what their roots will be.

c) Copy and complete the table and draw the graph of $y = x^2 - 1$ for $-4 \leqslant x \leqslant 4$.

x	–4	–3	–2	–1	0	1	2	3	4
y	15				–1			8	

d) Copy and complete the table and draw the graph of $y = x^2 - 5$ for $-4 \leqslant x \leqslant 4$.

x	–4	–3	–2	–1	0	1	2	3	4
y	11		–1					4	

e) Were your predictions correct?

3. a) Copy and complete the table and draw the graph of $y = x^2 + 4x$ for $-5 \leqslant x \leqslant 2$.

x	−5	−4	−3	−2	−1	0	1	2
x^2	25			4			1	
$+4x$	−20			−8			4	
y	5			−4			5	

b) Use your graph to find the roots of the equation $x^2 + 4x = 0$.

4. a) Copy and complete the table and draw the graph of $y = x^2 + 3x$ for $-5 \leqslant x \leqslant 3$.

x	−5	−4	−3	−2	−1	0	1	2	3
y	10			−2				10	

b) Use your graph to find the roots of the equation $x^2 + 3x = 0$.

5. a) Look at the equations of the graphs you drew in questions **3** and **4**. Is there a connection between the numbers in each equation and the roots?

b) Before you draw the graphs in parts **c)** and **d)**, try to predict what their roots will be.

c) Copy and complete the table and draw the graph of $y = x^2 - 3x$ for $-2 \leqslant x \leqslant 5$.

x	−2	−1	0	1	2	3	4	5
y	10			−2				10

d) Copy and complete the table and draw the graph of $y = x^2 + 5x$ for $-6 \leqslant x \leqslant 2$.

x	−6	−5	−4	−3	−2	−1	0	1	2
y	6			−6				6	

e) Were your predictions correct?

6. a) Copy and complete the table and draw the graph of $y = x^2 - 4x + 4$ for $-1 \leqslant x \leqslant 5$.

x	−1	0	1	2	3	4	5
y	9				1		

b) Use your graph to find the roots of the equation $x^2 - 4x + 4 = 0$.

c) What do you notice about the roots?

7. a) Copy and complete the table and draw the graph of $y = x^2 - 6x + 3$ for $-1 \leqslant x \leqslant 7$.

x	−1	0	1	2	3	4	5	6	7
y	10			−5			−2		

b) Use your graph to find the roots of the equation $x^2 - 6x + 3 = 0$.

8. a) Copy and complete the table and draw the graph of $y = 2x^2 + 5x - 6$ for $-5 \leqslant x \leqslant 2$.

x	−5	−4	−3	−2	−1	0	1	2
y								

b) Use your graph to find the roots of the equation $2x^2 + 5x - 6 = 0$.

9. Look back at questions **1** to **5**.

a) Write down the point of intersection of the graph with the y-axis for each one.

b) Write down the coordinates of the minimum point (vertex) of each graph for each one.

c) Explain the connection between these points and the original equation.

Graphs of functions

A **function** can be defined as a set of distinct parts, each with its own domain.

EXAMPLE 10

A function f(x) is defined as

$$f(x) = x^2 \quad 0 \leqslant x < 2$$
$$= 4 \quad 2 \leqslant x < 3$$
$$= 7 - x \quad 3 \leqslant x \leqslant 5$$

Draw the graph of $y = f(x)$ on the grid.

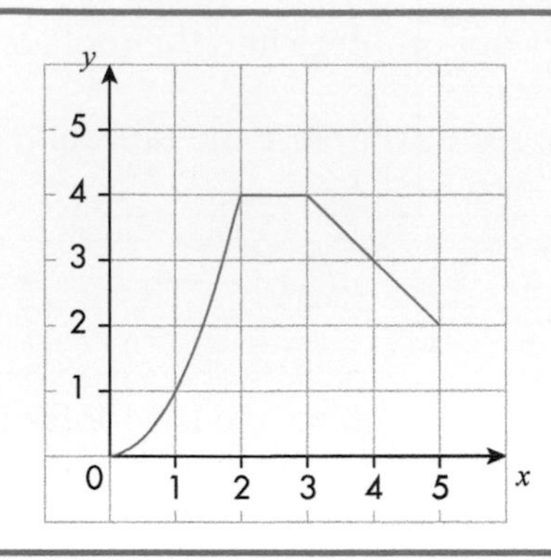

EXAMPLE 11

The graph of $y = f(x)$ is shown.

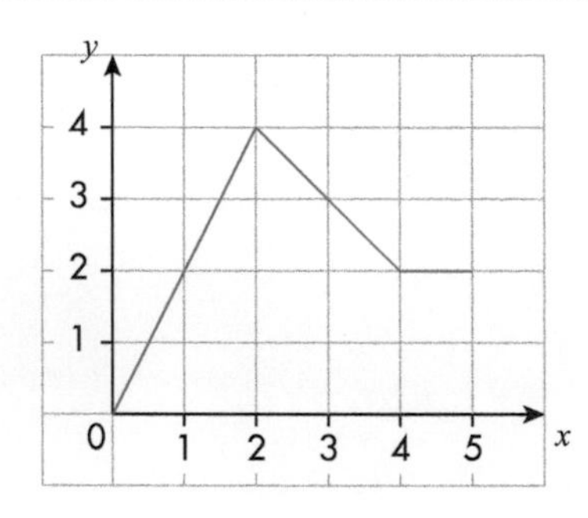

Define f(x), stating the domain for each part.

$$f(x) = 2x \quad 0 \leqslant x < 2$$
$$= 6 - x \quad 2 \leqslant x < 4$$
$$= 2 \quad 4 \leqslant x \leqslant 5$$

EXERCISE 5E

GRADE B

1. In each part sketch the graph of $y = f(x)$

a) $f(x) = 2 \quad 0 \leqslant x < 2$
$= x \quad 2 \leqslant x < 4$
$= 4 \quad 4 \leqslant x \leqslant 6$

b) $f(x) = x \quad 0 \leqslant x < 2$
$= 2 \quad 2 \leqslant x < 4$
$= 6 - x \quad 4 \leqslant x \leqslant 6$

c) $f(x) = x + 1 \quad 0 \leqslant x < 2$
$= 3 \quad 2 \leqslant x < 4$
$= x - 1 \quad 4 \leqslant x \leqslant 6$

d) $f(x) = x^2 \quad 0 \leqslant x < 2$
$= 8 - 2x \quad 2 \leqslant x \leqslant 4$

e) $f(x) = 2 + x \quad 0 \leqslant x < 2$
$= 4 \quad 2 \leqslant x < 4$
$= 6 - \frac{1}{2}x \quad 4 \leqslant x \leqslant 6$

f) $f(x) = x^2 \quad 0 \leqslant x < 1$
$= x \quad 1 \leqslant x < 3$
$= 2x - 3 \quad 3 \leqslant x \leqslant 6$

2. Write down the range of f(x) for each part of question 1.

3. Graphs of $y = \mathrm{f}(x)$ are shown.

Define $\mathrm{f}(x)$ stating the domain for each part of each graph.

a)

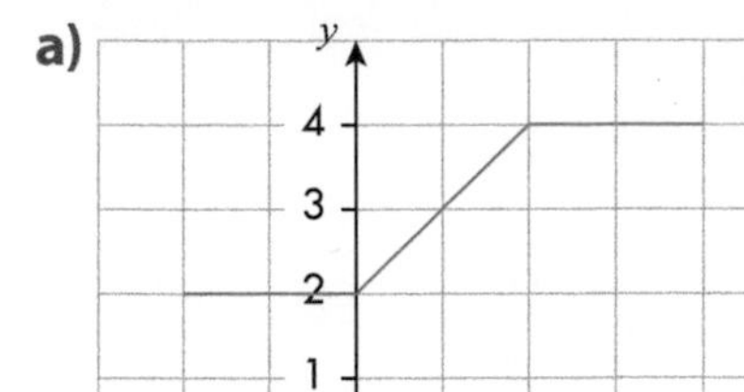

c)

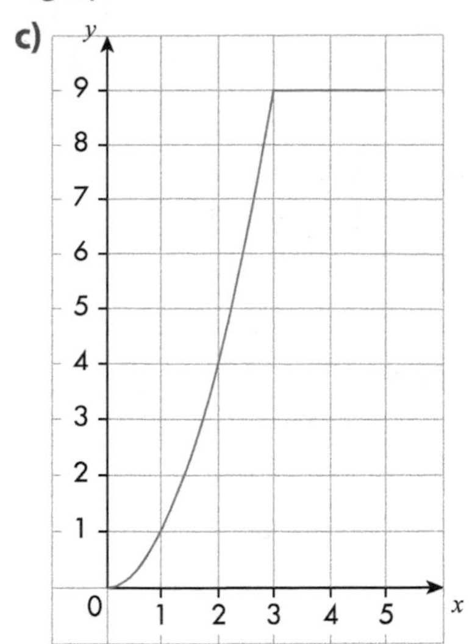

b)

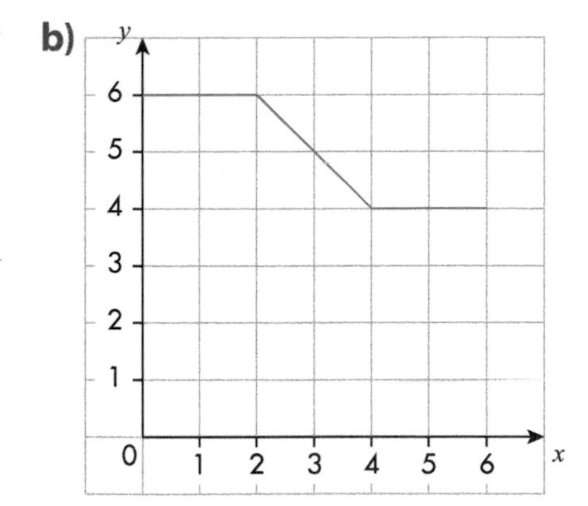

d)

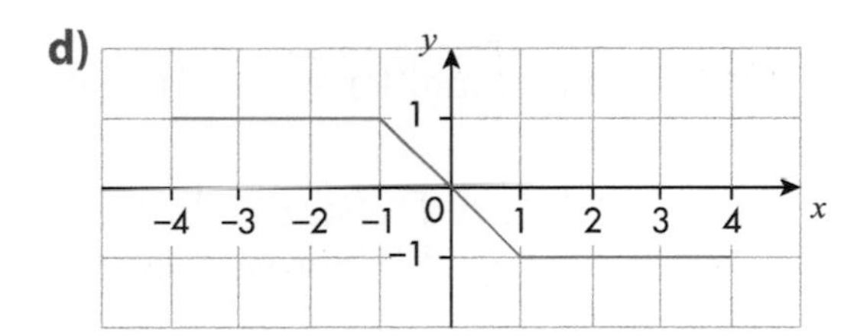

Exam-style questions

1. $\mathrm{f}(n) = n(n + 1)$ where n is an integer.

a) Find f(10) and f(–10).

b) Solve the equation $\mathrm{f}(n) = 0$. GRADE C

2. $\mathrm{f}(x) = 2x + x^2$ and $x \geqslant 3$.

a) Find f(3.5).

b) Write down the range of $\mathrm{f}(x)$. GRADE C

3. $\mathrm{g}(x) = \dfrac{x+2}{x+4}$ for all possible values of x.

a) What value of x must be excluded from the domain?

b) Solve the equation $\mathrm{g}(x) = 2$. GRADE B

6 Matrices

6.1 Introduction to matrices

THIS SECTION WILL SHOW YOU HOW TO ...

- ✓ multiply a 2×2 matrix by a 2×1 matrix
- ✓ multiply a 2×2 matrix by a 2×2 matrix
- ✓ multiply 2×2 and 2×1 matrices by a scalar
- ✓ understand that, in general, matrix multiplication is not commutative
- ✓ understand that matrix multiplication is associative

KEY WORDS

- ✓ matrix
- ✓ matrices
- ✓ row
- ✓ column
- ✓ order
- ✓ element
- ✓ scalar quantity
- ✓ matrix quantity

A **matrix** is a rectangular array of numbers. The plural of matrix is **matrices**.

Matrices can be used to store information. For example, suppose there are two ferries, A and B, across a river.

On Monday:

- Ferry A carries 70 adults and 42 children.
- Ferry B carries 52 adults and 30 children.

You can show this information in a matrix.

$$\mathbf{M} = \begin{pmatrix} 70 & 42 \\ 52 & 30 \end{pmatrix}$$

This matrix has two **rows** and two **columns**. It is a 2×2 matrix – read this as '2 by 2'.

2×2 is the **order** of the matrix. Note that the number of rows comes first, followed by the number of columns.

The numbers in the matrix are called the **elements** of the matrix.

Multiplying matrices by a scalar

The ferries run five days a week. The ferry manager estimates the number of passengers in one week by multiplying the numbers on Monday by five.

$$5\mathbf{M} = 5 \times \begin{pmatrix} 70 & 42 \\ 52 & 30 \end{pmatrix} = \begin{pmatrix} 350 & 210 \\ 260 & 150 \end{pmatrix}$$

To multiply the matrix by any number, multiply every element by that number.

The multiplying number (5 in the example above) is a **scalar quantity**. Quantities inside the matrix are called **matrix quantities**.

EXAMPLE 1

Here are two matrices. $\mathbf{A} = \begin{pmatrix} 12 \\ -7 \end{pmatrix}$ $\mathbf{B} = \begin{pmatrix} 13 & 6 \\ -4 & -5 \end{pmatrix}$

a) Write down the order of each matrix.

b) Write down matrices 2**A**, 3**B**, $\frac{1}{2}$**A**.

a) A has 2 rows and 1 column so the order is 2×1.

B has 2 rows and 2 columns so the order is 2×2.

b) $2\mathbf{A} = 2\begin{pmatrix} 12 \\ -7 \end{pmatrix} = \begin{pmatrix} 24 \\ -14 \end{pmatrix}$

$3\mathbf{B} = 3\begin{pmatrix} 13 & 6 \\ -4 & -5 \end{pmatrix} = \begin{pmatrix} 39 & 18 \\ -12 & -15 \end{pmatrix}$

$\frac{1}{2}\mathbf{A} = \frac{1}{2}\begin{pmatrix} 12 \\ -7 \end{pmatrix} = \begin{pmatrix} 6 \\ -3.5 \end{pmatrix}$

Multiplying matrices

Look again at the ferry example. This matrix gives the numbers of passengers on two different ferries on Monday.

$\mathbf{M} = \begin{pmatrix} 70 & 42 \\ 52 & 30 \end{pmatrix}$

The columns give the numbers of adults and children. The rows show ferries A and B.

The tickets cost £2 for adults and 50p for children. We can write this as a 2×1 matrix like this.

$\mathbf{C} = \begin{pmatrix} 2 \\ 0.5 \end{pmatrix}$

To find out how much money each ferry took, multiply **M** and **C** like this.

$\mathbf{MC} = \begin{pmatrix} 70 & 42 \\ 52 & 30 \end{pmatrix} \times \begin{pmatrix} 2 \\ 0.5 \end{pmatrix} = \begin{pmatrix} 140+21 \\ 104+15 \end{pmatrix} = \begin{pmatrix} 161 \\ 119 \end{pmatrix}$

Ferry A took £161 and ferry B took £119.

When you multiply two matrices, you combine each *row* of the first matrix with each *column* of the second to give a single number.

A 2×2 matrix multiplied by a 2×1 matrix gives a 2×1 matrix.

EXAMPLE 2

$\mathbf{A} = \begin{pmatrix} 2 & 1 \\ -2 & 3 \end{pmatrix}$ and $\mathbf{B} = \begin{pmatrix} 4 & -1 \\ 1 & 2 \end{pmatrix}$. Work out **AB** and **BA**.

$$\mathbf{AB} = \begin{pmatrix} 2 & 1 \\ -2 & 3 \end{pmatrix}\begin{pmatrix} 4 & -1 \\ 1 & 2 \end{pmatrix} = \begin{pmatrix} 8+1 & -2+2 \\ -8+3 & 2+6 \end{pmatrix} = \begin{pmatrix} 9 & 0 \\ -5 & 8 \end{pmatrix}$$

$$\mathbf{BA} = \begin{pmatrix} 4 & -1 \\ 1 & 2 \end{pmatrix}\begin{pmatrix} 2 & 1 \\ -2 & 3 \end{pmatrix} = \begin{pmatrix} 8+2 & 4+-3 \\ 2+-4 & 1+6 \end{pmatrix} = \begin{pmatrix} 10 & 1 \\ -2 & 7 \end{pmatrix}$$

Notice that $\mathbf{AB} \neq \mathbf{BA}$

EXERCISE 6A

GRADE B

1. $\mathbf{A} = \begin{pmatrix} 3 \\ -6 \end{pmatrix}$ $\mathbf{B} = \begin{pmatrix} -8 \\ 2 \end{pmatrix}$ $\mathbf{C} = \begin{pmatrix} 10 & -4 \\ -9 & 7 \end{pmatrix}$ $\mathbf{D} = \begin{pmatrix} 8 & 3 \\ -12 & -9 \end{pmatrix}$

Complete these scalar multiplications.

a) 5**A** **b)** 8**B** **c)** 2**C** **d)** 4**D**

e) –3**A** **f)** –6**C** **g)** $\frac{1}{3}$**A** **h)** $\frac{1}{4}$**B**

2. a) If $\mathbf{A} = \begin{pmatrix} 3 & 1 \\ 1 & 2 \end{pmatrix}$ find $\mathbf{A}^2$.

b) If $\mathbf{B} = \begin{pmatrix} 0 & 2 \\ 1 & 5 \end{pmatrix}$ find $\mathbf{B}^2$

C) If $\mathbf{C} = \begin{pmatrix} 1 & 2 \\ 2 & 1 \end{pmatrix}$ find $\mathbf{C}^2$ and $\mathbf{C}^3$.

HINTS AND TIPS

$\mathbf{A}^2$ means $\mathbf{A} \times \mathbf{A}$.

3. $\mathbf{A} = \begin{pmatrix} 5 & 1 \\ 2 & 3 \end{pmatrix}$ $\mathbf{B} = \begin{pmatrix} 1 & -2 \\ 2 & 3 \end{pmatrix}$ $\mathbf{C} = \begin{pmatrix} 1 & 2 \\ 2 & 1 \end{pmatrix}$ Complete these matrix multiplications.

a) **AB** **b)** **BA** **c)** **AC** **d)** **BC** **e)** $\mathbf{A}^2$ **f)** $\mathbf{B}^2$

4. Calculate: **a)** $\begin{pmatrix} 2 & 0 \\ 1 & 3 \end{pmatrix}\begin{pmatrix} 1 & 2 \\ 4 & 1 \end{pmatrix}$ **b)** $\begin{pmatrix} -1 & 2 \\ 3 & 1 \end{pmatrix}\begin{pmatrix} 2 & 2 \\ 3 & 4 \end{pmatrix}$

5. $\mathbf{A} = \begin{pmatrix} 2 & 1 \\ 0 & 2 \end{pmatrix}$ $\mathbf{B} = \begin{pmatrix} 2 & 1 \\ 3 & 1 \end{pmatrix}$

a) Calculate: **(i) AB** **(ii) BA.** **b)** Is **AB** the same as **BA**?

GRADE A

6. $\begin{pmatrix} 2 & 4 \\ 1 & 3 \end{pmatrix} \times \begin{pmatrix} 3 & x \\ y & 1 \end{pmatrix} = \begin{pmatrix} 14 & 0 \\ 9 & 1 \end{pmatrix}$

Find the values of x and y.

7. **A** and **B** are two 2×2 matrices. Is it always true, sometimes true or never true that **AB** = **BA**? Give a reason for your answer.

8. $\begin{pmatrix} 2 & -3 \\ 0 & 1 \end{pmatrix}\begin{pmatrix} 2 \\ x \end{pmatrix} = \begin{pmatrix} y \\ y+2 \end{pmatrix}$

Work out the values of x and y.

9. $\mathbf{A} = \begin{pmatrix} 3 & x \\ y & 2 \end{pmatrix}$ $\mathbf{B} = \begin{pmatrix} 1 & 2 \\ 1 & 0 \end{pmatrix}$

Work out the values of x and y, given that **AB** = **BA**.

6.2 The zero matrix and the identity matrix

THIS SECTION WILL SHOW YOU HOW TO ...

✓ understand that $\begin{pmatrix} 0 & 0 \\ 0 & 0 \end{pmatrix}$ is a **zero matrix**

✓ understand that $I = \begin{pmatrix} 1 & 0 \\ 0 & 1 \end{pmatrix}$ is the 2×2 **identity matrix**

✓ understand that **AI** = **IA** = **A**

KEY WORDS

✓ zero matrix
✓ identity matrix

The zero matrix

Given two matrices, $\mathbf{A} = \begin{pmatrix} 2 & 1 \\ -2 & 3 \end{pmatrix}$ and $\mathbf{Z} = \begin{pmatrix} 0 & 0 \\ 0 & 0 \end{pmatrix}$

$$\Rightarrow \mathbf{A} + \mathbf{Z} = \begin{pmatrix} 2 & 1 \\ -2 & 3 \end{pmatrix} + \begin{pmatrix} 0 & 0 \\ 0 & 0 \end{pmatrix} = \begin{pmatrix} 2 & 1 \\ -2 & 3 \end{pmatrix} = \mathbf{A}$$

and $\mathbf{AZ} = \begin{pmatrix} 2 & 1 \\ -2 & 3 \end{pmatrix}\begin{pmatrix} 0 & 0 \\ 0 & 0 \end{pmatrix} = \begin{pmatrix} 0+0 & 0+0 \\ 0+0 & 0+0 \end{pmatrix} = \begin{pmatrix} 0 & 0 \\ 0 & 0 \end{pmatrix} = \mathbf{Z}$

In general, if **M** is *any* 2×2 matrix, $\mathbf{M} + \mathbf{Z} = \mathbf{M}$ and $\mathbf{MZ} = \mathbf{Z}$.

You should see that **Z** is like 0 (zero) when you add and multiply numbers.

If a is any number, then $a + 0 = a$ and $a \times 0 = 0$.

For this reason $\begin{pmatrix} 0 & 0 \\ 0 & 0 \end{pmatrix}$ is called the 2×2 **zero matrix**.

The identity matrix

Now suppose $\mathbf{I} = \begin{pmatrix} 1 & 0 \\ 0 & 1 \end{pmatrix}$ and, as before, $\mathbf{A} = \begin{pmatrix} 2 & 1 \\ -2 & 3 \end{pmatrix}$.

$$\Rightarrow \mathbf{AI} = \begin{pmatrix} 2 & 1 \\ -2 & 3 \end{pmatrix}\begin{pmatrix} 1 & 0 \\ 0 & 1 \end{pmatrix} = \begin{pmatrix} 2+0 & 0+1 \\ -2+0 & 0+3 \end{pmatrix} = \begin{pmatrix} 2 & 1 \\ -2 & 3 \end{pmatrix} = \mathbf{A}$$

and $\mathbf{IA} = \begin{pmatrix} 1 & 0 \\ 0 & 1 \end{pmatrix}\begin{pmatrix} 2 & 1 \\ -2 & 3 \end{pmatrix} = \begin{pmatrix} 2+0 & 1+0 \\ 0+-2 & 0+3 \end{pmatrix} = \begin{pmatrix} 2 & 1 \\ -2 & 3 \end{pmatrix} = \mathbf{A}$

In general, if **M** is *any* 2×2 matrix, $\mathbf{MI} = \mathbf{IM} = \mathbf{M}$.

This is like multiplying any number a by 1: $a \times 1 = 1 \times a = a$.

The matrix $\begin{pmatrix} 1 & 0 \\ 0 & 1 \end{pmatrix}$ is the 2×2 **identity matrix**.

Remember: if you multiply a 2×2 matrix by 1, the answer is identical to the matrix you started with.

EXERCISE 6B

GRADE B

In this exercise $\mathbf{Z} = \begin{pmatrix} 0 & 0 \\ 0 & 0 \end{pmatrix}$ and $\mathbf{I} = \begin{pmatrix} 1 & 0 \\ 0 & 1 \end{pmatrix}$.

1. $\mathbf{M} = \begin{pmatrix} 4 & 1 \\ -2 & 1 \end{pmatrix}$ and $\mathbf{N} = \begin{pmatrix} -2 & 2 \\ 3 & 1 \end{pmatrix}$.

Find: **a)** $\mathbf{MN}$ **b)** $\mathbf{NM}$ **c)** $\mathbf{M}^2$ **d)** $\mathbf{N}^2$.

2. $\mathbf{P} = \begin{pmatrix} 3 & 4 \\ 5 & -2 \end{pmatrix}$ and $\mathbf{Q} = \begin{pmatrix} 8 & -5 \\ 9 & 2 \end{pmatrix}$.

Find: **a)** $\mathbf{PZ}$ **b)** $\mathbf{IQ}$ **c)** $\mathbf{QI}$ **d)** $\mathbf{ZQ}$.

3. Find: **a)** $\mathbf{Z}^2$ **b)** $\mathbf{I}^2$.

4. $\mathbf{C} = \begin{pmatrix} 4 & 1 \\ 1 & 2 \end{pmatrix}$ Find: **a)** $2\mathbf{C}$ **b)** $\mathbf{C}^2$.

5. $\mathbf{R} = \begin{pmatrix} 5 & 2 \\ 2 & 1 \end{pmatrix}$ and $\mathbf{S} = \begin{pmatrix} 1 & -2 \\ -2 & 5 \end{pmatrix}$.

Find: **a)** $\mathbf{RS}$ **b)** $\mathbf{SR}$.

c) What do you notice about your answers to **a** and **b**?

6. $\mathbf{A} = \begin{pmatrix} 4 & 2 \\ -2 & -1 \end{pmatrix}$ and $\mathbf{B} = \begin{pmatrix} -1 & -2 \\ 2 & 4 \end{pmatrix}$.

Find: **a)** $\mathbf{AB}$ **b)** $\mathbf{BA}$.

c) What do you notice about your answers to **a** and **b**?

7. $\mathbf{A} = \begin{pmatrix} 3 & 5 \\ 1 & 2 \end{pmatrix}$ and $\mathbf{B} = \begin{pmatrix} 2 & -5 \\ -1 & 3 \end{pmatrix}$.

a) Show that $\mathbf{AB} = \mathbf{I}$. **b)** Write down $\mathbf{BA}$.

8. $\mathbf{A} = \begin{pmatrix} -1 & 0 \\ 0 & -1 \end{pmatrix}$ Show that $\mathbf{A}^2 = \mathbf{I}$.

6.3 Transformations

THIS SECTION WILL SHOW YOU HOW TO ...

- ✓ work out the image of any point given a 2×2 transformation matrix
- ✓ work out the image of any vertex of the unit square given a 2×2 transformation matrix
- ✓ work out or recall the 2×2 transformation matrix for a given transformation

KEY WORD

- ✓ image

2×2 matrices can represent transformations. This means that you can use matrix multiplication to transform a point or several points.

Suppose $\mathbf{M} = \begin{pmatrix} 1 & 0 \\ 0 & -1 \end{pmatrix}$.

Then, for example, transforming the point $(2, -3)$ gives:

$$\begin{pmatrix} 1 & 0 \\ 0 & -1 \end{pmatrix}\begin{pmatrix} 2 \\ -3 \end{pmatrix} = \begin{pmatrix} 2 \\ 3 \end{pmatrix}$$

Using the transformation **M**, the **image** of the point $A(2, -3)$ is $(2, 3)$. The image point of A is written A′.

Any point transformed using the identity matrix will not move.

Some points do not move under other transformations.

EXAMPLE 3

Show that the point $B(3, 0)$ does not move under the transformation $\begin{pmatrix} 1 & 2 \\ 0 & 4 \end{pmatrix}$.

$$\begin{pmatrix} 1 & 2 \\ 0 & 4 \end{pmatrix}\begin{pmatrix} 3 \\ 0 \end{pmatrix} = \begin{pmatrix} 3 \\ 0 \end{pmatrix}$$

The image point B′ also has coordinates $(3, 0)$.

The unit square

The unit square has coordinates $O(0, 0)$, $A(1, 0)$, $B(1, 1)$ and $C(0, 1)$.

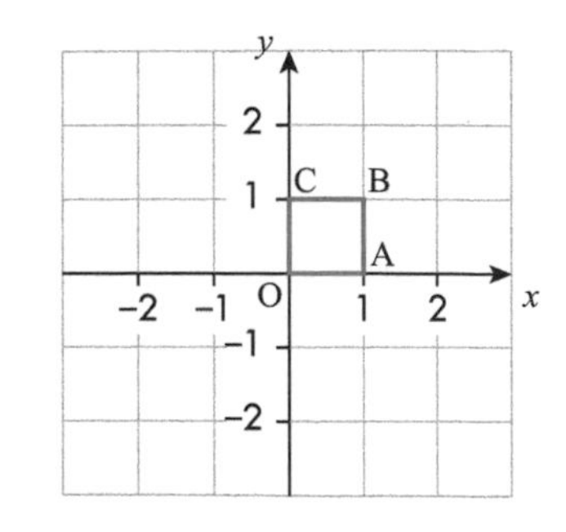

EXAMPLE 4

The unit square is mapped to OA′B′C′ under transformation matrix $\begin{pmatrix} 0 & 1 \\ -1 & 0 \end{pmatrix}$.

a) Work out the coordinates of A′, B′ and C′.

b) Describe geometrically the transformation represented by $\begin{pmatrix} 0 & 1 \\ -1 & 0 \end{pmatrix}$.

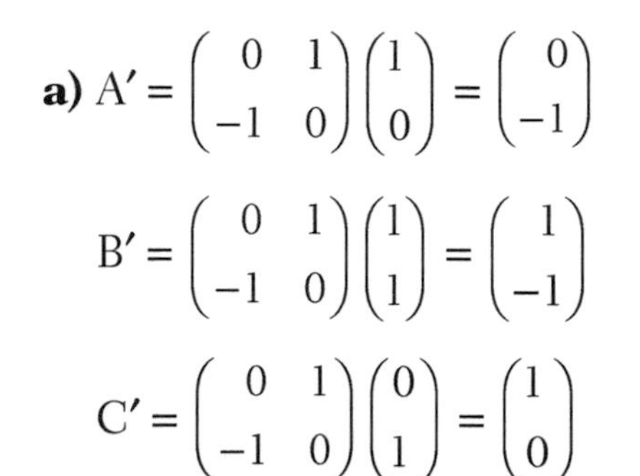

a) $A' = \begin{pmatrix} 0 & 1 \\ -1 & 0 \end{pmatrix}\begin{pmatrix} 1 \\ 0 \end{pmatrix} = \begin{pmatrix} 0 \\ -1 \end{pmatrix}$

$B' = \begin{pmatrix} 0 & 1 \\ -1 & 0 \end{pmatrix}\begin{pmatrix} 1 \\ 1 \end{pmatrix} = \begin{pmatrix} 1 \\ -1 \end{pmatrix}$

$C' = \begin{pmatrix} 0 & 1 \\ -1 & 0 \end{pmatrix}\begin{pmatrix} 0 \\ 1 \end{pmatrix} = \begin{pmatrix} 1 \\ 0 \end{pmatrix}$

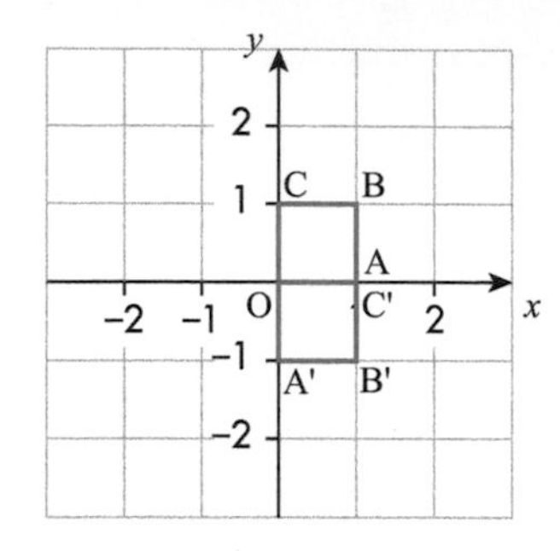

b) The transformation is a rotation 90° clockwise about O.

This table shows some common transformations.

You may spot a connection between the columns of each matrix and the coordinates of A′ and C′.

You may also spot a connection between the diagonals of each matrix and the coordinates of O and B′.

Transformation matrix	Geometrical description
$\begin{pmatrix} 1 & 0 \\ 0 & 1 \end{pmatrix}$	Identity matrix Points do not move
$\begin{pmatrix} 1 & 0 \\ 0 & -1 \end{pmatrix}$	Reflection in the x-axis
$\begin{pmatrix} -1 & 0 \\ 0 & 1 \end{pmatrix}$	Reflection in the y-axis
$\begin{pmatrix} 0 & 1 \\ 1 & 0 \end{pmatrix}$	Reflection in the line $y = x$
$\begin{pmatrix} 0 & -1 \\ -1 & 0 \end{pmatrix}$	Reflection in the line $y = -x$
$\begin{pmatrix} 0 & 1 \\ -1 & 0 \end{pmatrix}$	Rotation 90° clockwise about O

Transformation matrix	Geometrical description
$\begin{pmatrix} -1 & 0 \\ 0 & -1 \end{pmatrix}$	Rotation 180° about O
$\begin{pmatrix} 0 & -1 \\ 1 & 0 \end{pmatrix}$	Rotation 90° anticlockwise about O
$\begin{pmatrix} m & 0 \\ 0 & m \end{pmatrix}$	Enlargement scale factor m, centre O

EXERCISE 6C

GRADE B

1. $\mathbf{A} = \begin{pmatrix} 2 & 3 \\ 1 & 1 \end{pmatrix}$ Work out the image of the point P(3, –2) under the transformation matrix **A**.

2. $\mathbf{B} = \begin{pmatrix} 2 & 3 \\ 1 & 1 \end{pmatrix}$ Work out the image of the point Q(–2, 1) under the transformation matrix **B**.

3. The image of a point P(x, y) is (2, –5) under the matrix transformation $\begin{pmatrix} 5 & 1 \\ 1 & 2 \end{pmatrix}$.

Work out the values of x and y.

4. Copy the grid with the unit square. Transform the square, using each of these matrices.

Draw the result each time and describe fully the single transformation that the matrix represents.

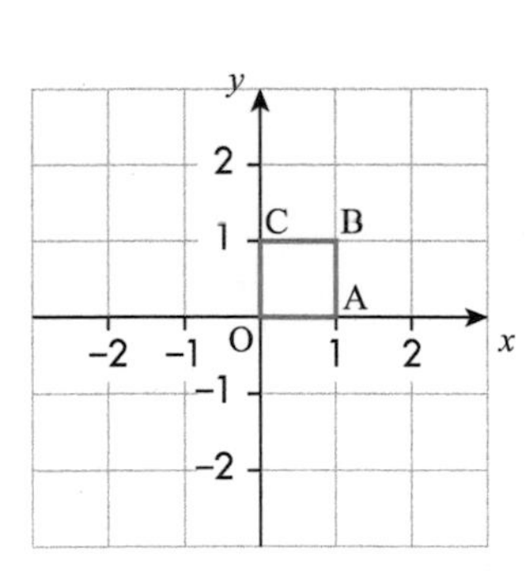

a) $\begin{pmatrix} 2 & 0 \\ 0 & 2 \end{pmatrix}$ **b)** $\begin{pmatrix} -1 & 0 \\ 0 & -1 \end{pmatrix}$ **c)** $\begin{pmatrix} 0 & -1 \\ -1 & 0 \end{pmatrix}$

d) $\begin{pmatrix} -1 & 0 \\ 0 & 1 \end{pmatrix}$ **e)** $\begin{pmatrix} 0 & 1 \\ 1 & 0 \end{pmatrix}$ **f)** $\begin{pmatrix} 1 & 0 \\ 0 & -1 \end{pmatrix}$

GRADE A

5. The matrix $\mathbf{M} = 4\mathbf{I}$.

Describe geometrically the transformation represented by **M**.

6. Work out the matrix that rotates the unit square through 90° anticlockwise.

7. A is the point (4, 2). B is the point (6, 0).

The points A and B are transformed using matrix $\mathbf{T} = \begin{pmatrix} 2 & 4 \\ 3 & -1 \end{pmatrix}$.

Work out the length of A′B′.

Give your answer as a surd in its simplest form.

6.4 Combinations of transformations

THIS SECTION WILL SHOW YOU HOW TO ...

- ✓ understand that a matrix product **AB** represents a transformation with matrix **B** followed by a transformation with matrix **A**
- ✓ work out the matrix which represents a combined transformation

KEY WORDS

- ✓ combined
- ✓ composite

A **combined** transformation or **composite** transformation can be represented by a single matrix. For example:

$\begin{pmatrix} 0 & 1 \\ 1 & 0 \end{pmatrix} \to$ reflection in $y = x$. $\begin{pmatrix} 0 & 1 \\ -1 & 0 \end{pmatrix} \to$ rotation 90° clockwise.

$\begin{pmatrix} 0 & 1 \\ 1 & 0 \end{pmatrix} \times \begin{pmatrix} 0 & 1 \\ -1 & 0 \end{pmatrix} = \begin{pmatrix} -1 & 0 \\ 0 & 1 \end{pmatrix}$ $\begin{pmatrix} -1 & 0 \\ 0 & 1 \end{pmatrix} \to$ reflection in the y-axis.

A rotation of 90° clockwise followed by a reflection in the line $y = x$ is the same as a reflection in the y-axis.

In general, matrix transformation **AB** is the same as matrix transformation **B** followed by matrix transformation **A**.

EXERCISE 6D

GRADE A*

1. $\mathbf{A} = \begin{pmatrix} 1 & 0 \\ 0 & -1 \end{pmatrix}$ $\mathbf{B} = \begin{pmatrix} -1 & 0 \\ 0 & 1 \end{pmatrix}$

Describe geometrically the transformations represented by:

a) **A** **b)** **B** **c)** **AB** **d)** **BA**
e) $\mathbf{A}^2$ **f)** $\mathbf{A}^4$ **g)** $\mathbf{B}^2$

2. These are three transformations in the x–y plane.

X: Reflection in the y-axis **Y**: Rotation 180° about O
Z: Reflection in the x-axis

a) Describe the combined (composite) transformation that represents:
i) **Y** followed by **X** **ii)** **Y** followed by **Z**
iii) **X** followed by **Y** **iv)** **X** followed by **Z**.

b) Work out the single matrix that represents each of transformations in part **a**.

3. Explain why $\mathbf{A}^4 = \mathbf{I}$, where $\mathbf{A} = \begin{pmatrix} 0 & 1 \\ -1 & 0 \end{pmatrix}$.

4. These are three transformations in the x–y plane.

X: Rotation 90° clockwise about O **Y**: Rotation 180° about O
Z: Rotation 90° anticlockwise about O

Use matrix multiplication to show that **Z** is the same as **X** followed by **Y**.

Exam-style questions

1. $\mathbf{C} = \begin{pmatrix} 2 & 3 \\ -2 & 1 \end{pmatrix}$ $\mathbf{D} = \begin{pmatrix} -1 & 2 \\ 4 & 0 \end{pmatrix}$ GRADE B

Work out: **a)** 4**C** **b) DC** **c) D**^{2}.

2. $\begin{pmatrix} 4 \\ x \end{pmatrix} + 2\begin{pmatrix} y \\ -5 \end{pmatrix} = \begin{pmatrix} 8 \\ 2 \end{pmatrix}$

Find the values of x and y. GRADE B

3. Work out the image of the point (3, –2) under the transformation represented by the matrix $\begin{pmatrix} 6 & -5 \\ -3 & 4 \end{pmatrix}$. GRADE A

4. $\mathbf{M} = \begin{pmatrix} 2 & 7 \\ 1 & 2 \end{pmatrix}$ $\mathbf{N} = \begin{pmatrix} 2 & -7 \\ -1 & 2 \end{pmatrix}$

Show that **MN** = k**I** where k is an integer. State the value of k. GRADE A

5. The unit square OABC has vertices O(0, 0), A(1, 0), B(1, 1) and C(0 ,1).

The square is mapped onto OA'B'C' by the matrix $\mathbf{M} = \begin{pmatrix} 0 & -1 \\ 1 & 0 \end{pmatrix}$. GRADE A*

a) Work out the coordinates of A', B' and C'.
b) Write down the transformation represented by the matrix **M**.
c) Find the transformation represented by the matrix **M**^{2}.

6. **M** is a matrix and $-\frac{1}{2}$ **M** = **I**.

Work out the transformation represented by the matrix **M**.

7. $\mathbf{R} = \begin{pmatrix} -1 & 0 \\ 0 & 1 \end{pmatrix}$ $\mathbf{S} = \begin{pmatrix} 0 & 1 \\ 1 & 0 \end{pmatrix}$ GRADE A*

a) Describe the transformations represented by **R** and **S**.
b) Work out the matrix **SR** and state the single transformation that it represents.
c) Use transformations to explain why **SR** and **RS** cannot be identical.

7 Algebra

7.1 Manipulation of rational expressions

THIS SECTION WILL SHOW YOU HOW TO ...

✓ simplify rational algebraic fractions

KEY WORDS

✓ reciprocal ✓ cancel ✓ single fractions

These five rules are used to work out the value of fractions.

Cancelling: $\frac{ac}{ad}=\frac{c}{d}$ a is a common factor of the numerator and the denominator and so can be removed from both.

Addition: $\frac{a}{b}+\frac{c}{d}=\frac{ad+bc}{bd}$

Subtraction: $\frac{a}{b}-\frac{c}{d}=\frac{ad-bc}{bd}$

Multiplication: $\frac{a}{b}\times\frac{c}{d}=\frac{ac}{bd}$

Division: $\frac{a}{b}\div\frac{c}{d}=\frac{a}{b}\times\frac{d}{c}=\frac{ad}{bc}$

To divide by a fraction, multiply by the **reciprocal**.

Note that a, b, c and d can be numbers, other letters or algebraic expressions. Remember:

- use brackets, if necessary
- factorise if you can
- cancel if you can.

EXAMPLE 1

Simplify the expression $\frac{1}{x}+\frac{x}{2y}$.

Using the addition rule: $\frac{1}{x}+\frac{x}{2y}=\frac{(1)(2y)+(x)(x)}{(x)(2y)}$

$$=\frac{2y+x^2}{2xy}$$

EXAMPLE 2

Write $\frac{3}{x-1}-\frac{2}{x+1}$ as a **single fraction** as simply as possible.

Using the subtraction rule: $\frac{3}{x-1}-\frac{2}{x+1}=\frac{3(x+1)-2(x-1)}{(x-1)(x+1)}$

$$=\frac{3x+3-2x+2}{(x-1)(x+1)}$$

$$=\frac{x+5}{(x-1)(x+1)}$$

EXAMPLE 3

Simplify this expression. $\frac{2x^2+x-3}{4x^2-9}$

Factorise the numerator and denominator: $\frac{(2x+3)(x-1)}{(2x+3)(2x-3)}$

The denominator is the difference of two squares.

Cancel any common factors: $\frac{\cancel{(2x+3)}(x-1)}{\cancel{(2x+3)}(2x-3)}$

The remaining fraction is the answer: $\frac{x-1}{2x-3}$

EXERCISE 7A

GRADE B

1. Simplify each expression.

a) $\frac{xy}{4}+\frac{2}{x}$ **b)** $\frac{2x+1}{2}+\frac{3x+1}{4}$ **c)** $\frac{x}{5}+\frac{2x+1}{3}$ **d)** $\frac{x-4}{4}+\frac{2x-3}{2}$

e) $\frac{xy}{4}-\frac{2}{y}$ **f)** $\frac{2x+1}{2}-\frac{3x+1}{4}$ **g)** $\frac{x}{5}-\frac{2x+1}{3}$ **h)** $\frac{x-4}{4}-\frac{2x-3}{2}$

i) $\frac{4y^2}{9x}\times\frac{3x^2}{2y}$ **j)** $\frac{x}{2}\times\frac{x-2}{5}$ **k)** $\frac{x-3}{15}\times\frac{5}{2x-6}$ **l)** $\frac{x-5}{10}\times\frac{5}{x^2-5x}$

m) $\frac{3x}{4}+\frac{x}{4}$ **n)** $\frac{3x}{4}\div\frac{x}{4}$ **o)** $\frac{3x+1}{2}-\frac{x-2}{5}$ **p)** $\frac{x^2-9}{10}\times\frac{5}{x-3}$

GRADE A

2. Write each expression as a single fraction as simply as possible.

a) $\frac{2}{x+1}+\frac{5}{x+2}$ **b)** $\frac{4}{x-2}+\frac{7}{x+1}$ **c)** $\frac{3}{4x+1}-\frac{4}{x+2}$

d) $\frac{2}{2x-1}-\frac{6}{x+1}$ **e)** $\frac{3}{2x-1}-\frac{4}{3x-1}$

3. An expression of the form $\frac{ax^2 + bx - c}{dx^2 - e}$ simplifies to $\frac{x-1}{2x-3}$.
What was the original expression?

4. Write each expression as a single fraction.

a) $\frac{4}{x+1} + \frac{5}{x+2}$

b) $\frac{18}{4x-1} - \frac{1}{x+1}$

c) $\frac{2x-1}{2} - \frac{6}{x+1}$

d) $\frac{3}{2x-1} - \frac{4}{3x-1}$

GRADE A*

5. Simplify each expression.

a) $\frac{x^2+2x-3}{2x^2+7x+3}$

b) $\frac{4x^2-1}{2x^2+5x-3}$

c) $\frac{6x^2+x-2}{9x^2-4}$

d) $\frac{4x^2+x-3}{4x^2-7x+3}$

e) $\frac{4x^2-25}{8x^2-22x+5}$

6. Simplify each expression.

a) $\frac{x^2+6x-7}{x^2-4} \div \frac{x+7}{x^2+2x}$

b) $\frac{x^2+x-30}{x^2-16} \div \frac{x+6}{x^2+4x}$

c) $\frac{x^2-7x+12}{x^2-25} \div \frac{x-3}{x^2+5x}$

d) $\frac{x^2+4x-60}{x^2-49} \div \frac{x+10}{x^2+7x}$

e) $\frac{x^2-4x-5}{x^2-9} \div \frac{x+1}{x^2-3x}$

f) $\frac{x^2-11x-42}{x^2-64} \div \frac{x-14}{x^2-8x}$

7. Simplify each expression.

a) $\frac{x^3+3x^2+2x}{x^2+x}$

b) $\frac{x^3+2x^2-3x}{x^2-x}$

c) $\frac{4x^3+x^2-5x}{x^2-x}$

d) $\frac{2x^3+3x^2-35x}{x^2+5x}$

e) $\frac{3x^3+6x^2+3x}{x^2+x}$

f) $\frac{x^3-4x}{x^2-2x}$

GRADE A**

8. Simplify each expression.

a) $\frac{6x^2-28x-10}{4x^2-9} \div \frac{x-5}{4x^2+6x}$

b) $\frac{9x^2+28x-32}{9x^2-49} \div \frac{x+4}{3x^2+7x}$

c) $\frac{3x^2+13x+4}{16x^2-9} \div \frac{3x+1}{8x^2+6x}$

d) $\frac{9x^2-3x-6}{25x^2-36} \div \frac{3x+2}{10x-12}$

e) $\frac{10x^2-4x-14}{64x^2-9} \div \frac{5x-7}{24x^2+9x}$

f) $\frac{8x^2-6x-5}{9x^2-16} \div \frac{2x+1}{3x^2-4x}$

7.2 Use and manipulation of formulae and expressions

THIS SECTION WILL SHOW YOU HOW TO ...

✓ rearrange formulae and expressions

KEY WORDS

✓ rearrange ✓ subject ✓ variable

The **subject** of a formula is the **variable** (letter) in the formula which stands on its own, usually on the left-hand side of the equals sign. For example, x is the subject of each of these formulae.

$x = 5t + 4$ $\qquad$ $x = 4(2y - 7)$ $\qquad$ $x = \frac{1}{t}$

To change the subject to a different variable, you have to **rearrange** the formula to get that variable on its own on the left-hand side.

EXAMPLE 4

From the formula $C = 2m^2 + 3$, make m the subject.

Subtract 3 from both sides to get $2m^2$ on its own: $C - 3 = 2m^2$

Divide both sides by 2: $\frac{C-3}{2} = \frac{2m^2}{2}$

Reverse the formula: $m^2 = \frac{C-3}{2}$

Take the square root on both sides: $m = \sqrt{\frac{C-3}{2}}$

EXERCISE 7B

GRADE C

1. $g = \frac{m}{v}$ Make m the subject.

2. $t = m^2$ Make m the subject.

3. $C = 2\pi r$ Make r the subject.

4. $A = bh$ Make b the subject.

5. $P = 2l + 2w$ Make l the subject.

6. $m = p^2 + 2$ Make p the subject.

7. $v = u + at$ Make a the subject.

8. $A = \frac{1}{4}\pi d^2$ Make d the subject.

9. $W = 3n + t$ **a)** Make n the subject. **b)** Express t in terms of n and W.

10. $k = 2p^2$ Make p the subject.

11. $v = u^2 - t$ **a)** Make t the subject. **b)** Make u the subject.

12. $k = m + n^2$ **a)** Make m the subject. **b)** Make n the subject.

13. $T = 5r^2$ Make r the subject.

14. $K = 5n^2 + w$ **a)** Make w the subject. **b)** Make n the subject.

EXAMPLE 5

Make y the subject of $a = c + \frac{d}{y^2}$

Subtract c from both sides: $a - c = \frac{d}{y^2}$

Multiply both sides by y^2: $y^2(a - c) = d$

Divide both sides by $a - c$: $y^2 = \frac{d}{a-c}$

Take the square root on both sides: $y = \sqrt{\frac{d}{a-c}}$

EXERCISE 7C

GRADE B

1. $a^2 + b^2 = c^2$ **a)** Make a the subject. **b)** Find a if $b = 6.0$ and $c = 6.5$.

2. $s = ut + \frac{1}{2}at^2$ is a formula used in mechanics.

a) Make a the subject. **b)** Find s if $u = -5, t = 4$ and $a = 10$.

3. $a = \frac{b+2}{c}$ **a)** Make c the subject. **b)** Make b the subject.

4. $p = \frac{r}{t-3}$ Make t the subject.

GRADE A

5. $d = \frac{12}{1+\sqrt{e}}$ Make e the subject.

6. $v^2 = u^2 + 2as$ **a)** Find v if $u = 3, a = 2$ and $s = 4$.

b) Make u the subject. **c)** Make s the subject.

7. $T = 2\pi\sqrt{\left(\frac{L}{G}\right)}$ **a)** Make L the subject. **b)** Show that $G = L\left(\frac{2\pi}{T}\right)^2$

8. $D = \pi R^2 - \pi r^2$

a) Make R the subject. **b)** Make r the subject. **c)** Make π the subject.

9. $3x^2 - 4y^2 = 11$

a) Find x if $y = 4$. **b)** Make x the subject. **c)** Make y the subject.

10. $T = 2\sqrt{\left(\frac{a}{c+3}\right)}$ **a)** Make a the subject. **b)** Make c the subject.

11. $a^2 = b^2 + c^2 - 2bcT$ Make T the subject.

12. $uv = fu + fv$ is a formula used in optics.

a) Make f the subject. **b)** Find f if $u = 20$ and $v = 30$.

c) Make u the subject. **d)** Make v the subject.

13. Make x the subject of $\frac{1}{x} = \frac{1}{y} + \frac{1}{z}$. **14.** Make s the subject of $\frac{r}{s} = \frac{t}{2} + \frac{1}{2}$.

15. Make h the subject of $5 = \frac{2g}{h} - \frac{1}{i}$.

GRADE A**

16. Make k the subject of $\frac{1}{j} = \frac{1}{k} \times \frac{1}{t}$. **17.** Make c the subject of $\frac{2a}{9} = \frac{3b}{c} \div \frac{2d}{3}$.

18. Make h the subject of $\frac{12e}{5} = \frac{1}{8f} + \frac{1}{9g} \times \frac{h}{i}$.

7.3 The factor theorem

THIS SECTION WILL SHOW YOU HOW TO ...

✓ understand and use the factor theorem to factorise polynomials up to cubics
✓ find integer roots of polynomial equations up to and including cubics
✓ show that $x - a$ is a factor of the function $f(x)$ by checking $f(a) = 0$
✓ solve equations up to cubics, where at least one of the roots is an integer

KEY WORDS

✓ polynomial ✓ cubic

A **polynomial** is the general name for an expression with terms that have positive integer powers of the variable and possibly a constant term.
Examples of polynomials are:

$x^2 - 2x + 3$ This is also called a quadratic, as the highest power is 2.

$x^3 - 3x^2 + 2x$ This is also called a **cubic**, as the highest power is 3.

The quadratic expression $x^2 - 2x + 3$ can be factorised to give $(x - 3)(x + 1)$.

$x^2 - 2x + 3 = 0 \Rightarrow (x - 3)(x + 1) = 0 \Rightarrow x - 3 = 0$ or $x + 1 = 0$.

So the solutions are $x = 3$ and $x = -1$.

Now suppose the solutions of a quadratic equation are $x = -4$ and $x = 5$.
This gives: $x + 4 = 0$ or $x - 5 = 0 \Rightarrow (x + 4)(x - 5) = 0 \Rightarrow x^2 - x - 20 = 0$.
So the quadratic equation is $x^2 - x - 20 = 0$ with solutions $x = -4$ and $x = 5$.

Using function notation: $f(x) = x^2 - x - 20$.

Since $f(4) = 0$ and $f(-5) = 0$, $x = 4$ and $x = -5$ are solutions to the quadratic equation $x^2 - x - 20 = 0$.
This can be extended to cubics and other polynomials.
In general, if $f(x)$ is a polynomial and a value a can be found so that $f(a) = 0$, then $x = a$ is a solution of the equation $f(x) = 0$ and $x - a$ is a factor of the polynomial.

EXAMPLE 6

Factorise $x^3 - 6x^2 + 11x - 6$.

Let $f(x) = x^3 - 6x^2 + 11x - 6$

$f(1) = 1^3 - 6(1)^2 + 11(1) - 6 = 1 - 6 + 11 - 6 = 0$

So, as $f(1) = 0$, $x - 1$ is a factor of $x^3 - 6x^2 + 11x - 6$.

The constant term is -6 so try the other factors of -6: 2, 3, -1, -2, -3 and -6.

$f(2) = 2^3 - 6(2)^2 + 11(2) - 6 = 8 - 24 + 22 - 6 = 0$

So, as $f(2) = 0$, $x - 2$ is a factor of $x^3 - 6x^2 + 11x - 6$.

$f(3) = 3^3 - 6(3)^2 + 11(3) - 6 = 27 - 54 + 33 - 6 = 0$

So, as $f(3) = 0$, $x - 3$ is a factor of $x^3 - 6x^2 + 11x - 6$.

$x^3 - 6x^2 + 11x - 6 = (x - 1)(x - 2)(x - 3)$

EXAMPLE 7

Solve $x^3 - 6x^2 + 11x - 6 = 0$.

From Example 6 you know that $(x - 1)(x - 2)(x - 3) = 0 \Rightarrow x - 1 = 0$ or $x - 2 = 0$ or $x - 3 = 0$.

So $x = 1$, $x = 2$ and $x = 3$ are the solutions.

EXERCISE 7D

GRADE A**

1. In each part, show that the expression is a factor of the polynomial.

a) $x + 1, x^3 + 6x^2 - 9x - 14$ **b)** $x - 3, x^3 + 3x^2 - 13x - 15$

c) $x - 4, x^3 - 7x^2 + 2x + 40$ **d)** $x + 6, x^3 + 13x^2 + 54x + 72$

e) $x + 7, x^3 - 37x + 84$ **f)** $2x - 3, 2x^3 - 5x^2 + x + 6$

2. Factorise each expression.

a) $x^3 + 3x^2 - 13x - 15$ **b)** $x^3 + 5x^2 + 2x - 8$

c) $x^3 + 6x^2 + 11x + 6$ **d)** $x^3 - 10x^2 - x - 30$

e) $2x^3 - 13x^2 + 17x + 12$ **f)** $6x^3 + 19x^2 + 11x - 6$

3. Solve each equation.

a) $x^3 - 4x^2 - 7x + 10 = 0$ **b)** $x^3 - 3x^2 + 3x - 1 = 0$

c) $x^3 + 2x^2 - 16x - 32 = 0$ **d)** $x^3 + 6x^2 + 5x - 12 = 0$

e) $x^3 + 15x^2 + 68x + 84 = 0$ **f)** $x^3 - 11x^2 - 10x + 200 = 0$

4. The equation $x^3 - 3x^2 + ax + b = 0$ has solutions $x = 1$ and $x = 4$.
Work out the third solution to the equation.

5. a) $x + 5$ is a factor of $x^3 + 3x^2 - 13x + c$. Show that $c = -15$.

b) Work out the other two factors of the cubic.

6. $x^3 + 3x^2 + ax + b$ factorises as $(x + c)^3$. Work the values of a, b and c.

7. $x^3 + 3x^2 - 16x - 48$ factorises as $(x^2 - a^2)(x + b)$. Work out the three linear factors of $x^3 + 3x^2 - 16x - 48$.

8. Work out the common factors of $x^3 - 5x^2 - 2x + 24$ and $x^3 - x^2 - 10x - 8$.

Exam-style questions

1. Write $\frac{2x+1}{x+2}-\frac{2x-1}{x+1}$ as a single fraction, as simply as possible. GRADE B

2. Simplify fully $\frac{4x^3-x}{x-3}\div\frac{4x^3+4x^2+x}{x^2-6x+9}$. GRADE B

3. Simplify fully $\frac{x^3-4x^2+4x}{x^4-8x^2+16}$. GRADE B

4. Make x the subject of $4(y-x)=2+5x$. GRADE B

5. Make v the subject of the formulae $\frac{1}{u}+\frac{1}{v}=\frac{1}{f}$. GRADE A

6. The formulae for the area and circumference of a circle are $A=\pi r^2$ and $C=2\pi r$.

Show that $A=\frac{C^2}{4\pi}$. GRADE A

7. $f(x)=x^3-x^2-14x+24$ GRADE A*

a) Show that $x+4$ is a factor of $f(x)$.

b) Factorise $f(x)$.

c) Solve the equation $f(x)=0$.

8. a) Show that $x-2$ is a factor of $x^3-6x^2+12x-8$.

b) Factorise $x^3-6x^2+12x-8$. GRADE A*

9. Factorise $x^3-4x^2-25x+100$. GRADE A*

10. Solve the equation $x^3+6x^2+11x+6=0$. GRADE A*

11. Solve these equations. GRADE A*

a) $x^2-3x-4=0$ **b)** $x^3-3x^2-4x=0$ **c)** $x^3-3x^2-4x+12=0$

8 Sequences

8.1 Number sequences

THIS SECTION WILL SHOW YOU HOW TO ...

✓ spot patterns in number sequences

KEY WORDS

✓ sequence ✓ term ✓ difference ✓ consecutive

A number **sequence** is an ordered set of numbers with a rule for finding every number in the sequence. Each number in a sequence is called a **term** and is in a certain position in the sequence.

Look at these sequences and their rules.

3, 6, 12, 24, …	doubling the previous term each time	… 48, 96, …
2, 5, 8, 11, …	adding 3 to the previous term each time	… 14, 17, …

Differences

For some sequences you need to look at the **differences** between **consecutive** terms to determine the pattern. The number n is the position of the term in the sequence.

EXERCISE 8A

GRADE C

1. State the next three terms in each sequence and explain how it is formed.

a) 4, 10, 16, 22, … **b)** 3, 8, 13, 18, … **c)** 2, 20, 200, 2000, …
d) 7, 10, 13, 16, … **e)** 10, 19, 28, 37, … **f)** 5, 15, 45, 135, …
g) 2, 6, 10, 14, … **h)** 1, 5, 25, 125, …

2. By considering differences, give the next two terms in each sequence.

a) 1, 2, 4, 7, 11, … **b)** 1, 2, 5, 10, 17, … **c)** 1, 3, 7, 13, 21, …
d) 1, 4, 10, 19, 31, … **e)** 1, 9, 25, 49, 81, … **f)** 1, 2, 4, 5, 7, 8, 10, …
g) 2, 3, 5, 9, 17, … **h)** 3, 8, 18, 33, 53, …

3. Find the rule for each sequence and write down its next three terms.

a) 3, 6, 12, 24, … **b)** 3, 9, 15, 21, 27, … **c)** 128, 64, 32, 16, 8, …
d) 50, 47, 44, 41, … **e)** 5, 6, 8, 11, 15, 20, … **f)** 4, 7, 10, 13, 16, …
g) 1, 3, 6, 10, 15, 21, … **h)** 1, 0.5, 0.25, 0.125, …

4. Find the next two numbers in each sequence and explain the pattern.

a) 1, 1, 2, 3, 5, 8, 13, … **b)** 1, 4, 9, 16, 25, 36, …
c) 3, 4, 7, 11, 18, 29, … **d)** 1, 8, 27, 64, 125, …

8.2 The nth term of a sequence

THIS SECTION WILL SHOW YOU HOW TO ...

✓ find the rule for a simple number sequence

KEY WORD

✓ nth term

When using a number sequence, you sometimes need to give, say, the 50th term, or even a higher term in the sequence. To do so, you need to find the rule that produces the sequence in its general form.

It may be helpful to look at the problem backwards. That is, take a rule and see how it produces a sequence. The rule is given for the general term, which is called the ***n*th term**.

EXAMPLE 1

The nth term of a sequence is $3n + 1$, where $n = 1, 2, 3, 4, 5, 6, \ldots$.
Write down the first five terms of the sequence.

Substituting $n = 1, 2, 3, 4, 5$ in turn:

$(3 \times 1 + 1), (3 \times 2 + 1), (3 \times 3 + 1), (3 \times 4 + 1), (3 \times 5 + 1), \ldots$

4 7 10 13 16

So the sequence is 4, 7, 10, 13, 16, ….

EXERCISE 8B

GRADE C

1. Write down the first five terms of the sequence that has as its nth term:

a) $n + 3$ **b)** $3n - 1$ **c)** $5n - 2$ **d)** $4n + 5$.

2. A haulage company uses this formula to calculate the cost of transporting n pallets.

For $n \leqslant 5$, the cost will be £$(40n + 50)$.

For $6 \leqslant n \leqslant 10$, the cost will be £$(40n + 25)$.

For $n \geqslant 11$, the cost will be £$40n$.

a) How much will the company charge to transport 7 pallets?
b) How much will the company charge to transport 15 pallets?
c) A company is charged £170 for transporting pallets.
How many pallets did they transport?
d) Another haulage company uses the formula £$50n$ to calculate the cost for transporting n pallets.
At what value of n do the two companies charge the same amount?

HINTS AND TIPS

Substitute numbers into the expressions until you can see how the sequence works.

8.3 The nth term of a linear sequence

THIS SECTION WILL SHOW YOU HOW TO ...

✓ find the nth term of a linear sequence

KEY WORDS

✓ linear sequence
✓ difference
✓ nth term

A **linear sequence** has the same **difference** between each term and the next. For example:

2, 5, 8, 11, 14, ... difference of 3

The ***n*th term** of this sequence is given by $3n - 1$.

Here is another linear sequence.

5, 7, 9, 11, 13, ... difference of 2

The nth term of this sequence is given by $2n + 3$.

So, you can see that the nth term of a linear sequence is *always* of the form $An + b$, where:

- A, the coefficient of n, is the difference between consecutive terms.
- b is the difference between the first term and A.

EXAMPLE 2

From the sequence 5, 12, 19, 26, 33, ... find:

a) the nth term **b)** the 50th term

c) the first term that is greater than 1000.

a) The difference between consecutive terms is 7. So the first part of the nth term is $7n$.

Subtract the difference 7 from the first term 5, which gives $5 - 7 = -2$.

So the nth term is given by $7n - 2$.

b) The 50th term is found by substituting $n = 50$ into the rule, $7n - 2$.

So 50th term $= 7 \times 50 - 2 = 350 - 2 = 348$

c) The first term that is greater than 1000 is given by:

$7n - 2 > 1000 \Rightarrow 7n > 1000 + 2 \Rightarrow n > \frac{1002}{7} \Rightarrow n > 143.14$

So the first term (which has to be a whole number) over 1000 is the 144th.

EXERCISE 8C

GRADE C

1. Find the next two terms and the nth term in each linear sequence.

a) 3, 5, 7, 9, 11, … **b)** 5, 9, 13, 17, 21, … **c)** 8, 13, 18, 23, 28, …

d) 2, 8, 14, 20, 26, … **e)** 5, 8, 11, 14, 17, … **f)** 2, 9, 16, 23, 30, …

g) 1, 5, 9, 13, 17, … **h)** 3, 7, 11, 15, 19, … **i)** 2, 5, 8, 11, 14, …

2. Find the nth term and the 50th term in each linear sequence.

a) 4, 7, 10, 13, 16, … **b)** 7, 9, 11, 13, 15, … **c)** 3, 8, 13, 18, 23, …

d) 1, 5, 9, 13, 17, … **e)** 2, 10, 18, 26, … **f)** 5, 6, 7, 8, 9, …

3. a) Which term of the sequence 5, 8, 11, 14, 17, … is the first one to be greater than 100?

b) Which term of the sequence 1, 8, 15, 22, 29, … is the first one to be greater than 200?

c) Which term of the sequence 4, 9, 14, 19, 24, … is the closest to 500?

HINTS AND TIPS

Remember to look at the differences and the first term.

4. For each sequence, find:

i) the nth term **ii)** the 100th term **iii)** the term closest to 100.

a) 5, 9, 13, 17, 21, … **b)** 3, 5, 7, 9, 11, 13, … **c)** 4, 7, 10, 13, 16, …

d) 8, 10, 12, 14, 16, … **e)** 9, 13, 17, 21, … **f)** 6, 11, 16, 21, …

GRADE B

5. A sequence of fractions is $\frac{3}{4}, \frac{5}{7}, \frac{7}{10}, \frac{9}{13}, \frac{11}{16}, \ldots$

a) Find the nth term in the sequence.

b) By changing each fraction to a decimal, can you see a pattern?

c) What, as a decimal, will be the value of:

i) the 100th term **ii)** the 1000th term?

d) Use your answers to part **c** to predict what the 10 000th term and the millionth term are. (Check these on your calculator.)

6. The formula for working out a series of fractions is $\frac{2n-1}{3n-1}$.

a) Work out the first three fractions in the series.

b) i) Work out the value of the fraction as a decimal when $n = 1\,000\,000$.

ii) What fraction is equivalent to this decimal?

c) How can you tell this from the original formula?

7. Look at this series of fractions. $\frac{31}{109}, \frac{33}{110}, \frac{35}{111}, \frac{37}{112}, \frac{39}{113}, \ldots$

a) Explain why the nth term of the numerators is $2n + 29$.

b) Write down the nth term of the denominators.

c) Explain why the terms of the series will eventually get very close to 2.

d) Which term of the series has a value equal to 1?

8.4 The nth term of a quadratic sequence

THIS SECTION WILL SHOW YOU HOW TO ...

✓ find the nth term of a quadratic sequence

KEY WORDS

✓ quadratic sequence ✓ first difference ✓ second difference

In a **quadratic sequence** the **first differences** between terms form a linear sequence. The **second differences** are always the same number.

EXAMPLE 3

Look at this sequence.

	2		12		28		50		78
First differences		10		16		22		28	
Second differences			6		6		6		

The second difference is 6, so $a = 3$.

Writing out the first five terms of the quadratic sequence $3n^2$ gives:

3 12 27 48 75

Subtracting these from the first five terms of the original sequence gives:

	2	12	28	50	78
−	3	12	27	48	75
=	−1	0	1	2	3

This is a linear sequence with nth term equal to $n - 2$.

So the nth term of 2 12 28 50 78 is $3n^2 + n - 2$.

EXAMPLE 4

Now look at this sequence.

	9		20		39		66		101
First differences		11		19		27		35	
Second differences			8		8		8		

The second difference is 8, so $a = 4$.

Writing out the first five terms of the quadratic sequence $4n^2$ gives:

4 16 36 64 100

Subtracting these from the first five terms of the original sequence gives:

	9	20	39	66	101
−	4	16	36	64	100
=	5	4	3	2	1

This is a linear sequence with nth term equal to $-n + 6$.

So the nth term of 2 12 28 50 78 is $4n^2 - n + 6$.

The nth term of a quadratic sequence is *always* of the form $an^2 + bn + c$ where:

- a, the coefficient of n, is half the second difference
- b *and* c can be worked out by subtracting the sequence with nth term an^2 from the original sequence to leave a linear sequence with nth term $bn + c$.

EXERCISE 8D

GRADE A**

1. Use the difference method to work out the next two terms of each quadratic sequence.

a) 1, 4, 9, 16, 25 **b)** 3, 6, 11, 18, 27 **c)** 2, 6, 12, 20, 30

d) 4, 10, 20, 34, 52 **e)** 3, 10, 21, 36. 55 **f)** 4, 13, 28, 49, 76

g) 3, 7, 13, 21, 31 **h)** 3, 18, 43, 78, 123 **i)** 4, 14, 30, 52, 80

j) 1, 6, 15, 28, 45

2. Use the difference method to work out the nth terms of the sequences in question 1.

3. The nth term of a sequence is $\frac{n^2-1}{4n+1}$.

a) Write down the first three terms. **b)** Which term of the sequence is $\frac{7}{5}$?

4. The nth term of a sequence is $\frac{4n^2}{3n-2}$.

Explain why all terms of this sequence are positive.

5. Work out the nth term of the sequence $\frac{1}{3}, \frac{4}{5}, \frac{9}{7}, \frac{16}{9}, \frac{25}{11}$.

6. a) Work out the nth term of the linear sequence 1, 4, 7, 10, 13 …

b) Work out the nth term of the linear sequence 2, 6, 10, 14, 18 …

c) Here is a quadratic sequence: 2, 24, 70, 140, 234

Using the difference method or your answers to parts a and b work out a formula for the nth term of this sequence.

8.5 The limiting value of a sequence as $n \to \infty$

THIS SECTION WILL SHOW YOU HOW TO ...

✓ find the limiting value of a sequence as the number of terms becomes very large

KEY WORDS

✓ limit ✓ limiting value ✓ infinity

When lots of terms of a sequence are written down, some sequences approach a fixed value.

For example, If you start with the number 1 and divide by 2 each time, you get this sequence:

$$1, \frac{1}{2}, \frac{1}{4}, \frac{1}{8}, \frac{1}{16}, \frac{1}{32}, \ldots$$

The terms of the sequence get smaller and smaller and approach 0.

The value that the sequence approaches is called the **limit** or **limiting value** as n tends to **infinity**. It is written $n \to \infty$.

This means that, as n gets very large, the nth term approaches zero. Zero is the limiting value.

The limiting value of a sequence as $n \to \infty$ is the value that the nth term of the sequence approaches as n becomes very large.

EXAMPLE 5

Work out the limiting value of $\frac{6n}{2n+1}$ as $n \to \infty$.

As n becomes very large, or as $n \to \infty$, the terms of greatest value are the terms with the highest power in the numerator ($6n$) and the denominator ($2n$).

You write this as $\frac{6n}{2n+1} \to \frac{6n}{2n} = 3$, so the limiting value is 3.

EXAMPLE 6

Work out the limiting value of $\frac{n^2+4n}{n+8}$ as $n \to \infty$.

$\frac{n^2+4n}{n^2+8} \to \frac{n^2}{n^2} = 1$, so the limiting value is 1.

EXAMPLE 7

Work out the limiting value of $\frac{n+7}{n^2+7}$ as $n \to \infty$.

$\frac{n+7}{n^2+7} \to \frac{n}{n^2} = \frac{1}{n}$, so as $n \to \infty$ the limiting value is 0.

EXAMPLE 8

Work out the limiting values of the sequence $\frac{2}{5}, \frac{3}{6}, \frac{4}{7}, \frac{5}{8}, \frac{6}{9}$.

nth term $= \frac{n+1}{n+4}$ and as $n \to \infty$ so $\frac{n+1}{n+4} \to \frac{n}{n} = 1$, so the limiting value is 1.

EXERCISE 8E

GRADE A*

1. Work out the limiting value of each sequence, as $n \to \infty$.

a) $\frac{2n}{n+3}$ **b)** $\frac{n}{3n-1}$ **c)** $\frac{12n+1}{3n+5}$

d) $\frac{5-n}{n+4}$ **e)** $\frac{6n+1}{n^2+n}$ **f)** $\frac{(2n+1)(2n-1)}{2n^2}$

2. Work out the limiting values of each sequence as $n \to \infty$.

a) $\frac{3}{5}, \frac{4}{6}, \frac{5}{7}, \frac{6}{8}, \frac{7}{9}$ **b)** $\frac{1}{3}, \frac{1}{5}, \frac{1}{7}, \frac{1}{9}, \frac{1}{11}$ **c)** $\frac{1}{4}, \frac{1}{9}, \frac{1}{16}, \frac{1}{25}, \frac{1}{36}$

d) $\frac{5}{4}, \frac{7}{9}, \frac{9}{16}, \frac{11}{25}, \frac{13}{36}$ **e)** $\frac{1}{6}, \frac{2}{11}, \frac{3}{16}, \frac{4}{21}, \frac{5}{26}$ **f)** $\frac{4}{5}, \frac{3}{6}, \frac{2}{7}, \frac{1}{8}, 0$

Exam-style questions

1. The first term that these two sequences have in common is 17.

What are the next two terms that the two sequences have in common?

8, 11, 14, 17, 20, ...
1, 5, 9, 13, 17, ...

GRADE B

2. Look at these two sequences.

Will the two sequences ever have a term in common? Yes or no? Justify your answer.

2, 5, 8, 11, 14, ...
3, 6, 9, 12, 15, ...

GRADE B

3. The nth term of a sequence is $\frac{8n}{n+1}$. GRADE A*

a) Work out the first three terms of the sequence.

b) Which term in the sequence is the first that is larger than 7?

c) Work out the limiting value of the sequence as $n \to \infty$.

4. A quadratic sequence starts 3, 8, 15, 24, 35, ….

Work out a formula for the nth term. GRADE A*

5. The nth terms of two sequences are $100 - 5n$ and $n^2 + 1$.

How many numbers occur in both sequences? GRADE A*

9 Pythagoras' theorem and trigonometry

9.1 Pythagoras' theorem

THIS SECTION WILL SHOW YOU HOW TO ...

- ✓ work out the length of the side of a right-angled triangle, given the lengths of the other two sides
- ✓ identify appropriate right-angled triangles in 2- and 3-dimensional shapes and apply Pythagoras' theorem
- ✓ recognise and use Pythagorean triples

KEY WORDS

- ✓ hypotenuse
- ✓ isosceles triangle
- ✓ Pythagoras' theorem
- ✓ 3D

Pythagoras, who was a philosopher as well as a mathematician, was born in 580BCE in Greece. He later moved to Italy, where he established the Pythagorean Brotherhood, which was a secret society devoted to politics, mathematics and astronomy.

This is his famous theorem.

Consider squares being drawn on each side of a right-angled triangle, with sides 3 cm, 4 cm and 5 cm.

Pythagoras' theorem

The usual description is: *In any right-angled triangle, the square of the* **hypotenuse** *is equal to the sum of the squares of the other two sides.*

Pythagoras' theorem can be written as a formula: $c^2 = a^2 + b^2$

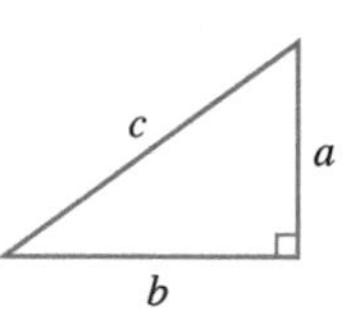

Remember that Pythagoras' theorem can only be used in right-angled triangles.

Finding the length of the hypotenuse

EXAMPLE 1

Find the length of the hypotenuse, marked x on the diagram.

Using Pythagoras' theorem gives:

$x^2 = 8^2 + 5.2^2 = 64 + 27.04 = 91.04 \Rightarrow x = \sqrt{91.04} = 9.5$

The length of the hypotenuse is 9.5 cm (1 decimal place)

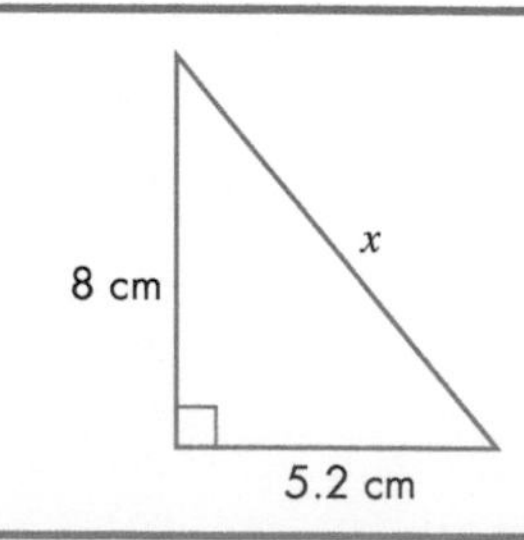

Finding the length of a shorter side

By rearranging the formula for Pythagoras' theorem, you can calculate the length of one of the shorter sides.

$c^2 = a^2 + b^2 \Rightarrow a^2 = c^2 - b^2$ or $b^2 = c^2 - a^2$

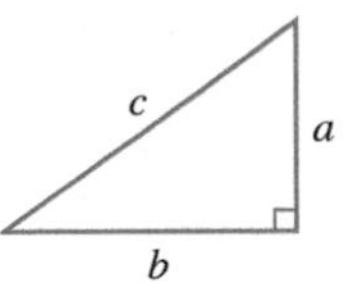

EXAMPLE 2

Find the length of the side marked x.

This is one of the shorter sides.

Pythagoras' theorem gives:

$x^2 = 15^2 - 11^2 = 225 - 121 = 104 \Rightarrow x = \sqrt{104}$

$\Rightarrow$ the length is 10.2 cm (1 decimal place)

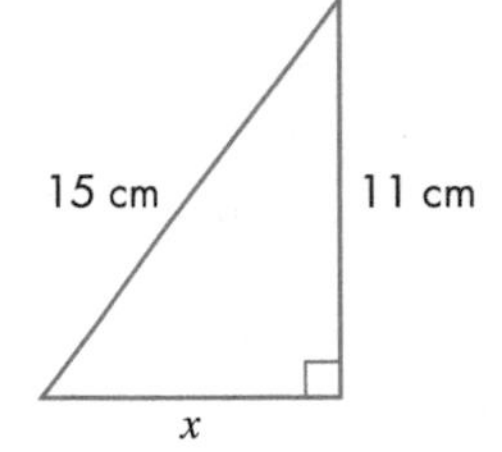

EXERCISE 9A

GRADE C

For each of the triangles in questions **1** to **3**, calculate the length of the hypotenuse, x, giving your answers to 1 decimal place.

1.

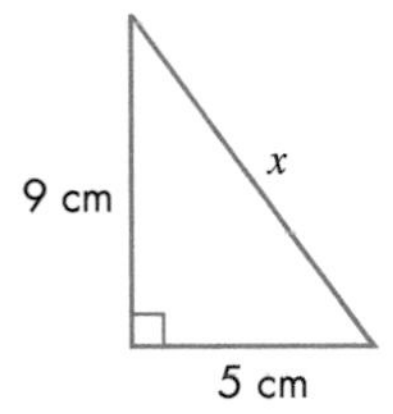

2.

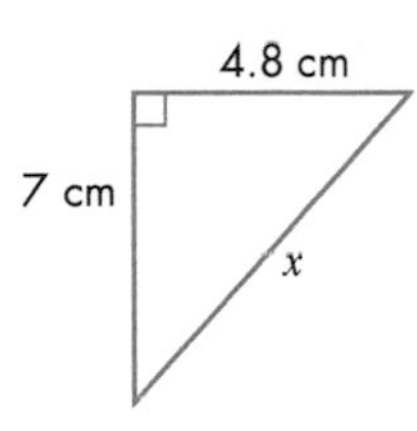

3.

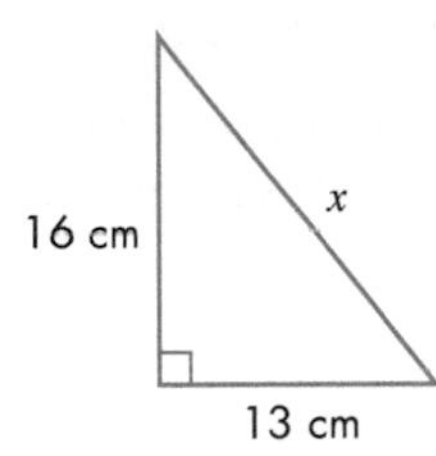

4. How does this diagram show that Pythagoras' theorem is true?

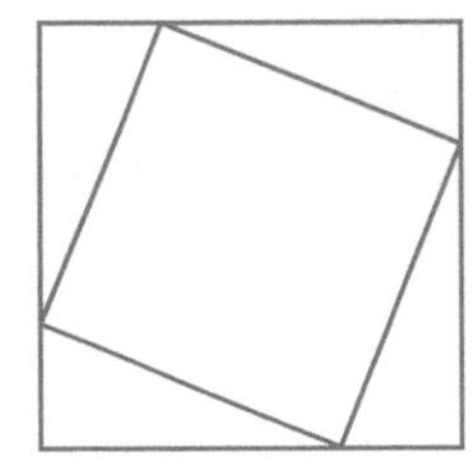

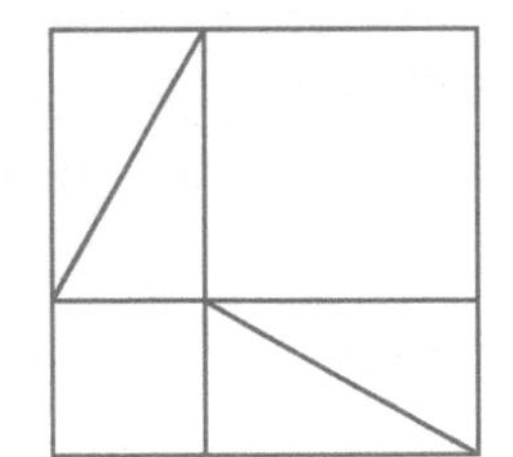

5. For each triangle, calculate the length marked x, giving your answers to 1 decimal place.

a)

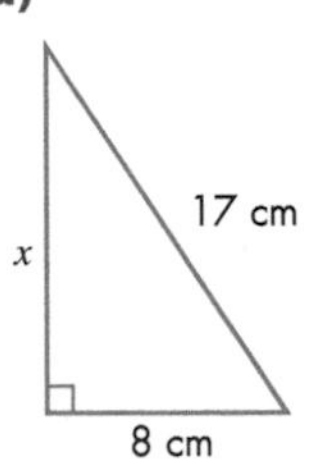

b)

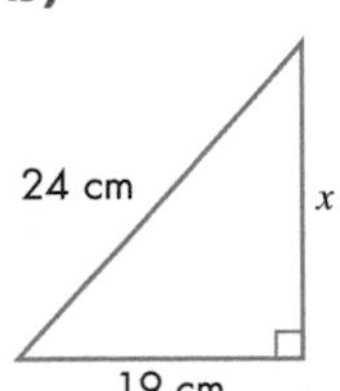

c)

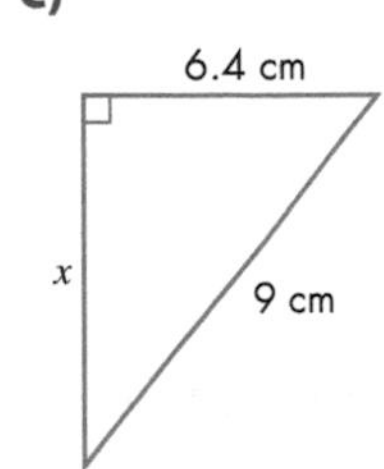

d)

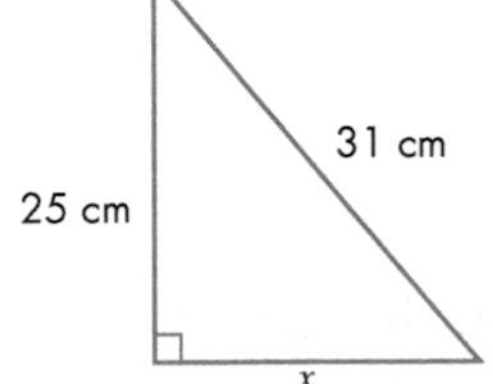

6. For each triangle, calculate the length x, giving your answers to 1 decimal place.

a)

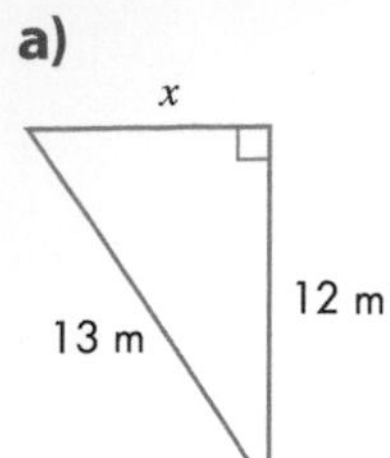

b)

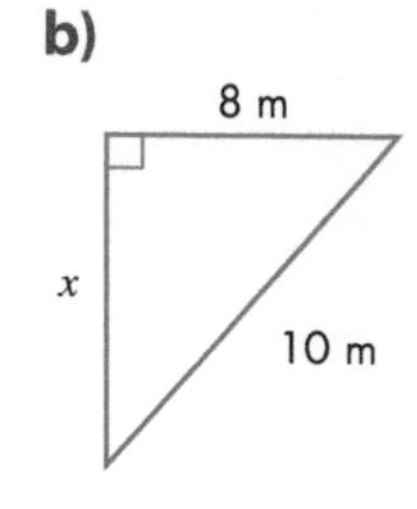

c)

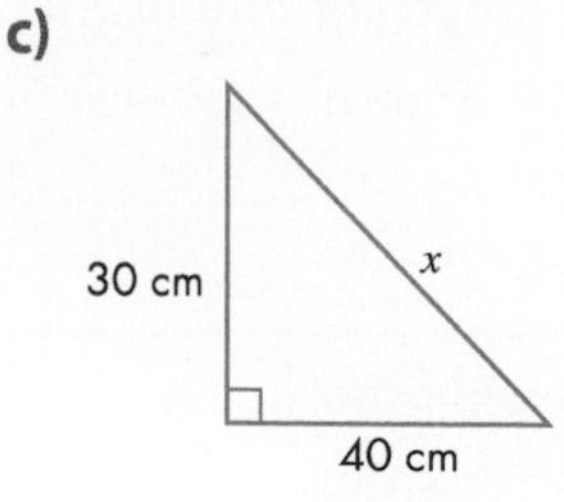

7. In question **6**, you found sets of three **whole** numbers that satisfy $a^2 + b^2 = c^2$. These are called Pythagorean triples. Can you find any more?

8. Calculate the value of x.

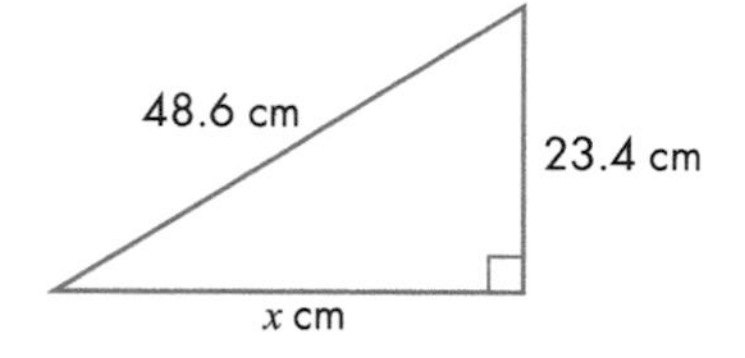

Remember the following tips when solving problems.

- Always sketch the right-angled triangle you need. Sometimes, the triangle is already drawn for you but some problems involve other lines and triangles that may confuse you. Identify which right-angled triangle you need and sketch it separately.
- Label the triangle with the necessary information, such as the length of its sides, taken from the question. Label the unknown side x.
- Set out your solution as in Example 2. Avoid short cuts, since they often cause errors. You gain marks in your examination for clearly showing how you are applying Pythagoras' theorem to the problem.
- Round your answer to a suitable degree of accuracy.

EXERCISE 9B

GRADE C

1. A ladder, 12 m long, leans against a wall. The ladder reaches 10 m up the wall. The ladder is safe if the foot of the ladder is about 2.5 m away from the wall.

   Is this ladder safe?

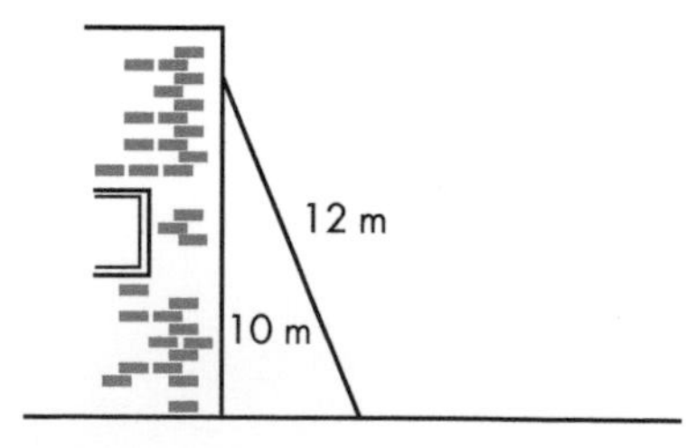

2. A ship going from a port to a lighthouse steams 15 km east and then 12 km north. The journey takes 1 hour. How much time would the ship save by travelling directly to the lighthouse in a straight line?

3. A and B are two points on a coordinate grid. They have coordinates (13, 6) and (1, 1). How long is the line that joins them?

4. The regulation for safe use of ladders states that: *the foot of a 5.00 m ladder must be placed between 1.20 m and 1.30 m from the foot of the wall.*
   a) What is the maximum height the ladder can safely reach up the wall?
   b) What is the minimum height the ladder can safely reach up the wall?

5. Is the triangle with sides 7 cm, 24 cm and 25 cm a right-angled triangle?
   Give a reason for your answer.

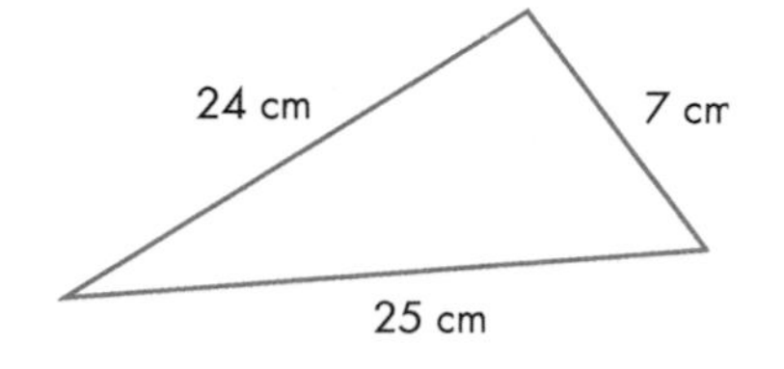

6. A 4 m long ladder is leaning against a wall. The foot of the ladder is 1 m from the wall. The foot of the ladder is not securely held and slips 20 cm further away from the wall.
   How far does the top of the ladder move down the wall?

Pythagoras' theorem and isosceles triangles

Every **isosceles triangle** has a line of symmetry that divides the triangle into two congruent right-angled triangles. So when you are faced with a problem involving an isosceles triangle, be aware that you are quite likely to have to split that triangle down the middle to create a right-angled triangle that will help you to solve the problem.

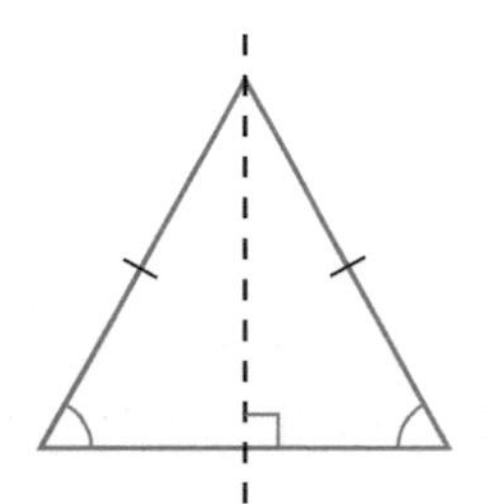

EXAMPLE 3

Calculate the area of this triangle.

It is an isosceles triangle and you need to calculate its height to find its area.

First split the triangle into two right-angled triangles to find its height.

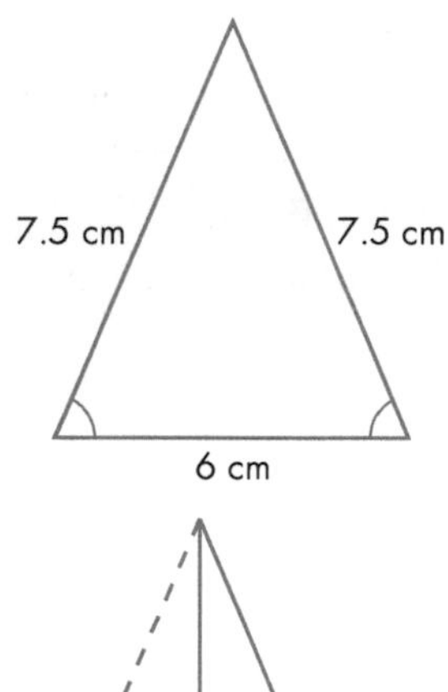

Let the height be x cm.

Then, using Pythagoras' theorem:

$x^2 = 7.5^2 - 3^2 = 56.25 - 9 = 47.25 \Rightarrow x = \sqrt{47.25} = 6.87$

Keep the accurate figure in the calculator memory.

The area of the triangle is $\frac{1}{2} \times 6 \times 6.87$ cm^2 (from the calculator memory) = 20.6 cm^2 (1 decimal place).

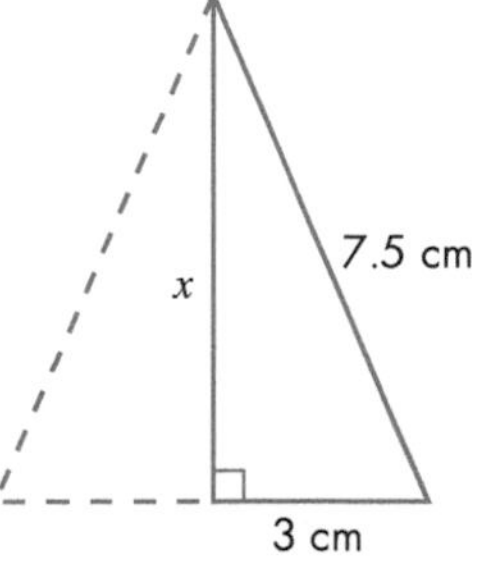

EXERCISE 9C

GRADE B

1. Calculate the areas of these isosceles triangles.

a)

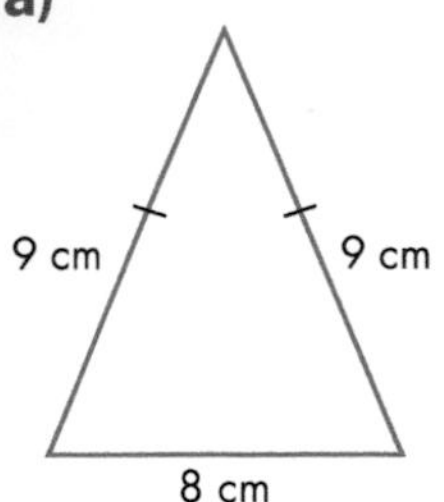

b)

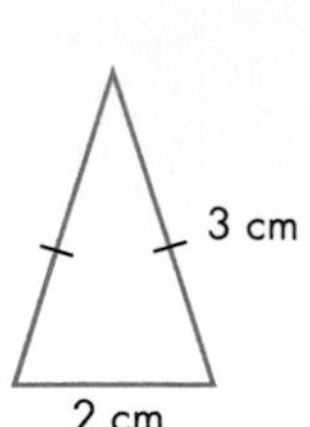

c)

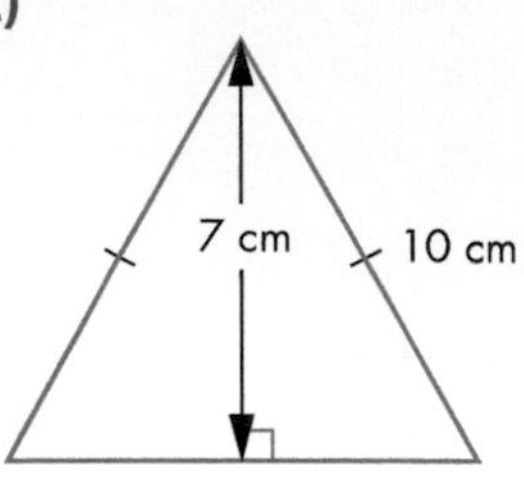

2. Calculate the area of an isosceles triangle with sides of 8 cm, 8 cm and 6 cm.

3. Calculate the area of an equilateral triangle of side 6 cm.

4. An isosceles triangle has sides of 5 cm and 6 cm.

a) Sketch the two different isosceles triangles that fit this data.

b) Which of the two triangles has the greater area?

5. Calculate the area of a regular hexagon of side 10 cm.

6. The diagram shows an isosceles triangle ABC.

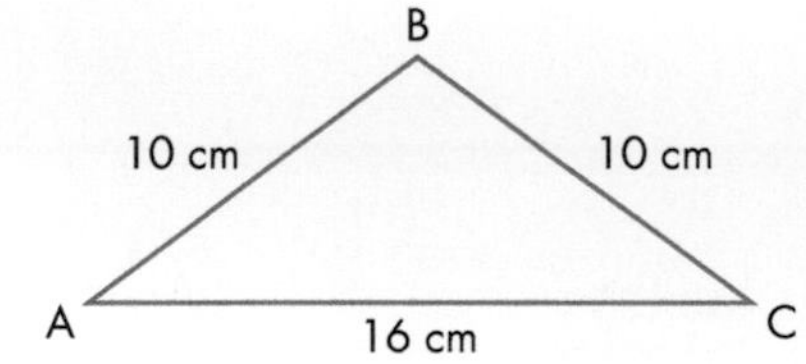

Calculate the area of triangle ABC.

State the units of your answer.

7. Calculate the lengths marked x in these isosceles triangles.

a)

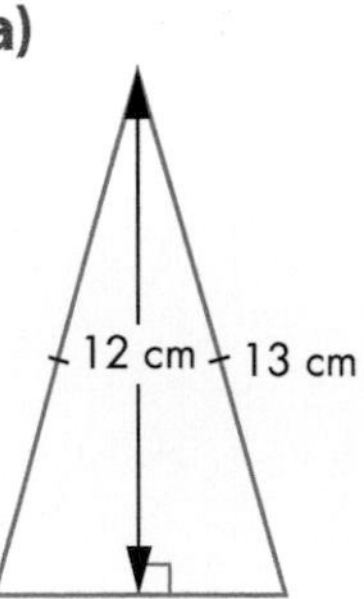

b)

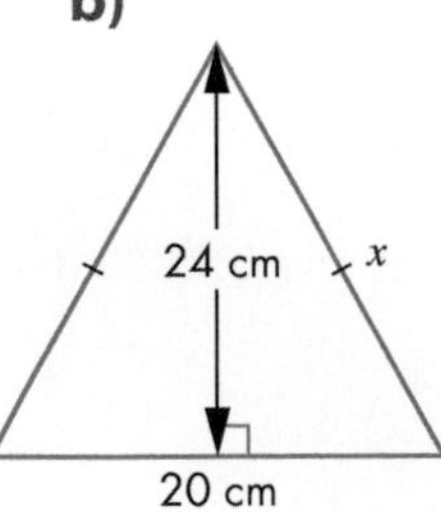

c)

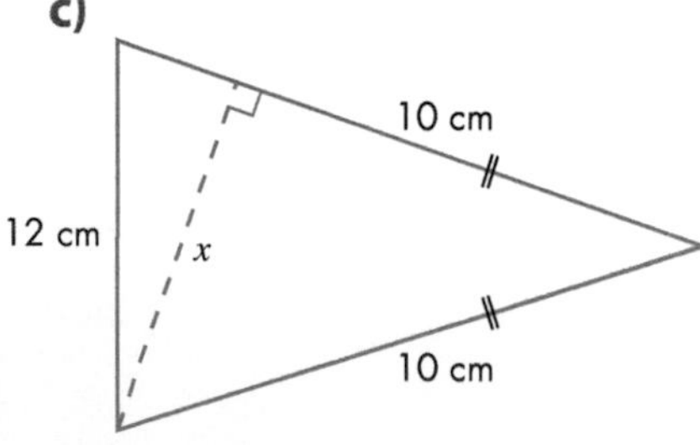

Pythagoras' theorem in three dimensions

You can use Pythagoras' theorem to solve problems in **3D** in exactly the same way as for 2D problems.

- Identify the right-angled triangle you need.
- Redraw this triangle and label it with the given lengths and the length to be found, usually x or y.
- From your diagram, decide whether you need to find the hypotenuse or one of the shorter sides.
- Solve the problem, rounding to a suitable degree of accuracy.

EXAMPLE 4

What is the longest piece of straight wire that can be stored in this box measuring 30 cm by 15 cm by 20 cm?

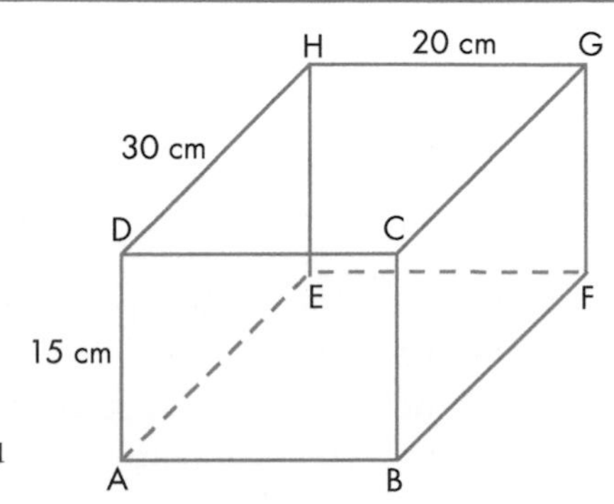

The longest distance across this box is any one of the diagonals AG, DF, CE or HB.

In this case, take AG.

First, identify a right-angled triangle containing AG and draw it.

This gives a triangle AFG, which contains two lengths you do not know, AG and AF.

Let AG = x cm and AF = y cm.

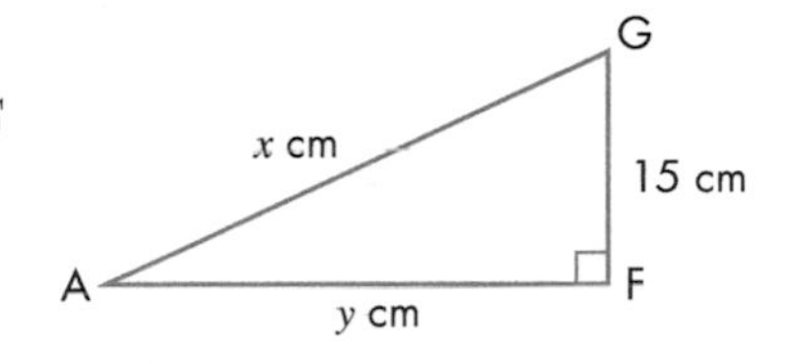

Next identify a right-angled triangle that contains the side AF and draw it.

This gives a triangle ABF. You can now find AF.

By Pythagoras' theorem:

$y^2 = 30^2 + 20^2 = 1300$ (there is no need to find y.)

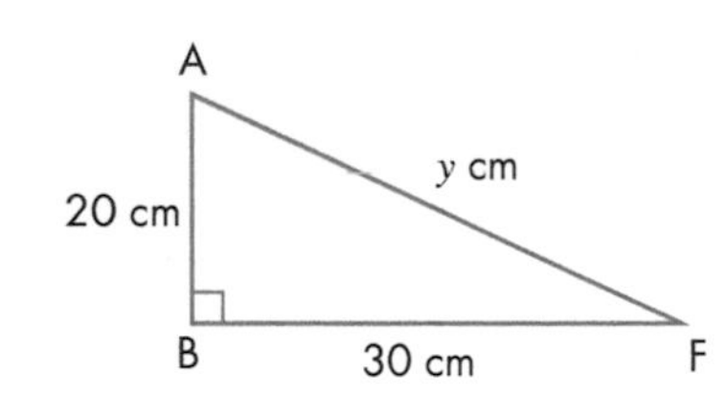

Now use triangle AFG to find AG.

By Pythagoras' theorem:

$x^2 = y^2 + 15^2 = 1300 + 225 = 1525$

So x cm = 39.1 cm (1 decimal place)

So, the longest straight wire that can be stored in the box is 39.1 cm.

EXERCISE 9D

GRADE A

1. A box measures 8 cm by 12 cm by 5 cm.

a) Calculate the length of: **i)** AC **ii)** BG **iii)** BE.

b) Calculate the diagonal distance BH.

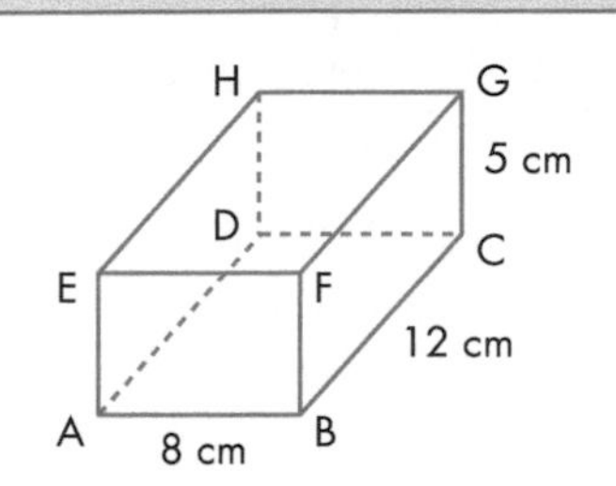

2. A garage is 5 m long, 3 m wide and 3 m high. Can a 7 m long pole be stored in it?

3. Spike, a spider, is at the corner S of the wedge shown in the diagram. Fred, a fly, is at the corner F of the same wedge.

   a) Calculate the shortest distance Spike would have to travel to get to Fred, if she used the edges of the wedge.

   b) Calculate the distance Spike would have to travel across the face of the wedge to get directly to Fred.

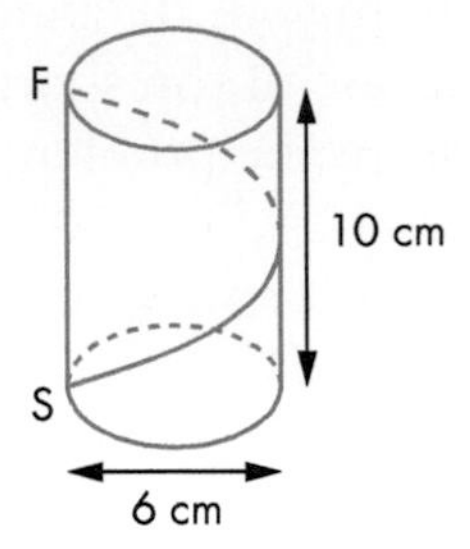

4. Fred is now at the top of a baked-beans can and Spike is directly below him on the base of the can. To catch Fred by surprise, Spike takes a diagonal route round the can.

   How far does Spike travel?

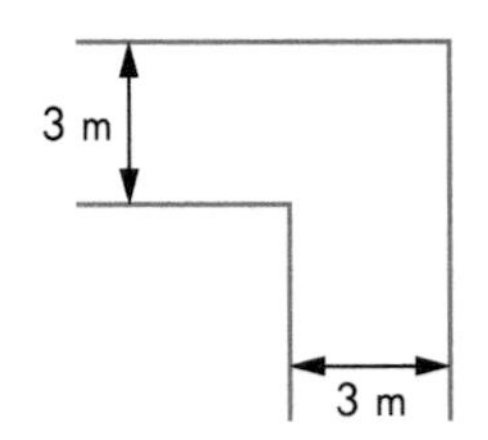

5. A corridor is 3 m wide and turns through a right angle, as in the diagram.

   a) What is the longest pole that can be carried along the corridor horizontally?

   b) If the corridor is 3 m high, what is the longest pole that can be carried along it in any direction?

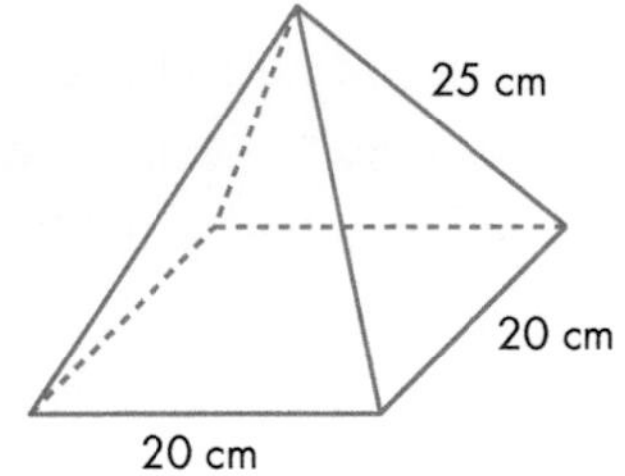

6. If each side of a cube is 10 cm long, how far will it be from one corner of the cube to the opposite one?

7. A pyramid has a square base of side 20 cm and each sloping edge is 25 cm long.

   How high is the pyramid?

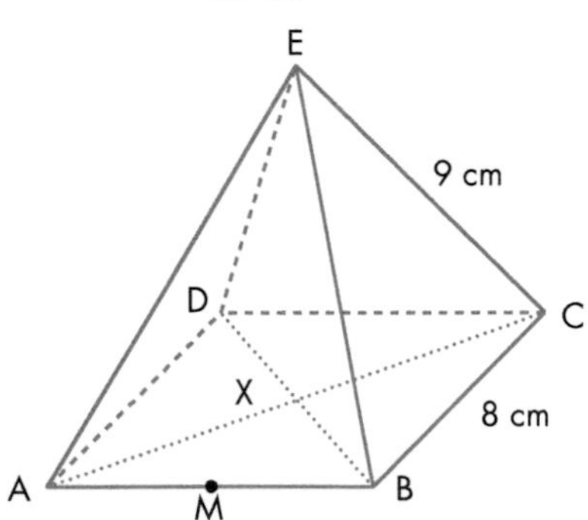

8. The diagram shows a square-based pyramid with base length 8 cm and sloping edges 9 cm.

   M is the midpoint of the side AB, X is the midpoint of the base, and E is directly above X.

   a) Calculate the length of the diagonal AC.

   b) Calculate EX, the height of the pyramid.

   c) Using triangle ABE, calculate the length EM.

E
9 cm
D
C
X
8 cm
A
M
B

9. The diagram shows a cuboid with sides of 40 cm, 30 cm and 22.5 cm.

   M is the midpoint of the side FG. Calculate (or write down) these lengths, giving your answers to 3 significant figures if necessary.

   a) AH b) AG c) AM d) HM

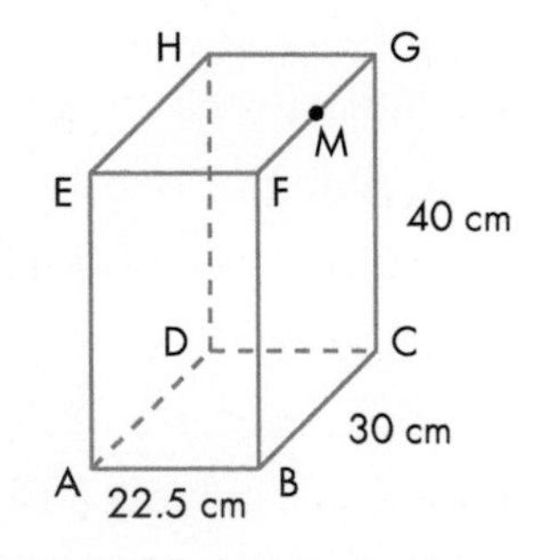

9.2 Trigonometry in right-angled triangles

THIS SECTION WILL SHOW YOU HOW TO ...

✓ understand and use trigonometry in right-angled triangles
✓ identify appropriate right-angled triangles in 2- and 3-dimensional shapes and apply trigonometry
✓ work out the angle between a line and a plane
✓ work out the angle between two planes

KEY WORDS

✓ ratio
✓ inverse
✓ sine
✓ cosine
✓ tangent
✓ opposite side
✓ adjacent side
✓ angle of elevation
✓ angle of depression

Trigonometry uses three important **ratios** to calculate sides and angles: **sine**, **cosine** and **tangent**.

These ratios are defined in terms of the sides of a right-angled triangle and an angle. The angle is often written as θ.

In a right-angled triangle:

- the side opposite the right angle is called the hypotenuse and is the longest side
- the side opposite the angle θ is called the **opposite side**
- the other side next to both the right angle and the angle θ is called the **adjacent side**.

In a right-angled triangle, sine, cosine and tangent ratios for θ are defined as:

$$\text{sine } \theta = \frac{\text{opposite}}{\text{hypotenuse}} \qquad \text{cosine } \theta = \frac{\text{adjacent}}{\text{hypotenuse}} \qquad \text{tangent } \theta = \frac{\text{opposite}}{\text{adjacent}}$$

These ratios are usually abbreviated as:

$$\sin \theta = \frac{\text{O}}{\text{H}} \qquad \cos \theta = \frac{\text{A}}{\text{H}} \qquad \tan \theta = \frac{\text{O}}{\text{A}}$$

These abbreviated forms are also used on calculator keys.

Using your calculator to calculate angles

What angle has a cosine of 0.6? You can use a calculator to find out.

'The angle with a cosine of 0.6' is written as $\cos^{-1} 0.6$ and is called the '**inverse** cosine of 0.6'.

Find out where $\cos^{-1}$ is on your calculator. You will probably find it on the same key as cos, but you will need to press SHIFT or INV or 2ndF first.

Look to see if $\cos^{-1}$ is written above the cos key.

Check that $\cos^{-1} 0.6 = 53.1301 \ldots = 53.1°$ (1 decimal place)

Check that $\cos 53.1° = 0.600$ (3 decimal places)

Check that you can find inverse sine and inverse tangent in the same way.

EXAMPLE 5

What angle has a sine of $\frac{3}{8}$?

You need to find $\sin^{-1}\frac{3}{8}$.

You could use the fraction button on your calculator or you could calculate $\sin^{-1}(3 \div 8)$.

If you use the fraction key you may not need brackets, or your calculator may put them in automatically.

Try to do it in both of these ways and then use whichever you prefer.

The answer should be 22.0°.

EXERCISE 9E

GRADE B

1. Find these values, rounding your answers to 3 significant figures.

   a) sin 43° **b)** sin 56° **c)** sin 67.2° **d)** sin 90°

2. **a) i)** What is sin 35°? **ii)** What is cos 55°?

   b) i) What is sin 12°? **ii)** What is cos 78°?

   c) i) What is cos 67°? **ii)** What is sin 23°?

   d) What connects the values in parts **a**, **b** and **c**?

   e) Copy and complete these sentences.

   i) sin 15° is the same as cos … **ii)** cos 82° is the same as sin …

   iii) sin x is the same as cos …

3. Use your calculator to work these out. Give your answers to three significant figures.

   a) tan 43° **b)** tan 56° **c)** tan 67.2° **d)** tan 90°

4. What is so different about tan compared with both sin and cos?

5. Use your calculator to work these out. Give your answers to three significant figures.

   a) 4 sin 63° **b)** 7 tan 52° **c)** $\frac{5}{\sin 63°}$ **d)** $\frac{6}{\cos 32°}$

6. Calculate sin x, cos x, and tan x for each triangle. Leave your answers as fractions.

   a)

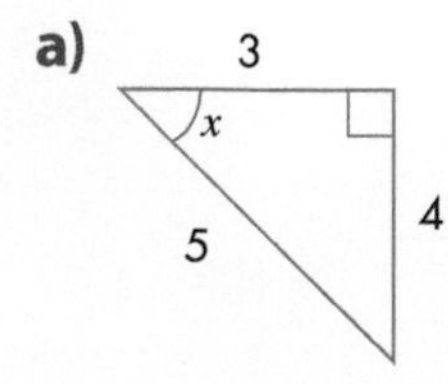

   b)

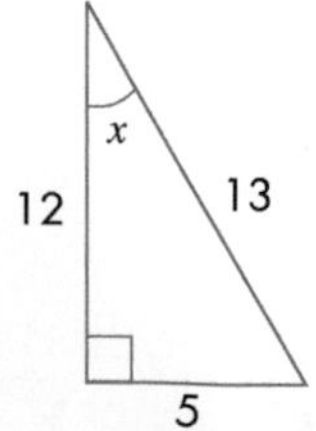

   c)

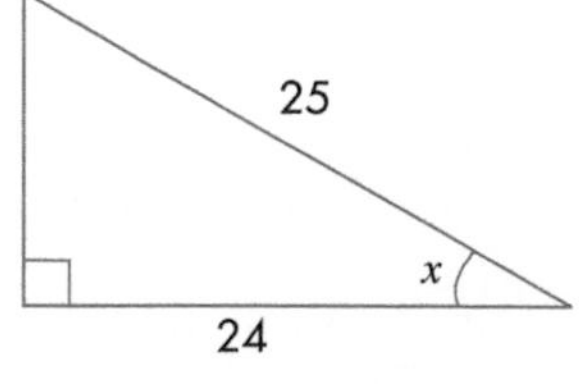

7. What angles have these sines? Give your answers to 1 decimal place.

   a) 0.5 **b)** 0.785 **c)** 0.64 **d)** 0.877 **e)** 0.999 **f)** 0.707

8. What angles have these cosines? Give your answers to 1 decimal place.

   a) 0.5 **b)** 0.64 **c)** 0.999 **d)** 0.707 **e)** 0.2 **f)** 0.7

9. What angles have these tangents? Give your answers to 1 decimal place.

a) 0.6 **b)** 0.38 **c)** 0.895 **d)** 1.05 **e)** 2.67 **f)** 4.38

10. What happens when you try to find the angle with a sine of 1.2? What is the largest value of sine you can put into your calculator without getting an error when you ask for the inverse sine? What is the smallest?

Using the sine function

Remember, sine $\theta = \dfrac{\text{opposite}}{\text{hypotenuse}}$

You can use the sine ratio to calculate the lengths of sides and angles in right-angled triangles.

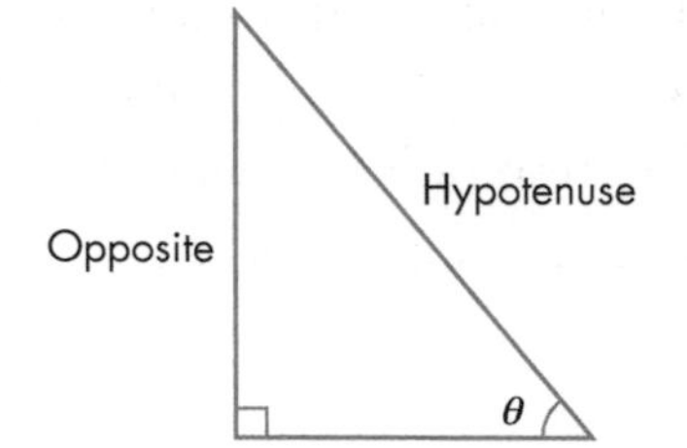

EXAMPLE 6

Find the angle θ, given that the opposite side is 7 cm and the hypotenuse is 10 cm.

Draw a diagram. (This is an essential step.)

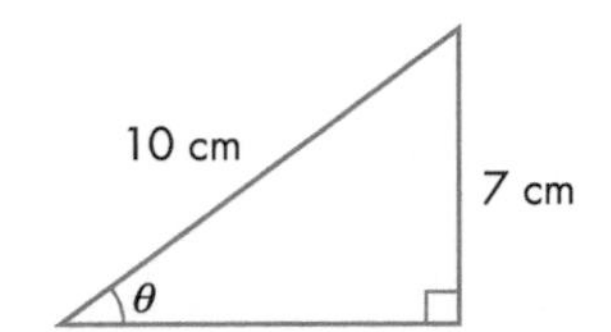

From the information given, use sine.

$\sin\theta = \dfrac{O}{H} = \dfrac{7}{10} = 0.7$ What angle has a sine of 0.7?

To find out, use the inverse sine function on your calculator.

$\sin^{-1} 0.7 = 44.4°$ (1 decimal place)

EXAMPLE 7

Find the length of the side marked a in this triangle.

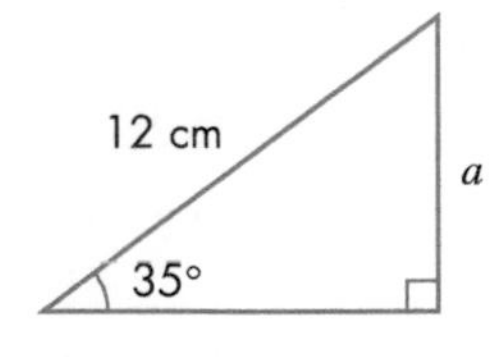

Side a is the opposite side, with 12 cm as the hypotenuse, so use sine.

$\sin\theta = \dfrac{O}{H} \Rightarrow \sin 35° = \dfrac{a}{12}$

So $a = 12 \sin 35° = 6.88$ cm (3 significant figures)

EXAMPLE 8

Find the length of the hypotenuse, h, in this triangle.

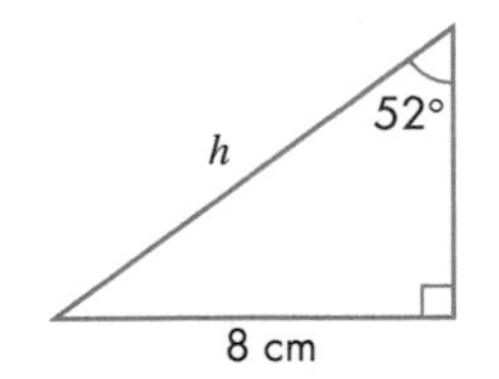

Note that although the angle is in the other corner, the opposite side is again given. So use sine.

$\sin\theta = \dfrac{O}{H} \Rightarrow \sin 52° = \dfrac{8}{h}$

So $h = \dfrac{8}{\sin 52°} = 10.2$ cm (3 significant figures)

Using the cosine function

Remember, cosine $\theta = \dfrac{\text{adjacent}}{\text{hypotenuse}}$

You can use the cosine ratio to calculate the lengths of sides and angles in right-angled triangles.

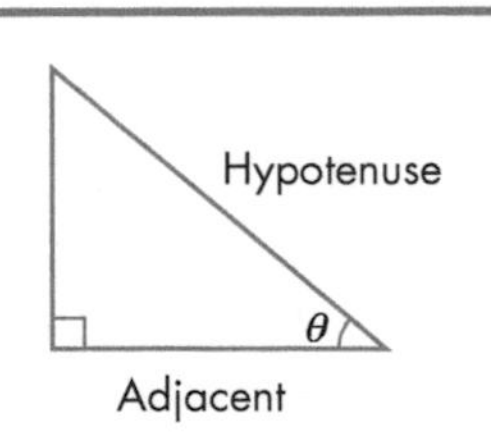

EXAMPLE 9

Find the angle θ, given that the adjacent side is 5 cm and the hypotenuse is 12 cm.

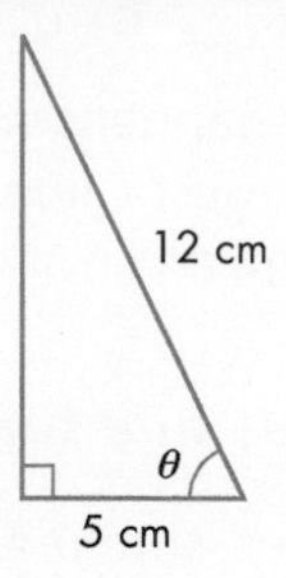

Draw a diagram. (This is an essential step.)

From the information given, use cosine.

$\cos\theta = \frac{A}{H} = \frac{5}{12}$

What angle has a cosine of $\frac{5}{12}$? To find out, use the inverse cosine function on your calculator.

$\cos^{-1} = 65.4°$ (1 decimal place)

EXAMPLE 10

Find the length of the hypotenuse, h, in this triangle.

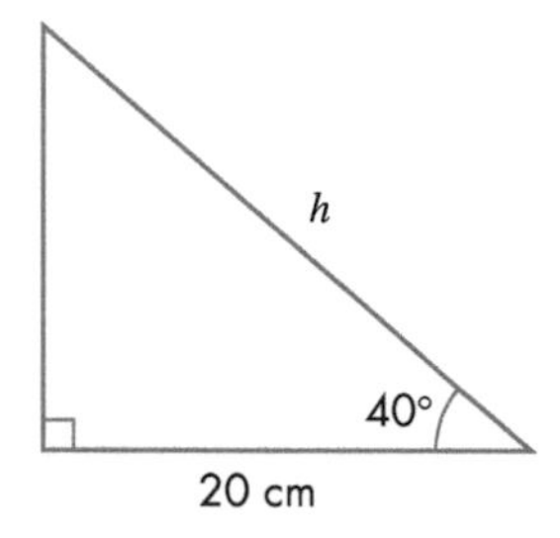

The adjacent side is given, so use cosine.

$\cos\theta = \frac{A}{H}$

$\cos 40° = \frac{20}{h} \Rightarrow h = \frac{20}{\cos 40°} = 26.1$ cm (3 significant figures)

Using the tangent function

Remember, tangent $\theta = \frac{\text{opposite}}{\text{adjacent}}$

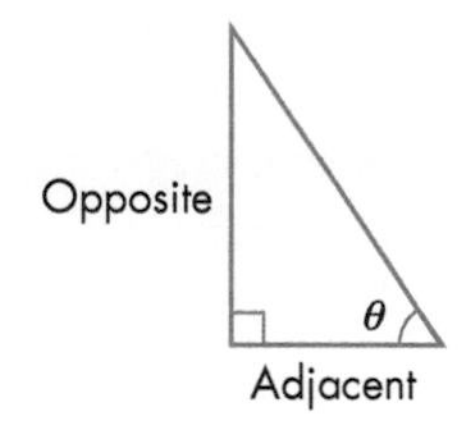

You can use the tangent ratio to calculate the lengths of sides and angles in right-angled triangles.

EXAMPLE 11

Find the length of the side marked x in this triangle.

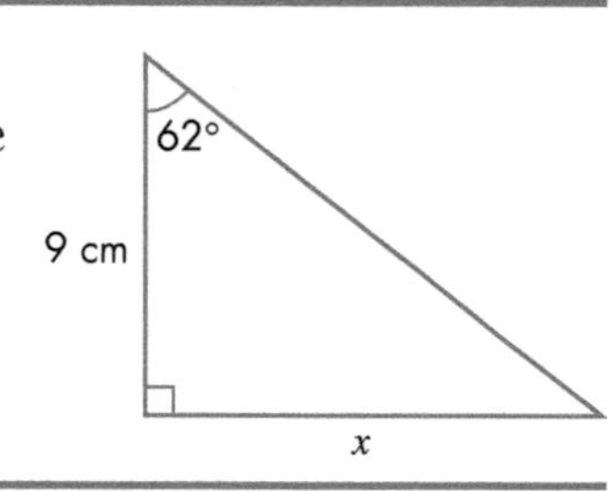

Side x is the opposite side, with 9 cm as the adjacent side, so use tangent.

$\tan\theta = \frac{O}{A}$

$\tan 62° = \frac{x}{9} \Rightarrow x = 9\tan 62° = 16.9$ cm (3 significant figures)

EXERCISE 9F

GRADE B

1. In each triangle, find the size of the angle marked x.

a)

10 cm, 3 cm, x

b)

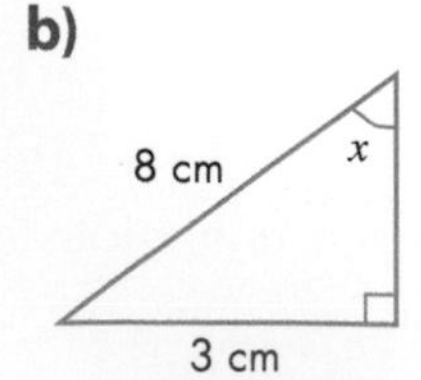

c)

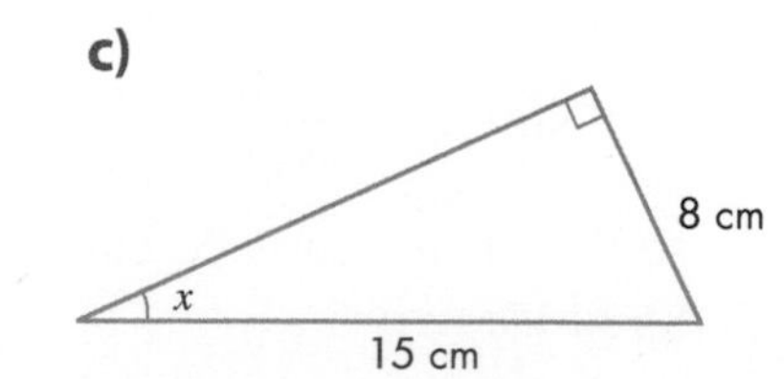

2. In each triangle, find the length of the side marked x.

a)

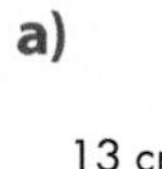
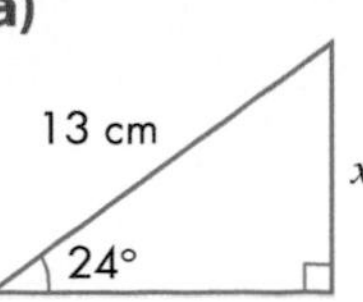

b)

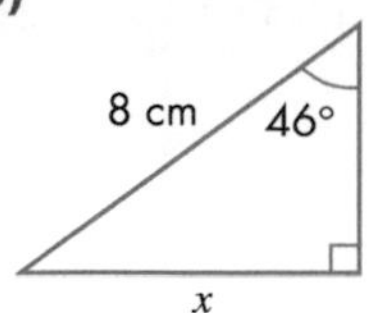

c)

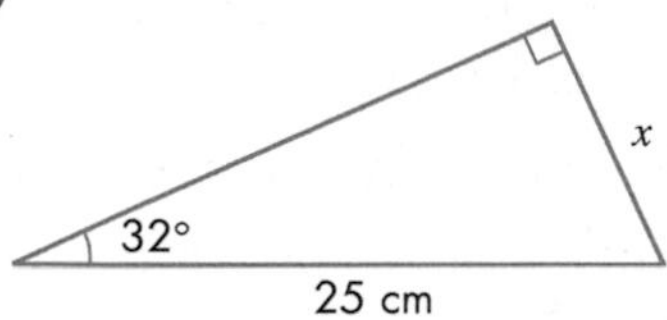

3. In each triangle, find the length of the side marked x.

a)

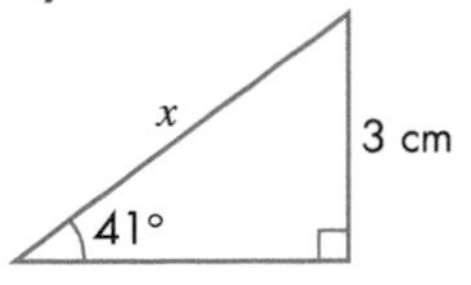

b)

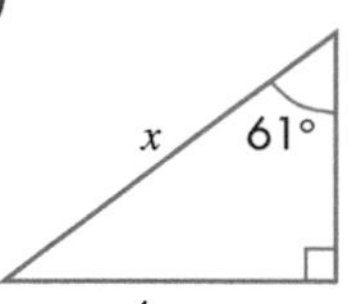

c)

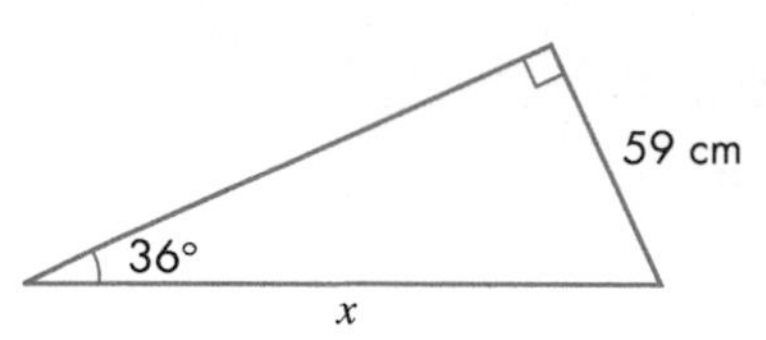

4. In each triangle, find the length of the side marked x.

a)

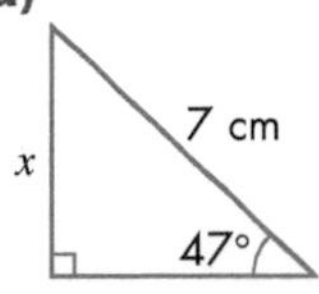

b)

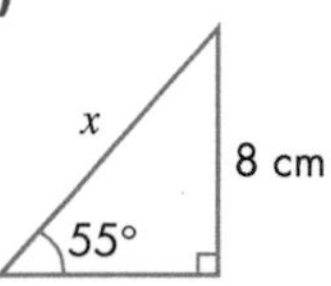

c)

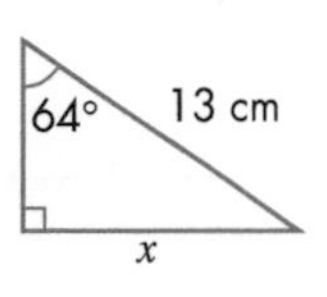

d)

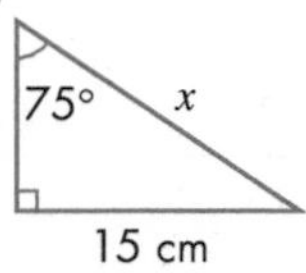

5. In each triangle, find the size of the angle marked x.

a)

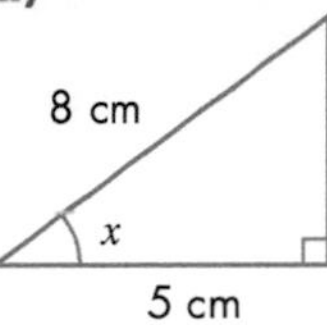

b)

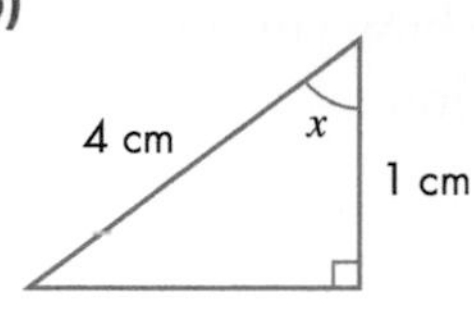

c)

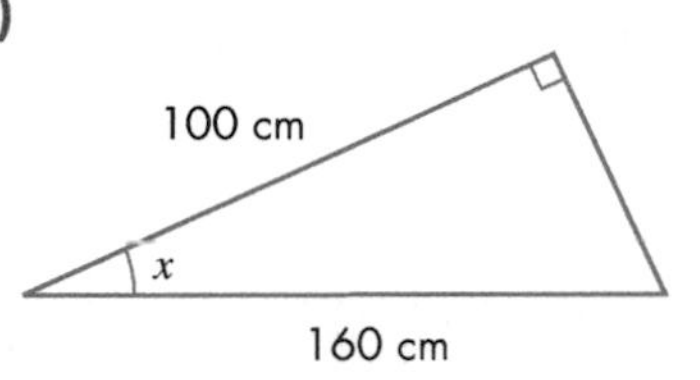

6. In each triangle, find the length of the side marked x.

a)

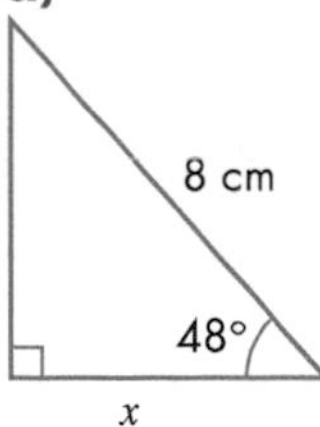

b)

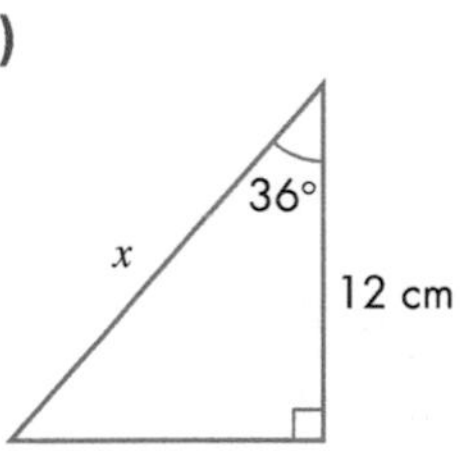

c)

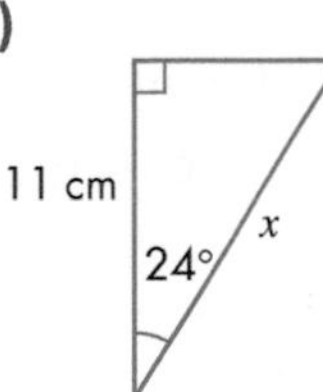

d)

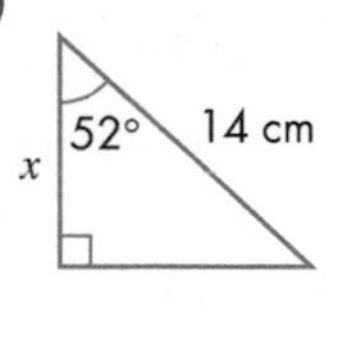

7. In each triangle, find the value of x.

a)

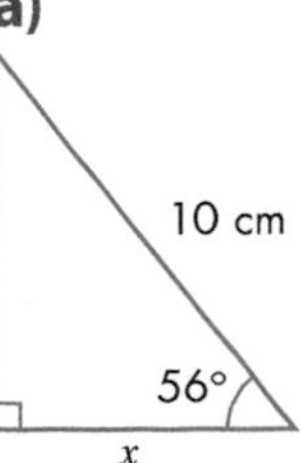

b)

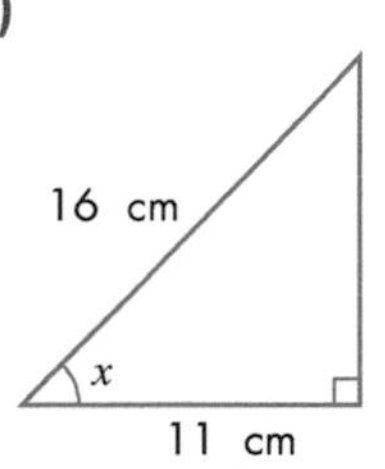

c)

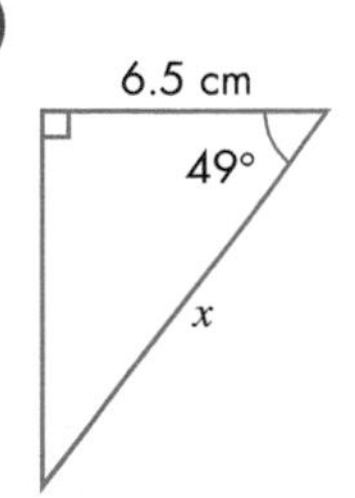

d)

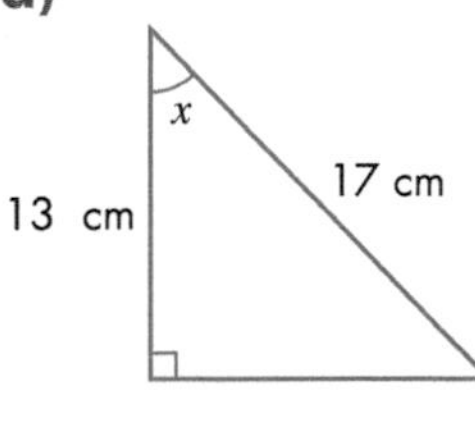

8. In each triangle, find the size of the angle marked x.

a)

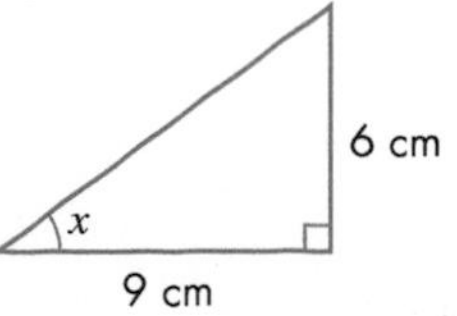

b)

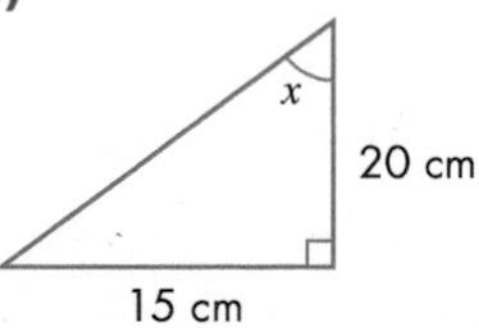

c)

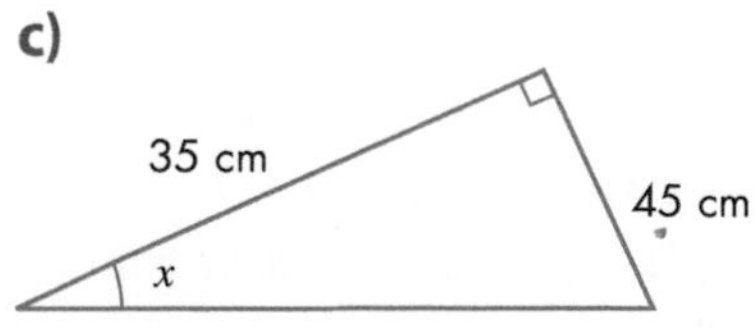

9. In each triangle, find the length of the side marked x.

a)

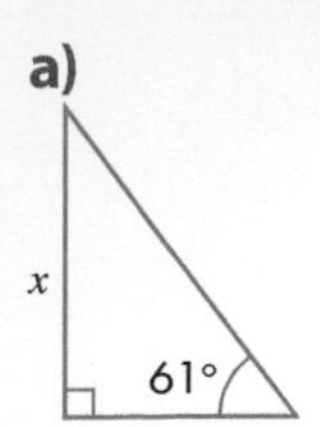

b)

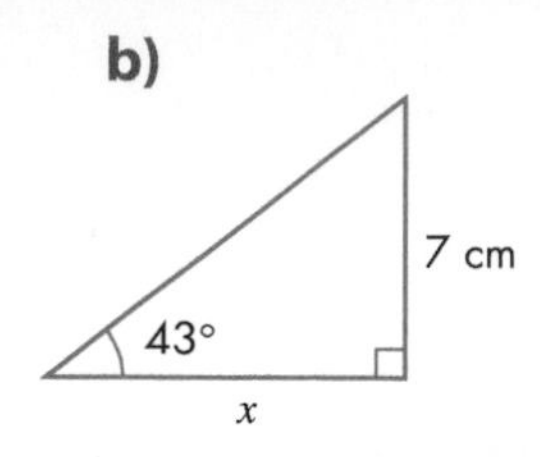

c)

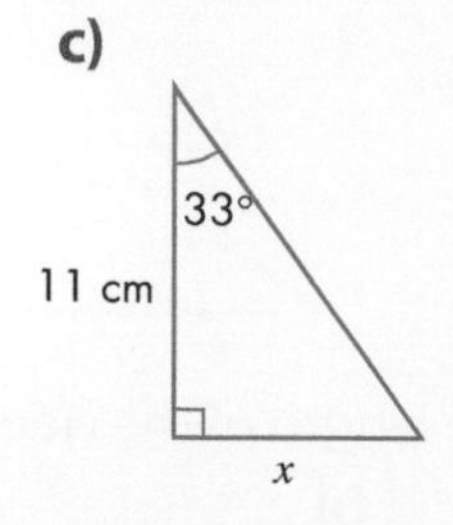

d)

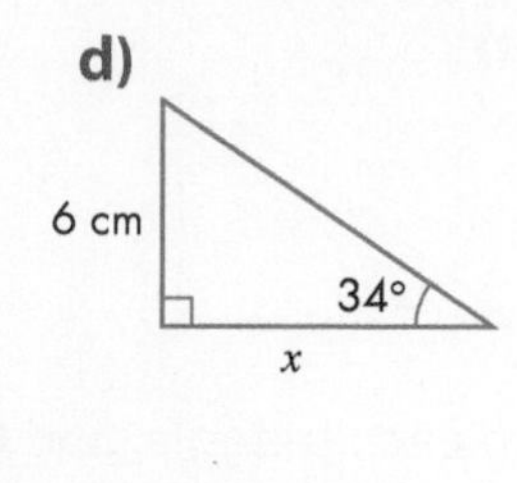

10. In each triangle, find the value of x.

a)

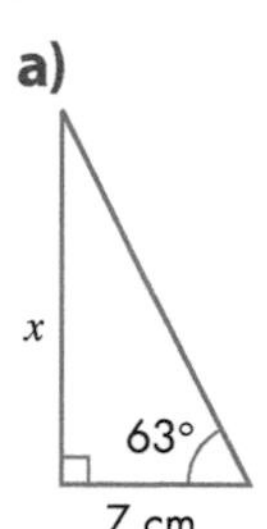

b)

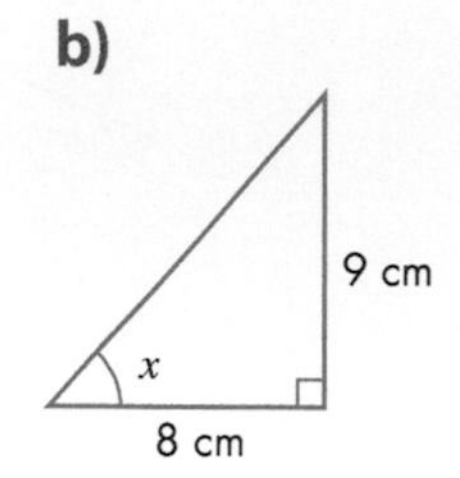

c)

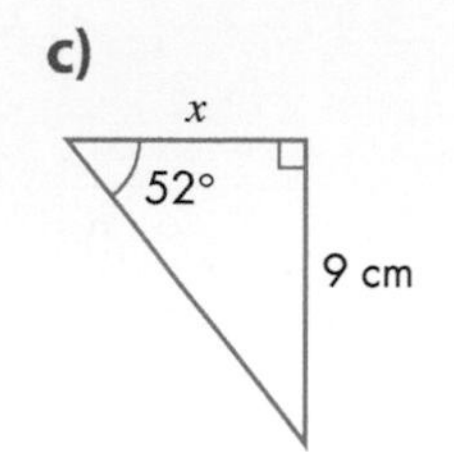

d)

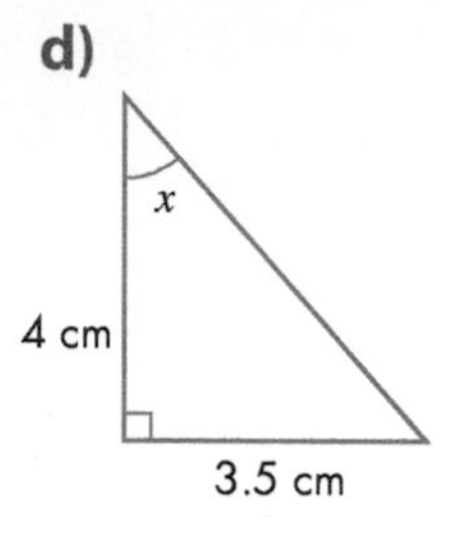

Which ratio to use

The difficulty with any trigonometric problem is knowing which ratio to use to solve it. The following examples show you how to determine which ratio you need in any given situation.

EXAMPLE 12

Find the length of the side marked x in this triangle.

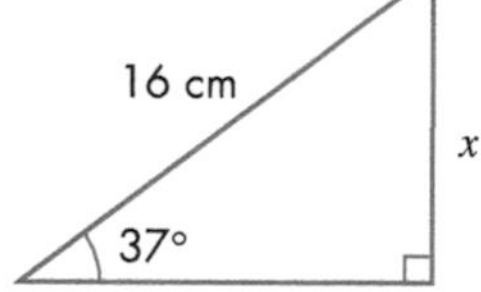

Step 1: Identify what information is given and what needs to be found. Namely, x is opposite the angle and 16 cm is the hypotenuse.

Step 2: Decide which ratio to use. Only one ratio uses opposite and hypotenuse: sine.

Step 3: Remember, $\sin \theta = \frac{\text{O}}{\text{H}}$.

Step 4: Put in the numbers and letters: $\sin 37° = \frac{x}{16}$

Step 5: Rearrange the equation and work out the answer:

$x = 16 \sin 37° = 9.629\ 040\ 371$ cm

Step 6: Give the answer to an appropriate degree of accuracy:
$x = 9.63$ cm (3 significant figures).

In reality, you do not write down every step as in Example 12. Step 1 can be done by marking the triangle. Steps 2 and 3 can be done in your head. Steps 4 to 6 are what you write down.

Examiners want to see evidence of working. Any reasonable attempt at identifying the sides and using a ratio will probably gain some method marks, but only if the fraction is the right way round. The next examples are set out in a way that requires the *minimum* amount of working but gets *maximum* marks.

EXAMPLE 13

Find the length of the side marked x in this triangle.

Mark on the triangle the side you know (H) and the side you want to find (A).

Recognise it is a cosine problem because you have A and H.

So $\cos 50° = \frac{x}{7} \Rightarrow x = 7 \cos 50° = 4.50$ cm (3 significant figures)

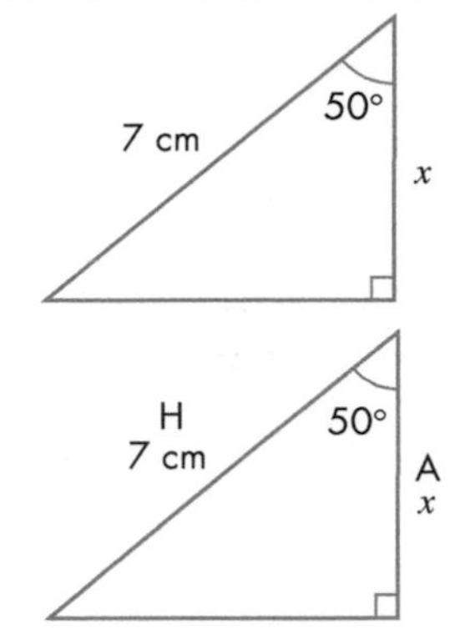

EXAMPLE 14

Find the size of the angle marked x in this triangle.

Mark on the triangle the sides you know.

Recognise it is a tangent problem because you have O and A.

So $\tan x = \frac{12}{7} \Rightarrow x = \tan^{-1} \frac{12}{7} = 59.7°$ (1 decimal place)

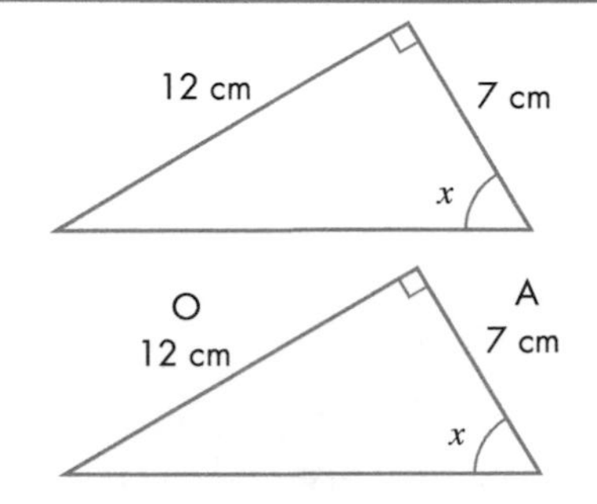

EXERCISE 9G

GRADE B

1. In each triangle, find the length of the side marked x.

a)

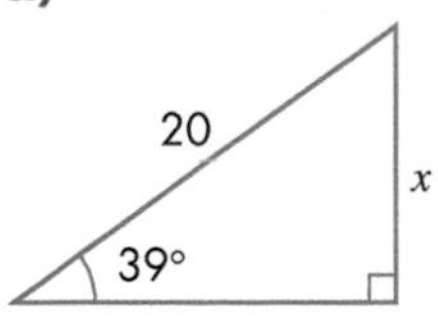

b)

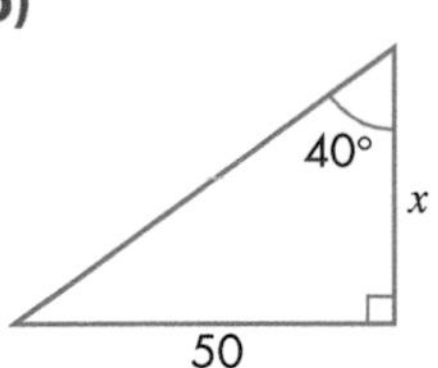

c)

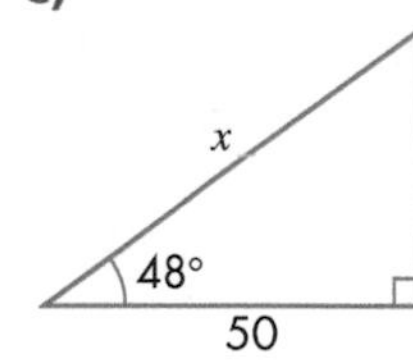

d)

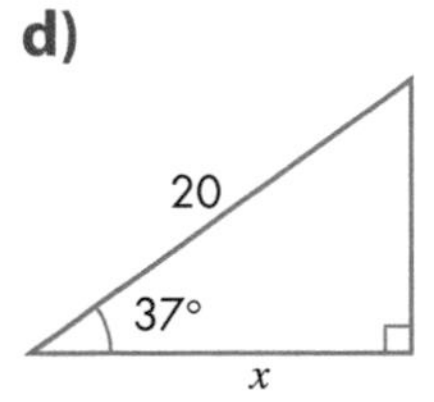

e)

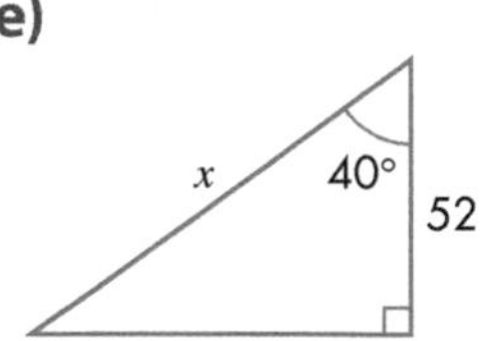

f)

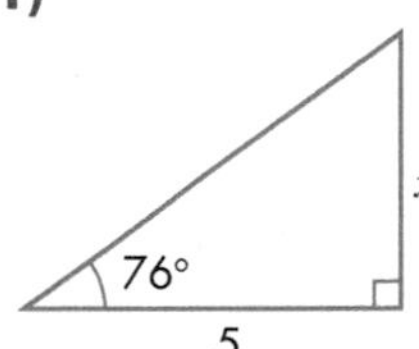

2. In each triangle, find the size of the angle marked x.

a)

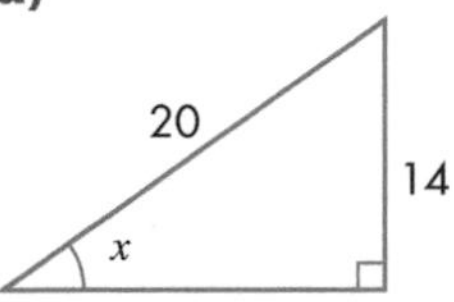

b)

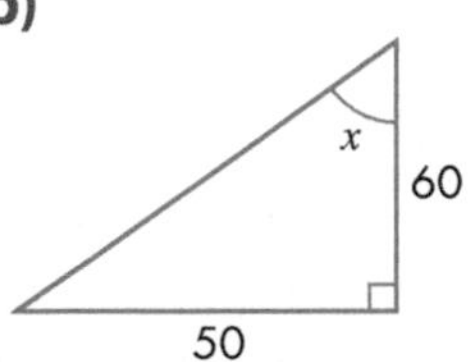

c)

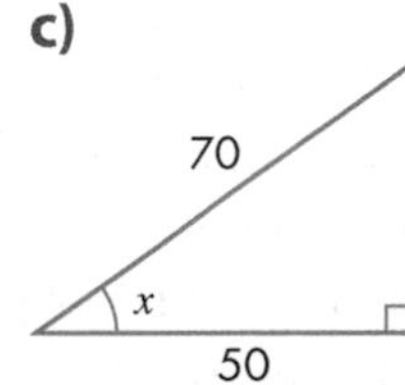

d)

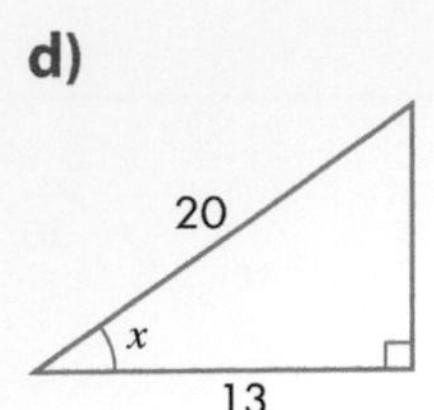

e)

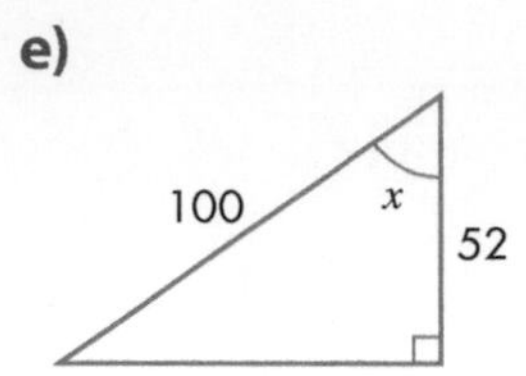

f)

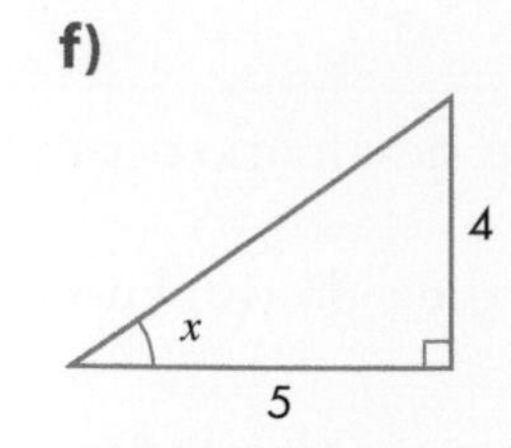

3. In each triangle, find the length or angle marked x.

a)

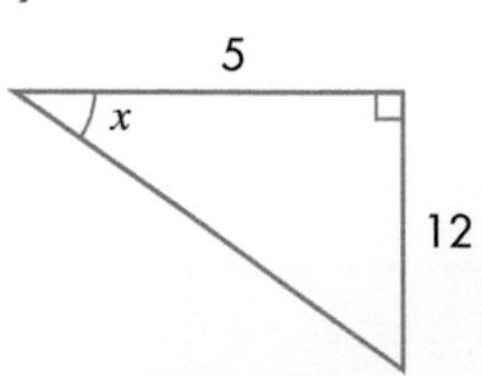

b)

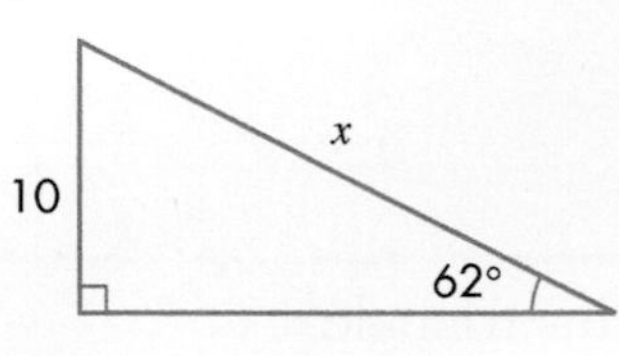

c)

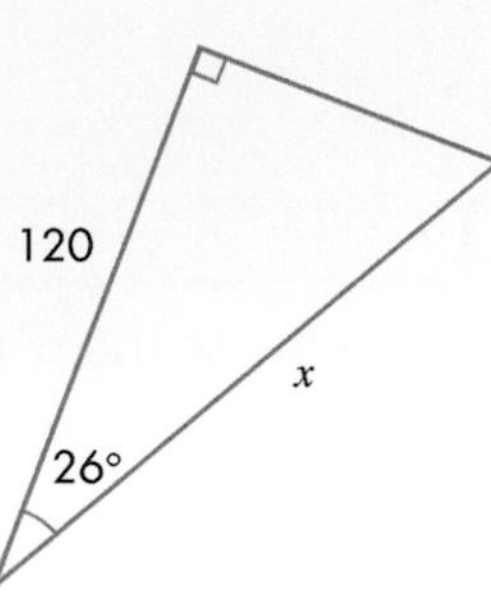

d)

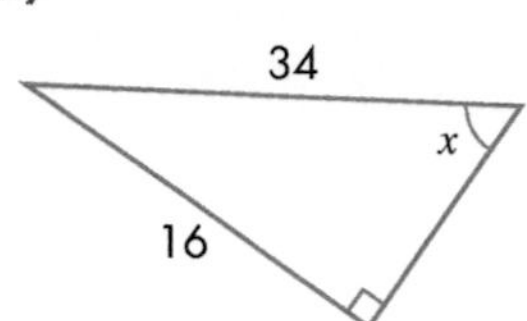

e)

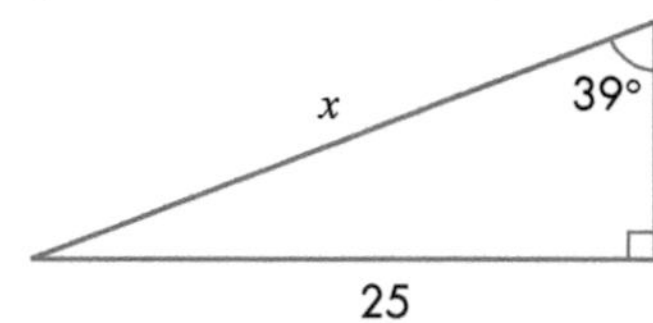

f)

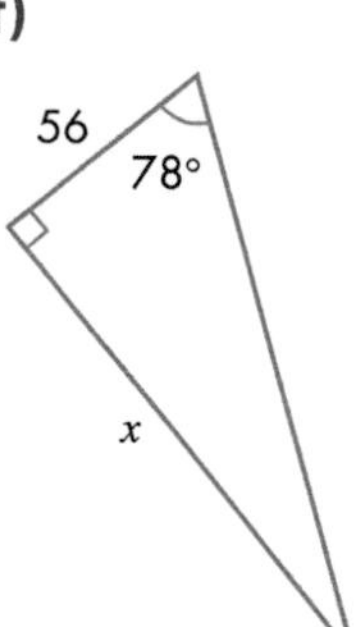

g)

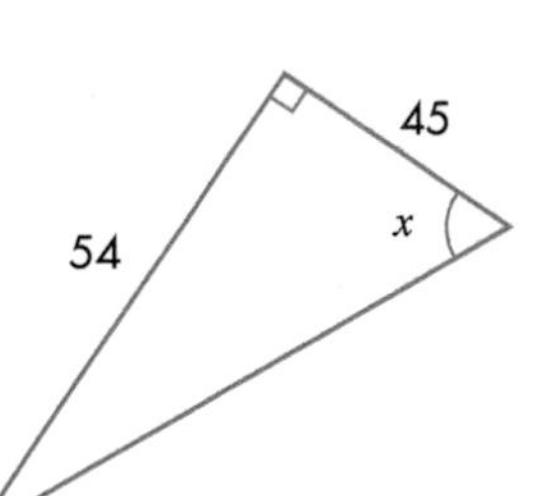

h)

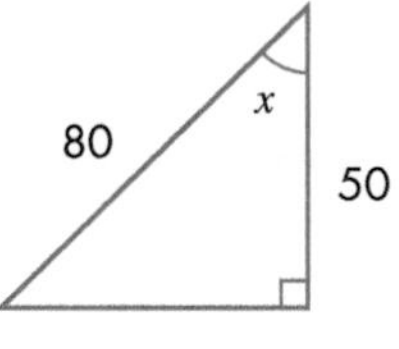

i)

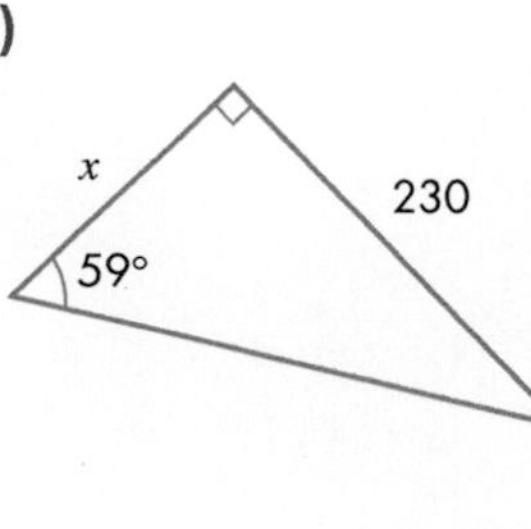

j)

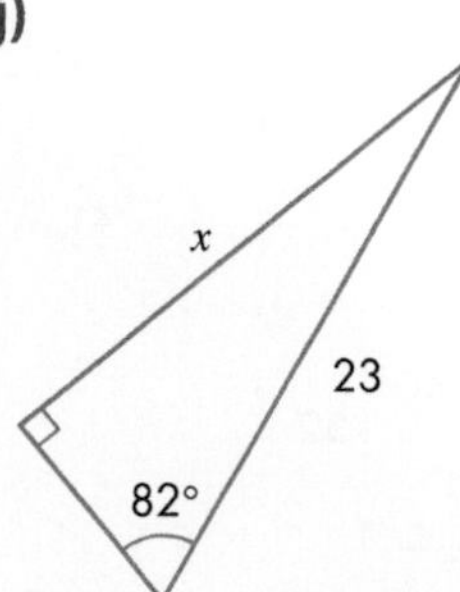

4. a) How does this diagram show that $\tan\theta = \dfrac{\sin\theta}{\cos\theta}$?

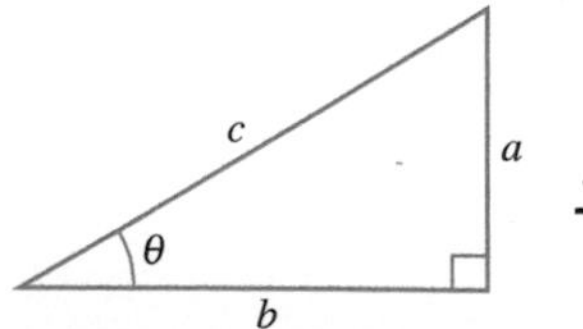

b) How does the diagram show that $(\sin\theta)^2 + (\cos\theta)^2 = 1$?

c) Choose a value for θ and check the two results in parts **a** and **b** are true.

EXERCISE 9H

GRADE A

In these questions, give answers involving angles to the nearest degree.

1. A ladder, 6 m long, rests against a wall. The foot of the ladder is 2.5 m from the base of the wall. What angle does the ladder make with the ground?
2. The ladder in question **1** has a 'safe angle' with the ground of between 70° and 80°. What are the safe limits for the distance of the foot of this ladder from the wall? How high up the wall does the ladder reach?
3. Calculate the angle that the diagonal makes with the long side of a rectangle that measures 10 cm by 6 cm.
4. Taipei is 800 km from Hong Kong on a bearing of 065°.
   a) How far north of Hong Kong is Taipei?
   b) How far west of Taipei is Hong Kong?

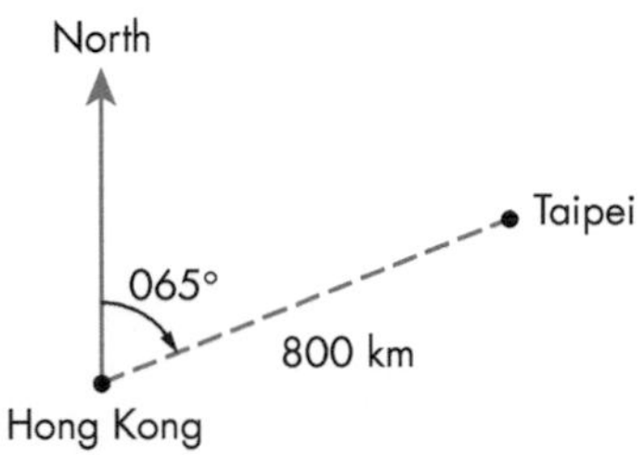

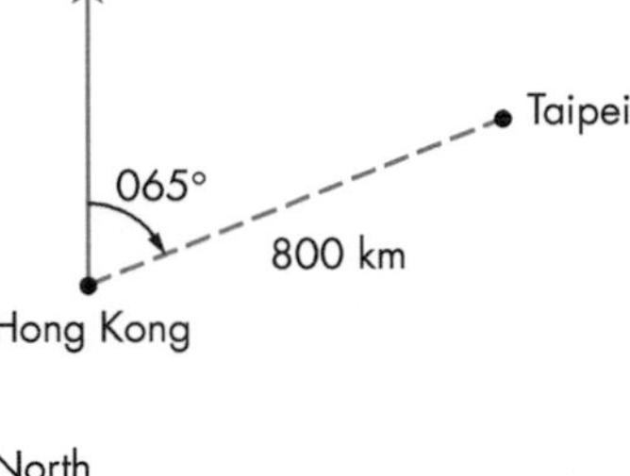

5. A ship is at S where it is 65 km from port P on a bearing of 132°.
   It sails north to port T where it is east of the port.
   What is the distance from S to T?

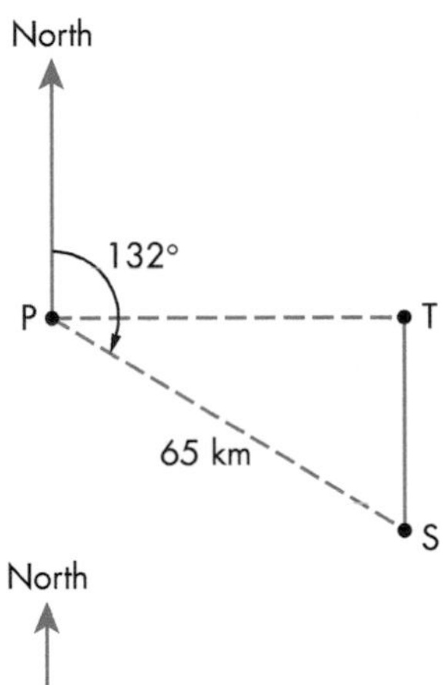

6. A plane at P is 160 km west of airport A.
   It flies on a bearing of 070° to Q which is north of A.
   How far does the plane fly?

North
Q
070°
P
A
160 km

7. Helena is standing on one bank of a wide river. She wants to find the width of the river. She cannot get to the other side. She asks if you can use trigonometry to find the width of the river.
   What can you suggest?

Angles of elevation and depression

When you look *up* at an aircraft in the sky, the angle through which your line of sight turns from looking straight ahead (the horizontal) is called the **angle of elevation**.

When you are standing on a high point and look *down* at a boat, the angle through which your line of sight turns from looking straight ahead (the horizontal) is called the **angle of depression**.

EXAMPLE 15

From the top of a vertical cliff, 100 m high, Ali sees a boat out at sea. The angle of depression from Ali to the boat is 42°. How far from the base of the cliff is the boat?

Draw a sketch.
Then you see that this is a tangent problem.

So $\tan 42° = \dfrac{100}{x} \Rightarrow x = \dfrac{100}{\tan 42°} = 111$ m

(3 significant figures)

EXERCISE 9I

GRADE A

In these questions, give any answers involving angles to the nearest degree.

1. Eric sees an aircraft in the sky. The aircraft is at a horizontal distance of 25 km from Eric. The angle of elevation is 22°. How high is the aircraft?
2. An aircraft is flying at an altitude of 4000 m and is 10 km from the airport. If a passenger can see the airport, what is the angle of depression?
3. A man standing 200 m from the base of a television transmitter looks at the top of it and notices that the angle of elevation of the top is 65°. How high is the tower?
4. **a)** From the top of a vertical cliff, 200 m high, a boat has an angle of depression of 52°.
   How far from the base of the cliff is the boat?

   b) The boat now sails away from the cliff so that the distance is doubled. Does that mean that the angle of depression is halved?
   Give a reason for your answer.

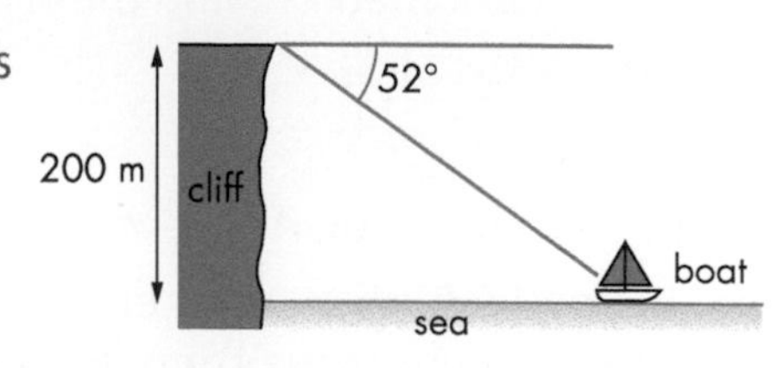

5. From a boat, the angle of elevation of the foot of a lighthouse on the edge of a cliff is 34°.
   a) If the cliff is 150 m high, how far from the base of the cliff is the boat?
   b) If the lighthouse is 50 m high, what would be the angle of elevation of the top of the lighthouse from the boat?

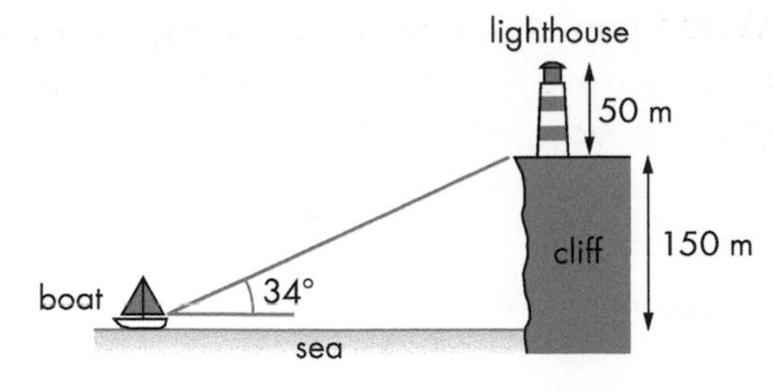

6. A bird flies from the top of a 12 m tall tree, at an angle of depression of 34°, to catch a worm on the ground.
   a) How far does the bird actually fly?
   b) How far was the worm from the base of the tree?

7. Sunil wants to work out the height of a building. He stands about 50 m away from a building. The angle of elevation from Sunil to the top of the building is about 15°.
   How tall is the building?

8. The top of a ski run is 100 m above the finishing line. The run is 300 m long. What is the angle of depression of the ski run?

9. Nessie and Cara are standing on opposite sides of a tree.
   Nessie is 14 m away and the angle of elevation of the top of the tree is 30°.
   Cara is 28 m away. She says the angle of elevation for her must be 15° because she is twice as far away.
   Is she correct?
   What do you think the angle of elevation is?

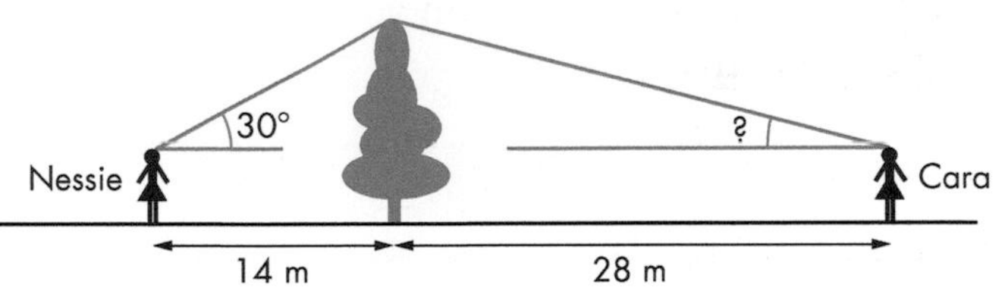

Problems in three dimensions

To find the value of an angle or side in a three-dimensional figure you need to find a right-angled triangle in the figure that contains it. This triangle also has to include two known values that you can use in the calculation.

You must redraw this triangle separately as a plain, right-angled triangle. Add the known values and the unknown value you want to find. Then use the trigonometric ratios and Pythagoras' theorem to solve the problem.

EXAMPLE 16

A, B and C are three points at ground level. They are in the same horizontal plane. C is 50 km east of B. B is north of A. C is on a bearing of 050° from A.

An aircraft, flying east, passes over B and over C at the same height. When it passes over B, the angle of elevation from A is 12°. Find the angle of elevation of the aircraft from A when it is over C.

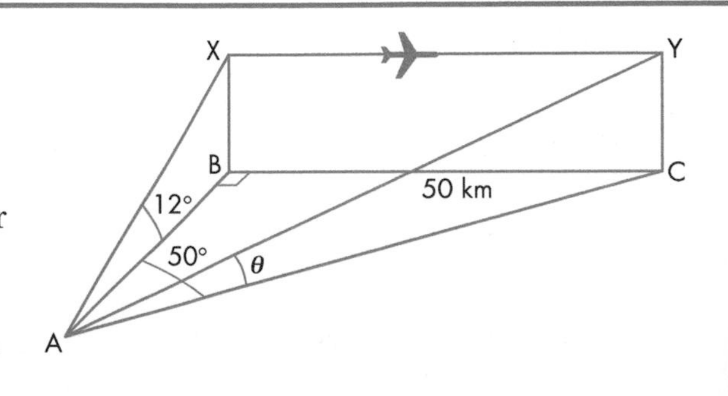

First, draw a diagram, including all the known information.

Next, use the right-angled triangle ABC to calculate AB and AC.

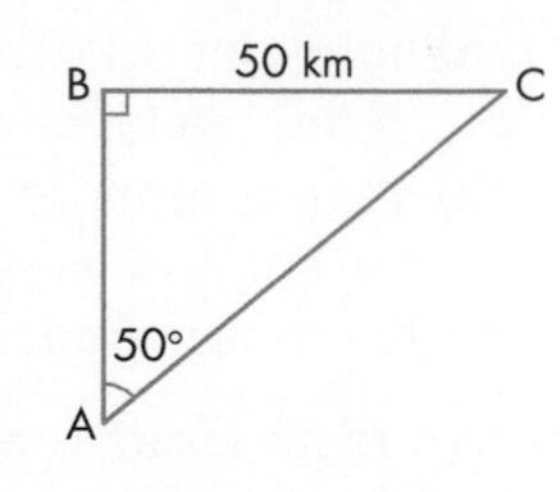

$$AB = \frac{50}{\tan 50°} = 41.95 \text{ km (4 significant figures)}$$

$$AC = \frac{50}{\sin 50°} = 65.27 \text{ km (4 significant figures)}$$

Then use the right-angled triangle ABX to calculate BX, and hence CY.

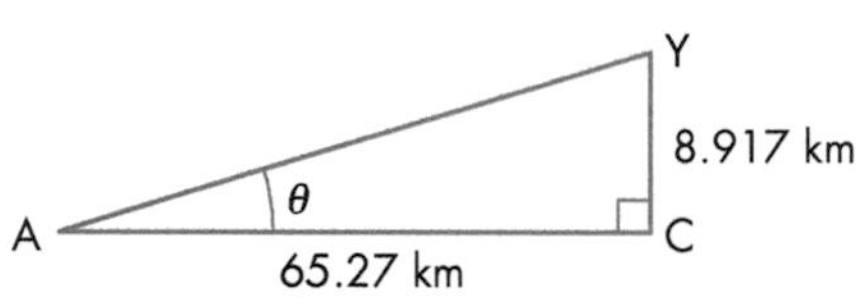

BX = 41.95 tan 12° = 8.917 km (4 significant figures)

Finally, use the right-angled triangle ACY to calculate the required angle of elevation, θ.

$$\tan \theta = \frac{8.917}{65.27} = 0.1366$$

$$\Rightarrow \theta = \tan^{-1} 0.1366 = 7.8° \text{ (1 decimal place)}$$

Always write down intermediate working values to at least 4 significant figures, or use the answer on your calculator display to avoid inaccuracy in the final answer.

EXERCISE 9J

GRADE A*

1. A vertical flagpole AP stands at the corner of a rectangular courtyard ABCD.

Calculate the angle of elevation of P from C.

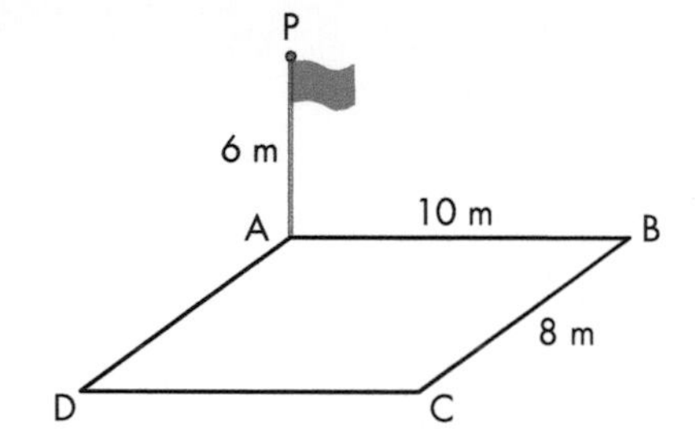

2. The diagram shows a pyramid. The base is a horizontal rectangle ABCD, 20 cm by 15 cm. The length of each sloping edge is 24 cm. The apex, V, is over the centre of the rectangular base. Calculate:

a) the length of the diagonal AC

b) the size of the angle VAC

c) the height of the pyramid.

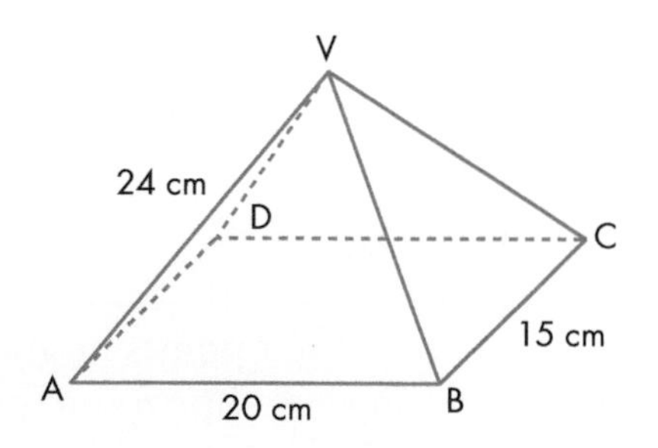

3. The diagram shows the roof of a building. The base ABCD is a horizontal rectangle 7 m by 4 m. The triangular ends are equilateral triangles. Each side of the roof is an isosceles trapezium. The length of the top of the roof, EF, is 5 m. Calculate:

   a) the length EM, where M is the midpoint of AB

   b) the size of angle EBC

   c) the size of the angle between planes EAB and the base ABCD (the angle between EM and ABCD).

4. ABCD is a vertical rectangular plane. EDC is a horizontal triangular plane. Angle CDE = 90°, AB = 10 cm, BC = 4 cm and ED = 9 cm. Calculate:

   a) angle AED **b)** angle DEC

   c) EC **d)** angle BEC.

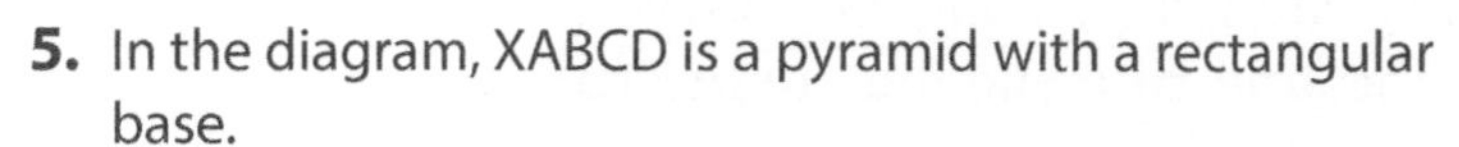

5. In the diagram, XABCD is a pyramid with a rectangular base.

   a) Revina says that the angle between the edge XD and the base ABCD is 56.3°. Work out the correct answer to show that Revina is wrong.

   b) Work out the angle between the planes ABCD and XDC.

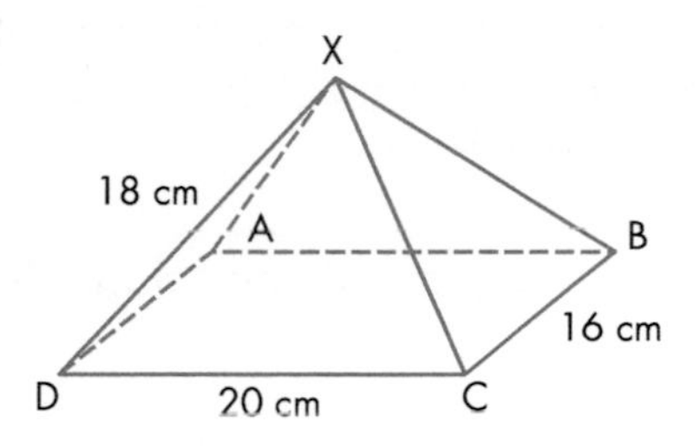

9.3 The sine rule and the cosine rule

THIS SECTION WILL SHOW YOU HOW TO ...

✓ understand and use the formulae for sine rule and cosine rule

KEY WORDS

✓ sine rule ✓ cosine rule ✓ included angle ✓ obtuse

Any triangle has six measurements: three sides and three angles. To find any unknown angles or sides, you need to know at least three of the measurements. Any combination of three measurements – except that of all three angles – is enough to work out the rest.

When you need to find a side or an angle in a triangle which contains no right angle, you can use one of two rules, depending on what you know about the triangle. These are the **sine rule** and the **cosine rule**.

The sine rule

Take a triangle ABC and draw the perpendicular from A to the opposite side BC.

From right-angled triangle ADB, $h = c \sin B$

From right-angled triangle ADC, $h = b \sin C$

Therefore:

$c \sin B = b \sin C$

which can be rearranged to give:

$$\frac{c}{\sin C} = \frac{b}{\sin B}$$

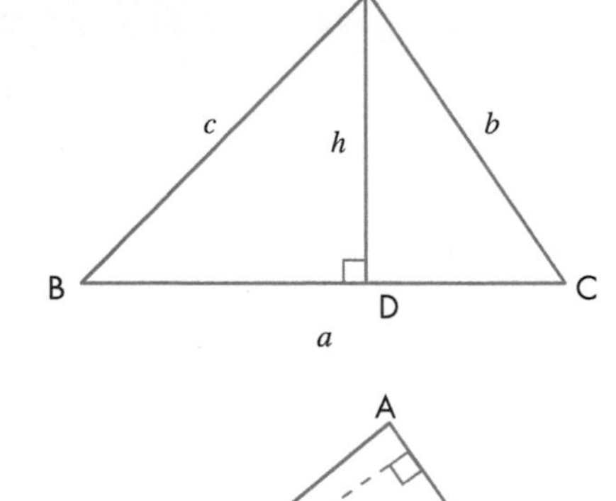

By drawing a perpendicular from each of the other two vertices to the opposite side (or by algebraic symmetry), you can see that:

$$\frac{a}{\sin A} = \frac{c}{\sin C} \text{ and } \frac{a}{\sin A} = \frac{b}{\sin B}$$

These are usually combined in the form:

$$\frac{a}{\sin A} = \frac{b}{\sin B} = \frac{c}{\sin C}$$

which can be inverted to give:

$$\frac{\sin A}{a} = \frac{\sin B}{b} = \frac{\sin C}{c}$$

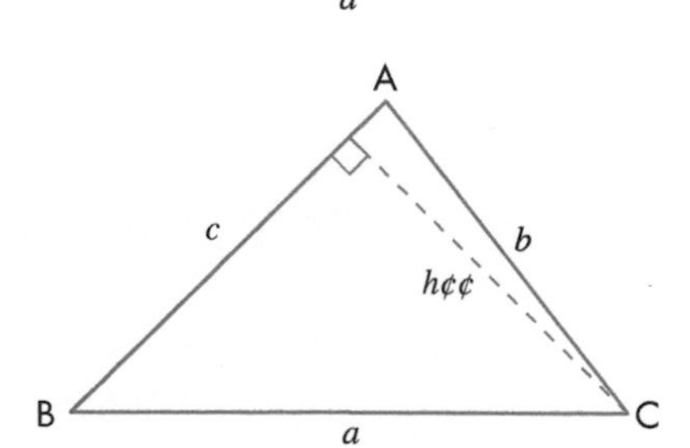

Remember, when using the sine rule: take each side in turn, divide it by the sine of the angle opposite and then equate the results.

Note:

- When you are calculating a *side*, use the rule with the *sides on top*.
- When you are calculating an *angle*, use the rule with the *sines on top*.

EXAMPLE 17

In triangle ABC, find the value of x.

Use the sine rule with sides on top, which gives:

$$\frac{x}{\sin 84^\circ} = \frac{25}{\sin 47^\circ}$$

$$\Rightarrow x = \frac{25 \sin 84^\circ}{\sin 47^\circ} = 34.0 \text{ cm (3 significant figures)}$$

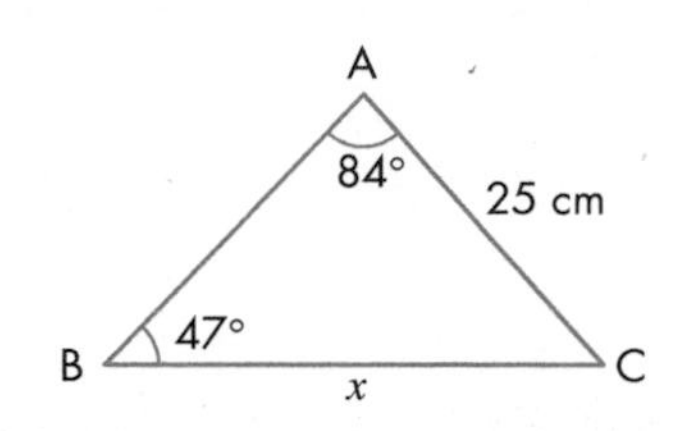

EXAMPLE 18

In the triangle ABC, find the value of the acute angle x.

Use the sine rule with sines on top, which gives:

$$\frac{\sin x}{7} = \frac{\sin 40^\circ}{6} \Rightarrow \sin x = \frac{7 \sin 40^\circ}{6} = 0.7499$$

$$\Rightarrow x = \sin^{-1} 0.7499 = 48.6^\circ \text{ (3 significant figures)}$$

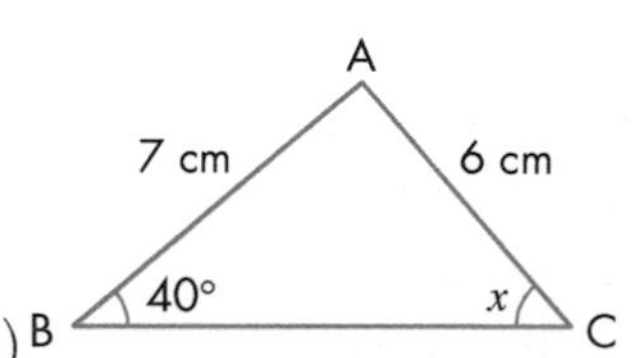

The sine rule works even if the triangle has an obtuse angle, because you can find the sine of an obtuse angle.

EXERCISE 9K

GRADE A

1. In each triangle, find the length of the side marked x.

a)

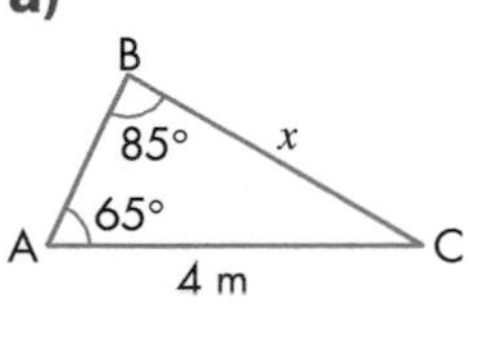

b)

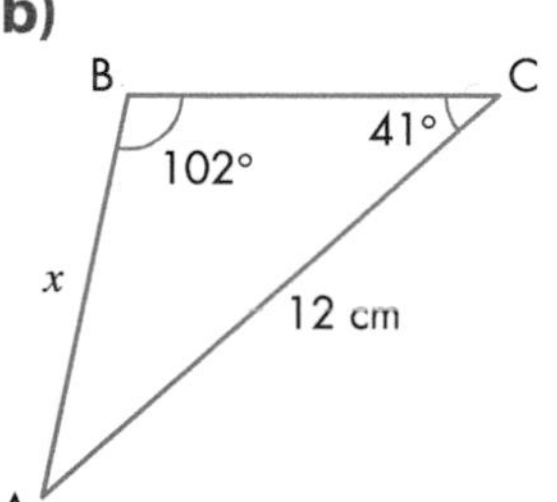

c)

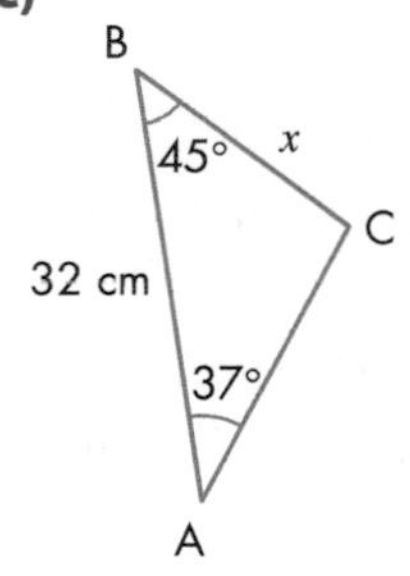

2. In each triangle, find the size of the angle marked x.

a)

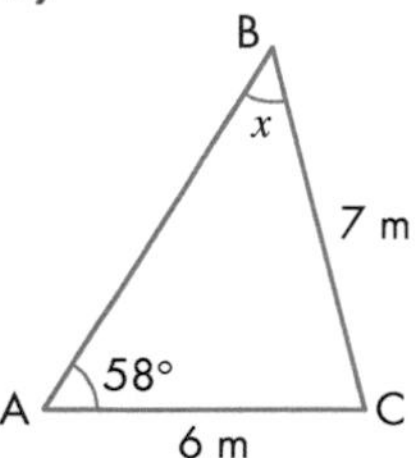

b)

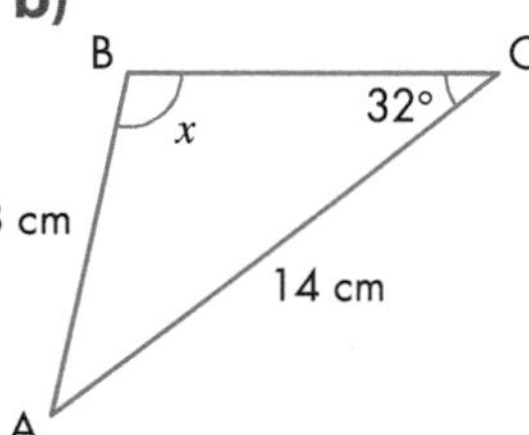

c)

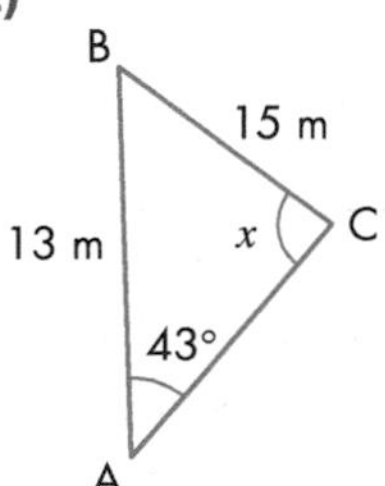

GRADE A*

3. A mass is hung from a horizontal beam by two strings. The shorter string is 2.5 m long and makes an angle of 71° with the horizontal. The longer string makes an angle of 43° with the horizontal. What is the length of the longer string?

4. To find the height of a tower standing on a small hill, Maria made some measurements (see diagram).

From a point B, the angle of elevation of C is 20°, the angle of elevation of A is 50°, and the distance BC is 25 m.

a) Calculate these angles.

i) ABC **ii)** BAC.

b) Using the sine rule and triangle ABC, calculate the height, h, of the tower.

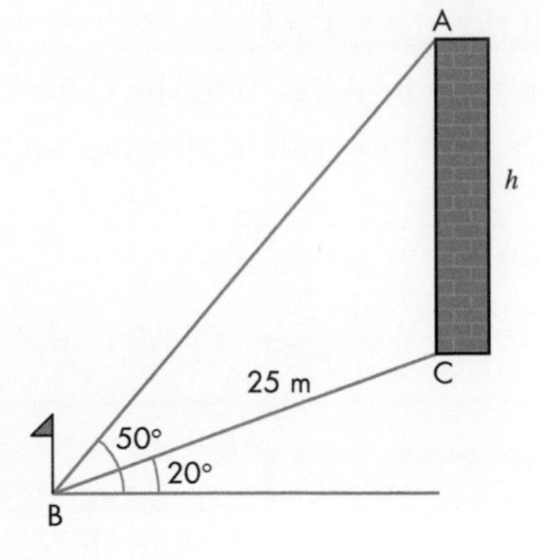

5. Use the information on this sketch to calculate the width, w, of the river.

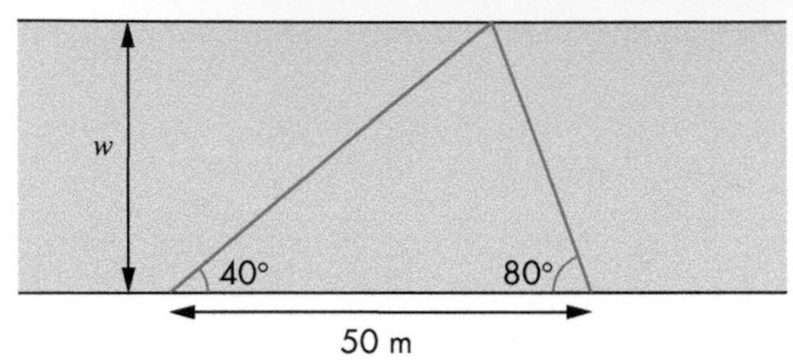

6. An old building is unsafe and is protected by a fence.

A company is going to demolish the building and has to work out its height BD, marked h on the diagram.

Use the given information to calculate the value of h.

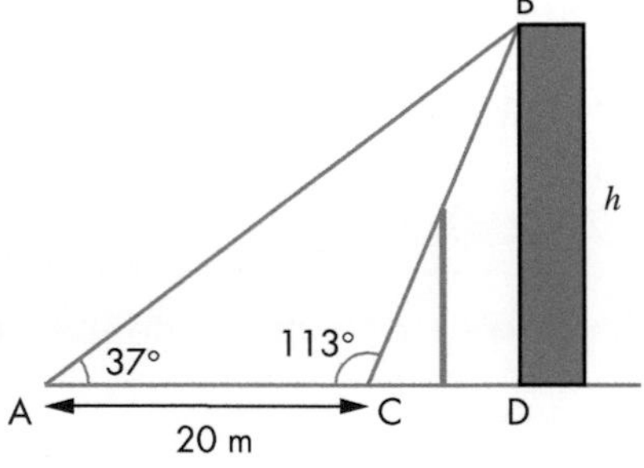

7. A rescue helicopter is based at an airfield at A.

It is sent out to rescue a man from a mountain at M, due north of A.

The helicopter then flies on a bearing of 145° to a hospital at H as shown on the diagram.

Calculate the direct distance from the mountain to the hospital.

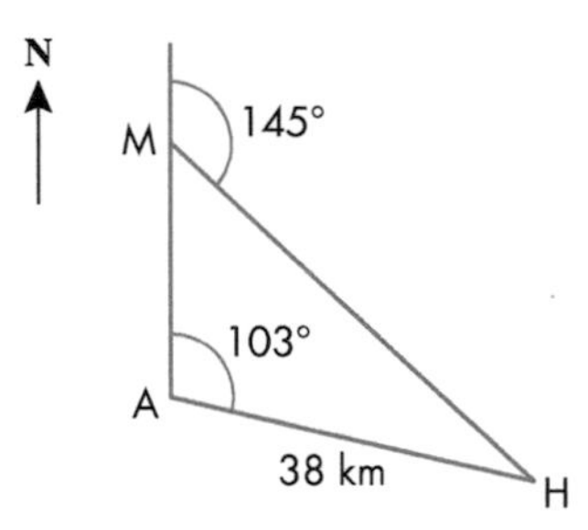

The cosine rule

Look at the triangle, shown on the right, where D is the foot of the perpendicular to BC from A.

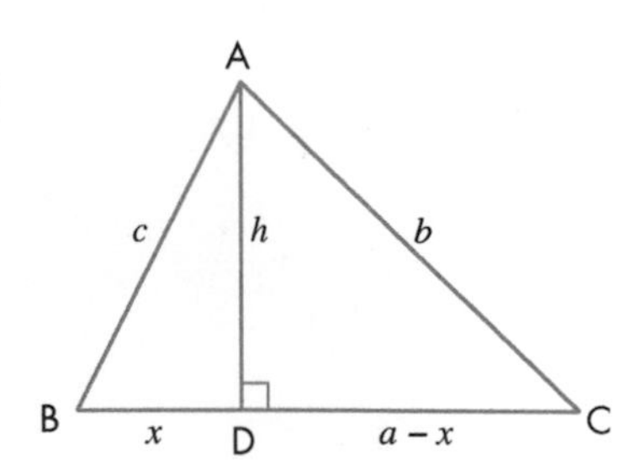

Using Pythagoras' theorem on triangle BDA:

$h^2 = c^2 - x^2$

Using Pythagoras' theorem on triangle ADC:

$h^2 = b^2 - (a - x)^2$

Therefore:

$c^2 - x^2 = b^2 - (a - x)^2 = b^2 - a^2 + 2ax - x^2 \Rightarrow c^2 = b^2 - a^2 + 2ax$

From triangle BDA, $x = c \cos B$.

$c^2 = b^2 - a^2 + 2ac \cos B \Rightarrow b^2 = a^2 + c^2 - 2ac \cos B$

By algebraic symmetry:

$a^2 = b^2 + c^2 - 2bc \cos A$ and $c^2 = a^2 + b^2 - 2ab \cos C$

This is the cosine rule.

The formula can be rearranged to find any of the three angles.

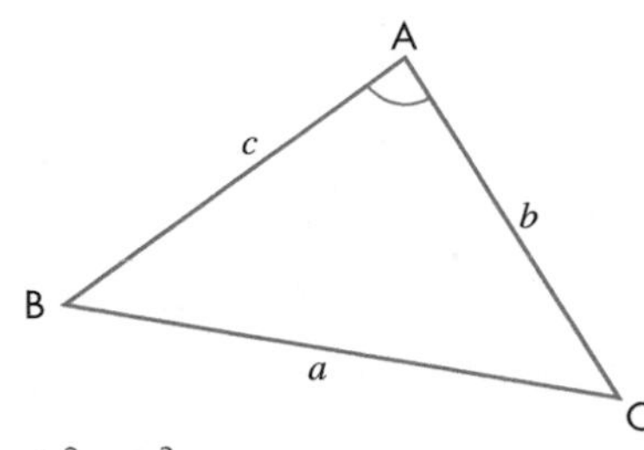

$$\cos A = \frac{b^2 + c^2 - a^2}{2bc} \qquad \cos B = \frac{a^2 + c^2 - b^2}{2ac} \qquad \cos C = \frac{a^2 + b^2 - c^2}{2ab}$$

EXAMPLE 19

Find the value of x in this triangle.

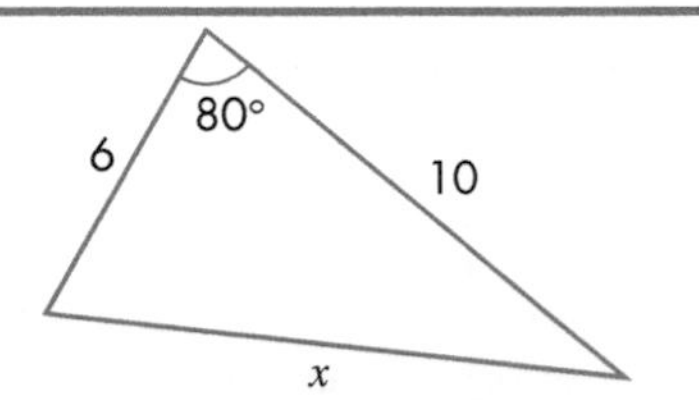

By the cosine rule:

$x^2 = 6^2 + 10^2 - 2 \times 6 \times 10 \times \cos 80° = 115.16$

$\Rightarrow x = 10.7$ (3 significant figures)

EXAMPLE 20

Find the value of x in this triangle.

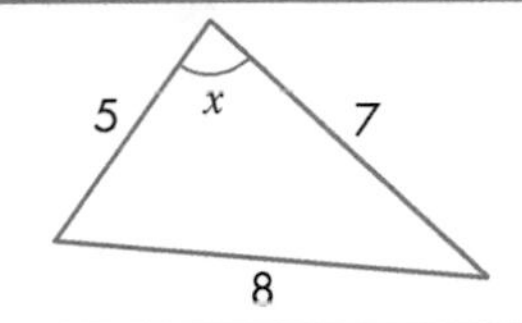

By the cosine rule:

$\cos x = \dfrac{5^2 + 7^2 - 8^2}{2 \times 5 \times 7} = 0.1428 \Rightarrow x = 81.8°$ (3 significant figures)

EXERCISE 9L

GRADE A

1. In each triangle, find the length of the side marked x.

a)

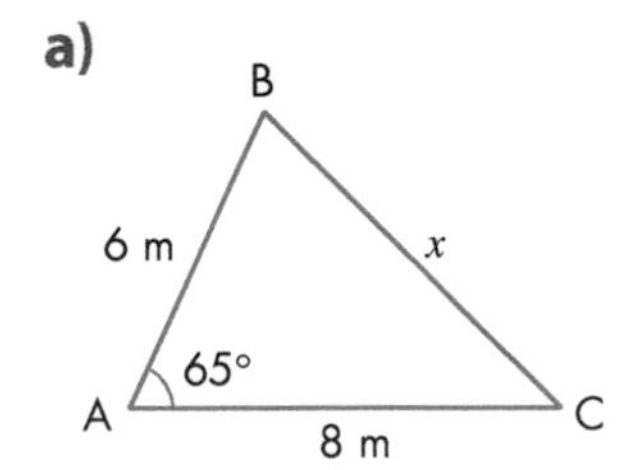

b)

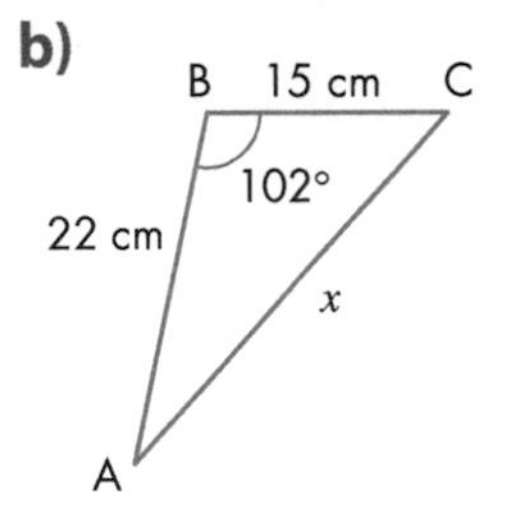

c)

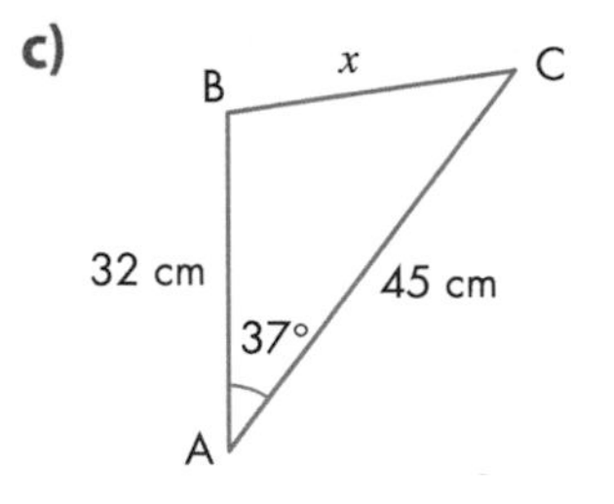

2. a) In each triangle, find the size of the angle marked x.

i)

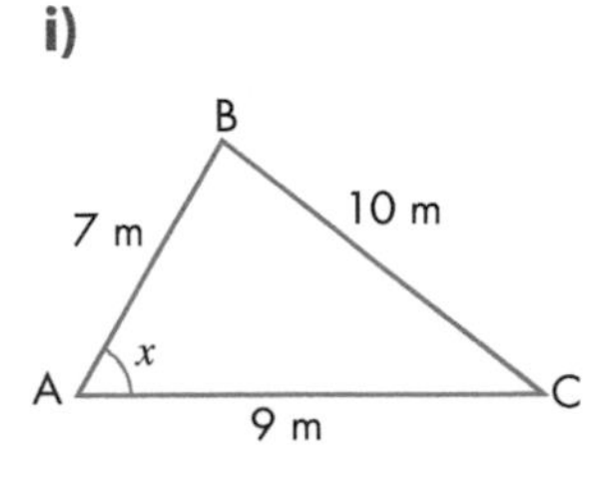

ii)

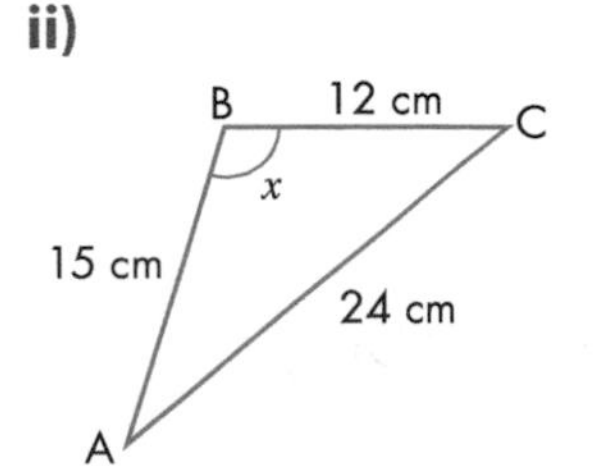

iii)

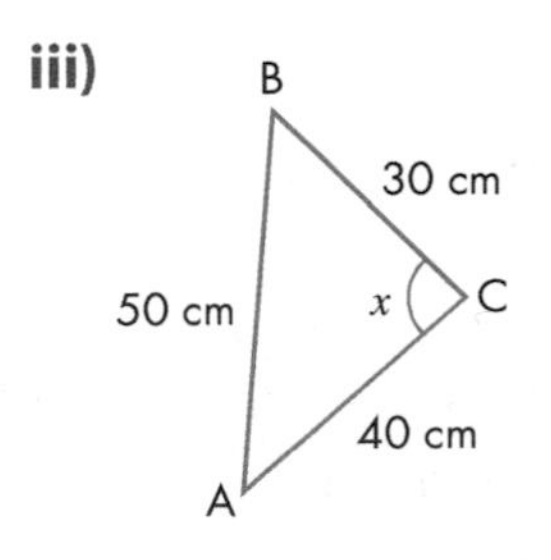

b) Explain the significance of the answer to part **a iii**.

GRADE A*

3. The diagram shows a trapezium ABCD.

   AB = 6.7 cm, AD = 7.2 cm, CB = 9.3 cm and angle DAB = 100°. Calculate:

   a) the length DB **b)** angle DBA

   c) angle DBC **d)** the length DC

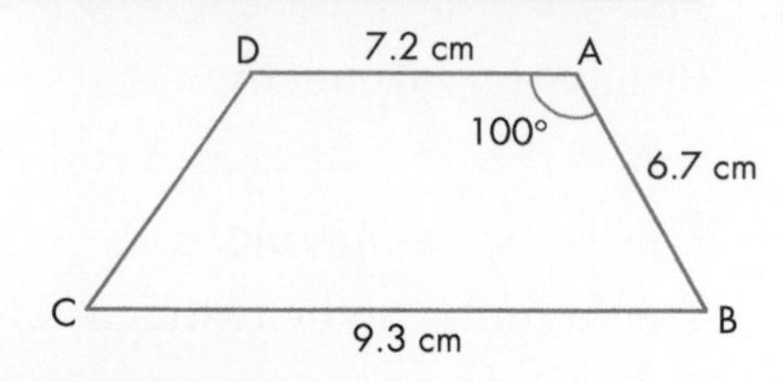

4. A ship sails from a port on a bearing of 050° for 50 km then turns onto a bearing of 150° for 40 km. A crewman is taken ill, so the ship drops anchor. What course and distance should a rescue helicopter from the port fly to reach the ship in the shortest possible time?

5. The three sides of a triangle are given as $3a$, $5a$ and $7a$. Calculate the smallest angle in the triangle.

6. Two ships, X and Y, leave a port at 9 am. Ship X travels at an average speed of 20 km/h on a bearing of 075° from the port. Ship Y travels at an average speed of 25 km/h on a bearing of 130° from the port.

   Calculate the distance between the two ships at 11 am.

7. Calculate the size of the largest angle in the triangle ABC.

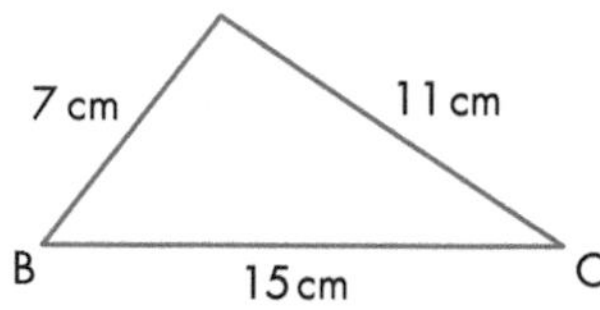

Choosing the correct rule

When you are solving a triangle, by calculating unknown measurements, there are three sets of information you may be given.

Two sides and the included angle

1. Use the cosine rule to find the third side.
2. Use the sine rule to find either of the other angles.
3. Use the sum of the angles in a triangle to find the third angle.

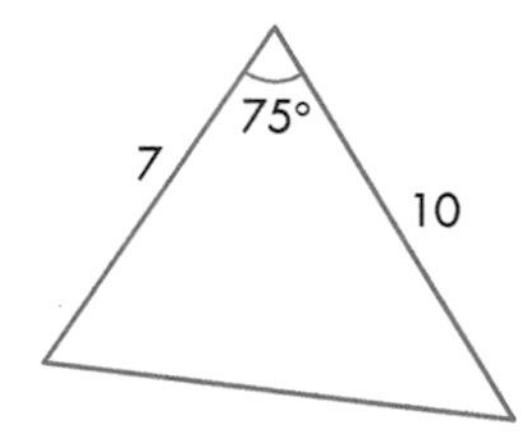

Two angles and a side

1. Use the sum of the angles in a triangle to find the third angle.

2, 3. Use the sine rule to find the other two sides.

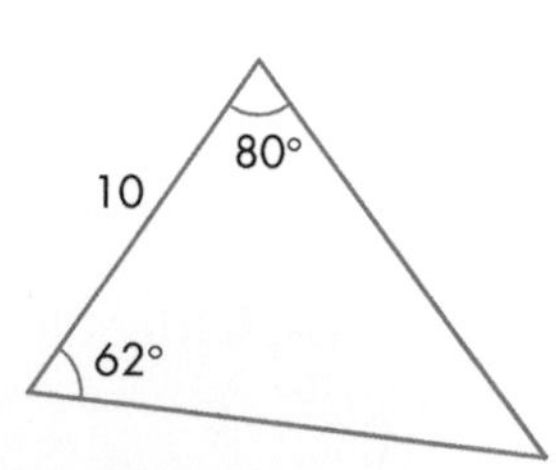

Three sides

1. Use the cosine rule to find one angle.
2. Use the sine rule to find another angle.
3. Use the sum of the angles in a triangle to find the third angle.

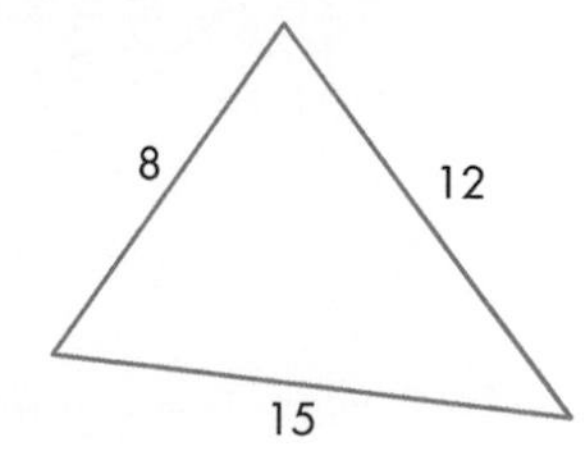

EXERCISE 9M

GRADE A

1. In each triangle, find the value of the length or angle marked x.

a)

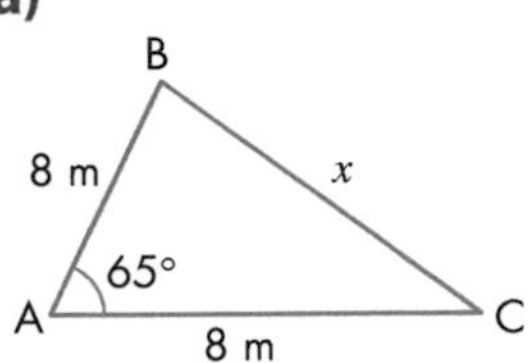

b)

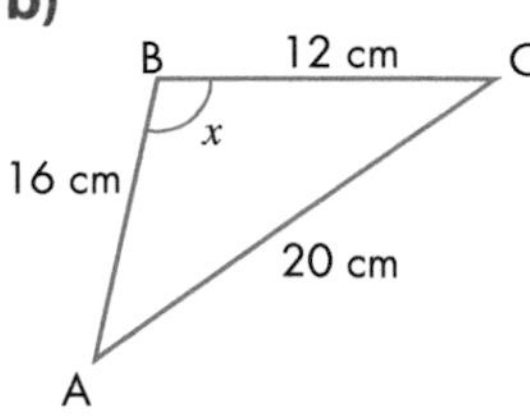

c)

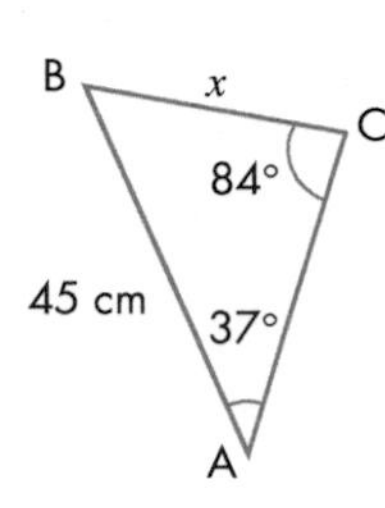

d)

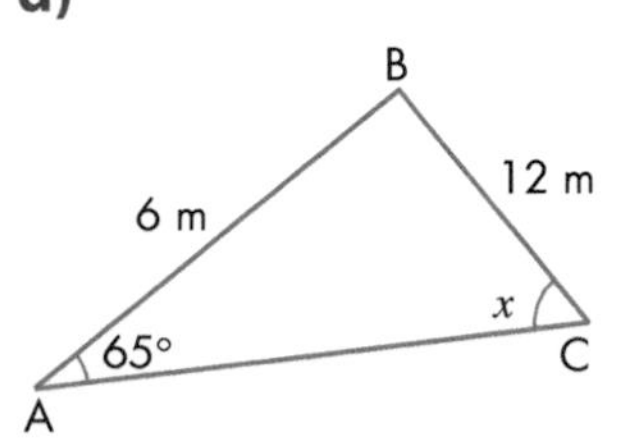

e)

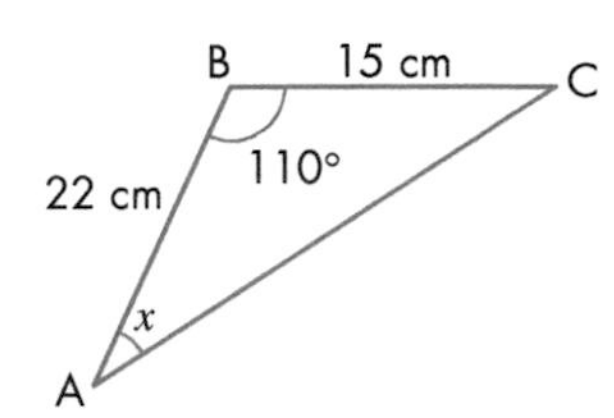

f)

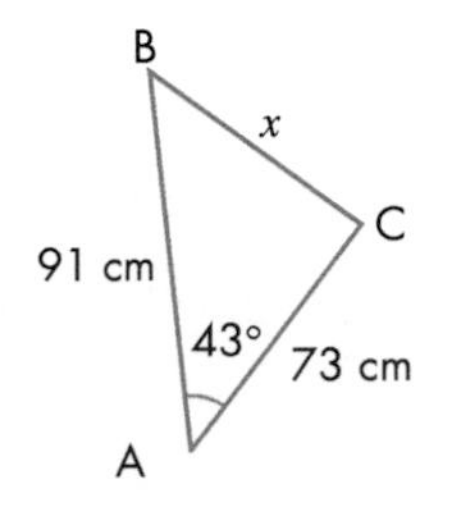

2. The hands of a clock have lengths 3 cm and 5 cm. Find the distance between the tips of the hands at 4 o'clock.

3. A spacecraft is seen hovering at a point which is in the same vertical plane as two towns, X and F, which are on the same level. Its distances from X and F are 8.5 km and 12 km respectively. The angle of elevation of the spacecraft when observed from F is 43°.

Calculate the distance between the two towns.

GRADE A*

4. Triangle ABC has sides with lengths a, b and c, as shown in the diagram.

a) What can you say about the angle BAC, if $b^2 + c^2 - a^2 = 0$?

b) What can you say about the angle BAC, if $b^2 + c^2 - a^2 > 0$?

c) What can you say about the angle BAC, if $b^2 + c^2 - a^2 < 0$?

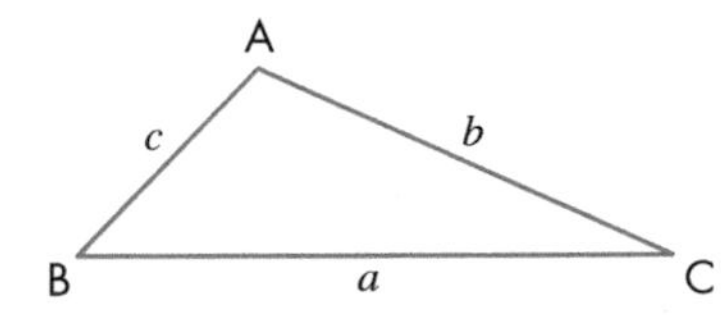

5. The diagram shows a sketch of a field ABCD.

A farmer wants to put a new fence round the perimeter of the field.

Calculate the perimeter of the field.

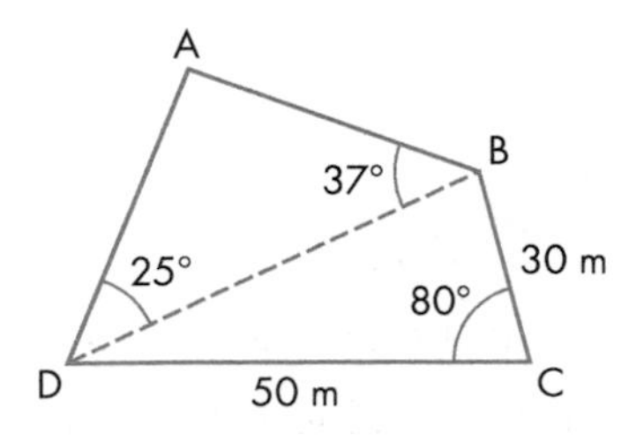

Exam-style questions

Do not use a calculator for questions 1 and 2.

1. In triangle ABC, the angle ADC is a right angle. GRADE B

AB = 13 cm, AD = 12 cm and AC = 15 cm.

Work out the area of triangle ABC.

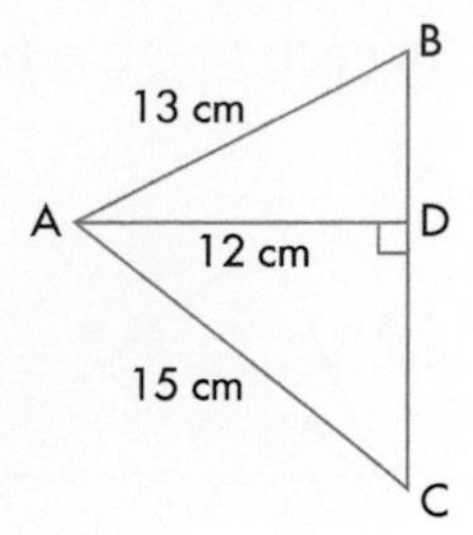

2. ABD and DBC are right-angled triangles.

Angle ABD = angle DBC = 30°.

The length of BC is 6 cm.

Work out the length of AB. GRADE B

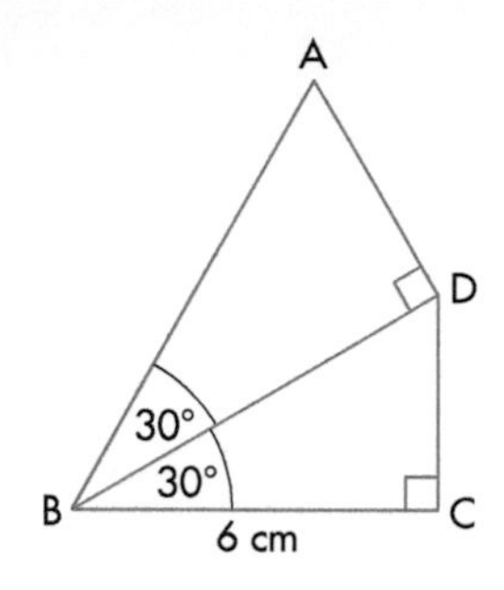

3. In triangle ABC, AC = AB = 4 cm and CB = 5 cm.

Work out the exact value of cos A. GRADE B

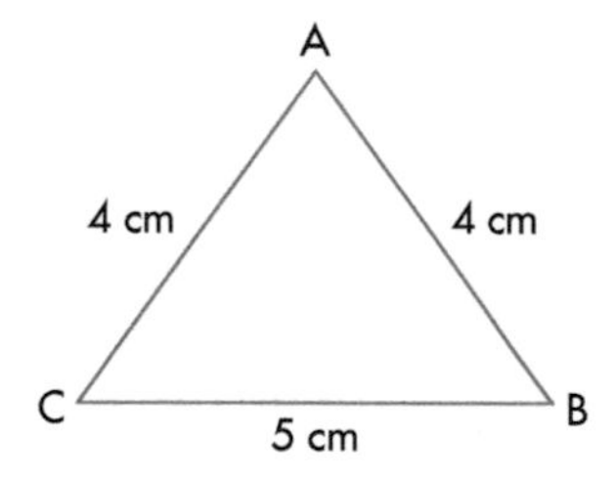

4. The lengths of the sides of a triangle are shown on the diagram. GRADE A

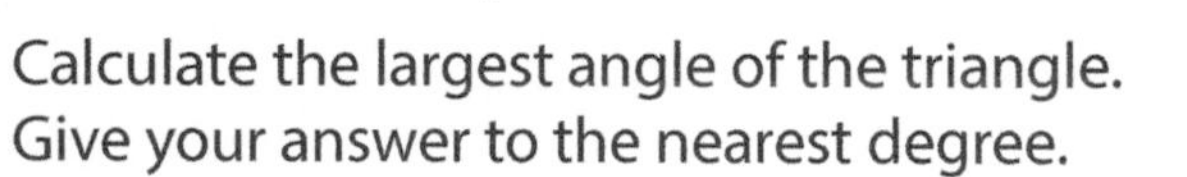
Calculate the largest angle of the triangle.
Give your answer to the nearest degree.

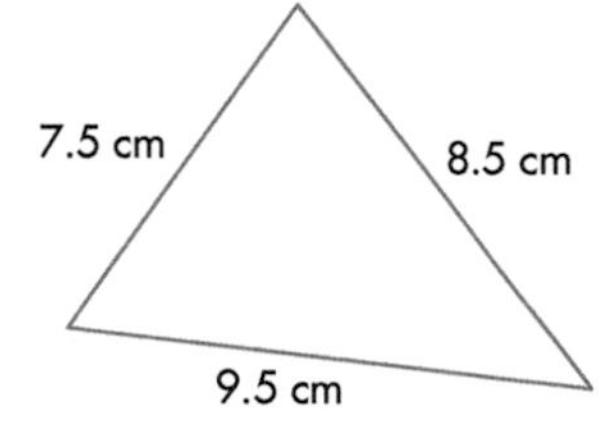

5. ABCD is a parallelogram with sides of length 8 cm and 12 cm.

Angle CAD = 25°

a) Calculate the size of angle ACD.

b) Calculate the length of the diagonal BD. GRADE A

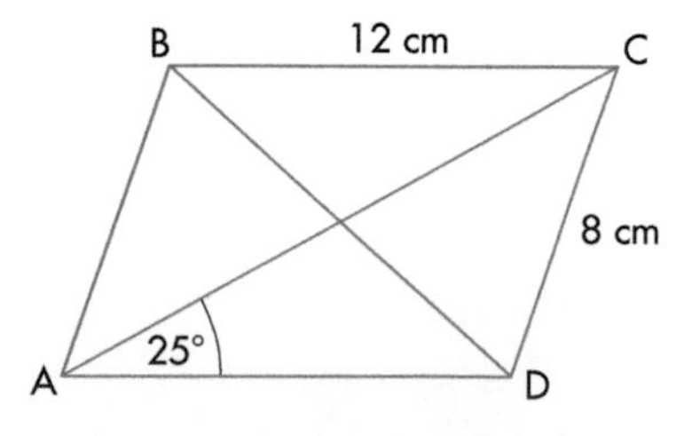

6. In triangle ABC, AB = 5 cm, BC = 6 cm and angle ABC = 55°.
Find the length of AC. GRADE A

7. A triangle has two sides of length 40 cm and an angle of 110°.
Work out the length of the third side of the triangle. GRADE A

8. The diagram shows an equilateral triangle of side 15 cm with a circle passing through the three vertices.

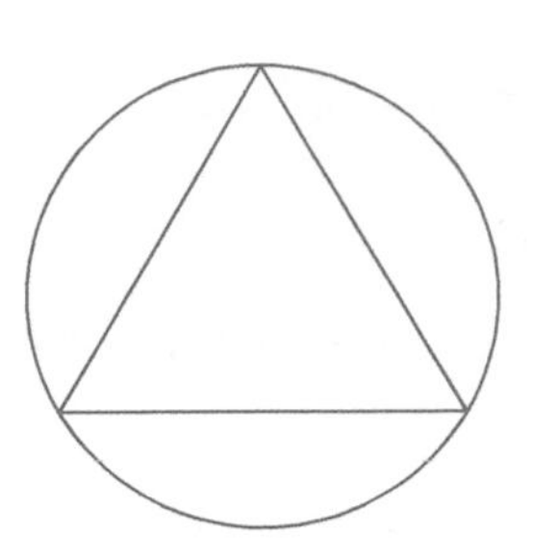

Calculate the diameter of the circle. GRADE A

9. A ship sails 30 km from X on a bearing of 050° and then 50 km on a bearing of 140° to arrive at Y.

Calculate the distance from X to Y. GRADE A

10. XC is a vertical flagpole standing on level ground.

C, A and B are three points on the ground and CAB is a straight line.

A and B are 20 m apart.

The angle of elevation of the top of the flagpole from A is 25°.

The angle of elevation of the top of the flagpole from B is 17°.

Calculate the height of the flagpole. GRADE A

11. ABCD is the top face of a cuboid and EFGH is the base.

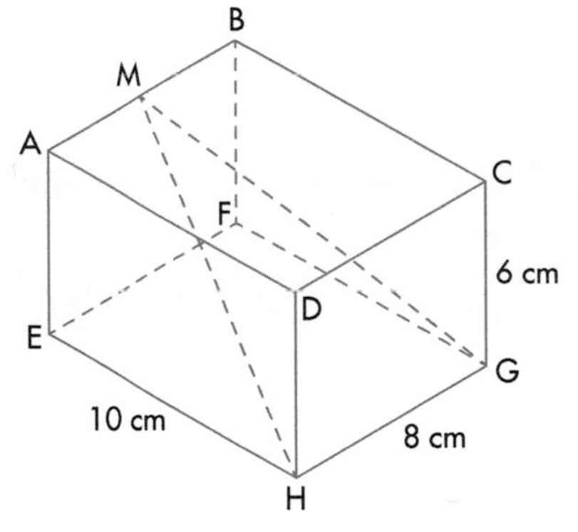

The lengths of the sides of the cuboid are 6 cm, 8 cm and 10 cm.

M is the midpoint of AB.

a) Calculate the length of MH.

b) Calculate the angle between MH and the base EFGH.

c) Calculate the angle between triangle MGH and face CDHG. GRADE A*

10 Solving equations

10.1 Solving linear equations

THIS SECTION WILL SHOW YOU HOW TO:

✓ solve linear equations

KEY WORDS

✓ equation
✓ variable
✓ solution
✓ solve
✓ brackets

A teacher gave these instructions to her class.

What algebraic expression represents the teacher's statement?

- Think of a number.
- Double it.
- Add 3.

Teacher

This is what two of her students said.

I chose the number 5.

Kim

My final answer was10.

Freda

Can you work out Kim's answer and the number that Freda started with?

Kim's answer will be:

$2 \times 5 + 3 = 13$.

Freda's answer can be set up as an **equation**.

An equation is formed when an expression is put equal to a number or another expression.

You are expected to deal with equations that have only one **variable** or letter.

The **solution** to an equation is the value of the variable that makes the equation true.

For example, the equation for Freda's answer is:

$2x + 3 = 10$

where x represents Freda's number.

The value of x that makes this true is $x = 3\frac{1}{2}$.

To **solve** an equation like $2x + 3 = 10$, do the same thing to each side of the equation to get x on its own.

$$2x + 3 = 10$$

Subtract 3 from both sides: $2x + 3 - 3 = 10 - 3 \Rightarrow 2x = 7$

Divide both sides by 2: $\frac{2x}{2} = \frac{7}{2} \quad \Rightarrow x = 3\frac{1}{2}$

EXAMPLE 1

Solve each of these equations by 'doing the same to both sides'.

a) $3x - 5 = 16$

Add 5 to both sides:

$$3x - 5 + 5 = 16 + 5$$

$$3x = 21$$

Divide both sides by 3:

$$\frac{3x}{3} = \frac{21}{3}$$

$$x = 7$$

Checking the answer gives:

$$3 \times 7 - 5 = 16$$

which is correct.

b) $\frac{x}{2} + 2 = 10$

Subtract 2 from both sides:

$$\frac{x}{2} + 2 - 2 = 10 - 2$$

$$\frac{x}{2} = 8$$

Multiply both sides by 2:

$$\frac{x}{2} \times 2 = 8 \times 2$$

$$x = 16$$

Checking the answer gives:

$$16 \div 2 + 2 = 10$$

which is correct.

Brackets

When you have an equation that contains **brackets**, you first must multiply out the brackets and then solve the resulting equation.

EXAMPLE 2

Solve $5(x + 3) = 25$

First multiply out the brackets to get:

$$5x + 15 = 25$$

Subtract 15: $5x = 25 - 15 = 10$

Divide by 5: $\frac{5x}{5} = \frac{10}{5} \Rightarrow x = 2$

An alternative method is to divide by the number outside the brackets.

Equations with the variable on both sides

When a variable appears on both sides of an equation, collect all the terms containing that variable on the left-hand side of the equation. But when there are more of the letters on the right-hand side, it is easier to turn the equation round. When an equation contains brackets, they must be multiplied out first.

EXAMPLE 3

Solve this equation. $5x + 4 = 3x + 10$

There are more xs on the left-hand side, so leave the equation as it is.

Subtract $3x$ from both sides: $2x + 4 = 10$

Subtract 4 from both sides: $2x = 6$

Divide both sides by 2: $x = 3$

EXAMPLE 4

Solve this equation. $3(2x + 5) + x = 2(2 - x) + 2$

Multiply out both brackets: $6x + 15 + x = 4 - 2x + 2$

Simplify both sides: $7x + 15 = 6 - 2x$

There are more xs on the left-hand side, so leave the equation as it is.

Add $2x$ to both sides: $9x + 15 = 6$

Subtract 15 from both sides: $9x = -9$

Divide both sides by 9: $x = -1$

EXERCISE 10A

GRADE C

1. Solve each equation by 'doing the same to both sides'. Check that each answer works for its original equation.

a) $3x - 7 = 11$ **b)** $5y + 3 = 18$ **c)** $7 + 3t = 19$ **d)** $5 + 4f = 15$

e) $3 + 6k = 24$ **f)** $4x + 7 = 17$ **g)** $\frac{m}{7} - 3 = 5$ **h)** $\frac{x}{5} + 3 = 3$

i) $\frac{h}{7} + 2 = 1$ **j)** $\frac{w}{3} + 10 = 4$ **k)** $\frac{x}{3} - 5 = 7$ **l)** $\frac{y}{2} - 13 = 5$

m) $\frac{2x + 1}{3} = 5$ **n)** $\frac{6y + 3}{9} = 1$ **o)** $\frac{2x - 3}{5} = 4$ **p)** $\frac{5t + 3}{4} = 1$

2. Solve each equation.

a) $2(x + 5) = 16$ **b)** $5(x - 3) = 20$ **c)** $3(t + 1) = 18$

d) $2(3x + 1) = 11$ **e)** $4(5y - 2) = 42$ **f)** $6(3k + 5) = 39$

g) $5(x - 4) = -25$ **h)** $3(t + 7) = 15$ **i)** $2(3x + 11) = 10$

3. Fill in values for a, b and c so that the answer to this equation is $x = 4$.

$a(bx + 3) = c$

4. Solve each equation.

a) $2x + 3 = x + 5$ **b)** $5y + 4 = 3y + 6$ **c)** $4a - 3 = 3a + 4$

d) $6k + 5 = 2k + 1$ **e)** $4m + 1 = m + 10$ **f)** $8s - 1 = 6s - 5$

5. Solve each equation.

a) $2(d + 3) = d + 12$ **b)** $5(x - 2) = 3(x + 4)$

c) $3(2y + 3) = 5(2y + 1)$ **d)** $3(h - 6) = 2(5 - 2h)$

e) $4(3b - 1) + 6 = 5(2b + 4)$ **f)** $2(5c + 2) - 2c = 3(2c + 3) + 7$

6. Explain why the equation $3(2x + 1) = 2(3x + 5)$ cannot be solved.

7. Explain why these are an infinite number of solutions to the equation:

$2(6x + 9) = 3(4x + 6)$

10.2 Setting up equations

THIS SECTION WILL SHOW YOU HOW TO …

✓ set up and solve linear equations
✓ solve more complex equations

Equations are used to represent situations, so that you can solve real-life problems. Many real-life problems can be solved by setting them up as linear equations and then solving the equation.

EXAMPLE 5

The rectangle shown has a perimeter of 40 cm.

Find the value of x.

$3x + 1$

$x + 3$

The perimeter of the rectangle is:

$$3x + 1 + x + 3 + 3x + 1 + x + 3 = 40$$

This simplifies to: $8x + 8 = 40$

Subtract 8 from both sides: $8x = 32$

Divide both sides by 8: $x = 4$

EXERCISE 10B

GRADE C

Set up an equation to represent each situation described below. Then solve the equation.

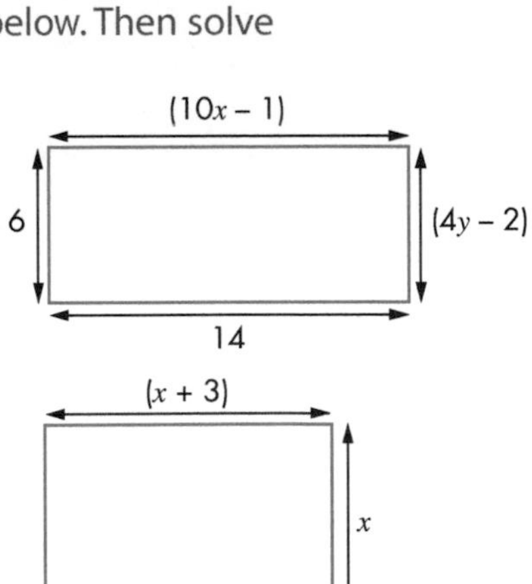

1. The diagram shows a rectangle.

   a) What is the value of x?

   b) What is the value of y?

2. In this rectangle, the length is 3 cm more than the width. The perimeter is 12 cm.

   a) What is the value of x?

   b) What is the area of the rectangle?

3. A rectangular room is 3 m longer than it is wide. The perimeter is 16 m.

   Floor tiles cost 9 dollars per square metre. How much will it cost to cover the floor?

4. A boy is Y years old. His father is 25 years older than he is. The sum of their ages is 31. How old is the boy?

5. Another boy is Y years old. His sister is twice as old as he is. The sum of their ages is 27. How old is the boy?

6. The diagram shows a square.

Find the value of x if the perimeter is 44 cm.

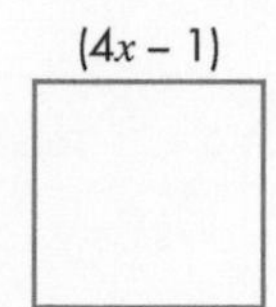

7. Max thought of a number. He then multiplied his number by 3. He added 4 to the answer. He then doubled that answer to get a final value of 38. What number did he start with?

8. The angles of a triangle, in degrees, are $2x$, $x+5$ and $x+35$.

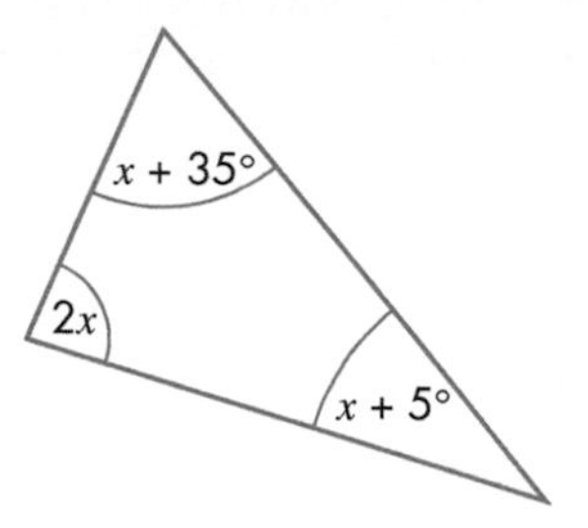

a) Write down an equation to show that the angles add up to 180 degrees.

b) Solve your equation to find the value of x.

9. Five friends went for a meal in a restaurant. The bill was \$$x$. They decided to add a \$10 tip and split the bill between them. Each person paid \$9.50.

a) Set up this problem as an equation.

b) Solve the equation to work out the bill before the tip was added.

10. A teacher asked her class to find three angles of a triangle that were consecutive even numbers.

Tammy wrote: $x+(x+2)+(x+4)=180$

$$3x+6=180$$
$$3x=174$$
$$x=58$$

So the angles are 58°, 60° and 62°.

The teacher then asked the class to find four angles of a quadrilateral that are consecutive even numbers.

Can this be done? Explain your answer.

HINTS AND TIPS

Use the same type of working as Tammy did for a triangle. Work out the value of x. What happens?

11. Maria has a large and a small bottle of cola. The large bottle holds 50 cl more than the small bottle.

From the large bottle she fills four cups and has 18 cl left over.

From the small bottle she fills three cups and has 1 cl left over.

How much does each bottle hold?

HINTS AND TIPS

Set up equations for both using x as the amount of cola in a cup. Put them equal but remember to add 50 to the small bottle equation to allow for the difference. Solve for x, then work out how much is in each bottle.

More complex equations

More complicated equations will require a number of steps to reach a solution. Often they can be solved in a number of different ways.

EXAMPLE 6

Solve the equation $\frac{x+1}{3} - \frac{x-3}{2} = x - 4$

Method 1

Write the expression on the left as a single fraction:

$$\frac{2(x+1) - 3(x-3)}{3 \times 2} = x - 4$$

Multiply by 6: $2(x+1) - 3(x-3) = 6(x-4)$

Remove the brackets: $2x + 2 - 3x + 9 = 6x - 24$

$11 - x = 6x - 24$

Rearrange: $35 = 7x \Rightarrow x = 5$

Method 2

The LCM of 2 and 3 is 6, so multiply throughout by 6.

$$6 \times \frac{x+1}{3} - 6 \times \frac{x-3}{2} = 6(x-4) \Rightarrow 2(x+1) - 3(x-3) = 6(x-4)$$

This gets rid of the fractions. Now continue as before.

EXERCISE 10C

GRADE B

1. Solve these equations.

a) $\frac{3x+5}{2} = x+6$ **b)** $\frac{12-x}{3} = x-8$ **c)** $\frac{2x-3}{2} = \frac{3x+8}{4}$

d) $\frac{3x+1}{2} = \frac{9x-5}{5}$ **e)** $\frac{6x+5}{x} = 8$ **f)** $10 - x = \frac{18-3x}{2}$

2. Solve these equations.

a) $\frac{x+1}{2} + \frac{x+2}{5} = 3$ **b)** $\frac{x+2}{4} + \frac{x+1}{7} = 3$ **c)** $\frac{4x+1}{3} - \frac{x+2}{4} = 2$

d) $\frac{2x-1}{3} + \frac{3x+1}{4} = 7$ **e)** $\frac{2x+1}{2} + \frac{x+1}{7} = 1$ **f)** $\frac{3x+1}{5} - \frac{5x-1}{7} = 0$

3. Solve these equations.

a) $\frac{x}{3} + \frac{x}{4} = \frac{x+1}{2}$ **b)** $\frac{12-x}{2} = \frac{11-x}{3}$ **c)** $\frac{x+1}{4} + \frac{x+2}{3} = 12 - x$

4. The perimeter of this triangle is 18.

Calculate the length of each side.

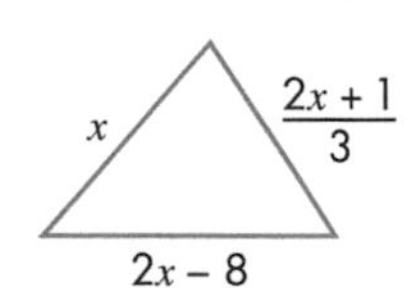

5. The angles of a triangle are x, $\frac{2x+10}{2}$, and $\frac{3x}{2}$ degrees.

Use the fact that the angles of a triangle add up to 180 degrees to calculate the size of each angle.

10.3 Solving quadratic equations by factorisation or the quadratic formula

THIS SECTION WILL SHOW YOU HOW TO ...

✓ solve quadratic equations by factorisation or by using the quadratic formula

KEY WORDS

✓ linear equation
✓ quadratic expression
✓ quadratic equation
✓ factor
✓ quadratic formula
✓ soluble
✓ coefficient
✓ constant

Solving quadratic equations by factorisation

All the equations in this chapter so far have been **linear equations**. We shall now look at equations which involve **quadratic expressions** such as $x^2 - 2x - 3$ which contain the square of the variable.

Solving the quadratic equation $x^2 + ax + b = 0$

To solve a **quadratic equation** such as $x^2 - 2x - 3 = 0$, you first have to be able to factorise it. Work through the examples below to see how this is done.

EXAMPLE 7

Solve $x^2 + 6x + 5 = 0$

This factorises into $(x + 5)(x + 1) = 0$.

The only way this expression can equal 0 is if the value of the expression in one of the brackets is 0.

Hence either $(x + 5) = 0$ or $(x + 1) = 0$

$\Rightarrow \quad x + 5 = 0$ or $x + 1 = 0$

$\Rightarrow \quad x = -5$ or $x = -1$

So the solution is $x = -5$ or $x = -1$.

There are *two* possible values for x.

EXAMPLE 8

Solve $x^2 - 6x + 9 = 0$

This factorises into $(x - 3)(x - 3) = 0$.

That is: $(x - 3)^2 = 0$

Hence, there is only one solution, $x = 3$.

EXERCISE 10D

GRADE B

Solve the equations in questions **1–12**.

1. $(x+2)(x+5)=0$
2. $(t+3)(t+1)=0$
3. $(a+6)(a+4)=0$
4. $(x+3)(x-2)=0$
5. $(x+1)(x-3)=0$
6. $(t+4)(t-5)=0$
7. $(x-1)(x+2)=0$
8. $(x-2)(x+5)=0$
9. $(a-7)(a+4)=0$
10. $(x-3)(x-2)=0$
11. $(x-1)(x-5)=0$
12. $(a-4)(a-3)=0$

First factorise, then solve the equations in questions **13–26**.

13. $x^2+5x+4=0$
14. $x^2+11x+18=0$
15. $x^2-6x+8=0$
16. $x^2-8x+15=0$
17. $x^2-3x-10=0$
18. $x^2-2x-15=0$
19. $t^2+4t-12=0$
20. $t^2+3t-18=0$
21. $x^2-x-2=0$
22. $x^2+4x+4=0$
23. $m^2+10m+25=0$
24. $t^2-8t+16=0$
25. $t^2+8t+12=0$
26. $a^2-14a+49=0$

27. A woman is x years old. Her husband is three years younger. The product of their ages is 550.
a) Set up a quadratic equation to represent this situation.
b) How old is the woman?

28. A rectangular field is 40 m longer than it is wide. The area is 48 000 square metres. The farmer wants to place a fence all around the field. How long will the fence be?

First rearrange the equations in questions **29–37**, then solve them.

29. $x^2+10x=-24$
30. $x^2-18x=-32$
31. $x^2+2x=24$
32. $x^2+3x=54$
33. $t^2+7t=30$
34. $x^2-7x=44$
35. $t^2-t=72$
36. $x^2=17x-72$
37. $x^2+1=2x$

Solving the general quadratic equation by factorisation

The general quadratic equation is of the form $ax^2+bx+c=0$ where a, b and c are positive or negative whole numbers. (It is easier to make sure that a is always positive.) Before you can solve any quadratic equation by factorisation, you must rearrange it to this form.

The method is similar to that used to solve equations of the form $x^2+ax+b=0$. That is, you have to find two **factors** of ax^2+bx+c with a product of 0.

EXAMPLE 9

Solve the quadratic equation $12x^2-28x=-15$.

First, rearrange the equation to the general form.

$12x^2-28x+15=0$

This factorises into $(2x-3)(6x-5)=0$.

The only way the product can equal 0 is if the value of the expression in one of the brackets is 0. Hence:

either $2x - 3 = 0$ or $6x - 5 = 0$

$\Rightarrow 2x = 3$ or $6x = 5$

$\Rightarrow x = \frac{3}{2}$ or $x = \frac{5}{6}$

So the solution is $x = 1\frac{1}{2}$ or $x = \frac{5}{6}$

Note: It is more accurate to give your answer as a fraction when it is appropriate. If you have to round a solution it becomes less accurate.

Special cases

Sometimes the values of b and c are zero. (Note that if a is zero the equation is no longer a quadratic equation but a linear equation.)

EXAMPLE 10

Solve these quadratic equations. **a)** $3x^2 - 4 = 0$ **b)** $6x^2 - x = 0$

a) Rearrange to get $3x^2 = 4$.

Divide both sides by 3: $x^2 = \frac{4}{3}$

Take the square root on both sides: $x = \pm\sqrt{\frac{4}{3}} = \pm\frac{2}{\sqrt{3}} = \pm\frac{2\sqrt{3}}{3}$

Note: A square root can be positive or negative. The symbol ± indicates that the square root has a positive and a negative value, *both* of which must be used in solving for x.

b) There is a common factor of x, so factorise as $x(6x - 1) = 0$.

There is only one set of brackets this time but each factor can be equal to zero, so $x = 0$ or $6x - 1 = 0$.

Hence, $x = 0$ or $\frac{1}{6}$.

EXERCISE 10E

GRADE A

Give your answers either in rational form or as mixed numbers.

1. Solve these equations.

a) $3x^2 + 8x - 3 = 0$ **b)** $6x^2 - 5x - 4 = 0$ **c)** $5x^2 - 9x - 2 = 0$

d) $4t^2 - 4t - 35 = 0$ **e)** $18t^2 + 9t + 1 = 0$ **f)** $3t^2 - 14t + 8 = 0$

g) $6x^2 + 15x - 9 = 0$ **h)** $12x^2 - 16x - 35 = 0$ **i)** $15t^2 + 4t - 35 = 0$

j) $28x^2 - 85x + 63 = 0$ **k)** $24x^2 - 19x + 2 = 0$ **l)** $16t^2 - 1 = 0$

m) $4x^2 + 9x = 0$ **n)** $25t^2 - 49 = 0$ **o)** $9m^2 - 24m - 9 = 0$

2. Rearrange these equations into the general form and then solve them.

a) $x^2 - x = 12$ **b)** $8x(x + 1) = 30$ **c)** $(x + 1)(x - 2) = 40$

d) $13x^2 = 11 - 2x$ **e)** $(x + 1)(x - 2) = 4$ **f)** $10x^2 - x = 2$

g) $8x^2 + 6x + 3 = 2x^2 + x + 2$ **h)** $25x^2 = 10 - 45x$ **i)** $8x - 16 - x^2 = 0$

j) $(2x + 1)(5x + 2) = (2x - 2)(x - 2)$ **k)** $5x + 5 = 30x^2 + 15x + 5$

l) $2m^2 = 50$ **m)** $6x^2 + 30 = 5 - 3x^2 - 30x$

n) $4x^2 + 4x - 49 = 4x$ **o)** $2t^2 - t = 15$

3. Here are three equations.

A: $(x - 1)^2 = 0$ B: $3x + 2 = 5$ C: $x^2 - 4x = 5$

a) Give some mathematical fact that equations A and B have in common.

b) Give a mathematical reason why equation B is different from equations A and C.

4. Pythagoras' theorem states that the sum of the squares of the two short sides of a right-angled triangle equals the square of the long side (hypotenuse).

A right-angled triangle has sides of length $5x - 1$, $2x + 3$ and $x + 1$ cm.

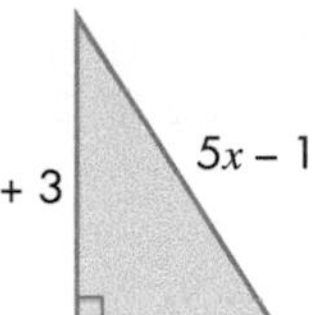

a) Show that $20x^2 - 24x - 9 = 0$.

b) Find the area of the triangle.

Solving quadratic equations by the quadratic formula

Many quadratic equations cannot be solved by factorisation because they do not have simple factors. Try to factorise, for example, $x^2 - 4x - 3 = 0$ or $3x^2 - 6x + 2 = 0$. You will find it is impossible.

One way to solve this type of equation is to use the **quadratic formula**. This formula can be used to solve *any* quadratic equation that can be solved (is **soluble**).

The solution of the equation $ax^2 + bx + c = 0$ is given by:

$$x = \frac{-b \pm \sqrt{b^2 - 4ac}}{2a}$$

where a and b are the **coefficients** of x^2 and x respectively and c is the **constant** term.

Real solutions

If $b^2 - 4ac > 0$ the quadratic equation will have two real solutions.

If $b^2 - 4ac = 0$ the quadratic equation will have one solution.

If $b^2 - 4ac < 0$ the quadratic equation will have no real solutions.

EXAMPLE 11

Solve $5x^2 - 11x - 4 = 0$, giving solutions correct to 2 decimal places.

Take the quadratic formula:

$$x = \frac{-b \pm \sqrt{b^2 - 4ac}}{2a}$$

and put $a = 5$, $b = -11$ and $c = -4$, which gives:

$$x = \frac{-(11) \pm \sqrt{(-11)^2 - 4(5)(-4)}}{2(5)}$$

Note that the values for a, b and c have been put into the formula in brackets.

This is to avoid mistakes in calculation. It is a very common mistake to get the sign of b wrong or to think that -11^2 is -121. Using brackets will help you do the calculation correctly.

$$x = \frac{11 \pm \sqrt{121 + 80}}{10} = \frac{11 \pm \sqrt{201}}{10} \Rightarrow x = 2.52 \text{ or } -0.32$$

Note: The calculation has been done in stages. With a calculator it is possible just to work out the answer, but make sure you can use your calculator properly. If not, break the calculation down. Remember the rule 'if you try to do two things at once, you will probably get one of them wrong'.

Examination tip: If you are asked to solve a quadratic equation to one or two decimal places, you can be sure that it can be solved *only* by the quadratic formula.

EXERCISE 10F

GRADE A

Use the quadratic formula to solve the equations in questions **1–15**. Give your answers to 2 decimal places.

1. $2x^2 + x - 8 = 0$

2. $3x^2 + 5x + 1 = 0$

3. $x^2 - x - 10 = 0$

4. $5x^2 + 2x - 1 = 0$

5. $7x^2 + 12x + 2 = 0$

6. $3x^2 + 11x + 9 = 0$

7. $4x^2 + 9x + 3 = 0$

8. $6x^2 + 22x + 19 = 0$

9. $x^2 + 3x - 6 = 0$

10. $3x^2 - 7x + 1 = 0$

11. $2x^2 + 11x + 4 = 0$

12. $4x^2 + 5x - 3 = 0$

13. $4x^2 - 9x + 4 = 0$

14. $7x^2 + 3x - 2 = 0$

15. $5x^2 - 10x + 1 = 0$

16. A rectangular lawn is 2 m longer than it is wide.

The area of the lawn is 21 m². The gardener wants to edge the lawn with edging strips, which are sold in lengths of $1\frac{1}{2}$ m. How many will she need to buy?

17. Shaun is solving a quadratic equation, using the formula.
He correctly substitutes values for a, b and c to get:
$x = \frac{3 \pm \sqrt{37}}{2}$
What is the equation Shaun is trying to solve?

18. Explain why the equation $2x^2 + 4x + 3 = 0$ has no real solutions.

19. Work out the number of real solutions for each equation.
a) $x^2 - 10x + 25 = 0$ **b)** $2x^2 + x + 1 = 0$ **c)** $3x^2 - 7x - 2 = 0$

EXAMPLE 12

Solve this equation. $\frac{3}{x-1} - \frac{2}{x+1} = 1$

Use the rule for combining fractions, and cross-multiply to take the denominator of the left-hand side to the right-hand side. Use brackets to help with expanding and to avoid problems with minus signs.

$3(x + 1) - 2(x - 1) = (x - 1)(x + 1)$

$3x + 3 - 2x + 2 = x^2 - 1$ (Right-hand side is the difference of two squares.)

Rearrange into the general quadratic form.

$x^2 - x - 6 = 0$

Factorise and solve $(x - 3)(x + 2) = 0 \Rightarrow x = 3$ or -2

EXERCISE 10G

GRADE A

1. Show that each algebraic fraction simplifies to the given expression.
   a) $\frac{2}{x+1} + \frac{5}{x+2} = 3$ simplifies to $3x^2 + 2x - 3 = 0$
   b) $\frac{4}{x-2} + \frac{7}{x+1} = 3$ simplifies to $3x^2 - 14x + 4 = 0$
   c) $\frac{3}{4x+1} - \frac{4}{x+2} = 2$ simplifies to $8x^2 + 31x + 2 = 0$
   d) $\frac{2}{2x-1} - \frac{6}{x+1} = 11$ simplifies to $22x^2 + 21x - 19 = 0$
   e) $\frac{3}{2x-1} - \frac{4}{3x-1} = 1$ simplifies to $x^2 - x = 0$

2. Solve these equations.
   a) $\frac{4}{x+1} + \frac{5}{x+2} = 2$ **b)** $\frac{18}{4x-1} - \frac{1}{x+1} = 1$ **c)** $\frac{2x-1}{2} - \frac{6}{x+1} = 1$
   d) $\frac{3}{2x-1} - \frac{4}{3x-1} = 1$

3. **a)** Solve the equation $x^2 - 7x + 10 = 0$.
   b) Use your answer to part **a** to solve the equation $x - 7\sqrt{x} + 10 = 0$.

4. **a)** Solve the equation $x^2 + x - 2 = 0$.
   b) Use your answer to part **a** to solve the equation $x + \sqrt{x} - 2 = 0$.

10.4 Solving quadratic equations by completing the square

THIS SECTION WILL SHOW YOU HOW TO ...

✓ complete the square in a quadratic expression
✓ solve quadratic equations by completing the square

KEY WORDS

✓ completing the square root

Another method for solving quadratic equations is **completing the square**. This method can be used as an alternative to the quadratic formula.

You will remember that:

$(x + a)^2 = x^2 + 2ax + a^2$

which can be rearranged to give:

$x^2 + 2ax = (x + a)^2 - a^2$

This is the principle behind completing the square.

There are three basic steps in rewriting $x^2 + px + q$ in the form $(x + a)^2 + b$.

Step 1: Ignore q and just look at the first two terms, $x^2 + px$.

Step 2: Rewrite $x^2 + px$ as $\left(x + \frac{p}{2}\right)^2 - \left(\frac{p}{2}\right)^2$.

Step 3: Bring q back to get $x^2 + px + q = \left(x + \frac{p}{2}\right)^2 - \left(\frac{p}{2}\right)^2 + q$.

EXAMPLE 13

Rewrite the following in the form $(x \pm a)^2 \pm b$.

a) $x^2 + 6x - 7$ **b)** $x^2 - 8x + 3$

a) Ignore -7 for the moment.

Rewrite $x^2 + 6x$ as $(x + 3)^2 - 9$.

(Expand $(x + 3)^2 - 9 = x^2 + 6x + 9 - 9 = x^2 + 6x$. The 9 is subtracted to get rid of the constant term when the brackets are expanded.)

Now bring the -7 back, so $x^2 + 6x - 7 = (x + 3)^2 - 9 - 7$.

Combine the constant terms to get the final answer:
$x^2 + 6x - 7 = (x + 3)^2 - 16$.

b) Ignore $+3$ for the moment.

Rewrite $x^2 - 8x$ as $(x - 4)^2 - 16$.

(Note that you still subtract $(-4)^2$, as $(-4)^2 = +16$)

Now bring the +3 back, so $x^2 - 8x + 3 = (x - 4)^2 - 16 + 3$.
Combine the constant terms to get the final answer:
$x^2 - 8x + 3 = (x - 4)^2 - 13$

EXAMPLE 14

Rewrite $x^2 + 4x - 7$ in the form $(x + a)^2 - b$.
Hence solve the equation $x^2 + 4x - 7 = 0$, giving your answers to 2 decimal places.

Note that:

$x^2 + 4x = (x + 2)^2 - 4$

So:

$x^2 + 4x - 7 = (x + 2)^2 - 4 - 7 = (x + 2)^2 - 11$

When $x^2 + 4x - 7 = 0$, you can rewrite the equations completing the square as:

$(x + 2)^2 - 11 = 0$

Rearranging gives $(x + 2)^2 = 11$.

Taking the **square root** of both sides gives $x + 2 = \pm\sqrt{11} \Rightarrow x = -2 \pm \sqrt{11}$.

This answer could be left like this, but you are asked to calculate it to 2 decimal places.

$\Rightarrow x = 1.32$ or -5.32 (to 2 decimal places)

To solve $ax^2 + bx + c = 0$ when a is not 1, start by dividing through by a.

EXAMPLE 15

Rewrite $3x^2 + 12x - 6$ in the form $a(x + b)^2 + c$.

Start by ignoring the '−6' term.

Factorise $3x^2 + 12x$ as $3(x^2 + 4x)$.

Rewrite $x^2 + 4x$ as $(x + 2)^2 - 4$.

So $3x^2 + 12x = 3(x^2 + 4x)$ which is the same as $3((x + 2)^2 - 4)$.

$3((x + 2)^2 - 4) = 3(x + 2)^2 - 12$

Now bring back the '−6' term.

Then $3x^2 + 12x - 6 \equiv 3(x + 2)^2 - 12 - 6 = 3(x + 2)^2 - 18$

EXAMPLE 16

Solve by completing the square.

$2x^2 - 6x - 7 = 0$

Divide by 2: $x^2 - 3x - 3.5 = 0$

$x^2 - 3x = (x - 1.5)^2 - 2.25$

So: $x^2 - 3x - 3.5 = (x - 1.5)^2 - 5.75$

When $x^2 - 3x - 3.5 = 0$

Then: $(x - 1.5)^2 = 5.75 \Rightarrow x - 1.5 = \pm\sqrt{5.75} \Rightarrow x = 1.5 \pm \sqrt{5.75}$.

EXERCISE 10H

GRADE A*

1. Write an equivalent expression in the form $(x \pm a)^2 - b$.

a) $x^2 + 4x$ **b)** $x^2 + 14x$ **c)** $x^2 - 6x$ **d)** $x^2 + 6x$

e) $x^2 - 3x$ **f)** $x^2 - 9x$ **g)** $x^2 + 13x$ **h)** $x^2 + 10x$

i) $x^2 + 8x$ **j)** $x^2 - 2x$ **k)** $x^2 + 2x$

2. Write an equivalent expression in the form $(x \pm a)^2 - b$.

Question **1** will help with **a** to **h**.

a) $x^2 + 4x - 1$ **b)** $x^2 + 14x - 5$ **c)** $x^2 - 6x + 3$ **d)** $x^2 + 6x + 7$

e) $x^2 - 3x - 1$ **f)** $x^2 + 6x + 3$ **g)** $x^2 - 9x + 10$ **h)** $x^2 + 13x + 35$

i) $x^2 + 8x - 6$ **j)** $x^2 + 2x - 1$ **k)** $x^2 - 2x - 7$ **l)** $x^2 + 2x - 9$

3. Solve each equation by completing the square. Leave a square root sign in your answer where appropriate. The answers to question **2** will help.

a) $x^2 + 4x - 1 = 0$ **b)** $x^2 + 14x - 5 = 0$ **c)** $x^2 - 6x + 3 = 0$

d) $x^2 + 6x + 7 = 0$ **e)** $x^2 - 3x - 1 = 0$ **f)** $x^2 + 6x + 3 = 0$

g) $x^2 - 9x + 10 = 0$ **h)** $x^2 + 13x + 35 = 0$ **i)** $x^2 + 8x - 6 = 0$

j) $x^2 + 2x - 1 = 0$ **k)** $x^2 - 2x - 7 = 0$ **l)** $x^2 + 2x - 9 = 0$

4. Solve by completing the square. Give your answers to 2 decimal places.

a) $x^2 + 2x - 5 = 0$ **b)** $x^2 - 4x - 7 = 0$ **c)** $x^2 + 2x - 9 = 0$

5. Rewrite each equation in the form $a(x + b)^2 + c$.

a) $2x^2 + 4x + 7$ **b)** $3x^2 + 12x + 3$ **c)** $6x^2 + 12x + 4$

d) $5x^2 - 30x + 12$ **e)** $8x^2 - 32x + 10$ **f)** $9x^2 + 9x + 9$

g) $12x^2 - 36x + 14$ **h)** $5x^2 + 10x + 6$ **i)** $7x^2 + 14x + 5$

j) $7x^2 + 7x + 2$ **k)** $10x^2 - 20x + 5$ **l)** $11x^2 + 22x + 6$

6. Work out the values of a, b and c such that $4x^2 + ax - 5 \equiv b(x + 3)^2 + c$.

7. Work out the values of a, b and c such that $ax^2 + bx + 10 \equiv (2x + 1)^2 + c$.

8. Work out the values of a and b such that $5 - 4x - x^2 \equiv a - (x + b)^2$.

9. Solve these equations by completing the square.

a) $2x^2 - 6x - 3 = 0$ **b)** $4x^2 - 8x + 1 = 0$ **c)** $2x^2 + 5x - 10 = 0$

d) $0.5x^2 - 7.5x + 8 = 0$

10. Ahmed rewrites the expression $x^2 + px + q$ by completing the square. He correctly does this and gets $(x - 7)^2 - 52$. What are the values of p and q?

11. Jorge writes the steps to solve $x^2 + 6x + 7 = 0$ by completing the square on sticky notes. Unfortunately he drops them and they get out of order. Can you put the notes in the correct order?

Add 2 to both sides	Subtract 3 from both sides	Write $x^2 + 6x + 7 = 0$ as $(x + 3)^2 - 2 = 0$	Take the square root of both sides

EXAM-STYLE QUESTIONS

1. ABCD is a trapezium. GRADE A

The area of the trapezium is 110 cm².

a) Set up an equation in x.

b) Solve the equation and hence find the length of CD.

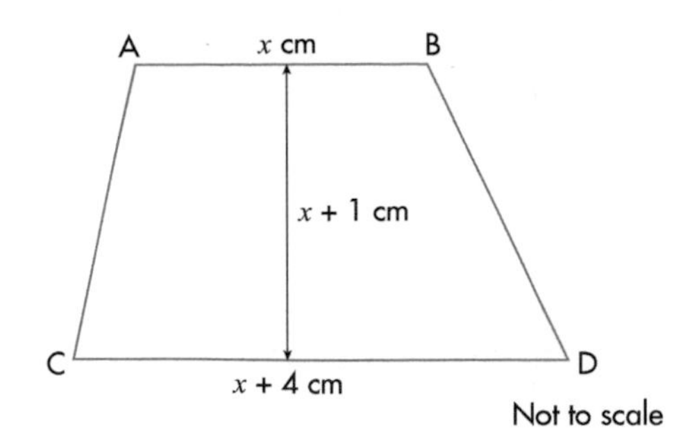

2. The diagrams show a rectangle and a square.

a) i) If the rectangle and the square have the same *perimeter*, use this fact to set up an equation in x.

ii) Solve your equation and hence find the length of the side of the square.

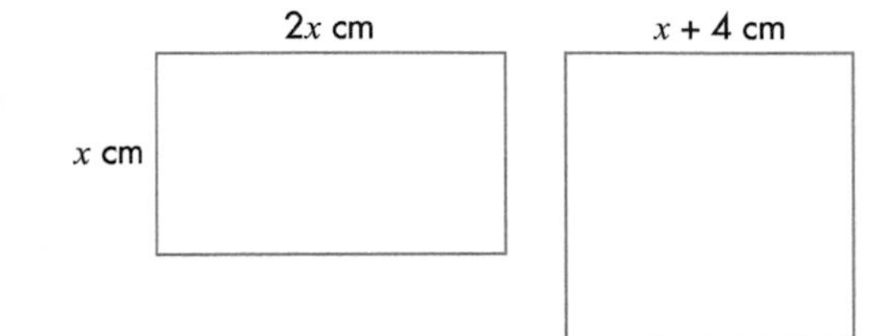

b) i) If the rectangle and the square have the same *area*, use this fact to set up an equation in x.

ii) Solve your equation and hence find the length of the side of the square in this case.

3. Find the values of a, b and c such that $6x^2 + 6x - 1 \equiv a(x + b)^2 + c$. GRADE A*

4. a) Find the values of a, b and c such that $3x^2 - 12x + 4 \equiv a(x + b)^2 + c$.

b) Hence, or otherwise, solve the equation $3x^2 - 12x + 4 = 0$, giving your solutions correct to 2 decimal places.

5. ABC is a straight line.

The length of BC is 1 m and the length of AB is x m.

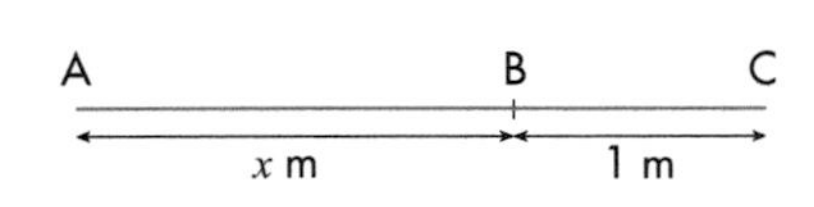

a) AB : BC = AC : AB

Use this fact to set up an equation in x.

b) Solve your equation and hence write the length of AB in metres in the form $a + b\sqrt{5}$.

11 Simultaneous equations

11.1 Simultaneous linear equations

THIS SECTION WILL SHOW YOU HOW TO ...

✓ solve simultaneous linear equations in two variables

KEY WORDS

✓ simultaneous equations
✓ simultaneous linear equations
✓ elimination
✓ substitution

Simple simultaneous linear equations

Sometimes a problem might include more than one unknown variable. In that case you will have several **simultaneous equations** to solve.

You can solve simultaneous equations by **elimination** or by **substitution**.

The elimination method

Here, you solve simultaneous equations by eliminating one of the variables. There are six steps in this method.

Step 1: Make the coefficients of one of the variables the same.

Step 2: **Eliminate** this variable by adding or subtracting the equations.

Step 3: Solve the resulting linear equation in the other variable.

Step 4: **Substitute** the value found back into one of the original equations.

Step 5: Solve the resulting equation.

Step 6: Check that the two values found satisfy the original equations.

EXAMPLE 1

Solve the equations: $6x + y = 15$ and $4x + y = 11$.

Label the equations so that the method is clear.

$6x + y = 15 \quad (1)$

$4x + y = 11 \quad (2)$

Step 1: Since the y-term in both equations has the same coefficient there is no need to balance them.

Step 2: Subtract one equation from the other. (1) minus (2) will give positive values.

(1) – (2) $\qquad 2x = 4$

Step 3: Solve this equation. $\quad x = 2$

Step 4: Substitute $x = 2$ into one of the original equations. (Choose the one with the smallest numbers.)

So substitute into: $4x + y = 11$

which gives: $8 + y = 11$

Step 5: Solve this equation. $y = 3$

Step 6: Test the solution in the original equations. Substitute $x = 2$ and $y = 3$ into $6x + y$, which gives $12 + 3 = 15$ and into $4x + y$, which gives $8 + 3 = 11$. These are correct, so you can confidently say the solution is $x = 2$ and $y = 3$.

EXAMPLE 2

Solve these equations. $5x + y = 22$ (1)

$2x - y = 6$ (2)

Step 1: Both equations have the same y-coefficient but with *different* signs so there is no need to balance them.

Step 2: As the signs are different, *add* the two equations, to eliminate the y-terms.

(1) + (2) $7x = 28$

Step 3: Solve this equation. $x = 4$

Step 4: Substitute $x = 4$ into one of the original equations, $5x + y = 22$, which gives:

$$20 + y = 22$$

Step 5: Solve this equation. $y = 2$

Step 6: Test the solution by putting $x = 4$ and $y = 2$ into the original equations, $2x - y = 6$, which gives $8 - 2 = 6$ and $5x + y = 22$ which gives $20 + 2 = 22$.

These are correct, so the solution is $x = 4$ and $y = 2$.

The substitution method

There are five steps in the substitution method.

Step 1: Rearrange one of the equations into the form $y = \ldots$ or $x = \ldots$.

Step 2: Substitute the right-hand side of this equation into the other equation in place of the variable on the left-hand side.

Step 3: Expand and solve this equation.

Step 4: Substitute the value into the $y = \ldots$ or $x = \ldots$ equation.

Step 5: Check that the values work in both original equations.

The method you use depends very much on the coefficients of the variables and the way that the equations are written in the first place.

EXAMPLE 3

Solve the simultaneous equations: $y = 2x + 3$, $3x + 4y = 1$.

Because the first equation is in the form $y = \ldots$ it suggests that the substitution method should be used.

Again label the equations to help follow through the method.

$y = 2x + 3$ (1)

$3x + 4y = 1$ (2)

Step 1: As equation (1) is in the form $y = \ldots$ there is no need to rearrange an equation.

Step 2: Substitute the right-hand side of equation (1) into equation (2) for the variable y.

$3x + 4(2x + 3) = 1$

Step 3: Expand and solve the equation.

$3x + 8x + 12 = 1 \quad \Rightarrow 11x = -11 \quad \Rightarrow x = -1$

Step 4: Substitute $x = -1$ into $y = 2x + 3$: $y = -2 + 3 = 1$

Step 5: Test the values in $y = 2x + 3$ which gives $1 = -2 + 3$ and $3x + 4y = 1$, which gives $-3 + 4 = 1$. These are correct so the solution is $x = -1$ and $y = 1$.

Balancing coefficients in one equation only

You have seen how to solve simultaneous equations by adding or subtracting the equations in each pair, or by substituting without rearranging. This is not always possible. The next examples show what to do when there are no identical terms, or when you need to rearrange.

EXAMPLE 4

Solve these equations. $3x + 2y = 18$ (1)

$2x - y = 5$ (2)

Step 1: Multiply equation (2) by 2. There are other ways to balance the coefficients but this is the easiest and leads to less work later. With practice, you will get used to which will be the best way to balance the coefficients.

$2 \times (2)$ $\quad 4x - 2y = 10$ (3)

Be careful to multiply every term and not just the y-term.

Step 2: As the signs of the y-terms are opposite, add the equations.

$(1) + (3)$ $\quad 7x = 28$

Be careful to add the correct equations. This is why labelling them is useful.

Step 3: Solve this equation: $x = 4$

Step 4: Substitute $x = 4$ into any equation, say $2x - y = 5 \Rightarrow 8 - y = 5$

Step 5: Solve this equation: $y = 3$

Step 6: Check: (1), $3 \times 4 + 2 \times 3 = 18$ and (2), $2 \times 4 - 3 = 5$, which are correct so the solution is $x = 4$ and $y = 3$.

EXAMPLE 5

Solve the simultaneous equations: $3x + y = 5$ (1)

$5x - 2y = 12$ (2)

Step 1: Multiply the first equation by 2:

$6x + 2y = 10$ (3)

Step 2: Add (2) + (3): $11x = 22$

Step 3: Solve: $x = 2$

Step 4: Substitute back: $3 \times 2 + y = 5$

Step 5: Solve: $y = -1$

Step 6: Check: (1) $3 \times 2 - 1 = 5$ and (2) $5 \times 2 - 2 \times (-1) = 10 + 2 = 12$, which are correct.

Balancing coefficients in both equations

There are also cases where you need to change *both* equations to obtain identical terms.

The next example shows you how this is done.

Note: The substitution method is not suitable for these types of equation as you end up with fractional terms.

EXAMPLE 6

Solve these equations. $4x + 3y = 27$ (1)

$5x - 2y = 5$ (2)

Both equations have to be changed to obtain identical terms in either x or y. However, you can see that if you make the y-coefficients the same, you will add the equations. This is always safer than subtraction, so this is obviously the better choice. You do this by multiplying the first equation by 2 (the y-coefficient of the second equation) and the second equation by 3 (the y-coefficient of the first equation).

Step 1: (1) × 2 or $2 \times (4x + 3y = 27) \Rightarrow 8x + 6y = 54$ (3)

(2) × 3 or $3 \times (5x - 2y = 5) \Rightarrow 15x - 6y = 15$ (4)

Label the new equations (3) and (4).

Step 2: Eliminate one of the variables: (3) + (4) $\Rightarrow 23x = 69$

Step 3: Solve the equation: $x = 3$

Step 4: Substitute into equation (1): $12 + 3y = 27$

Step 5: Solve the equation: $y = 5$

Step 6: Check: (1), $4 \times 3 + 3 \times 5 = 12 + 15 = 27$, and (2), $5 \times 3 - 2 \times 5 = 15 - 10 = 5$, which are correct so the solution is $x = 3$ and $y = 5$.

EXERCISE 11A

GRADE B

1. Solve each pair of simultaneous equations.

a) $x + y = 15$
$y = 2x$

b) $x = 3y$
$x + y = 24$

c) $x + y = 60$
$y = 4x$

d) $y = x + 12$
$y = 3x$

e) $y = x - 10$
$x = 5y$

f) $x + 4 = y$
$y = 9x$

2. Solve each pair of simultaneous equations.

a) $x + y = 20$
$x - y = 6$

b) $y + x = 23$
$y - x = 5$

c) $x + y = 6$
$x - y = 14$

d) $y = 2x + 3$
$y = 8x$

e) $x + y = 20$
$y = 3x - 2$

f) $y = 2x + 4$
$y = 10 - x$

3. Solve these simultaneous equations.

a) $4x + y = 17$
$2x + y = 9$

b) $5x + 2y = 13$
$x + 2y = 9$

c) $2x + y = 7$
$5x - y = 14$

d) $2x + 5y = 37$
$y = 11 - 2x$

e) $4x - 3y = 7$
$x = 13 - 3y$

f) $4x - y = 17$
$x = 2 + y$

4. Solve these simultaneous equations.

a) $5x + 2y = 4$
$4x - y = 11$

b) $4x + 3y = 37$
$2x + y = 17$

c) $3x + 5y = 15$
$x + 3y = 7$

d) $5x - 2y = 24$
$3x + y = 21$

e) $5x - 2y = 4$
$3x - 6y = 6$

f) $3x - 2y = 3$
$5x + 6y = 12$

5. Solve these simultaneous equations.

a) $2x + 5y = 15$
$3x - 2y = 13$

b) $2x + 3y = 30$
$5x + 7y = 71$

c) $2x - 3y = 15$
$5x + 7y = 52$

d) $2x + y = 4$
$x - y = 5$

e) $5x + 2y = 11$
$3x + 4y = 8$

f) $x - 2y = 4$
$3x - y = -3$

g) $3x - y = 5$
$x + 3y = -20$

h) $3x - 4y = 4.5$
$2x + 2y = 10$

i) $x - 5y = 15$
$3x - 7y = 17$

6. Here are four equations.

A: $5x + 2y = 1$ B: $4x + y = 9$ C: $3x - y = 5$ D: $3x + 2y = 3$

Here are four sets of (x, y) values.

$(1, -2)$, $(-1, 3)$, $(2, 1)$, $(3, -3)$

Match each pair of (x, y) values to a pair of equations.

7. Find the area of the triangle enclosed by these three equations.

$y - x = 2$ $x + y = 6$ $3x + y = 6$

8. Find the area of the triangle enclosed by these three equations.

$x - 2y = 6$ $x + 2y = 6$ $x + y = 3$

HINTS AND TIPS

Find the points of intersection of each pair of equations, plot the points on a grid and use any method to work out the area of the resulting triangle.

11.2 Linear and non-linear simultaneous equations

THIS SECTION WILL SHOW YOU HOW TO ...

✓ solve simultaneous equations, one linear and one non-linear

KEY WORDS

✓ linear ✓ non-linear

You have already seen the method of substitution for solving **linear** simultaneous equations.

You can use a similar method when you need to solve a pair of equations, one of which is linear and the other of which is **non-linear**. But you must always substitute from the linear into the non-linear.

EXAMPLE 7

Solve these simultaneous equations.

$x^2 + y^2 = 5$

$x + y = 3$

Call the equations (1) and (2):

$x^2 + y^2 = 5 \quad (1)$

$x + y = 3 \quad (2)$

Rearrange equation (2) to obtain: $x = 3 - y$

Substitute this into equation (1), which gives: $(3 - y)^2 + y^2 = 5$

Expand and rearrange into the general form of the quadratic equation:

$$9 - 6y + y^2 + y^2 = 5 \Rightarrow 2y^2 - 6y + 4 = 0$$

Cancel by 2: $y^2 - 3y + 2 = 0$

Factorise: $(y - 1)(y - 2) = 0 \Rightarrow y = 1$ or 2

Substitute for y in equation (2).
When $y = 1$, $x = 2$ and when $y = 2$, $x = 1$.

Note: You should always give answers as a pair of values in x and y.

EXAMPLE 8

Find the solutions of the pair of simultaneous equations:
$y = x^2 + x - 2$ and $y = 2x + 4$.

This example is slightly different, as both equations are given in terms of y, so substituting for y gives:

$2x + 4 = x^2 + x - 2$

Rearranging into the general quadratic: $x^2 - x - 6 = 0$

Factorising and solving gives: $(x + 2)(x - 3) = 0 \Rightarrow x = -2$ or 3

Substitute back to find y.

When $x = -2$, $y = 0$. When $x = 3$, $y = 10$.

So the solutions are $(-2, 0)$ and $(3, 10)$.

EXERCISE 11B

GRADE B

1. Use the substitution method to solve each pair of linear simultaneous equations.

a) $2x + y = 9$
$x - 2y = 7$

b) $3x - 2y = 10$
$4x + y = 17$

c) $x - 2y = 10$
$2x + 3y = 13$

GRADE A

2. Solve each pair of simultaneous equations.

a) $xy = 2$
$y = x + 1$

b) $xy = -4$
$2y = x + 6$

3. Solve each pair of simultaneous equations.

a) $x^2 + y^2 = 25$
$x + y = 7$

b) $x^2 + y^2 = 9$
$y = x + 3$

c) $x^2 + y^2 = 13$
$5y + x = 13$

4. Solve each pair of simultaneous equations.

a) $y = x^2 + 2x - 3$
$y = 2x + 1$

b) $y = x^2 - 2x - 5$
$y = x - 1$

c) $y = x^2 - 2x$
$y = 2x - 3$

5. Solve these pairs of simultaneous equations.

a) $y = x^2 + 3x - 3$ and $y = x$

b) $x^2 + y^2 = 13$ and $x + y = 1$

c) $x^2 + y^2 = 5$ and $y = x + 1$

d) $y = x^2 - 3x + 1$ and $y = 2x - 5$

e) $y = x^2 - 3$ and $y = x + 3$

f) $y = x^2 - 3x - 2$ and $y = 2x - 6$

g) $x^2 + y^2 = 41$ and $y = x + 1$

11.3 Using graphs to solve simultaneous linear equations

THIS SECTION WILL SHOW YOU HOW TO ...

✓ use graphs to solve linear simultaneous equations

Consider the equations $x + 2y = 10$ and $x + y = 8$.

The simultaneous solution is $x = 6$ and $y = 2$.

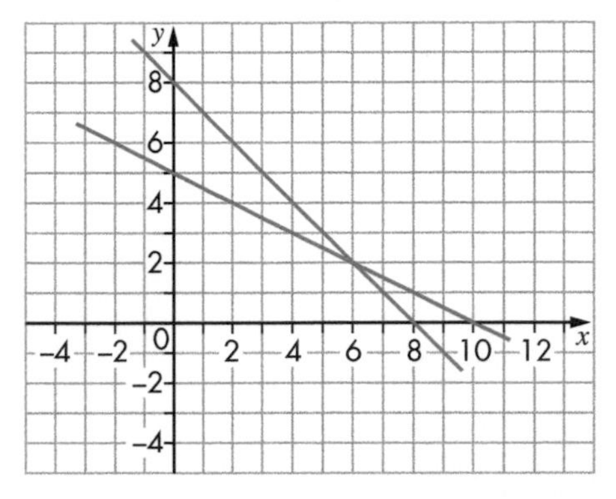

This is a graph of the lines with equations $x + 2y = 10$ and $x + y = 8$.

They cross at (6, 2).

The solution of a pair of simultaneous equations can always be interpreted as the coordinates of the point where the corresponding lines cross.

EXAMPLE 9

By drawing their graphs on the same grid, find the solution of these simultaneous equations.

$3x + y = 6$ $\qquad$ $y = 4x - 1$

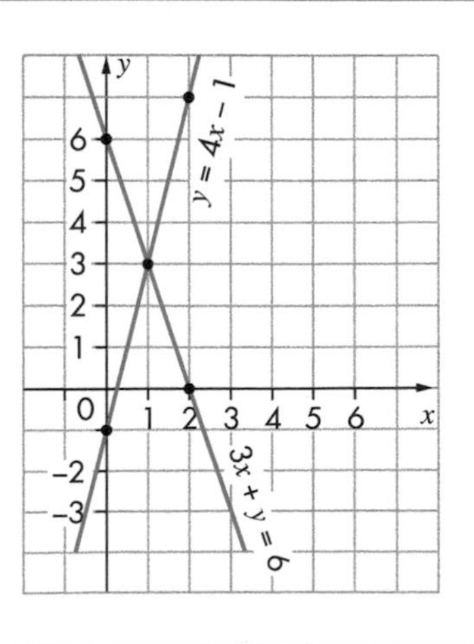

The first graph is drawn using the cover-up method. It crosses the x-axis at (2, 0) and the y-axis at (0, 6).

The second graph can be drawn by finding some points or by the gradient-intercept method. If you use the gradient-intercept method, you find the graph crosses the y-axis at -1 and has a gradient of 4.

The point where the graphs intersect is (1, 3). So the solution to the simultaneous equations is $x = 1$, $y = 3$.

EXERCISE 11C

GRADE B

1. *Use the graph* to solve these simultaneous equations.

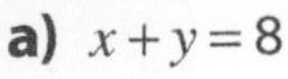

a) $x + y = 8$
$y - x = 4$

b) $x + 3y = 12$
$x + y = 8$

c) $x + y = 8$
$3x - y = 8$

d) $y - x = 4$
$x + 3y = 12$

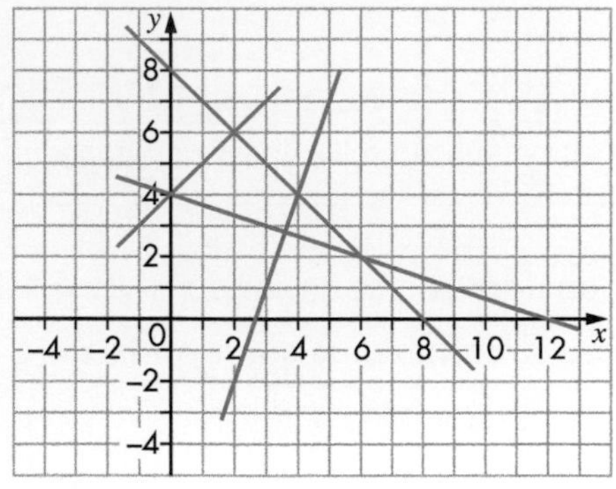

2. Draw graphs to solve these pairs of simultaneous equations

a) $x + 2y = 20$
$2x + y = 16$

b) $x - y = 2$
$2x + 3y = 24$

c) $2x - y = 12$
$3x + 2y = 18$

In questions **3–14**, draw the graphs to find the solution of each pair of simultaneous equations.

3. $x + 4y = 8$
$x - y = 3$

4. $y = 2x - 1$
$3x + 2y = 12$

5. $y = 2x + 4$
$y = x + 7$

6. $y = x$
$x + y = 10$

7. $y = 2x + 3$
$5x + y = 10$

8. $y = 5x + 1$
$y = 2x + 10$

9. $y = x + 8$
$x + y = 4$

10. $y - 3x = 9$
$y = x - 3$

11. $y = -x$
$y = 4x - 5$

12. $3x + 2y = 18$
$y = 3x$

13. $y = 3x + 2$
$y + x = 10$

14. $y = \frac{x}{3} + 1$
$x + y = 11$

15. The graph shows four lines.

P: $y = 4x + 1$ Q: $y = 2x + 2$

R: $y = x - 2$ S: $x + y + 1 = 0$

Which pairs of lines intersect at the following points?

a) $(-1, -3)$ **b)** $(\frac{1}{2}, -1\frac{1}{2})$ **c)** $(\frac{1}{2}, 3)$ **d)** $(-1, 0)$

e) Solve the simultaneous equations P and S to find the exact solution.

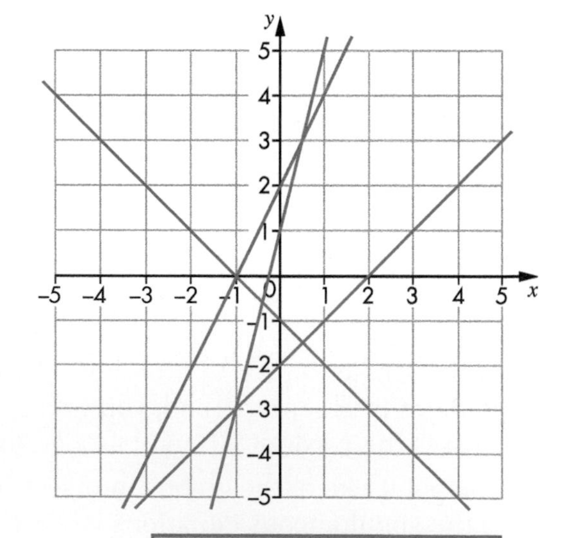

16. Four lines have the following equations.

A: $y = x$ B: $y = 2$

C: $x = -3$ D: $y = -x$

These lines intersect at six different points.

Without drawing the lines accurately, write down the coordinates of the six intersection points.

HINTS AND TIPS

Sketch the lines.

11.4 Using graphs to solve simultaneous equations, one linear and one non-linear

THIS SECTION WILL SHOW YOU HOW TO ...

✓ use graphs to solve equations, one linear and one non-linear

KEY WORDS

✓ simultaneous equations
✓ linear
✓ non-linear

In section 11.2 you saw how to use an algebraic method for solving a pair of **simultaneous equations** where one is **linear** (a straight line) and one is **non-linear** (a curve). In this section, you will learn how to do this graphically. In section 11.3 you saw how to find the solution to a pair of linear simultaneous equations. The same principle applies here. The point where the graphs cross gives the solution. However, in most cases, there are two solutions, because the straight line will cross the curve twice.

Most of the non-linear graphs will be quadratic graphs, but there is one other type you can meet. This is an equation of the form $x^2 + y^2 = r^2$, which is a circle, with the centre as the origin and a radius of r.

EXAMPLE 10

Find the approximate solutions of the pair of equations $y = x^2 + x - 2$ and $y = 2x + 3$ by graphical means.

Set up a table for the quadratic.

x	−4	−3	−2	−1	0	1	2	3	4
y	10	4	0	−2	−2	0	4	10	18

Draw both graphs on the same set of axes.

From the graph, the approximate solutions can be seen to be (−1.8, −0.6) and (2.8, 8.6).

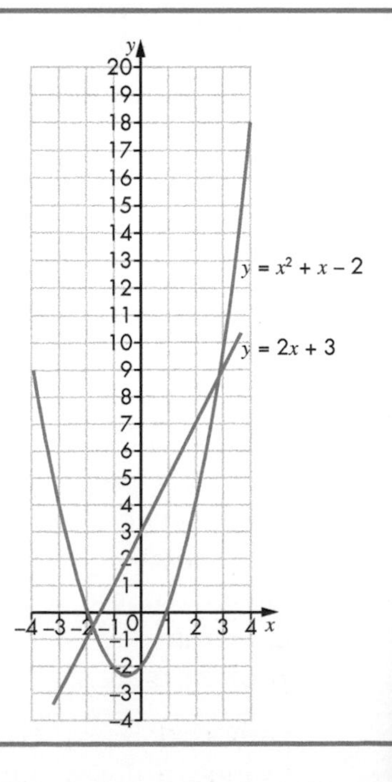

EXAMPLE 11

Find the approximate solutions of the pair of equations $x^2 + y^2 = 25$ and $y = x + 2$ by graphical means.

The curve is a circle of radius 5, centred on the origin.

From the graph, the approximate solutions can be seen to be (−4.4, −2.4), (2.4, 4.4).

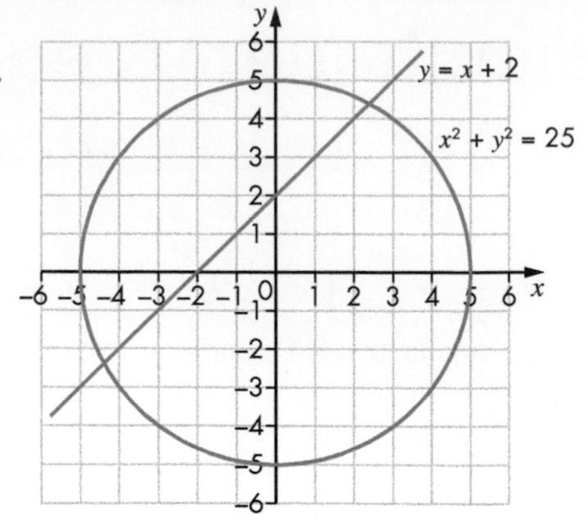

EXERCISE 11D

GRADE A

1. Use graphical methods to find the approximate or exact solutions to these pairs of simultaneous equations. In this question, suitable ranges for the axes are given. In an examination a grid will be supplied.

a) $y = x^2 + 3x - 2$ and $y = x$ $(-5 < x < 5, -5 < y < 5)$
b) $y = x^2 - 3x - 6$ and $y = 2x$ $(-4 < x < 8, -10 < y < 20)$
c) $x^2 + y^2 = 25$ and $x + y = 1$ $(-6 < x < 6, -6 < y < 6)$
d) $x^2 + y^2 = 4$ and $y = x + 1$ $(-5 < x < 5, -5 < y < 5)$
e) $y = x^2 - 3x + 1$ and $y = 2x - 1$ $(0 < x < 6, -4 < y < 12)$
f) $y = x^2 - 3$ and $y = x + 3$ $(-5 < x < 5, -4 < y < 8)$
g) $y = x^2 - 3x - 2$ and $y = 2x - 3$ $(-5 < x < 5, -5 < y < 10)$
h) $x^2 + y^2 = 9$ and $y = x - 1$ $(-5 < x < 5, -5 < y < 5)$

GRADE A*

2. a) Solve the simultaneous equations $y = x^2 + 3x - 4$ and $y = 5x - 5$ $(-5 < x < 5, -8 < y < 8)$.
b) What is special about the intersection of these two graphs?
c) Show that $5x - 5 = x^2 + 3x - 4$ can be rearranged to $x^2 - 2x + 1 = 0$.
d) Factorise and solve $x^2 - 2x + 1 = 0$.
e) Explain how the solution in part **d** relates to the intersection of the graphs.

3. a) Solve the simultaneous equations $y = x^2 + 2x + 3$ and $y = x - 1$ $(-5 < x < 5, -5 < y < 8)$.
b) What is special about the intersection of these two graphs?
c) Rearrange $x - 1 = x^2 + 2x + 3$ into the general quadratic form $ax^2 + bx + c = 0$.
d) Work out the discriminant $b^2 - 4ac$ for the quadratic in part **c**.
e) Explain how the value of the discriminant relates to the intersection of the graphs.

11.5 Solving equations by the method of intersection

THIS SECTION WILL SHOW YOU HOW TO ...

✓ solve equations by the method of intersection

Simultaneous equations can often be solved by drawing two intersecting graphs on the same axes and using the coordinates of their points of intersection. You may be given the graph of one line, drawn on Cartesian axes, and asked to draw a straight-line graph to solve the two equations simultaneously.

EXAMPLE 12

Show how each equation given below can be solved, using the graph of $y = x^3 - 2x - 2$ and its intersection with another graph. In each case, give the equation of the other graph and the solution(s).

a) $x^3 - 2x - 4 = 0$ **b)** $x^3 - 3x - 1 = 0$

a) This method will give the required graph.

Step 1: Write down the original (given) equation. $y = x^3 - 2x - 2$

Step 2: Write down the (new) equation to be solved in reverse. $0 = x^3 - 2x - 4$

Step 3: Subtract these equations. $y = \quad + 2$

Step 4: Draw this line on the original graph to solve the new equation.

The graphs of $y = x^3 - 2x - 2$ and $y = 2$ are drawn on the same axes.

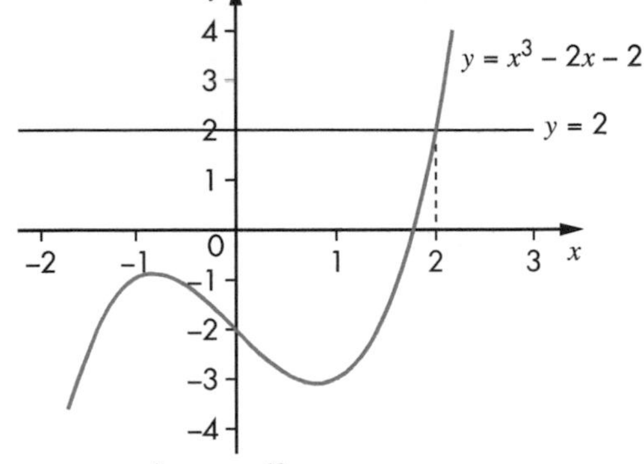

The intersection of these two graphs is the solution of $x^3 - 2x - 4 = 0$.

The solution is $x = 2$.

This works because you are drawing a straight line on the same axes as the original graph, and solving for x and y where they intersect.

At the points of intersection the y-values will be the same and so will the x-values.

So you can say: original equation = straight line

Rearranging this gives: (original equation) – (straight line) = 0

You have been asked to solve: (new equation) = 0

So: (original equation) – (straight line) = (new equation)

Rearranging this again gives: (original equation) – (new equation) = straight line

b) Write down given graph: $y = x^3 - 2x - 2$

Write down new equation: $0 = x^3 - 3x - 1$

Subtract: $y = \quad + x - 1$

The graphs of $y = x^3 - 2x - 2$ and $y = x - 1$ are then drawn on the same axes.

The intersection of the two graphs is the solution of $x^3 - 3x - 1 = 0$.

The solutions are $x = -1.5, -0.3$ and 1.9.

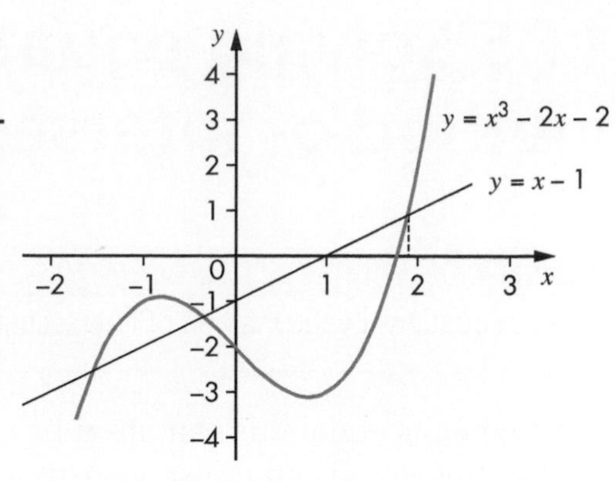

EXAMPLE 13

The graph shows the curve $y = x^2 + 3x - 2$.

By drawing a suitable straight line, solve these equations.

a) $x^2 + 3x - 1 = 0$ **b)** $x^2 + 2x - 3 = 0$

a) Given graph: $y = x^2 + 3x - 2$

New equation: $0 = x^2 + 3x - 1$

Subtract: $y = \quad - 1$

Draw: $y = -1$

Solutions: $x = 0.3, -3.3$

b) Given graph: $y = x^2 + 3x - 2$

New equation: $0 = x^2 + 2x - 3$

Subtract: $y = \quad + x \quad + 1$

Draw: $y = x + 1$

Solutions: $x = 1, -3$

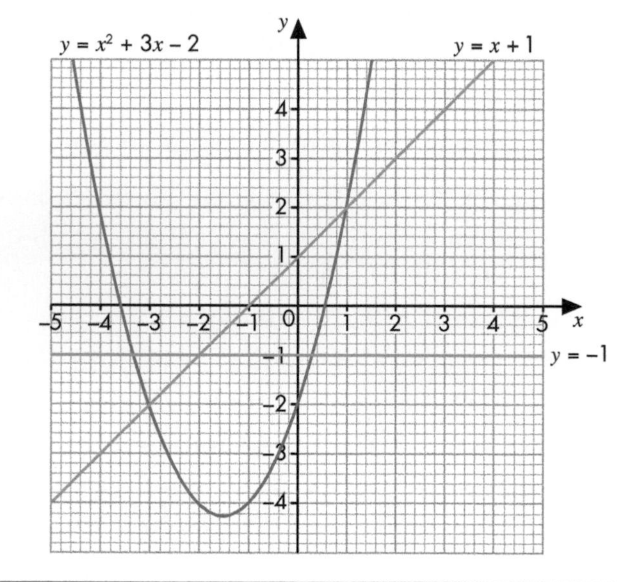

EXERCISE 11E

GRADE A*

In questions **1** to **5**, use the graphs given here. Trace the graphs or place a ruler over them in the position of the line. Solution values only need to be given to 1 decimal place. In questions **6** to **10**, either draw the graphs yourself or use a graphics calculator to draw them.

1. Below is the graph of $y = x^2 - 3x - 6$.

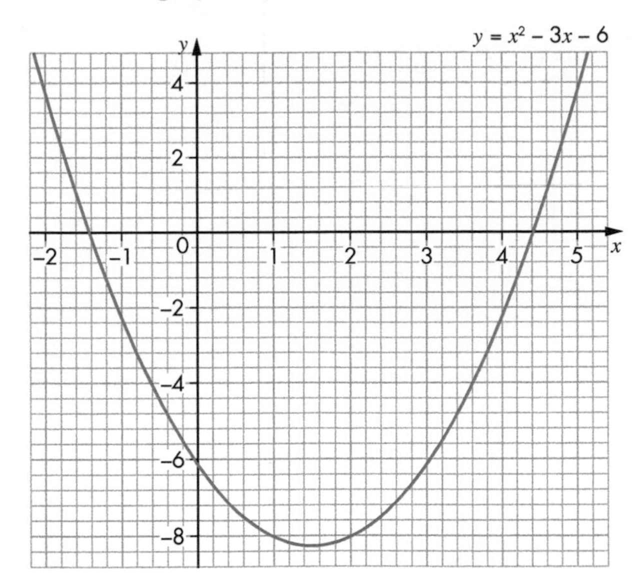

a) Solve these equations.

i) $x^2 - 3x - 6 = 0$ **ii)** $x^2 - 3x - 6 = 4$ **iii)** $x^2 - 3x - 2 = 0$

b) By drawing a suitable straight line solve $2x^2 - 6x + 2 = 0$.

2. Below is the graph of $y = x^2 + 4x - 5$.

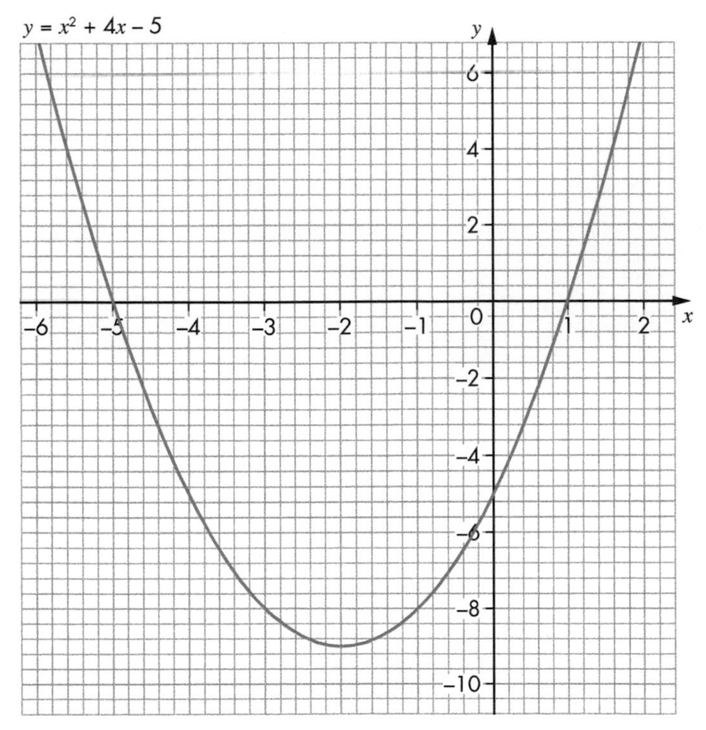

a) Solve $x^2 + 4x - 5 = 0$.

b) By drawing suitable straight lines solve these equations.

i) $x^2 + 4x - 5 = 2$ **ii)** $x^2 + 4x - 4 = 0$ **iii)** $3x^2 + 12x + 6 = 0$

3. Below are the graphs of $y = x^2 - 5x + 3$ and $y = x + 3$.

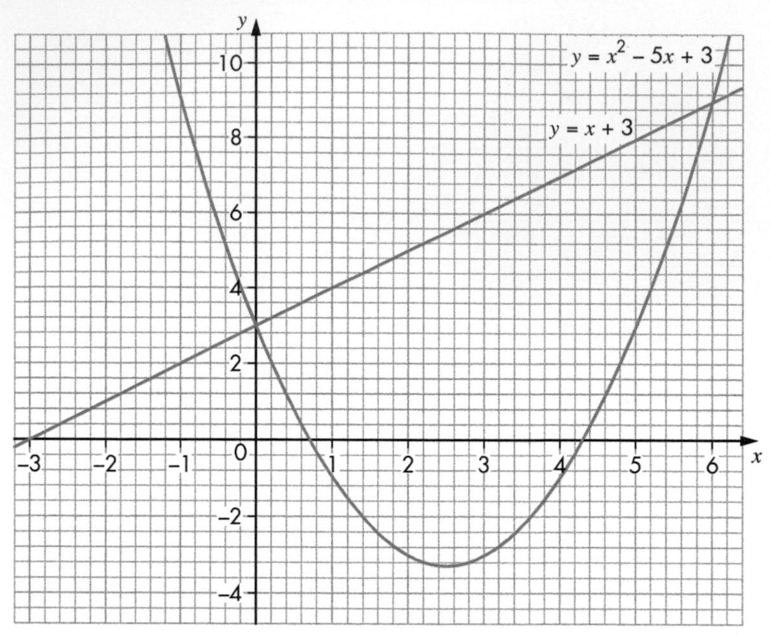

a) Solve these equations.

i) $x^2 - 6x = 0$ **ii)** $x^2 - 5x + 3 = 0$

b) By drawing suitable straight lines solve these equations.

i) $x^2 - 5x + 3 = 2$ **ii)** $x^2 - 5x - 2 = 0$

4. Below are the graphs of $y = x^2 - 2$ and $y = x + 2$.

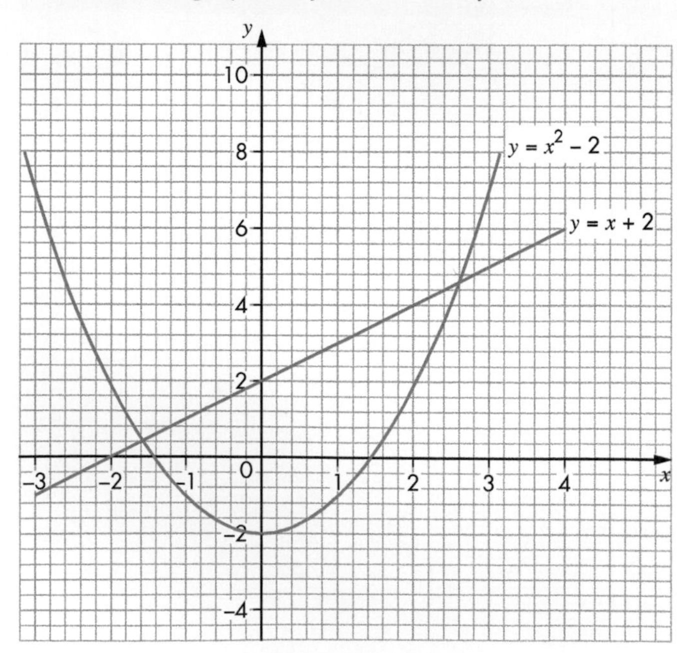

a) Solve these equations.

i) $x^2 - x - 4 = 0$ **ii)** $x^2 - 2 = 0$

b) By drawing suitable straight lines solve these equations.

i) $x^2 - 2 = 3$ **ii)** $x^2 - 4 = 0$

5. Below are the graphs of $y = x^3 - 2x^2$, $y = 2x + 1$ and $y = x - 1$.

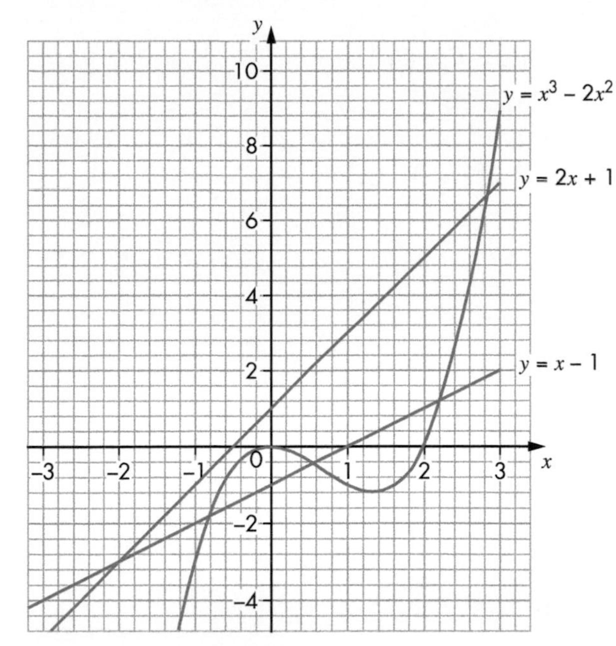

Solve these equations.

a) $x^3 - 2x^2 = 0$ **b)** $x^3 - 2x^2 = 3$ **c)** $x^3 - 2x^2 + 1 = 0$

d) $x^3 - 2x^2 - 2x - 1 = 0$ **e)** $x^3 - 2x^2 - x + 1 = 0$

6. Draw the graph of $y = x^2 - 4x - 2$.

a) Solve $x^2 - 4x - 2 = 0$.

b) By drawing a suitable straight line solve $x^2 - 4x - 5 = 0$.

7. Draw the graph of $y = 2x^2 - 5$.

a) Solve $2x^2 - 5 = 0$.

b) By drawing a suitable straight line solve $2x^2 - 3 = 0$.

8. Draw the graphs of $y = x^2 - 3$ and $y = x + 2$ on the same axes. Use the graphs to solve these equations.

a) $x^2 - 5 = 0$ **b)** $x^2 - x - 5 = 0$

9. Draw the graphs of $y = x^2 - 3x - 2$ and $y = 2x - 3$ on the same axes. Use the graphs to solve these equations.

a) $x^2 - 3x - 1 = 0$ **b)** $x^2 - 5x + 1 = 0$

10. Draw the graphs of $y = x^3 - 2x^2 + 3x - 4$ and $y = 3x - 1$ on the same axes. Use the graphs to solve these equations.

a) $x^3 - 2x^2 + 3x - 6 = 0$ **b)** $x^3 - 2x^2 - 3 = 0$

11. The graph shows the lines A: $y = x^2 + 3x - 2$; B: $y = x$; C: $y = x + 2$; D: $y + x = 3$ and E: $y + x + 1 = 0$.

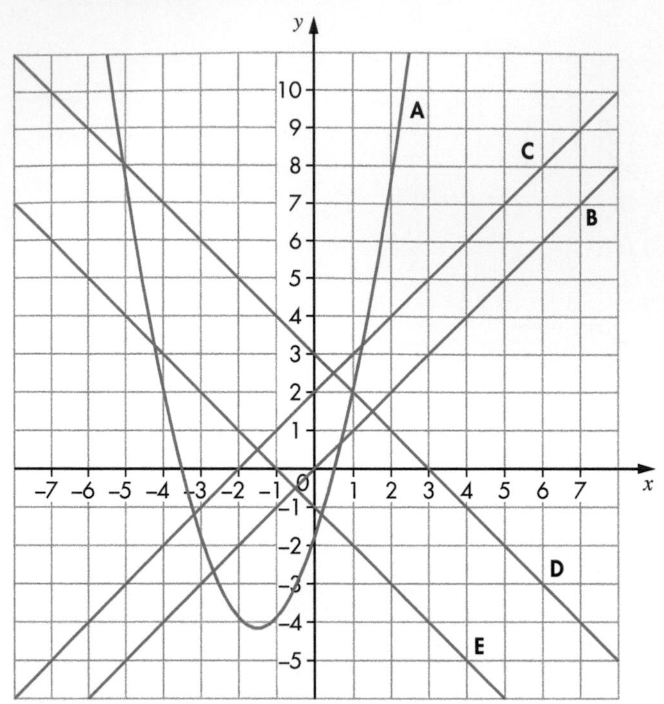

a) Which pair of lines has a common solution of (0.5, 2.5)?

b) Which pair of lines has the solutions of (1, 2) and (–5, 8)?

c) What quadratic equation has an approximate solution of (–4.2, 3.2) and (0.2, –1.2)?

d) The minimum point of the graph $y = x^2 + 3x - 2$ is at (–1.5, –4.25).
What is the minimum point of the graph $y = x^2 + 3x - 8$?

12. Jamil was given a sketch of the graph $y = x^2 + 3x + 5$ and asked to draw an appropriate straight line to solve $x^2 + x - 2 = 0$.

This is Jamil's working.

Original $\quad y = x^2 + 3x + 5$

New $\quad \underline{0 = x^2 + \ \ x - 2}$

$\quad y = \quad 2x - 7$

When Jamil drew the line $y = 2x - 7$, it did not intersect with the parabola $y = x^2 + 3x + 5$. He concluded that the equation $x^2 + x - 2 = 0$ did not have any solutions.

a) Show by factorisation that the equation $x^2 + x - 2 = 0$ has solutions –2 and 1.

b) Explain the error that Jamil made.

c) What line should Jamil have drawn?

Exam-style questions

1. Solve these simultaneous equations. GRADE B

$3x + 2y = 11$

$2x - 2y = 14$

2. Solve these simultaneous equations. GRADE B

$3x - 4y = 17$

$x - 4y = 3$

3. Solve these simultaneous equations. GRADE B

$2x + 3y = 19$

$6x + 2y = 22$

4. Solve these simultaneous equations. GRADE B

$5x - 2y = 26$

$3x - y = 15$

5. Solve these simultaneous equations. GRADE B

$3x - 2y = 15$

$2x - 3y = 5$

6. Solve these simultaneous equations. GRADE B

$5x - 3y = 14$

$4x - 5y = 6$

7. Solve these simultaneous equations. GRADE A*

$x = 3 + 2y$

$x^2 + 2y^2 = 27$

8. Find the points of intersection of the line $x + 3y = 5$ and the circle $x^2 + y^2 = 25$. GRADE A*

9. Find the points of intersection of the line $x = 4$ and the curve $y^2 = 9x$. GRADE A*

10. Solve these simultaneous equations. GRADE A*

$y = 2x + 1$

$y = x^2 + x$

Give your answers to 2 decimal places.

12 Inequalities

12.1 Linear inequalities

THIS SECTION WILL SHOW YOU HOW TO ...

✓ solve linear inequalities

KEY WORD

✓ inequality

Inequalities behave similarly to equations. You use the same rules to solve linear inequalities as you use for linear equations.

There are four inequality signs:

$<$ means 'less than'

$>$ means 'greater than'

$\leqslant$ means 'less than or equal to'

$\geqslant$ means 'greater than or equal to'.

EXAMPLE 1

Solve $2x + 3 < 14$.

Rewrite this as: $2x < 14 - 3 \Rightarrow 2x < 11$

Divide both sides by 2: $\frac{2x}{2} < \frac{11}{2} \Rightarrow x < 5.5$

This means that x can take any value below 5.5 but *not* the value 5.5.

EXAMPLE 2

Solve $\frac{x}{2} + 4 \geqslant 13$.

Solve just like an equation but leave the inequality sign in place of the equals sign.

Subtract 4 from both sides: $\frac{x}{2} \geqslant 9$

Multiply both sides by 2: $x \geqslant 18$

This means that x can take any value above and including 18.

If you divide by a negative number when you are solving an inequality you must change the sign.

Then:

- 'less than' becomes 'more than'
- 'more than' becomes 'less than'.

EXAMPLE 3

Solve the inequality $10 - 2x \geqslant 3$.

Subtract 10 from both sides: $-2x \geqslant -7$

Divide both sides by -2: $x \leqslant 3.5$

You must reverse inequality signs when multiplying or dividing both sides by a negative number. So the inequality has changed in the last line.

You could approach this example in a different way.

$10 - 2x \geqslant 3$

Add $2x$ to both sides: $10 \geqslant 2x + 3$

Subtract 3 from both sides: $7 \geqslant 2x$

Divide both sides by 2: $3.5 \geqslant x$

The sign does not change this time. $3.5 \geqslant x$ is equivalent to $x \leqslant 3.5$

EXERCISE 12A

GRADE C

1. Solve these linear inequalities.

a) $x + 4 < 7$ **b)** $t - 3 > 5$ **c)** $p + 2 \geqslant 12$

d) $2x - 3 < 7$ **e)** $4y + 5 \leqslant 17$ **f)** $3t - 4 > 11$

g) $\frac{x}{2} + 4 < 7$ **h)** $\frac{y}{5} + 3 \leqslant 6$ **i)** $\frac{t}{3} - 2 \geqslant 4$

j) $3(x - 2) < 15$ **k)** $5(2x + 1) \leqslant 35$ **l)** $2(4t - 3) \geqslant 34$

2. In each part, write down the largest integer value of x that satisfies the inequality.

a) $x - 3 \leqslant 5$, where x is positive

b) $x + 2 < 9$, where x is positive and even

c) $3x - 11 < 40$, where x is a square number

d) $5x - 8 \leqslant 15$, where x is positive and odd

e) $2x + 1 < 19$, where x is positive and prime

3. In each part, write down the smallest integer value of x that satisfies the inequality.

a) $x - 2 \geqslant 9$, where x is positive

b) $x - 2 > 13$, where x is positive and even

c) $2x - 11 \geqslant 19$, where x is a square number

4. a) Explain why you cannot make a triangle with three sticks of length 3 cm, 4 cm and 8 cm.

b) Three sides of a triangle are x, $x + 2$ and 10 cm.
x is a whole number.
What is the smallest value x can take?

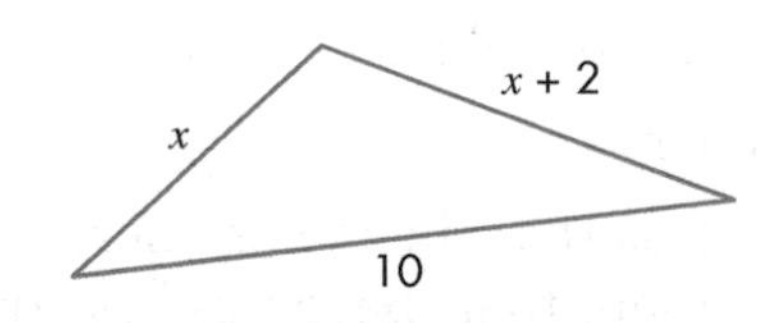

GRADE B

5. Five cards have inequalities and equations marked on them.

$x > 0$ $x < 3$ $x \geqslant 4$ $x = 2$ $x = 6$

The cards are shuffled and then turned over, one at a time. If two consecutive cards have any numbers in common, then a point is scored. If they do not have any numbers in common, then a point is deducted.

a) The first two cards below score −1 because $x = 6$ and $x < 3$ have no numbers in common.

Explain why the total for this combination scores 0.

$x = 6$ $x < 3$ $x > 0$ $x = 2$ $x \geqslant 4$

b) What does this combination score?

$x > 0$ $x = 6$ $x \geqslant 4$ $x = 2$ $x < 3$

c) Arrange the cards to give a maximum score of 4.

6. Solve these linear inequalities.

a) $4x + 1 \geqslant 3x - 5$ **b)** $5t - 3 \leqslant 2t + 5$ **c)** $3y - 12 \leqslant y - 4$

d) $2x + 3 \geqslant x + 1$ **e)** $5w - 7 \leqslant 3w + 4$ **f)** $2(4x - 1) \leqslant 3(x + 4)$

7. Solve these linear inequalities.

a) $\frac{x + 4}{2} \leqslant 3$ **b)** $\frac{x - 3}{5} > 7$ **c)** $\frac{2x + 5}{3} < 6$

d) $\frac{4x - 3}{5} \geqslant 3$ **e)** $\frac{2t - 2}{7} > 4$ **f)** $\frac{5y + 3}{5} \leqslant 2$

8. In this question n is always an integer.

a) Find the largest possible value of n if $2n + 3 < 12$.

b) Find the largest possible value of n if $\frac{n}{5} < 20$.

c) Find the smallest possible value of n if $3(n - 7) \geqslant 10$.

d) Find the smallest possible value of n if $\frac{6n - 2}{7} \geqslant 9$.

e) Find the smallest possible value of n if $3n + 14 \leqslant 8n - 13$.

9. a) If $20 - x > 4$, which of these numbers are possible values of x?

−10 0 10 20 30

b) Solve the inequality $20 - x > 4$.

10. Solve these inequalities.

a) $15 - x > 6$ **b)** $18 - x \leqslant 7$ **c)** $6 \geqslant 9 - x$

11. Solve these inequalities.

a) $20 - 2x \leqslant 5$ **b)** $3 - 4x \geqslant 11$ **c)** $25 - 3x > 7$

d) $2(6 - x) < 9$ **e)** $\frac{10 - 2x}{5} \leqslant 4$ **f)** $\frac{8 - 4x}{3} > 2$

12. Solve these linear inequalities.

a) $7 < 2x + 1 < 13$ **b)** $5 < 3x - 1 < 14$ **c)** $-1 \leqslant 5x + 4 \leqslant 19$

d) $1 \leqslant 4x - 3 < 13$ **e)** $11 \leqslant 3x + 5 < 17$ **f)** $-3 \leqslant 2x - 3 \leqslant 7$

12.2 Quadratic inequalities

THIS SECTION WILL SHOW YOU HOW TO …

✓ solve quadratic inequalities

KEY WORD

✓ solution set

Suppose $x^2 > 16$. What can you say about x?

First, change the inequality to an equals sign. $x^2 = 16$

There are two solutions, $x = 4$ and $x = -4$.

These divide the number line into three sections.

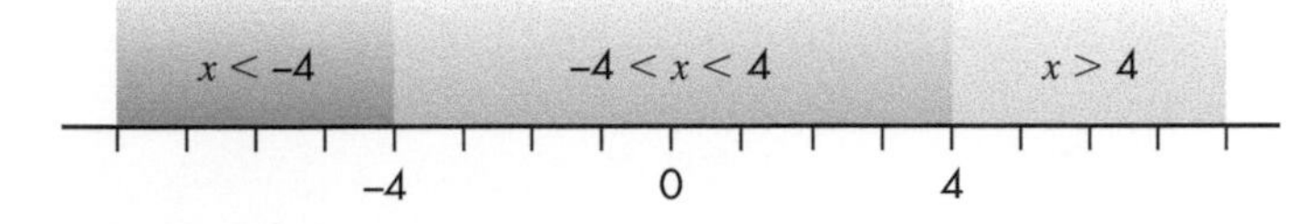

By choosing a value in each section in turn, and squaring it, you can see that:

if $x > 4$ then $x^2 > 16$ (e.g. $5^2 = 25 > 16$)

if $-4 < x < 4$ then $x^2 < 16$ (e.g. $2^2 = 4 < 16$)

if $x < -4$ then $x^2 > 16$ (e.g. $(-5)^2 = 25 > 16$)

So the **solution set** for $x^2 > 16$ is in two parts: $x < -4$ and $x > 4$.

There are two solutions, $x > 4$ or $x < -4$.

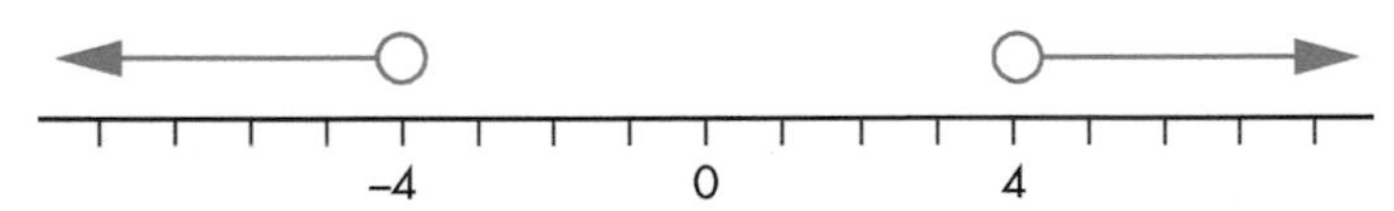

Notice that the boundary values are not included in this case because $4^2 = (-4)^2 = 16$.

Neither of them is less than 16.

The solution to $x^2 \geqslant 16$ is similar but in this case the boundary values are included because the inequality includes 16 as a possible value.

So the **solution set** for $x^2 \geqslant 16$ is in two parts: $x \leqslant -4$ and $x \geqslant 4$.

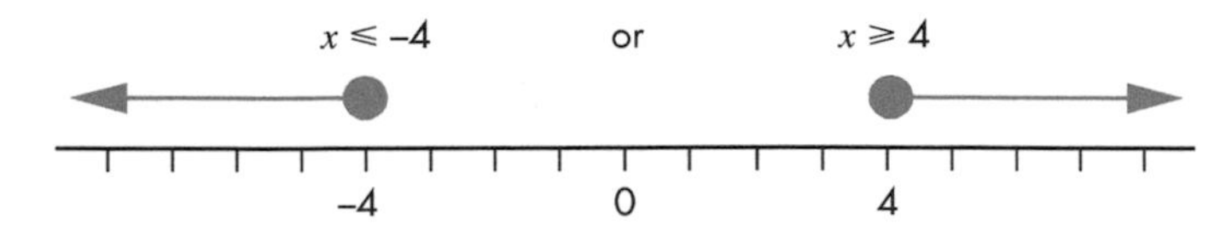

EXAMPLE 4

Solve the inequality $x^2 - x - 6 < 0$.

Factorising the quadratic gives $(x + 2)(x - 3) < 0$.

Now use a table to solve the inequality.

The key values are $x = -2$ (when $x + 2 = 0$) and $x = 3$ (when $x - 3 = 0$).

–2 3

	$x < -2$	$-2 < x < 3$	$x > 3$
$x + 2$	negative	positive	positive
$x - 3$	negative	negative	positive
$(x + 2)(x - 3)$	positive	negative	positive
	> 0	< 0	> 0

So $(x + 2)(x - 3) < 0$ when $-2 < x < 3$.

EXERCISE 12B

GRADE A

1. Solve these inequalities.

a) $x^2 \leqslant 16$ **b)** $x^2 < 4$ **c)** $x^2 > 6.25$ **d)** $x^2 \geqslant 1$

2. Solve these inequalities.

a) $2x^2 < 18$ **b)** $3x^2 > 75$ **c)** $4x^2 \geqslant 9$ **d)** $4x^2 \leqslant 1$

3. Solve these inequalities.

a) $x^2 - 4 \leqslant 0$ **b)** $x^2 - 12.25 > 0$ **c)** $8x^2 - 50 < 0$ **d)** $9 - x^2 \geqslant 0$

4. Solve each inequality.

a) $x^2 + 2x - 3 \leqslant 0$ **b)** $3x^2 + 7x > 6$ **c)** $9x^2 + 4x < 5$
d) $5x^2 + 22x \leqslant 15$ **e)** $3x^2 \leqslant 11x + 4$

5. Work out the integer values of x that satisfy the inequality $3x^2 - 19x + 20 < 0$.

6. Work out the integer values of x that satisfy the inequality $4x^2 - 12x + 5 < 0$.

12.3 Graphical inequalities

THIS SECTION WILL SHOW YOU HOW TO ...

✓ use graphs to solve inequalities

KEY WORDS

✓ region ✓ boundary ✓ origin

GRAPHICAL INEQUALITIES

A linear inequality can be plotted on a graph. The result is a **region** that lies on one side or the other of a straight line. You will recognise an inequality by the fact that it looks like an equation but instead of the equals sign it has an inequality sign: $<$, $>$, $\leqslant$ or $\geqslant$.

These are examples of linear inequalities that can be represented on a graph.

$y < 3$ $x > 7$ $-3 \leqslant y < 5$ $y \geqslant 2x + 3$ $2x + 3y < 6$ $y \leqslant x$

The method for graphing an inequality is to draw the **boundary** line that defines the inequality. This is found by replacing the inequality sign with an equals sign.

A common convention is to use a solid line when the boundary is included and a broken line when the boundary is not included. This means a solid line for $\leqslant$ and $\geqslant$ and a broken one for $<$ and $>$.

After drawing the boundary line, shade the *required region*.

To confirm on which side of the line the region lies, choose any point that is not on the boundary line and test it in the inequality. If it satisfies the inequality, that is the side required. If it doesn't, the other side is required.

Work through the six inequalities in the following example to see how the procedure is applied.

EXAMPLE 5

Show each inequality on a graph.

a) $y \leqslant 3$ **b)** $x > 7$ **c)** $-3 \leqslant y < 5$

d) $y \leqslant 2x + 3$ **e)** $2x + 3y < 6$ **f)** $y \leqslant x$

a) $y \leqslant 3$

Draw the line $y = 3$. Test a point that is not on the line. The **origin** is always a good choice if possible, as 0 is easy to test.

Putting 0 into the inequality gives $0 \leqslant 3$. The inequality is satisfied and so the region containing the origin is the side you want.

Shade it in.

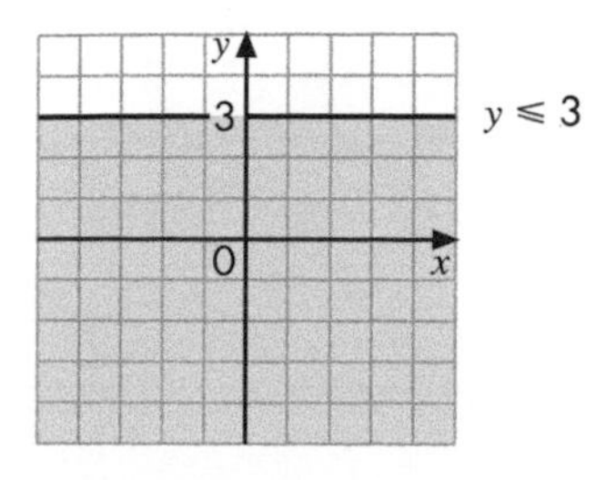

b) $x > 7$

Draw the line $x = 7$.

Test the origin (0, 0), which gives $0 > 7$. This is not true, so you want the other side of the line from the origin. Shade it in.

In this case the boundary line is not included.

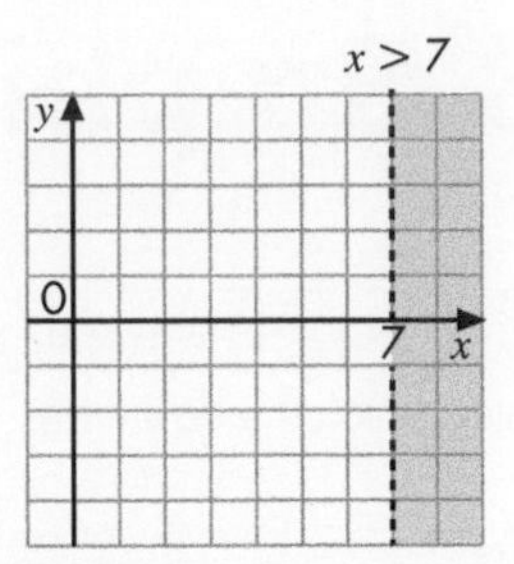

c) $-3 \leqslant y < 5$

Draw the lines $y = -3$ and $y = 5$.

Test a point that is not on either line, say (0, 0). Zero is between −3 and 5, so the required region lies between the lines. Shade it in.

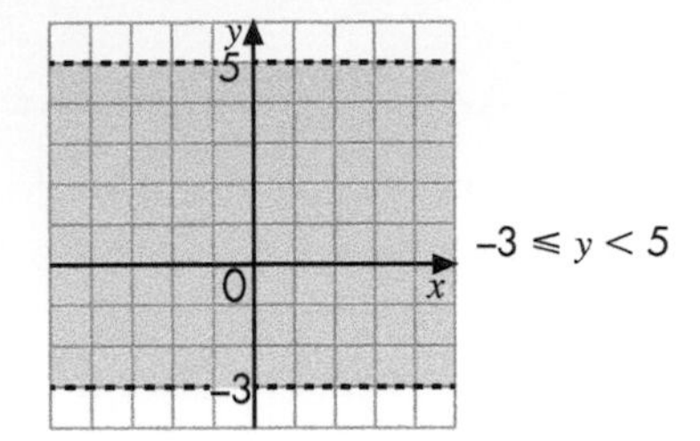

d) $y \leqslant 2x + 3$

Draw the line $y = 2x + 3$.

Test a point that is not on the line, (0, 0). Putting these x- and y-values in the inequality gives $0 \leqslant 2(0) + 3$, which is true. So the region that includes the origin is what you want. Shade it in.

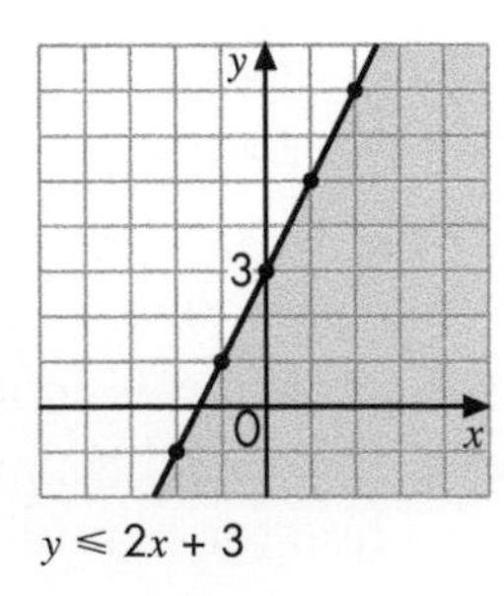

e) $2x + 3y < 6$

Draw the line $2x + 3y = 6$.

The easiest way is to find out where it crosses the axes.

If $x = 0$, $3y = 6 \Rightarrow y = 2$. Crosses y-axis at (0, 2).

If $y = 0$, $2x = 6 \Rightarrow x = 3$. Crosses x-axis at (3, 0).

Draw the line through these two points.

Test a point that is not on the line, say (0, 0). Is it true that $2(0) + 3(0) < 6$? The answer is yes, so the origin is in the region that you want. Shade it in.

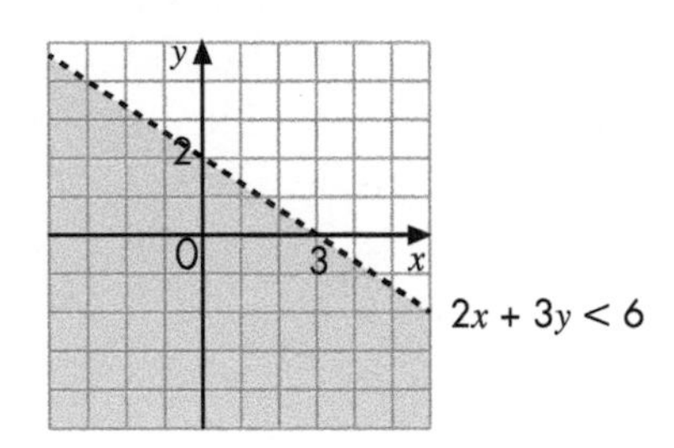

f) $y \leqslant x$

Draw the line $y = x$.

This time the origin is on the line, so pick any other point, say (1, 3). Putting $x = 1$ and $y = 3$ in the inequality gives $3 \leqslant 1$. This is not true, so the point (1, 3) is not in the region you want.

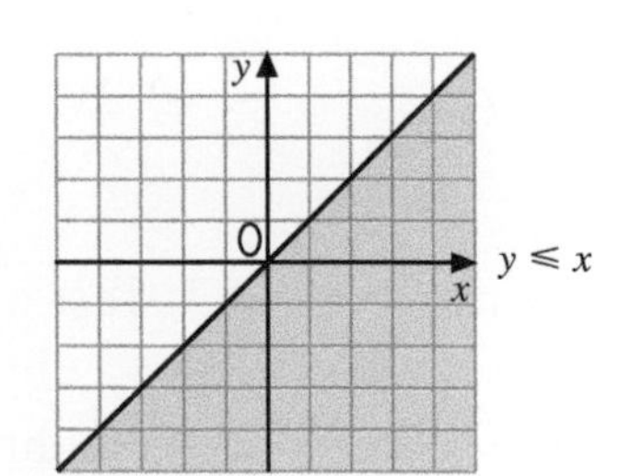

EXERCISE 12C

GRADE C

1. a) Draw the line $x = 2$. **b)** Shade the region defined by $x \leqslant 2$.

2. a) Draw the line $y = -3$. **b)** Shade the region defined by $y > -3$.

3. a) Draw the line $x = -2$. **b)** Draw the line $x = 1$ on the same grid.

c) Shade the region defined by $-2 \leqslant x \leqslant 1$.

4. a) Draw the line $y = -1$. **b)** Draw the line $y = 4$ on the same grid.

c) Shade the region defined by $-1 < y \leqslant 4$.

GRADE B

5. a) On the same grid, draw the regions defined by these inequalities.

i) $-3 \leqslant x \leqslant 6$ **ii)** $-4 < y \leqslant 5$

b) Are these points in the region defined by both inequalities?

i) (2, 2) **ii)** (1, 5) **iii)** (–2, –4)

6. a) Draw the line $y = 2x - 1$. **b)** Shade the region defined by $y < 2x - 1$.

7. a) Draw the line $x + y = 4$. **b)** Shade the region defined by $x + y \leqslant 4$.

8. a) Draw the line $y = \frac{1}{2}x + 3$. **b)** Shade the region defined by $y \geqslant \frac{1}{2}x + 3$.

9. Shade the region defined by $x + y \geqslant 3$.

More than one inequality

When several inequalities are given they define a **region**.

EXAMPLE 6

Shade the region where $x \geqslant 1$, $y \geqslant 2$ and $x + y \leqslant 8$.

Start by drawing the lines $x = 1$, $y = 2$ and $x + y = 8$.

Check which side of each line is required. Points inside the shaded triangle satisfy all three inequalities.

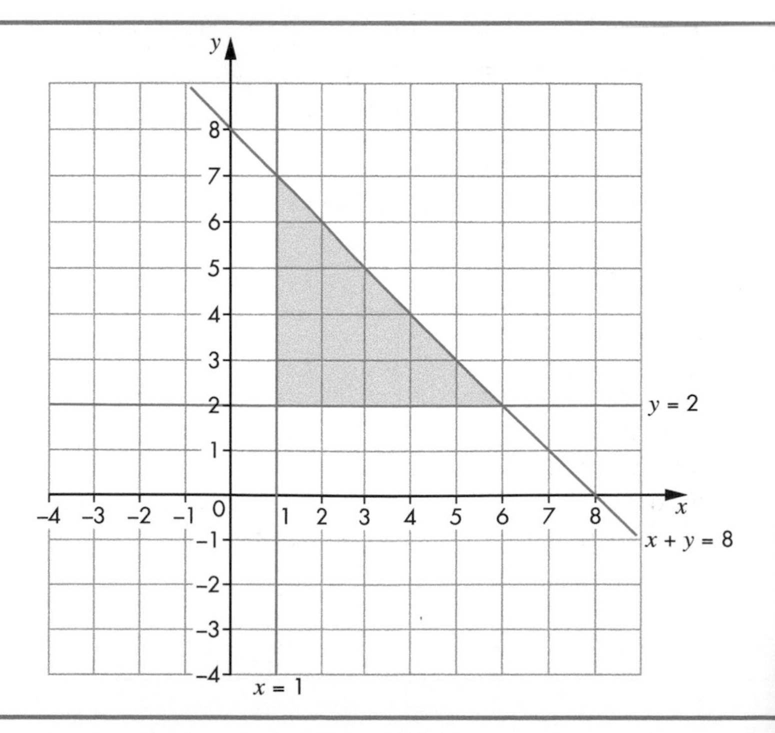

EXERCISE 12D

GRADE B

> **HINTS AND TIPS**
>
> Draw a set of Cartesian axes for each question.

1. Shade the region where $y > 4$ and $x > 5$.
2. Shade the region where $x > 0$, $y \geqslant 3$ and $x + y \leqslant 7$.
3. Shade the region where $x \leqslant 7$, $y \leqslant 6$ and $x + y \geqslant 3$.
4. Shade the region where $x \leqslant 10$, $y \geqslant 0$ and $y \leqslant x$.
5. Shade the region where $x + y > 4$, $x + y < 8$.
6. Shade the region where $x + y \geqslant 0$, $y \geqslant 3$.
7. Write down two inequalities to describe this region.

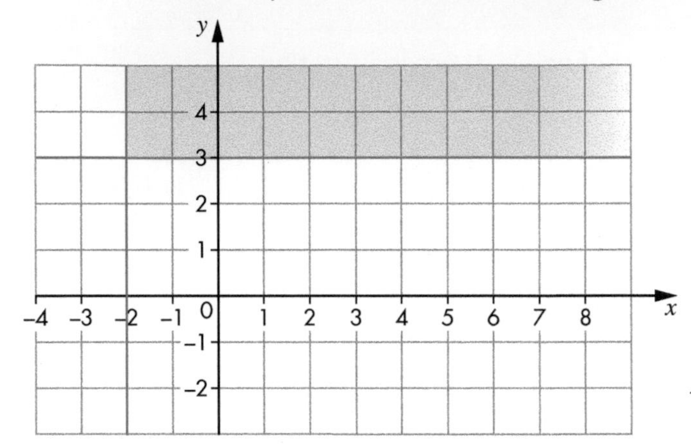

8. Write down three inequalities to define this region.

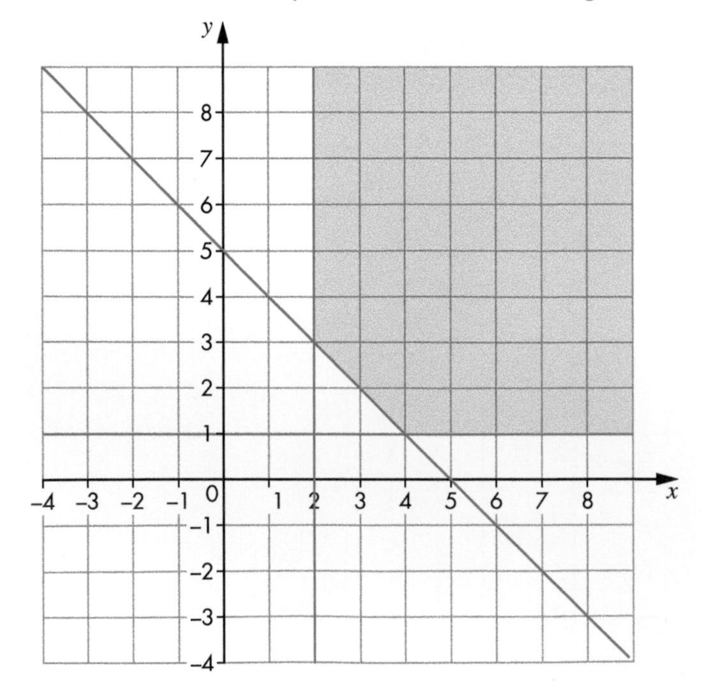

9. Show by shading the region where $y \geqslant x$, $x + y \leqslant 10$ and $x \geqslant 2$.
10. Show by shading the region where $y < x$ and $x < 0$.

More complex graphical inequalities

When more complex inequalities are involved you can use the same procedure. Draw the boundary lines first then decide on the required region.

EXAMPLE 7

Shade the region defined by $y \geqslant 2x - 4$, $x + 2y \leqslant 12$ and $3x + y \geqslant 6$.

The line $y = 2x - 4$ has gradient 2 and intercept -4 on the y-axis.

The inequality indicates the region above this line.

To draw the line $x + 2y = 12$, find where it crosses the axes.

If $x = 0$, $2y = 12 \Rightarrow y = 6$ so it goes through $(0, 6)$.

If $y = 0$, $x = 12$ so it goes through $(12, 0)$.

You want the region below this line.

The line $3x + y = 6$ passes through $(0, 6)$ and $(2, 0)$.

You want the region above this line.

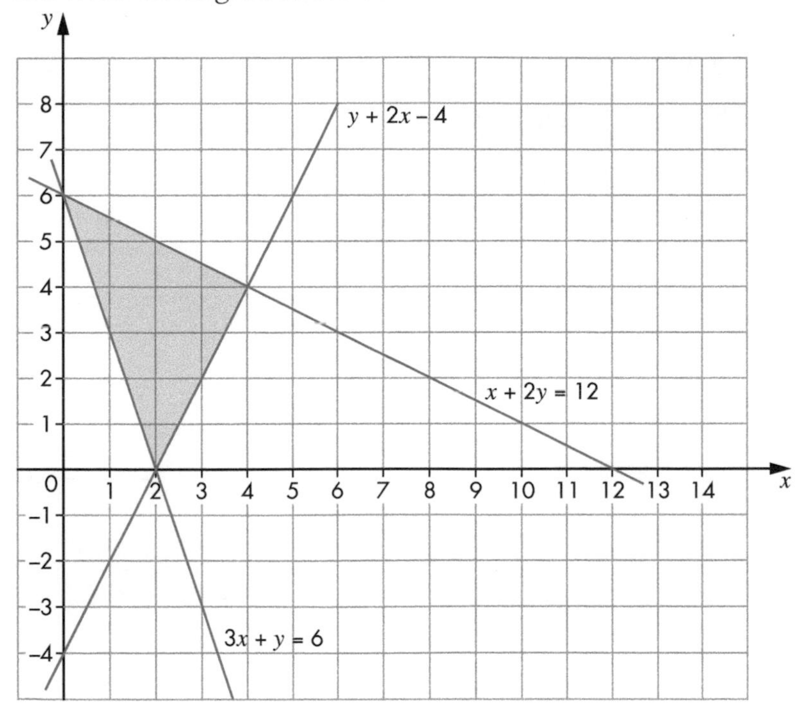

The required region is shaded.

EXERCISE 12E

GRADE B

1. **a)** Draw the line $y = x$ on a Cartesian grid.

   b) Draw the line $2x + 5y = 10$ on the same diagram.

   c) Draw the line $2x + y = 6$ on the same diagram.

   d) Shade the region defined by $y \geqslant x$, $2x + 5y \geqslant 10$ and $2x + y \leqslant 6$.

   e) Are the following points in the region defined by these inequalities?

   i) $(1, 1)$ **ii)** $(2, 2)$ **iii)** $(1, 3)$

2. **a)** On the same grid, shade the regions defined by these inequalities.

   i) $y > x - 3$ **ii)** $3y + 4x \leqslant 24$ **iii)** $x \geqslant 2$

   b) Are these points in the region defined by all three inequalities?

   i) (1, 1) **ii)** (2, 2) **iii)** (3, 3) **iv)** (4, 4)

HINTS AND TIPS

Find the points where the line crosses each axis.

3. **a)** On a graph draw the lines $y = x$, $x + y = 8$ and $y = 2$.

   b) Label the region R where $y \leqslant x$, $x + y \leqslant 8$ and $y \geqslant 2$. Shade the region that is required.

4. **a)** On a graph draw the lines $y = x - 4$, $y = 0.5x$ and $y = -x$.

   b) Show the region S where $y \geqslant x - 4$, $y \leqslant 0.5x$ and $y \geqslant -x$. Shade the region that is required.

   c) What is the largest y-coordinate of a point in S?

   d) What is the smallest y-coordinate of a point in S?

   e) What is the largest value of $x + y$ for a point in S?

5. Explain how you would find which side of the line represents the inequality $y < x + 2$.

6. The region marked R is bounded by the lines $x + y = 3$, $y = \frac{1}{2}x + 3$ and $y = 5x - 15$.

   a) What three inequalities are satisfied in region R?

   b) What is the greatest value of $x + y$ in region R?

   c) What is the greatest value of $x - y$ in region R?

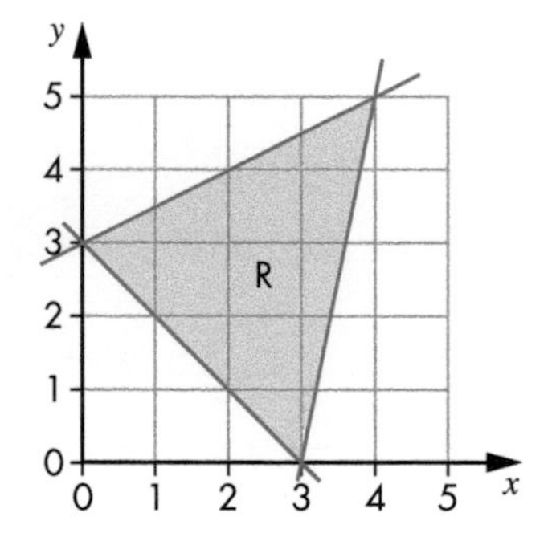

7. **a)** Copy the grid on the right.

   On your grid, sketch a graph of the equations $y = x^2 - 3$ and $y = 2x$.

   b) On your sketch, shade the region that satisfies both inequalities $y \geqslant x^2 - 3$ and $y \leqslant 2x$.

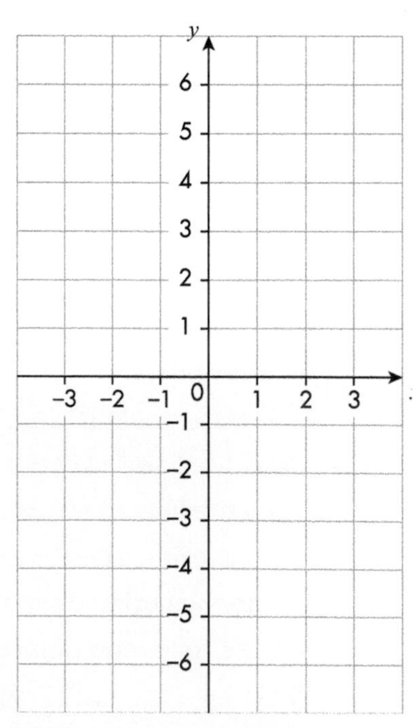

Exam-style questions

1. Solve the inequality $2(x - 3) \geqslant 16 - 3x$. GRADE B

2. Solve the inequality $(x + 2)^2 < (x - 6)^2$. GRADE B

3. Solve the inequality $x^2 + 6x + 5 > 0$. GRADE A

4. Find the integer solutions of the inequality $N^2 - 2N < 8$. GRADE A

5. a) On the same set of axes, sketch the graphs of $y = x^2 - 3$ and $y = 2x$. GRADE A

b) Use the graphs to solve the inequality $x^2 - 3 \leqslant 2x$.

6. ABCD is a rectangle. GRADE A

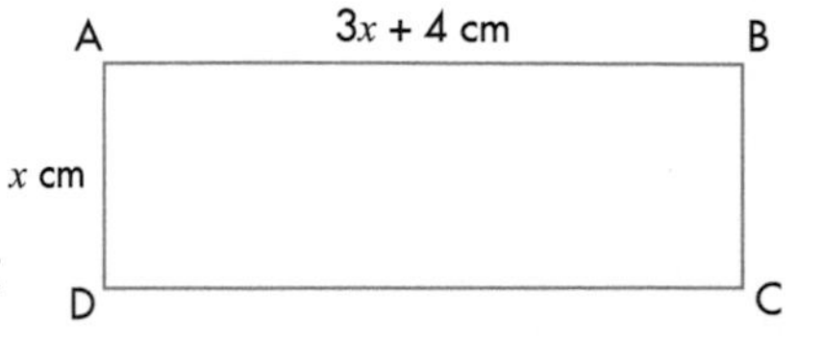

Not to scale

a) The area of the rectangle is at least 160 cm^2.

Use this fact to set up an inequality for x.

b) Solve the inequality and hence find the smallest possible length of AD.

7. Each side of this equilateral triangle is N cm long, where N is an integer.

The area of the triangle is at least 100 cm^2.

a) Set up an inequality for N.

b) Find the smallest positive integer value of N. GRADE A

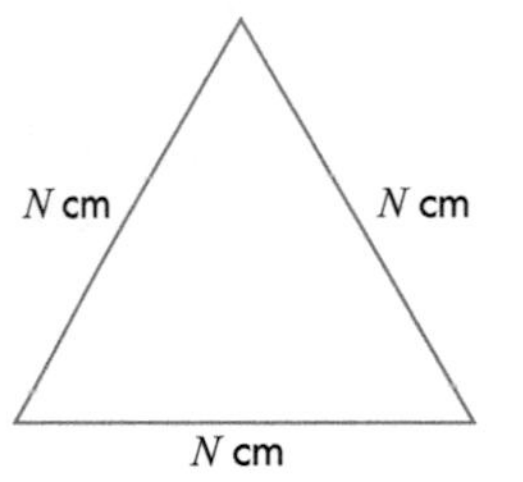

13 Coordinate geometry

13.1 Cartesian grids and straight-line graphs

THIS SECTION WILL SHOW YOU HOW TO ...

✓ work out the midpoint of the line between two points
✓ draw straight-line graphs
✓ calculate and use the gradient of a line

KEY WORDS

✓ Cartesian coordinates
✓ equation
✓ Cartesian plane
✓ midpoint
✓ line segment
✓ straight-line graph
✓ scale
✓ gradient

Using coordinates

Coordinates are sometimes called **Cartesian coordinates** after their inventor, René Descartes. The grid is sometimes called the **Cartesian plane**.

Note: The **equation** of the x-axis is $y = 0$ and the equation of the y-axis is $x = 0$.

Midpoints

The **midpoint** of a **line segment** is the same distance from each end.

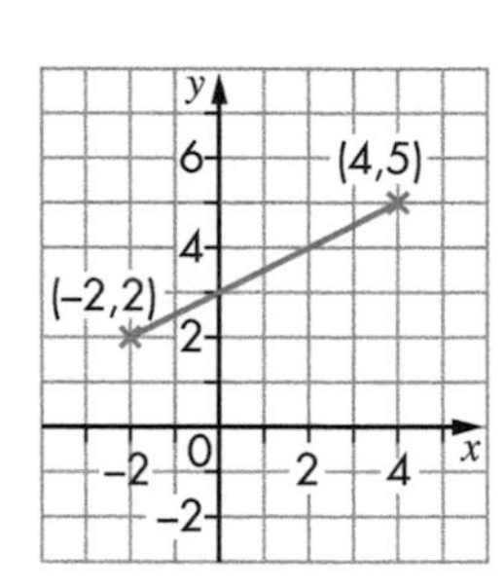

To find the coordinates of the midpoint, add the coordinates of the end points and divide by 2.

In the example shown the coordinates of the midpoint are

$$\left(\frac{-2+4}{2}, \frac{2+5}{2}\right) = (1, 3.5).$$

EXERCISE 13A

GRADE C

1. a) Write down the equation of the straight line through:
  i) A and C   **ii)** B and D   **iii)** D and E.
  b) Which point is on the line with equation $y = 5$?
  c) Which point is on the line with equation $x = -3$?

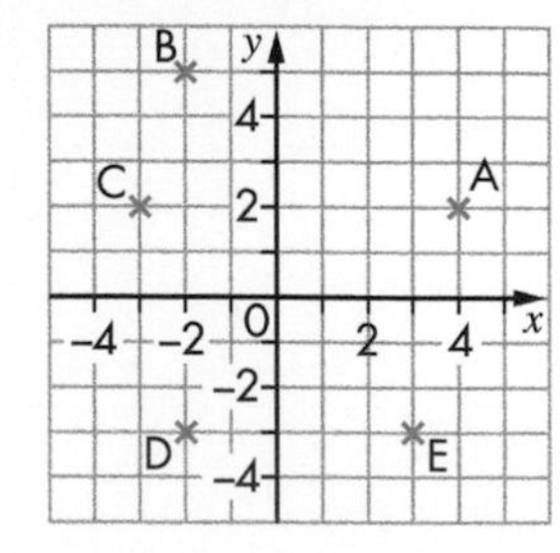

2. **a)** Find the coordinates of the midpoint of PQ.
b) Find the coordinates of the midpoint of:
i) QR **ii)** RS **iii)** SP.
c) The line $y = -2$ has three points on one side and one point on the other.
Which is the point on its own?

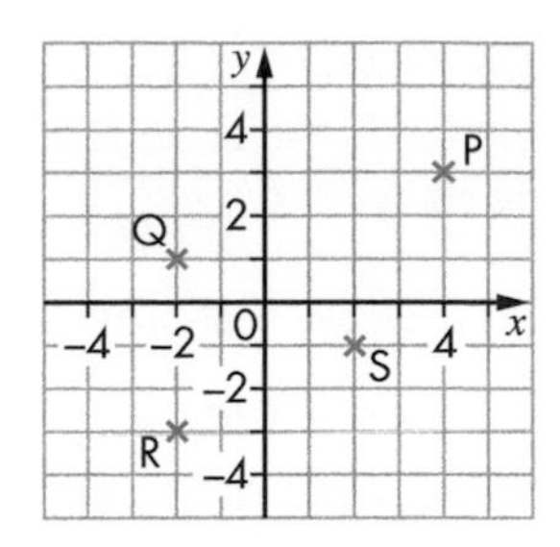

3. **a)** On the grid in question **1**, find the midpoints of AE and CD.
b) Find the equation of the line through both midpoints.

Drawing straight-line graphs

An equation of the form $y = mx + c$ where m and c are numbers will give a **straight-line graph**.

EXAMPLE 1

Draw the graph of $y = 4x - 5$ for values of x from 0 to 5.
This is usually written as $0 \leqslant x \leqslant 5$.
Choose three values for x: these should be the highest and lowest x-values and one in between. Work out the y-values by substituting the x-values into the equation.
Keep a record of your calculations in a table, as shown below.

x	0	3	5
y			

When $x = 0$, $y = 4(0) - 5 = -5$. This gives the point $(0, -5)$.
When $x = 3$, $y = 4(3) - 5 = 7$. This gives the point $(3, 7)$.
When $x = 5$, $y = 4(5) - 5 = 15$. This gives the point $(5, 15)$.

x	0	3	5
y	–5	7	15

You now have to decide the extent (range) of the axes.
You can find this out by looking at the coordinates that you have so far.
The smallest x-value is 0, the largest is 5. The smallest y-value is –5, the largest is 15.
Now draw the axes, plot the points and complete the graph.
It is nearly always a good idea to choose 0 as one of the x-values.

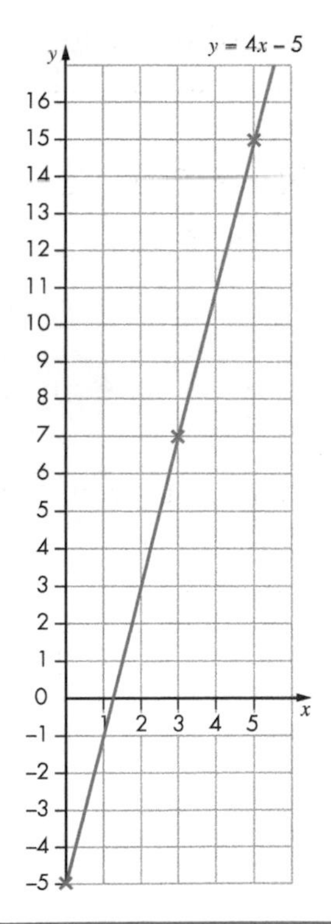

EXERCISE 13B

GRADE C

1. Draw the graph of $y = \frac{x}{2} - 3$ for $0 \leq x \leq 10$.
2. Draw the graph of $y = x + 4$ for $0 \leq x \leq 6$.
3. **a)** On the same set of axes, draw the graphs of $y = 3x - 2$ and $y = 2x + 1$ for $0 \leq x \leq 5$.
   b) At which point do the two lines intersect?
4. **a)** On the same axes, draw the graphs of $y = 4x - 5$ and $y = 2x + 3$ for $0 \leq x \leq 5$.
   b) At which point do the two lines intersect?
5. **a)** On the same axes, draw the graphs of $y = \frac{x}{3} - 1$ and $y = \frac{x}{2} - 2$ for $0 \leq x \leq 12$.
   b) At which point do the two lines intersect?
6. **a)** On the same axes, draw the graphs of $y = 3x + 1$ and $y = 3x - 2$ for $0 \leq x \leq 4$.
   b) Do the two lines intersect? If not, why not?

More straight-line graphs

The equation of a straight line can take a number of different forms.

EXAMPLE 2

Draw the straight line with the equation $2x + 3y = 12$.

You need to find pairs of numbers that satisfy this equation.

It is always useful to find where the line crosses the axes.

If $x = 0$, then $0 + 3y = 12 \Rightarrow y = 4$ so $(0, 4)$ is on the line.

If $y = 0$, then $2x + 0 = 12 \Rightarrow x = 6$ so $(6, 0)$ is on the line.

Use these points to draw the line.

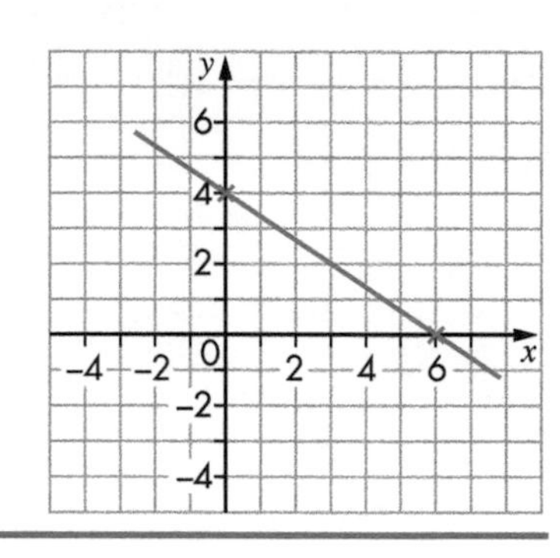

EXERCISE 13C

GRADE C

1. A line has equation $x + 2y = 6$.
   a) Where does it cross the x-axis? **b)** Where does it cross the y-axis?
   c) Draw a graph of $x + 2y = 6$.
2. Draw a graph of the line $2x + y = 8$.
3. Draw a graph of the line $x + 3y = 9$.
4. Draw a graph of the line $5x + 2y = 10$.
5. Draw a graph of the line $2y - x = 4$.

Gradient

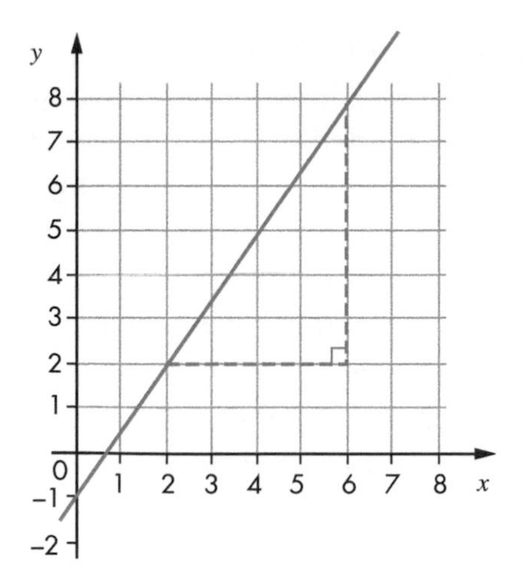

The slope of a line is called its gradient. The steeper the slope of the line, the larger the value of the gradient.

The gradient of the line shown here can be measured by drawing, as large as possible, a right-angled triangle that has part of the line as its hypotenuse (sloping side). The gradient is then given by:

$$\text{gradient} = \frac{\text{distance measured up}}{\text{distance measured along}}$$

$$= \frac{\text{difference on } y\text{-axis}}{\text{difference on } x\text{-axis}}$$

For example, to measure the steepness of the line in the diagram draw a right-angled triangle of which the hypotenuse is part of this line. Count how many squares there are on the vertical side. This is the difference between your *y*-coordinates. In the graph above, this is 6. Count how many squares there are on the horizontal side. This is the difference between your *x*-coordinates. In the case above, this is 4. Then:

$$\text{gradient} = \frac{\text{difference in the } y\text{-coordinates}}{\text{difference in the } x\text{-coordinates}} = \frac{6}{4} = \frac{3}{2} \text{ or } 1.5$$

You can use the method of counting squares in cases like this, where the **scale** is one square to one unit.

Remember: When a line slopes down from left to right, the gradient is negative, so you must place a minus sign in front of the calculated fraction.

EXAMPLE 3

Find the gradient of each of these lines.

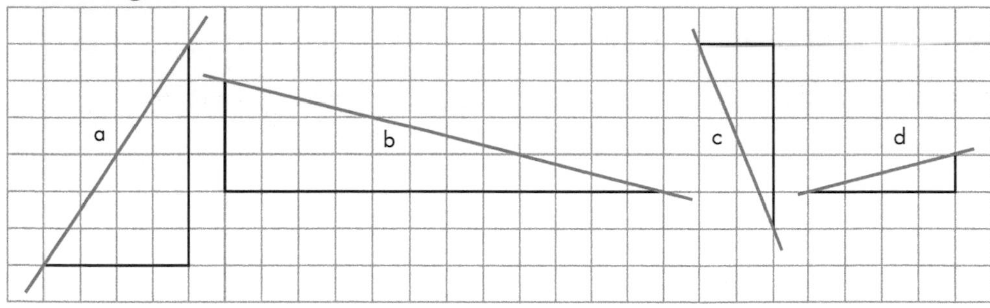

In each case, a sensible choice of triangle has already been made.

a) *y*-difference = 6, *x*-difference = 4	Gradient = 6 ÷ 4 = $\frac{3}{2}$ = 1.5
b) *y*-difference = 3, *x*-difference = 12	Line slopes down from left to right, so gradient = –(3 ÷ 12) = $-\frac{1}{4}$ = –0.25
c) *y*-difference = 5, *x*-difference = 2	Line slopes down from left to right, so gradient = –(5 ÷ 2) = $-\frac{5}{2}$ = –2.5
d) *y*-difference = 1, *x*-difference = 4	Gradient = 1 ÷ 4 = $\frac{1}{4}$ = 0.25

Drawing a line with a given gradient

To draw a line with a certain gradient, you reverse the process described above. That is, you first draw the right-angled triangle, using the given gradient. For example, take a gradient of 2.

Start at a convenient point (A in the diagrams below). A gradient of 2 means that for an x-step of 1 the y-step must be 2 (because 2 is the fraction). So, move one square across and two squares up, and mark a dot.

Repeat this as many times as you like and draw the line. You can also move one square back and two squares down, which gives the same gradient, as the third diagram shows.

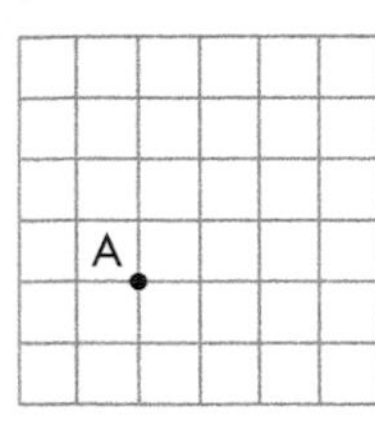

Stage 1

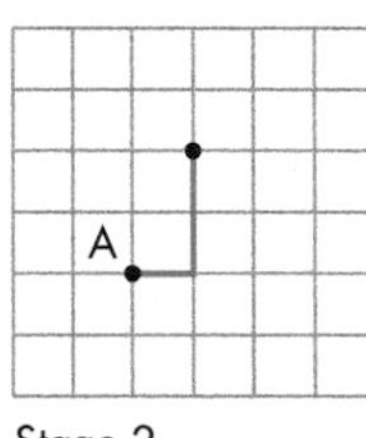

Stage 2

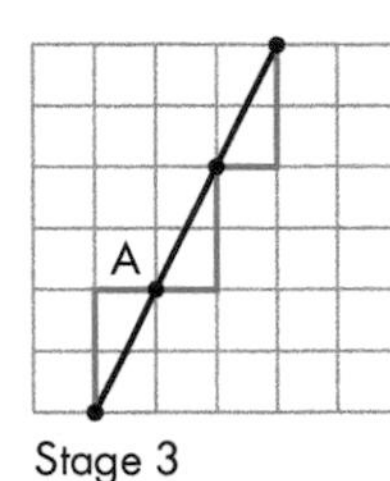

Stage 3

EXAMPLE 4

Draw lines with these gradients.

a) $\frac{1}{3}$ **b)** -3 **c)** $-\frac{1}{4}$

a) This is a fractional gradient which has a y-step of 1 and an x-step of 3. Move three squares across and one square up every time.

b) This is a negative gradient, so for every one square across, move three squares down.

c) This is also a negative gradient and it is a fraction. So for every four squares across, move one square down.

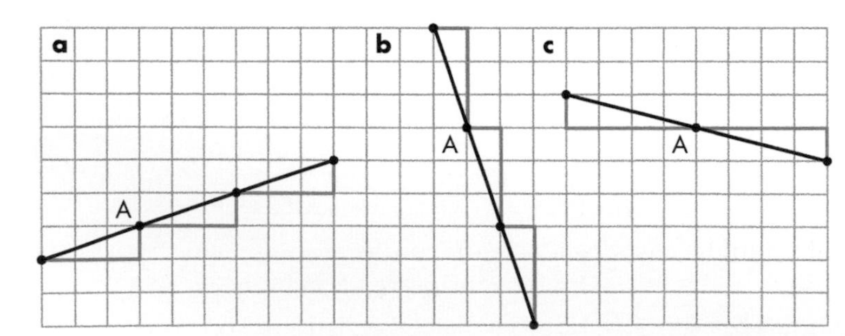

EXERCISE 13D

GRADE C

1. Find the gradient of each line.

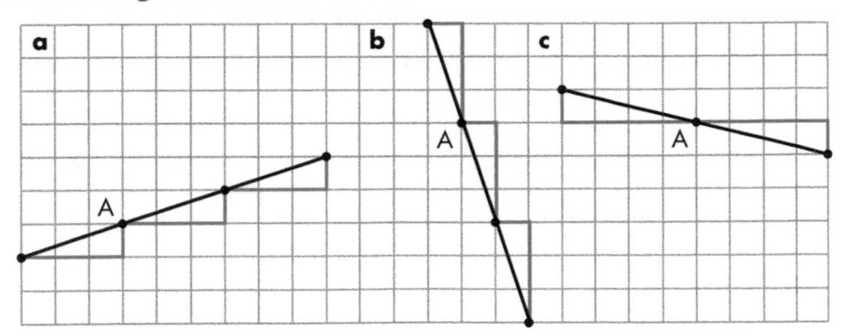

2. Draw lines with these gradients.

a) 4 **b)** $\frac{2}{3}$ **c)** -2 **d)** $-\frac{4}{5}$ **e)** 6 **f)** -6

3. Find the gradient of each line. What is special about these lines?

a)

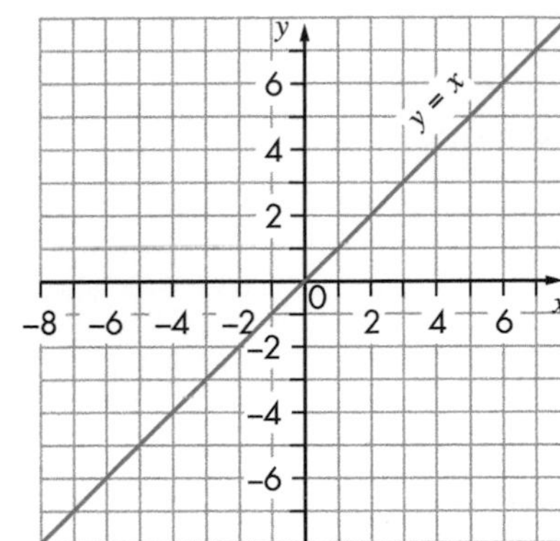

b)

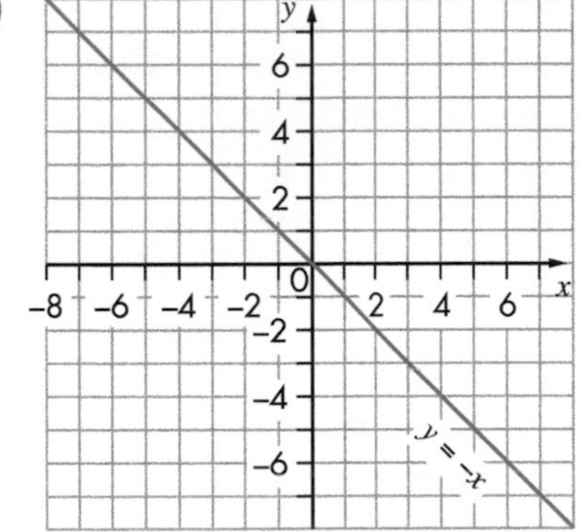

4. This graph shows the profile of a fell race. The horizontal axis shows the distance, in miles, of the race. The vertical axis is the height above sea level throughout the race. There are 5280 feet in a mile.

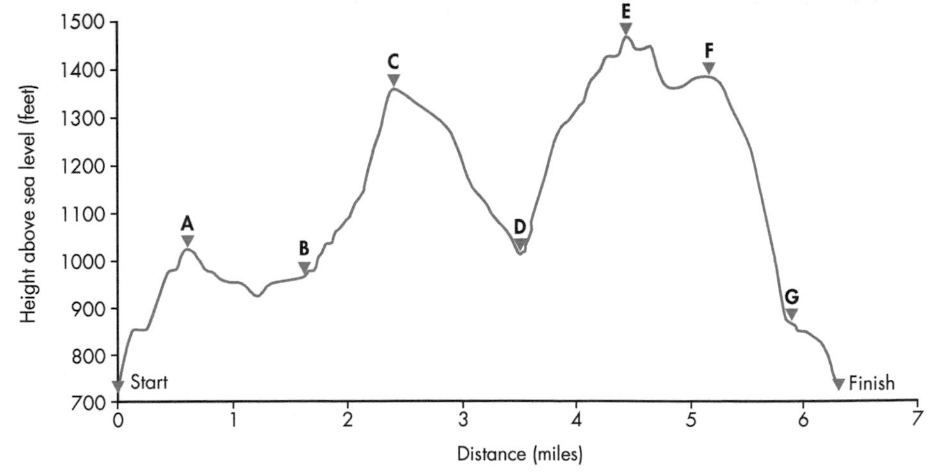

a) Work out the approximate gradient of the race from the start to point A.

b) The steepest part of the race is from F to G.

i) How can you tell this from the graph?

ii) Work out the approximate gradient from F to G.

c) Fell races are classified in terms of distance and amount of ascent.

Distance:	Short (S)	Less than 6 miles
	Medium (M)	Between 6 and 12 miles
	Long (L)	Over 12 miles
Ascent	C	An average of 100 to 125 feet per mile
	B	An average of 125 to 250 feet per mile
	A	An average of 250 or more feet per mile

So, for example, an AL race would be over 12 miles and have at least 250 feet of ascent, on average, per mile.

What category is the race above?

5. These lines are drawn on grids in which each square is equivalent to one unit. The line on grid **e** is horizontal. The lines on grids **a** to **d** get nearer and nearer to the horizontal.

a)

b)

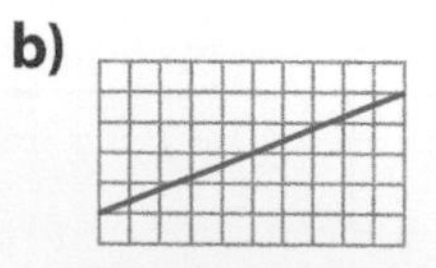

c)

d)

e)

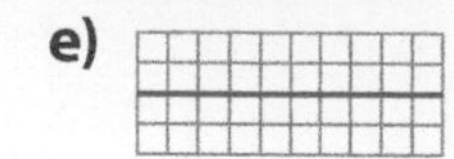

Find the gradient of each line in grids **a** to **d**. By looking at the values you obtain, what do you think the gradient of a horizontal line is?

6. These lines are drawn on grids in which each square is equivalent to one unit. The line on grid **e** is vertical. The lines on grids **a** to **d** get nearer and nearer to the vertical.

a)

b)

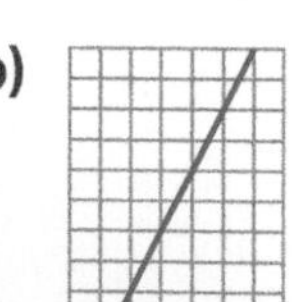

c)

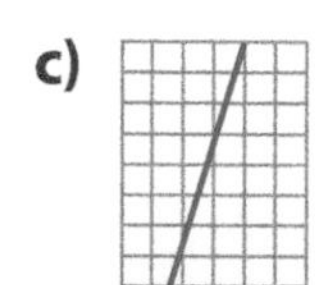

d)

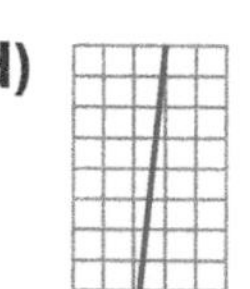

e)

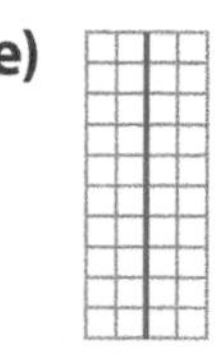

Find the gradient of each line in grids **a** to **d**. By looking at the values you obtain, what do you think the gradient of a vertical line is?

7. Find the gradients of the straight lines through each pair of points.

a) (0, 0) and (2, 8) **b)** (0, 0) and (8, 2) **c)** (3, 0) and (5, 5)

d) (4, 0) and (5, 10) **e)** (0, 8) and (4, 0) **f)** (0,0) and (10, –2)

g) (0, 5) and (7, 5) **h)** (0, 9) and (6, 0)

13.2 The equation of a straight line in the forms $y = mx + c$ and $y - y_1 = m(x - x_1)$

THIS SECTION WILL SHOW YOU HOW TO …

✓ use the gradient–intercept method to draw straight-line graphs
✓ find the equation of a straight line from its gradient and the y-intercept

KEY WORDS

✓ coefficient
✓ constant term
✓ cover-up method
✓ gradient
✓ gradient–intercept
✓ intercept

Drawing graphs by the gradient–intercept method

The ideas that you have already discovered lead to another way of plotting lines, known as the **gradient–intercept** method.

EXAMPLE 5

Draw the graph of $y = 3x - 1$, using the gradient–intercept method.

- Because the **constant term** is -1, you know that the graph goes through the y-axis at -1. Mark this point with a dot or a cross (**A** on diagram **i**).
- The number in front of x (called the **coefficient** of x) gives the relationship between y and x. 3 is the coefficient of x and this tells you that the y-value is 3 times the x-value, so the gradient of the line is 3. For an x-step of one unit, there is a y-step of three. Starting at -1 on the y-axis, move one square across and three squares up and mark this point with a dot or a cross (**B** on diagram **i**).

i)

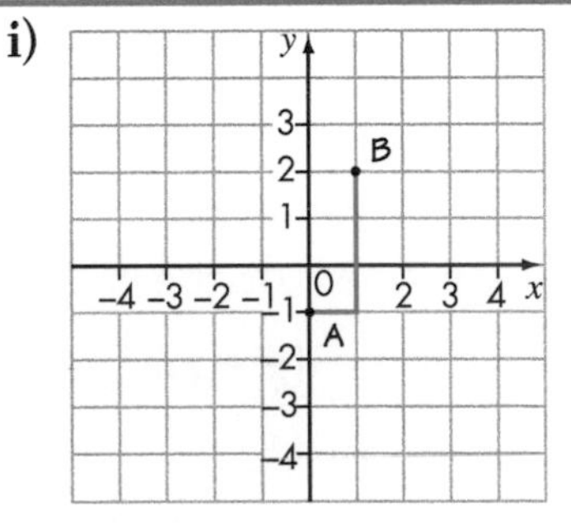

ii)

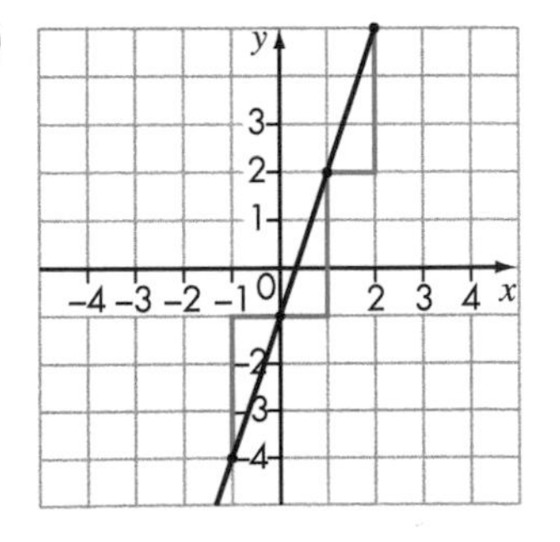

Repeat this from every new point. You can also move one square back and three squares down. When you have marked enough points, join the dots (or crosses) to make the graph (diagram **ii**). Note that if the points are not in a straight line, you have made a mistake.

In any equation of the form $\boldsymbol{y = mx + c}$, the constant term, c, is the **intercept** on the y-axis and the coefficient of x, m, is the **gradient** of the line.

EXERCISE 13E

GRADE C

1. Draw these lines, using the gradient–intercept method. Use the same grid, taking x from –10 to 10 and y from –10 to 10. If the grid gets too 'crowded', draw another one.

a) $y = 2x + 6$ **b)** $y = 3x - 4$ **c)** $y = \frac{1}{2}x + 5$

d) $y = x + 7$ **e)** $y = 4x - 3$ **f)** $y = 2x - 7$

g) $y = \frac{1}{4}x - 3$ **h)** $y = \frac{2}{3}x + 4$ **i)** $y = 6x - 5$

j) $y = x + 8$ **k)** $y = \frac{4}{5}x - 2$ **l)** $y = 3x - 9$

2. a) Using the gradient–intercept method, draw these lines on the same grid. Use axes with ranges $-6 \leqslant x \leqslant 6$ and $-8 \leqslant y \leqslant 8$.

i) $y = 3x + 1$ **ii)** $y = 2x + 3$

b) Where do the lines cross?

3. a) Using the gradient–intercept method, draw these lines on the same grid. Use axes with ranges $-14 \leqslant x \leqslant 4$ and $-2 \leqslant y \leqslant 6$.

i) $y = \frac{x}{3} + 3$ **ii)** $y = \frac{x}{4} + 2$

b) Where do the lines cross?

4. a) Using the gradient–intercept method, draw these lines on the same grid. Use axes with ranges $-4 \leqslant x \leqslant 6$ and $-6 \leqslant y \leqslant 8$.

i) $y = x + 3$ **ii)** $y = 2x$

b) Where do the lines cross?

5. Here are the equations of three lines.

A: $y = 3x - 1$ B: $2y = 6x - 4$ C: $y = 2x - 2$

a) State a mathematical property that lines A and B have in common.
b) State a mathematical property that lines B and C have in common.
c) Which of the following points is the intersection of lines A and C?
(1, –4) (–1, –4) (1, 4)

GRADE B

6. a) What is the gradient of line A?
b) What is the gradient of line B?
c) What angle is there between lines A and B?
d) What relationship do the gradients of A and B have with each other?
e) Another line C has a gradient of 3. What is the gradient of a line perpendicular to C?

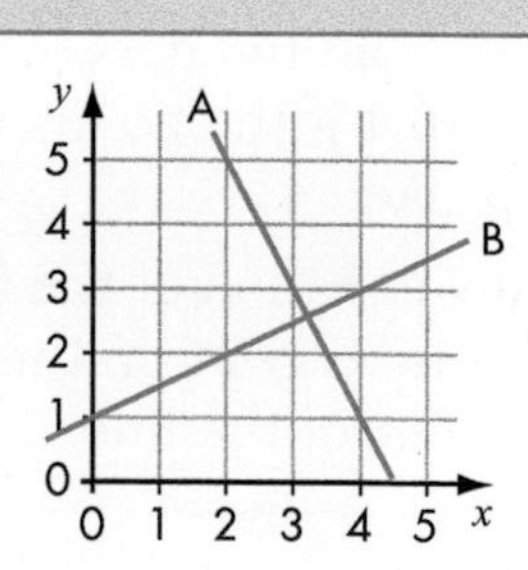

The x-axis has the equation $y = 0$. This means that all points on the x-axis have a y-value of 0.

The y-axis has the equation $x = 0$. This means that all points on the y-axis have an x-value of 0.

You can use these facts to draw any line that has an equation of the form $ax + by = c$.

EXAMPLE 6

Draw the graph of $4x + 5y = 20$.

Because the value of x is 0 on the y-axis, you can solve the equation for y.

$4(0) + 5y = 20 \quad \Rightarrow 5y = 20 \quad \Rightarrow y = 4$

Hence, the line passes through the point (0, 4) on the y-axis (diagram **A**).

Because the value of y is 0 on the x-axis, you can also solve the equation for x.

$4x + 5(0) = 20 \quad \Rightarrow 4x = 20 \quad \Rightarrow x = 5.$

Hence, the line passes through the point (5, 0) on the x-axis (diagram **B**). You need only two points to draw a line. (Normally, you would like a third point but in this case you can accept two.) Draw the graph by joining the points (0, 4) and (5, 0) (diagram **C**).

A

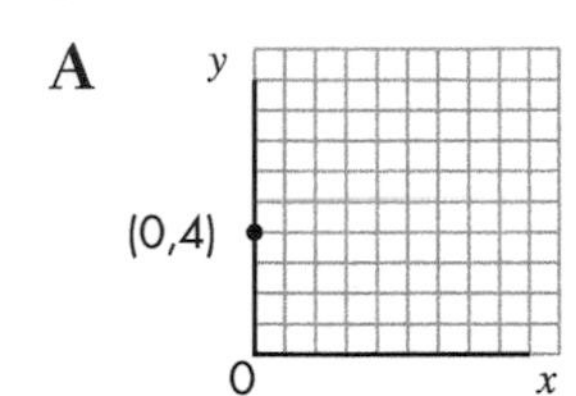

B

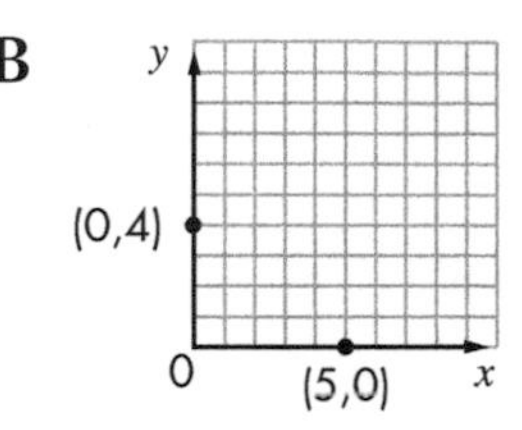

C

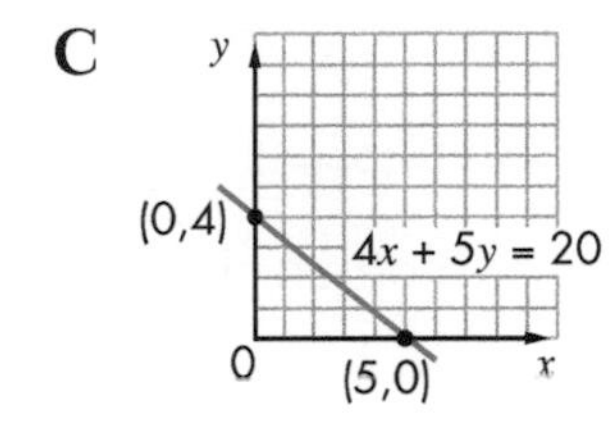

This type of equation can be drawn very easily, without much working at all, using the **cover-up method**.

Start with the equation: $4x + 5y = 20$

Cover up the x-term: $+ 5y = 20$

Solve the equation (when $x = 0$): $y = 4$

Now cover up the y-term: $4x +$ $= 20$

Solve the equation (when $y = 0$): $x = 5$

This gives the points (0, 4) on the y-axis and (5, 0) on the x-axis.

EXAMPLE 7

Draw the graph of $2x - 3y = 12$.

Start with the equation: $2x - 3y = 12$

Cover up the x-term: $- 3y = 12$

Solve the equation (when $x = 0$): $y = -4$

Now cover up the y-term: $2x +$ $=12$

Solve the equation (when $y = 0$): $x = 6$

This gives the points (0, −4) on the y-axis and (6, 0) on the x-axis.

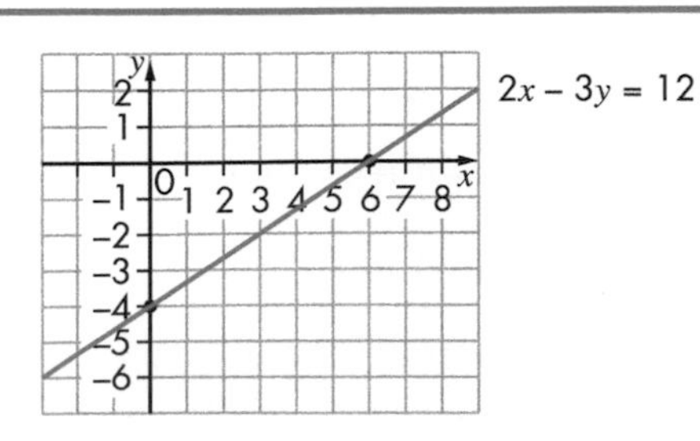

EXERCISE 13F

GRADE B

1. Draw these lines using the cover-up method. Use the same grid, taking x from −10 to 10 and y from −10 to 10. If the grid gets too 'crowded', draw another.

a) $3x + 2y = 6$ **b)** $4x + 3y = 12$ **c)** $4x - 5y = 20$
d) $x + y = 10$ **e)** $3x - 2y = 18$ **f)** $x - y = 4$
g) $5x - 2y = 15$ **h)** $2x - 3y = 15$ **i)** $6x + 5y = 30$
j) $x + y = -5$ **k)** $x + y = 3$ **l)** $x - y = -4$

2. a) Using the cover-up method, draw the following lines on the same grid.

Use axes with ranges $-2 \leqslant x \leqslant 6$ and $-2 \leqslant y \leqslant 6$.

i) $2x + y = 4$ **ii)** $x - 2y = 2$

b) Where do the lines cross?

3. a) Using the cover-up method, draw the following lines on the same grid.

Use axes with ranges $-2 \leqslant x \leqslant 6$ and $-3 \leqslant y \leqslant 6$.

i) $x + 2y = 6$ **ii)** $2x - y = 2$

b) Where do the lines cross?

4. a) Using the cover-up method, draw the following lines on the same grid.

Use axes with ranges $-6 \leqslant x \leqslant 8$ and $-2 \leqslant y \leqslant 8$.

i) $x + y = 6$ **ii)** $x - y = 2$

b) Where do the lines cross?

5. These are the equations of three lines.

A: $2x + 6y = 12$ B: $x - 2y = 6$ C: $x + 3y = -9$

a) State a mathematical property that lines A and B have in common.
b) State a mathematical property that lines B and C have in common.
c) State a mathematical property that lines A and C have in common.
d) The line A crosses the y-axis at (0, 2). The line C crosses the x-axis at (−9, 0). Find values of a and b so that this line passes through these two points.

$ax + by = 18$

6. The diagram shows an octagon ABCDEFGH.

The equation of the line through A and B is $y = 3$.

The equation of the line through B and C is $x + y = 4$.

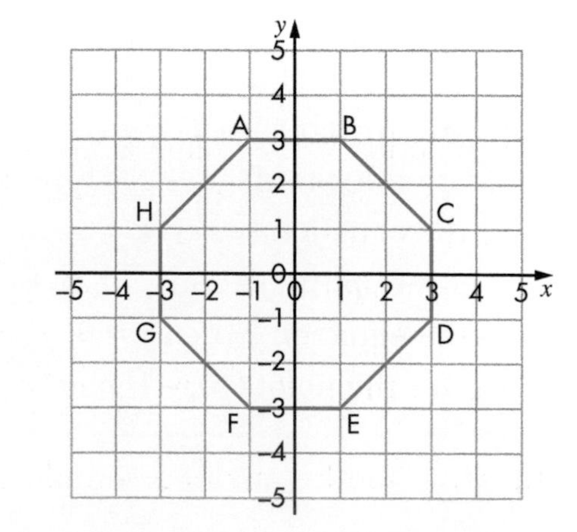

a) Write down the equation of the lines through:
i) C and D **ii)** D and E **iii)** E and F
iv) F and G **v)** G and H **vi)** H and A.

b) The gradient of the line through F and B is 3.
Write down the gradient of the lines through:
i) A and E **ii)** G and C **iii)** H and D.

Finding the equation of a line from its graph

When a graph can be expressed in the form $y = mx + c$, the **coefficient** of x, m, is the **gradient**, and the constant term, c, is the **intercept** on the y-axis.

This means that if you know the gradient, m, of a line and its intercept, c, on the y-axis, you can write down the equation of the line immediately.

For example, if $m = 3$ and $c = -5$, the equation of the line is $y = 3x - 5$.

All linear graphs can be expressed in the form $y = mx + c$.

This gives a method of finding the equation of any line drawn on a pair of coordinate axes.

EXAMPLE 8

Find the equation of the line shown in diagram **A**.

A

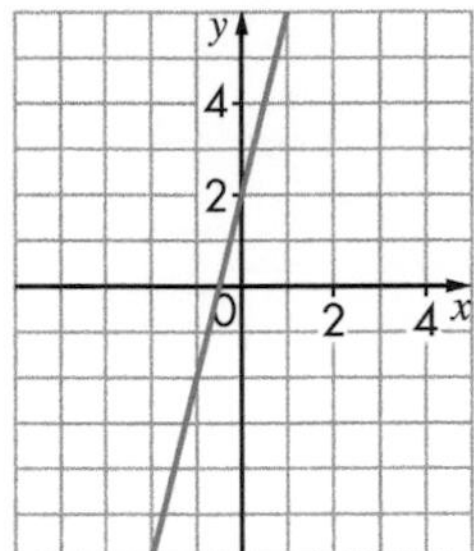

B

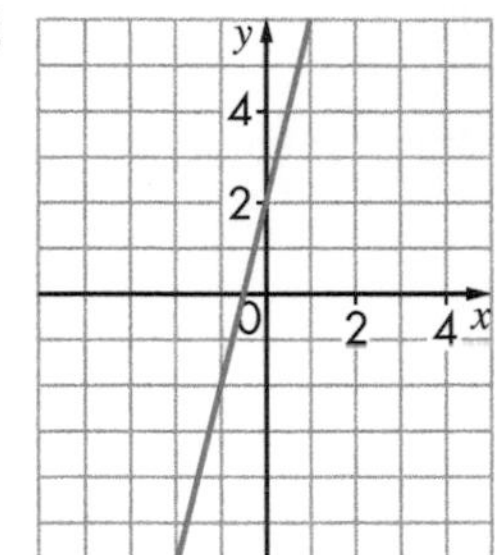

C

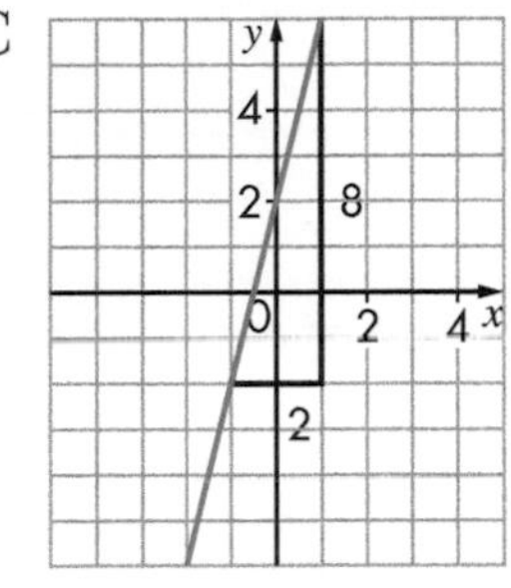

First, find where the graph crosses the y-axis (diagram **B**).

This is at (0, 2) so $c = 2$

Next, measure the gradient of the line (diagram **C**).

y-step = 8

x-step = 2

Gradient = $8 \div 2 = 4$

So $m = 4$

Finally, write down the equation of the line: $y = 4x + 2$.

The equation $y - y_1 = m(x - x_1)$

When you know the gradient, m, of a straight line and the coordinates, (x_1, y_1), of a point on the line, you can write down the equation of the graph in the form $y - y_1 = m(x - x_1)$. Note that the coefficient of x is the gradient, m.

For example, if the gradient $m = 3$ and the line passes through the point (2, 7), the equation of the line is $y - 7 = 3(x - 2)$, which simplifies to $y = 3x + 1$.

This gives another method of finding the equation of a straight line.

EXERCISE 13G

GRADE B

1. Give the equation of each line, all of which have positive gradients. (Each square represents one unit.)

a)

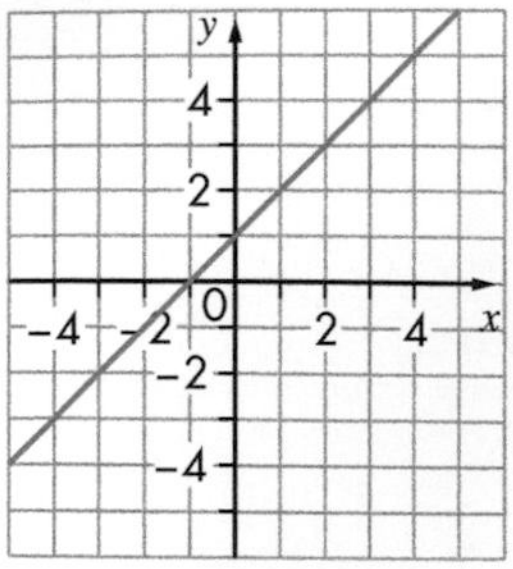

b)

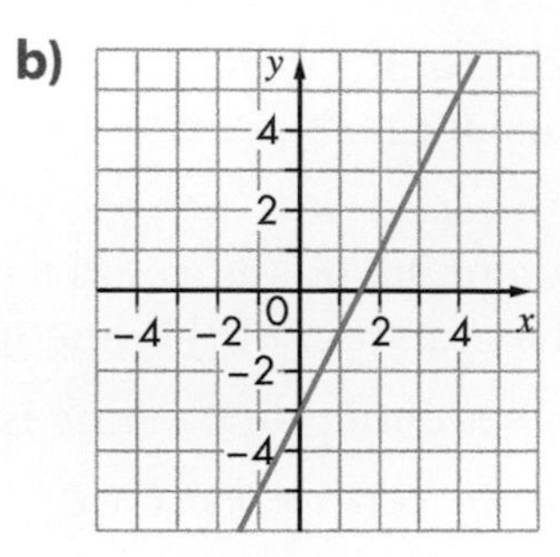

c)

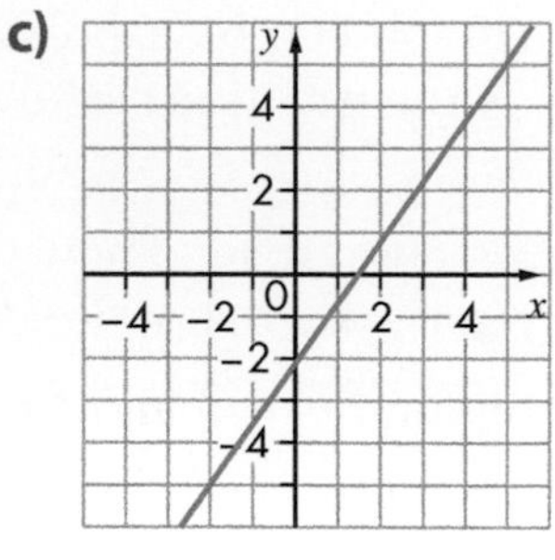

2. In each grid there are two lines. (Each square represents one unit.)

a)

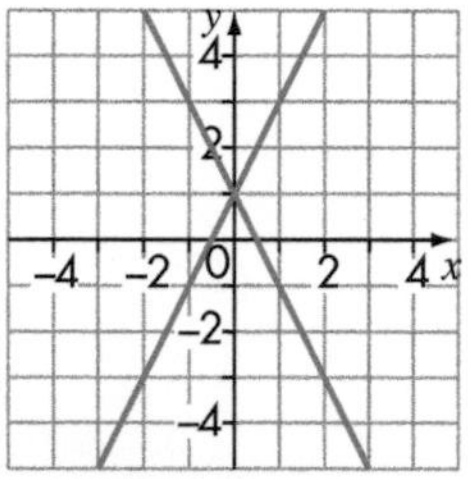

b)

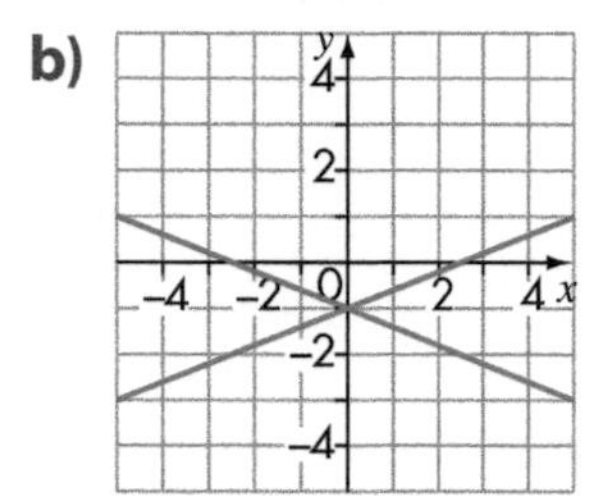

c)

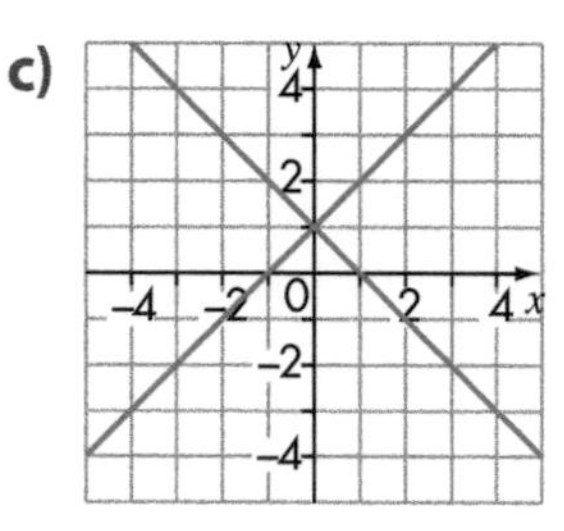

For each grid:

i) find the equation of each of the lines

ii) describe any symmetries that you can see

iii) describe any connection between the gradients of each pair of lines.

3. A straight line passes through the points (1, 3) and (2, 5).

a) Explain how you can tell that the line also passes through (0, 1).

b) Explain how you can tell that the line has a gradient of 2.

c) Work out the equation of the line that passes through (1, 5) and (2, 8).

4. Give the equation of each line, all of which have negative gradients. (Each square represents one unit.)

a)

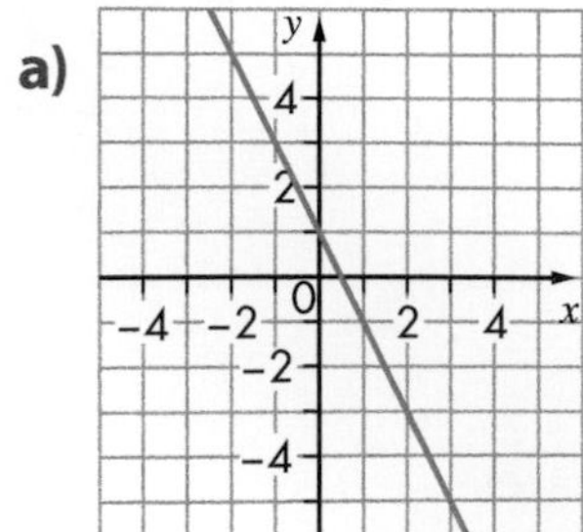

b)

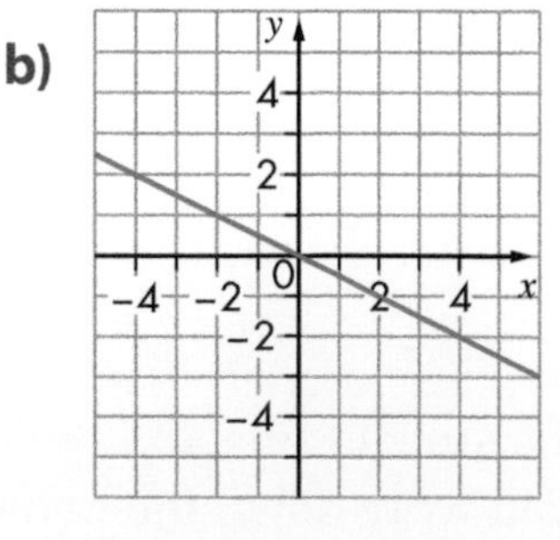

c)

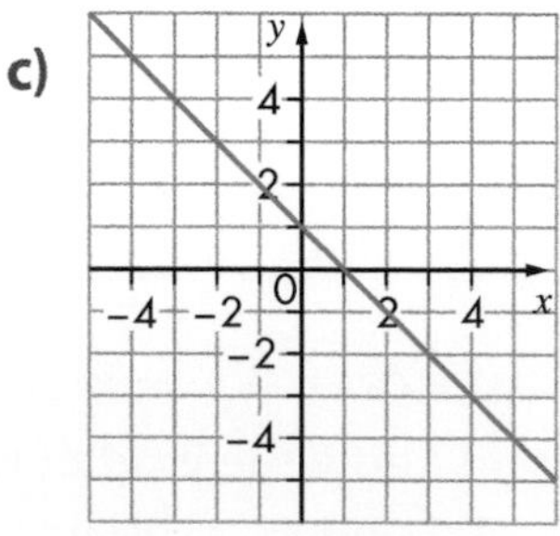

d)

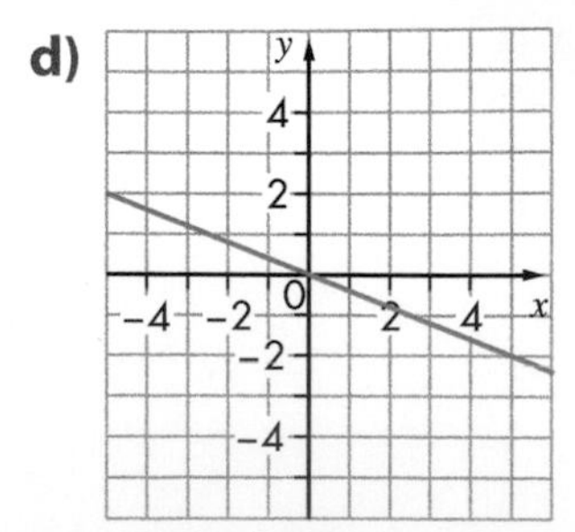

e)

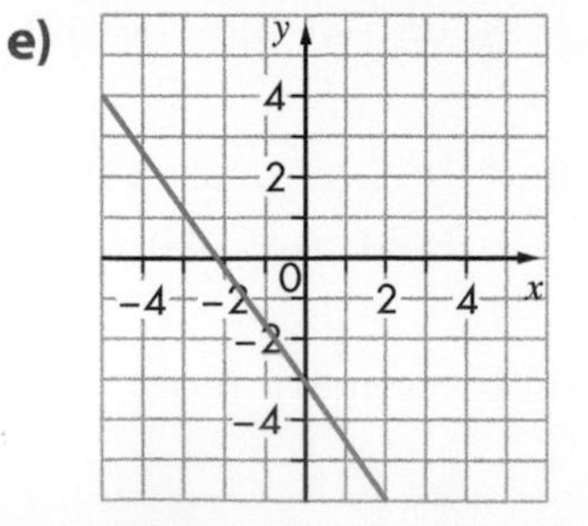

5. In each grid there are three lines. One of them is $y = x$.
(Each square represents one unit.)

a) 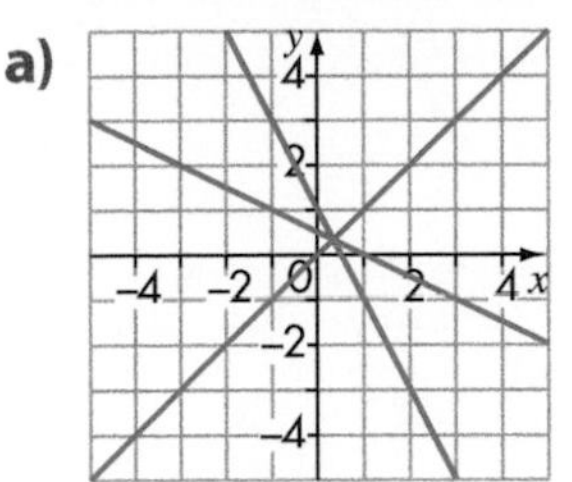

b) 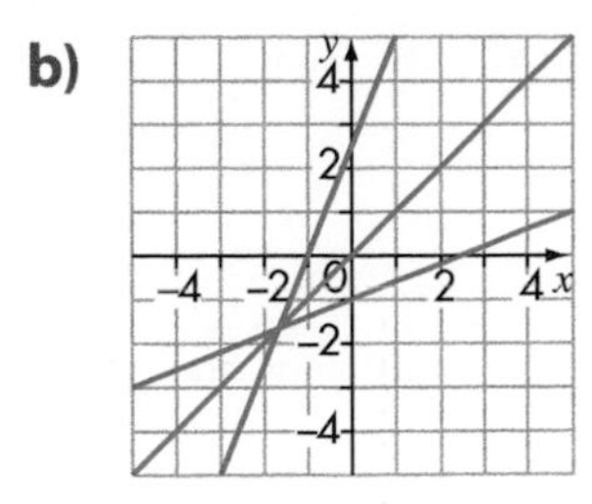

c) 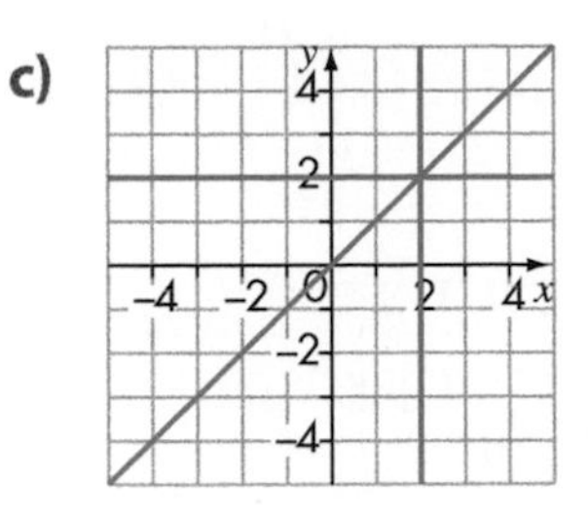

For each grid:

i) find the equation of each of the other two lines
ii) describe any symmetries that you can see
iii) describe any connection between the gradients of each group of lines.

6. Find the gradient of the lines through these pairs of points.
a) (4, 0) and (6, 6) **b)** (0, 3) and (8, 7) **c)** (2, –2) and (4, 6)
d) (1, 5) and (5, 1) **e)** (–4, 6) and (6, 1) **f)** (–5, –3) and (4, 3)

7. Find the equations of the lines joining these pairs of points.
a) (0, –3) and (4, 5) **b)** (–4, 2) and (2, 5)
c) (–1, –6) and (2, 6) **d)** (1, 5) and (4, –4)

8. Find the midpoints of the line segments joining the points in question **6**.

9. A is (–3, 5), B is (1, 1) and C is (5, 9).
a) Draw the triangle ABC on a coordinate grid.
b) Find the equation of the straight line through A and C.
c) Find the midpoint of AB.
d) Find the equation of the straight line through the midpoints of AC and BC.

10. Find the equations of the lines joining these pairs of points.
a) (2, 2) and (6, 5) **b)** (–3, 2) and (9, 8)
c) (1, 5) and (5, –3) **d)** (–6, –4) and (2, 4)

13.3 Parallel and perpendicular lines

THIS SECTION WILL SHOW YOU HOW TO …

✓ draw linear graphs parallel or perpendicular to other lines and passing through a specific point

KEY WORDS

✓ negative reciprocal ✓ perpendicular ✓ parallel

EXAMPLE 9

In each of these grids, there are two lines.

For each grid:

i) find the equation of each line

ii) describe the geometrical relationship between the lines

iii) describe the numerical relationships between their gradients.

a)

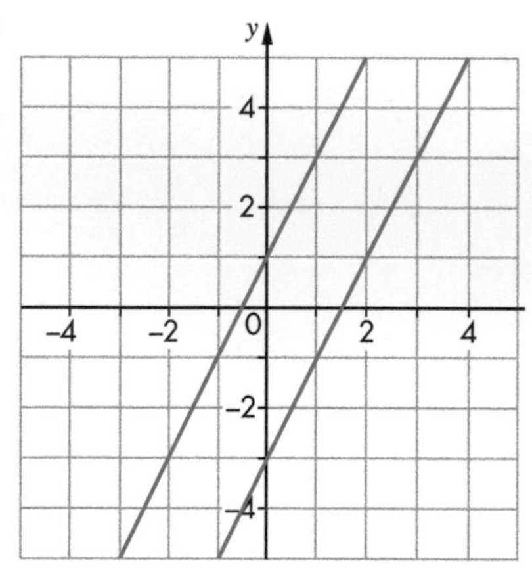

b)

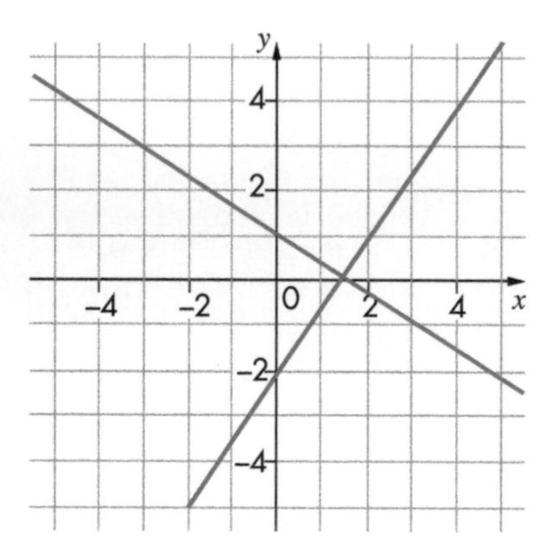

i) Grid **a**: the lines have equations $y = 2x + 1$, $y = 2x - 3$.

Grid **b**: the lines have equations $y = \frac{3}{2}x - 2$, $y = -\frac{2}{3}x + 1$.

ii) Grid **a**: the lines are parallel.

Grid **b**: the lines are perpendicular (at right angles).

iii) Grid **a**: the gradients are equal.

Grid **c**: the gradients are reciprocals of each other but with different signs.

Note: If two lines are **parallel**, then their gradients are equal.

If two lines are **perpendicular**, their gradients are **negative reciprocals** of each other.

EXAMPLE 10

Two points A and B are A(0, 1) and B(2, 4).

a) Work out the equation of the line AB.
b) Write down the equation of the line parallel to AB and passing through the point (0, 5).
c) Write down the gradient of a line perpendicular to AB.
d) Write down the equation of a line perpendicular to AB and passing through the point (0, 2).

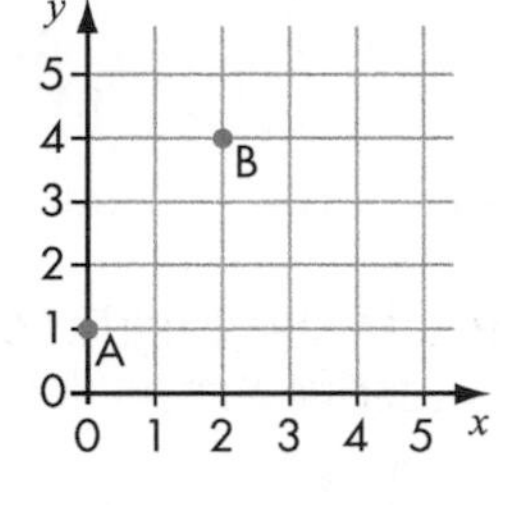

a) The gradient of AB is $\frac{3}{2}$ and passes through (0, 1) so the equation is $y = \frac{3}{2}x + 1$.

b) The gradient is the same and the intercept is (0, 5) so the equation is $y = \frac{3}{2}x + 5$.

c) The perpendicular gradient is the negative reciprocal $-\frac{2}{3}$.

d) The gradient is $-\frac{2}{3}$ and the intercept is (0, 2) so the equation is $y = -\frac{2}{3}x + 2$.

EXAMPLE 11

Find the equation of the line that is perpendicular to the line $y = \frac{1}{2}x - 3$ and passes through (0, 5).

The gradient of the new line will be the negative reciprocal of $\frac{1}{2}$ which is −2.

The point (0, 5) is the intercept on the *y*-axis so the equation of the line is:
$y = -2x + 5$.

EXAMPLE 12

The point A is (2, 1) and the point B is (4, 4).

a) Find the equation of the line parallel to AB and passing through (6, 11).
b) Find the equation of the line parallel to AB and passing through (8, 0).

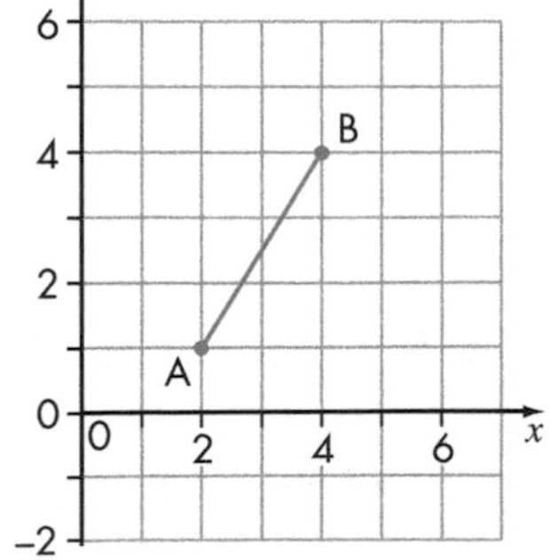

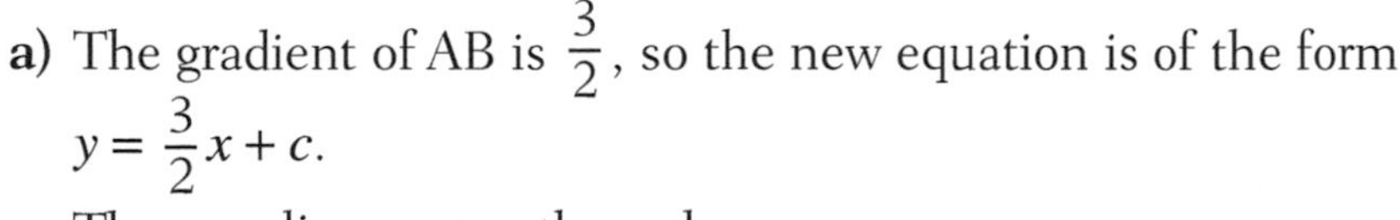
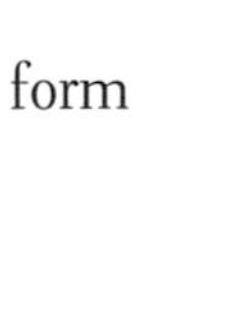

a) The gradient of AB is $\frac{3}{2}$, so the new equation is of the form $y = \frac{3}{2}x + c$.

The new line passes through (6, 11), so $11 = \frac{3}{2} \times 6 + c \Rightarrow c = 2$.

Hence the new line is $y = \frac{3}{2}x + 2$.

b) The gradient of AB is $\frac{3}{2}$, so the new equation is of the form $y = \frac{3}{2}x + c$.

The new line passes through (8, 0), so $0 = \frac{3}{2} \times 8 + c \Rightarrow c = -12$.
Hence the new line is $y = x - 12$.

EXAMPLE 13

The point A is (2, –1) and the point B is (4, 5).

a) Find the equation of the line parallel to AB and passing through (2, 8).

b) Find the equation of the line perpendicular to the midpoint of AB.

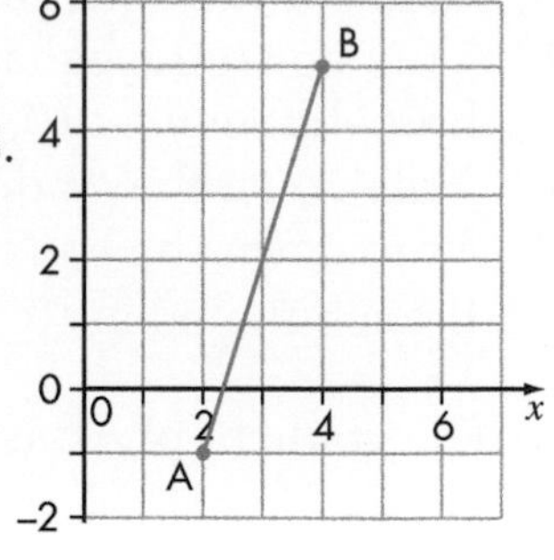

a) The gradient of AB is 3, so the new equation is of the form $y = 3x + c$.

The new line passes through (2, 8), so $8 = 3 \times 2 + c \Rightarrow c = 2$.
Hence the line is $y = 3x + 2$.

b) The midpoint of AB is (3, 2).

The gradient of the perpendicular line is the negative reciprocal of 3, which is $-\frac{1}{3}$.

You could find c as in part **a** but you can also do a sketch on the grid.
This will show that the perpendicular line passes through (0, 3).
Hence the equation of the line is $y = -\frac{1}{3}x + 3$.

EXERCISE 13H

GRADE A

1. Here are the equations of three lines.

Line A: $y = 3x - 2$ Line B: $y = 3x + 1$ Line C: $y = -\frac{1}{3}x + 1$

a) Give a reason why line A is the odd one out of the three.
b) Give a reason why line C is the odd one out of the three.
c) Which of the following would be a reason why line B is the odd one out of the three?
  i) Line B is the only one that intersects the negative x-axis.
  ii) Line B is not parallel to either of the other two lines.
  iii) Line B does not pass through (0, –2).

2. Write down the negative reciprocals of the following numbers.

a) 2 **b)** –3 **c)** 5 **d)** –1
e) $\frac{1}{2}$ **f)** $\frac{1}{4}$ **g)** $-\frac{1}{3}$ **h)** $-\frac{2}{3}$
i) 1.5 **j)** 10 **k)** –6 **l)** $\frac{4}{3}$

3. Write down the equation of the line perpendicular to each line and that passes through the same point on the y-axis.

a) $y = 2x - 1$ **b)** $y = -3x + 1$ **c)** $y = x + 2$ **d)** $y = -x + 2$
e) $y = \frac{1}{2}x + 3$ **f)** $y = \frac{1}{4}x - 3$ **g)** $y = -\frac{1}{3}x$ **h)** $y = -\frac{2}{3}x - 5$

4. Write down the equations of these lines.

a) Parallel to $y = 4x - 5$ and passes through (0, 1)
b) Parallel to $y = \frac{1}{2}x + 3$ and passes through (0, –2)
c) Parallel to $y = -x + 2$ and passes through (0, 3)

5. Write down the equations of these lines.
   a) Perpendicular to $y = 3x + 2$ and passes through (0, –1)
   b) Perpendicular to $y = -\frac{1}{3}x - 2$ and passes through (0, 5)
   c) Perpendicular to $y = x - 5$ and passes through (0, 1)

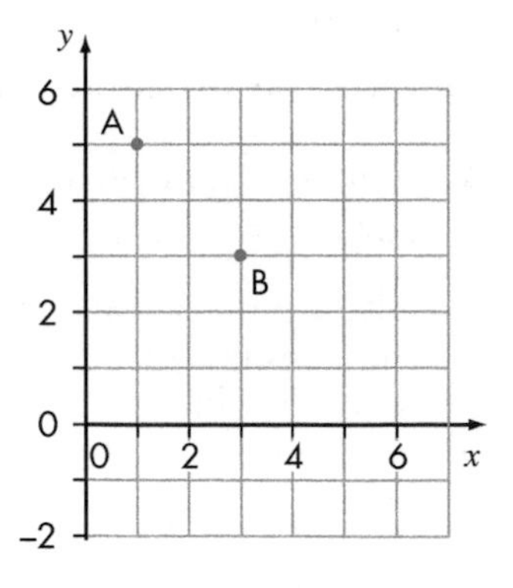

6. A is the point (1, 5). B is the point (3, 3).
   a) Find the equation of the line parallel to AB and passing through (5, 9).
   b) Find the equation of the line perpendicular to AB and passing through the midpoint of AB.

7. Find the equation of the line that passes through the midpoint of AB, where A is (–5, –3) and B is (–1, 3), and has a gradient of 2.

8. Find the equation of the line perpendicular to $y = 4x - 3$, passing though (–4, 3).

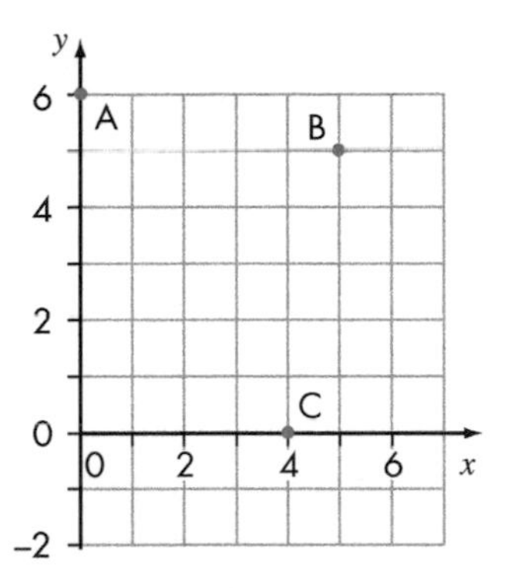

9. A is the point (0, 6), B is the point (5, 5) and C is the point (4, 0).
   a) Write down the coordinates of the point where the line BC intercepts the y-axis.
   b) Work out the equation of the line AB.
   c) Write down the equation of the line BC.

10. Find the equation of the perpendicular bisector of the points A(1, 2) and B(3, 6).

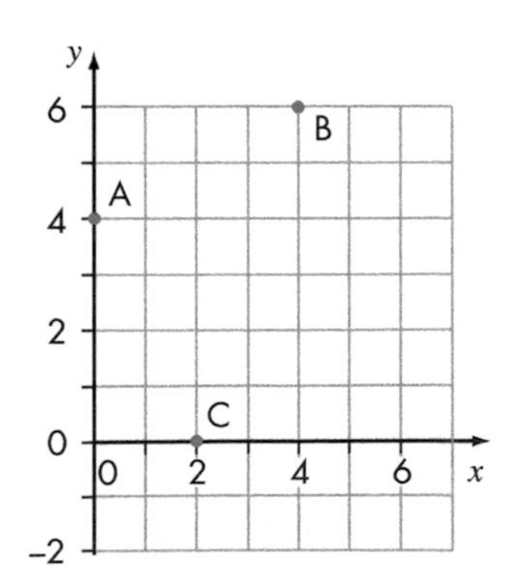

11. A is the point (0, 4), B is the point (4, 6) and C is the point (2, 0).
    a) Find the equation of the line BC.
    b) Show that the point of intersection of the perpendicular bisectors of AB and AC is (3, 3).
    c) Show algebraically that this point lies on the line BC.

12. A is the point (–2, 3) and B is the point (0, 2). Find the equation of the line that is perpendicular to AB and passes through the midpoint of AB.

13.4 Applications of coordinate geometry

THIS SECTION WILL SHOW YOU HOW TO ...

✓ apply what you know to problems involving coordinate geometry

You are likely to be asked to solve problems involving coordinates. This section gives some examples of what you might be asked to do.

EXAMPLE 14

Calculate the distance between the points A(1, 4) and B(5, 7).

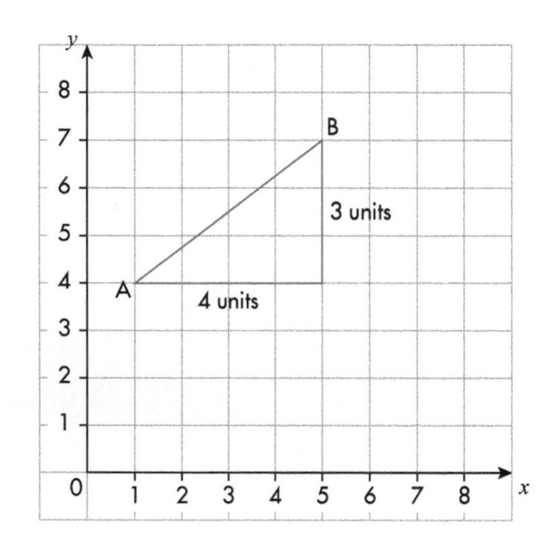

Plot the points on a grid and draw a right-angled triangle.

Using Pythagoras' theorem:

$AB^2 = 3^2 + 4^2 = 9 + 16 = 25$

So the length of AB is 5 units.

EXAMPLE 15

Work out the coordinates of the midpoint of the line joining A(3, 5) and B(7, 3).

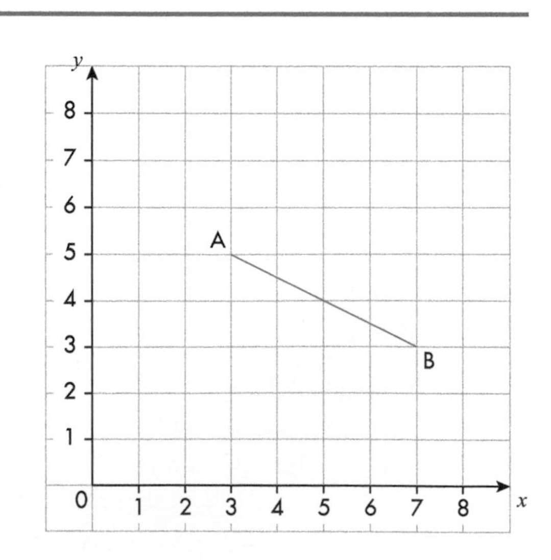

Plot the points on a grid.

From plotting the points you can see straight away that the coordinates of the midpoint are (5, 4).

However, it is better to work out answers by calculation.

You can work out the midpoint by calculating the average (mean) of the *x*-coordinates and the average of the *y*-coordinates.

So the coordinates of the midpoint are

$(\frac{3+7}{2}, \frac{5+3}{3}) = (5, 4)$.

EXAMPLE 16

The point A has coordinates (–3, 1) and point B has coordinates (2, 16). Point C is such that AC : CB = 2 : 3.

Work out the coordinates of C.

Look at the x-coordinates first.

From –3 to 2 is 5 units and 5 units divided in the ratio 2 : 3 gives 2 units and 3 units. So the x-coordinate of C is $-3 + 2 = -1$.

Now look at the y-coordinates.

From 1 to 16 is 15 units and 15 units divided in the ratio 2 : 3 gives 6 units and 9 units. So the y-coordinate of C is $1 + 6 = 7$.

The point C has coordinates (–1, 7).

EXAMPLE 17

A is the point (3, 4), B is the point (7, 2), C is the point (10, –3) and D is the point (2, 1). What type of quadrilateral is ABCD?

Work out the gradient for each side.

The gradient of AB $= \frac{4-2}{3-7} = -\frac{1}{2}$

The gradient of BC $= \frac{2--3}{7-10} = -\frac{5}{3}$

The gradient of CD $= \frac{-3-1}{10-2} = -\frac{1}{2}$

The gradient of DA $= \frac{1-4}{2-3} = 3$

The gradient of AB is equal to the gradient of CD so two opposite sides are parallel.

Therefore ABCD is a trapezium.

EXAMPLE 18

A is the point (1, 3), B is the point (5, 7) and C is the point (4, –2).

Work out the area of triangle ABC.

Draw the triangle.

Now draw a rectangle around the outside of the triangle.

Area of triangle ABC = area of rectangle – area of three small right-angled triangles

$$\text{Area of triangle ABC} = (4 \times 9) - \left(\frac{1}{2} \times 3 \times 4\right) - \left(\frac{1}{2} \times 1 \times 9\right) - \left(\frac{1}{2} \times 4 \times 5\right)$$

$$= 36 - 6 - 4.5 - 10$$

$$= 15.5 \text{ square units}$$

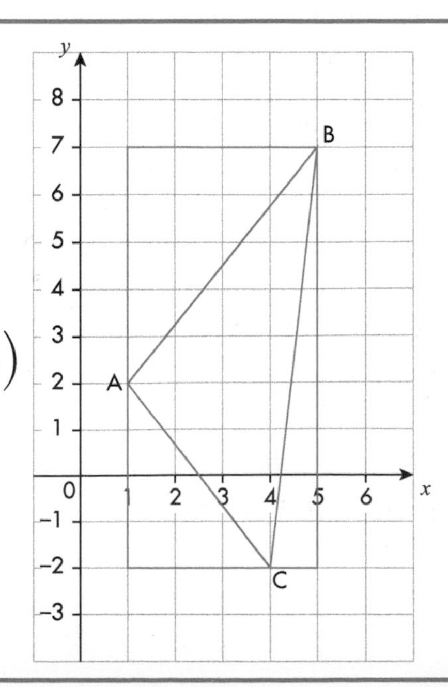

EXERCISE 13I

GRADE B

1. Use Pythagoras' theorem to calculate the distance between the points in each pair.

a) (3, 6) and (6, 10) **b)** (0, 0) and (5, 12) **c)** (–2, 3) and (6, 9)
d) (–2, –3) and (7, 9) **e)** (–5, 4) and (3, –11) **f)** (6, –1) and (–1, 23)

2. Work out the distance between points in each pair. Give your answers in the form $a\sqrt{b}$.

a) (0, 0) and (3, 3) **b)** (1, 5) and (3, 9) **c)** (–2, 3) and (2, 9)
d) (–1, –5) and (0, 2) **e)** (–4, –5) and (4, –1) **f)** (6, –3) and (–2, –1)

3. Work out the coordinates of the midpoints of the points in each pair.

a) (1, 11) and (3, –5) **b)** (0, 6) and (7, –2) **c)** (–1, –3) and (–5, 9)
d) (2, –4) and (8, 8) **e)** (–3, 2) and (3, –2) **f)** (7, –4) and (–3, 10)

4. A is the point (3, 8). The midpoint of the line segment AC is the point B(5, –2). Work out the coordinates of C.

5. D is the point (–1, 5). The midpoint of the line segment DF is the point E(–6, 8). Work out the coordinates of F.

6. Point A has coordinates (4, 9) and point B has coordinates (–2, 18).

Point C is on the line segment AB such that AC : CB = 2 : 1.

Work out the coordinates of C.

7. Point X has coordinates (–5, 5) and point Y has coordinates (3, 1).

The points X, Y and Z are in a straight line such that XY : YZ = 2 : 3.

Work out the coordinates of Z.

8. Point P has coordinates (4, –3) and point R has coordinates (–12, 21).

The points P, Q and R are in a straight line such that PR : PQ = 4 : 3.

Work out the coordinates of Q.

9. A is the point (4, –7), B is (10, –5), C is (8, –8) and D is (2, –10).

What type of quadrilateral is ABCD?

Explain your answer.

10. A is the point (2, 5), B is the point (6, 4) and C is the point (3, –5).

Work out the area of triangle ABC.

Exam-style questions

1. Work out the gradient of the line shown in the diagram. GRADE B

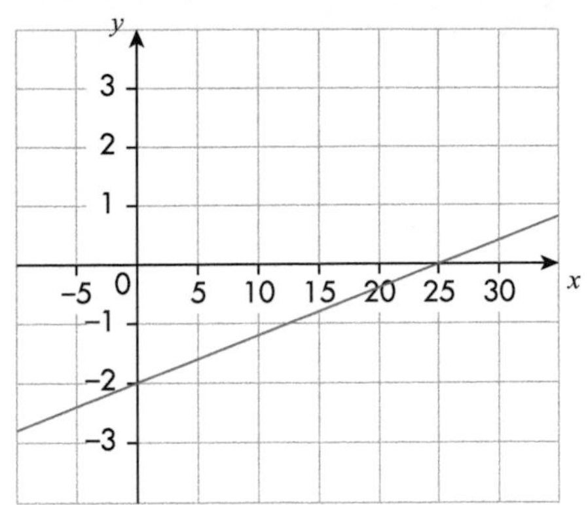

2. A line with gradient 6 passes through the point (5, 10).
   Where does the line cross the y-axis? GRADE B

3. Work out the equation of the straight line through the points (7, –4) and (–3, 1). GRADE A

4. On a grid, draw a line that is parallel to $2x + 3y = 6$ and that passes through the point (4, 4). GRADE A

5. Two lines are perpendicular.
   The equation of one is $3x + y = 10$.
   The lines cross at the point (2, 4).
   Work out the equation of the other line. GRADE A*

6. A quadrilateral has its vertices at A(4, 6), B(6, –2), C(–6, –5) and D(–8, 3).
   Show that ABCD is a rectangle. GRADE A*

7. The coordinates of point P are (–3, –2) and the coordinates of point Q are (6, 4).
   PQR is a straight line and PQ : QR = 4 : 1
   Work out the coordinates of Q. GRADE A**

8. **a)** A triangle has its vertices at A(5, 2), B(1, –2) and C(–1, 4). GRADE A**
   b) Show that ABC is an isosceles triangle.
   c) Find the equation of the line of symmetry of the triangle.
   d) Find the area of the triangle.

14 The equation of a circle

14.1 The equation of a circle centred on the origin (0, 0)

THIS SECTION WILL SHOW YOU HOW TO …

- ✓ recognise the equation of a circle, centre (0, 0), radius r
- ✓ write down the equation of a circle, given the centre (0, 0) and the radius
- ✓ work out coordinates of points of intersection of a given circle centred on the origin and a given straight line
- ✓ calculate the length of a chord formed by the intersection of a given circle centred on the origin and a given straight line

KEY WORDS

- ✓ centre
- ✓ radius

The general equation of a circle with **centre** (0, 0) and **radius** r is $x^2 + y^2 = r^2$.

For example, the equation of a circle with centre (0, 0) and radius 5 is $x^2 + y^2 = 25$.

The point (3, 4) lies on the circle as the values of x and y satisfy the equation of the circle, $3^2 + 4^2 = 25$.

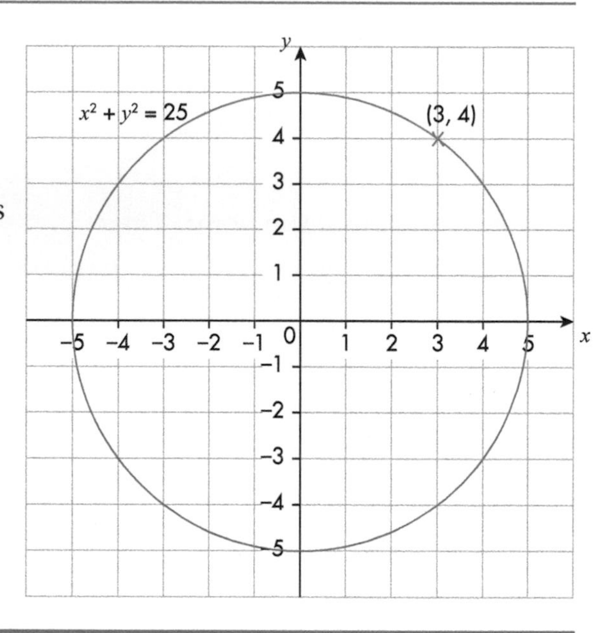

EXAMPLE 1

Write down the equation of a circle with centre (0, 0) and radius 3.

The equation is $x^2 + y^2 = 3^2$ or $x^2 + y^2 = 9$.

EXAMPLE 2

Show that the point (5, −2) lies on the circle with centre (0, 0) and radius $\sqrt{29}$.

The equation of a circle, centre (0, 0) and radius $\sqrt{29}$, is $x^2 + y^2 = 29$.

At the point (5, −2), $x = 5$ and $y = -2$.

$5^2 + (-2)^2 = 25 + 4 = 29$

So the point (5, −2) lies on the circle.

EXAMPLE 3

The circle $x^2 + y^2 = 12$ intersects the line $y = x + 1$ at the points A and B.

a) Work out the coordinates of A and B.

b) Work out the length of the chord AB.

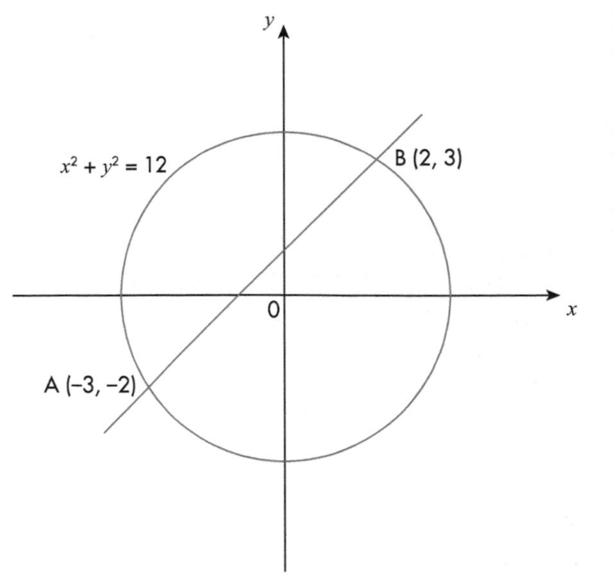

a) Substituting $y = x + 1$ into the equation $x^2 + y^2 = 13$ gives:

$x^2 + (x + 1)^2 = 13$

$\Rightarrow x^2 + x^2 + 2x + 1 = 13$

$\Rightarrow 2x^2 + 2x - 12 = 0$

$\Rightarrow x^2 + x - 6 = 0 \Rightarrow (x + 3)(x - 2) = 0$

$\Rightarrow x = -3$ or $x = 2$

When $x = -3$, $y = -2$ and when $x = 2$, $y = 3$.

So from the diagram the point A has coordinates (–3, –2) and point B has coordinates (2, 3).

b) Using Pythagoras' theorem:

$AB^2 = (3 - -2)^2 + (2 - -3)^2 \Rightarrow AB^2 = 5^2 + 5^2 \Rightarrow AB^2 = 25 + 25$

$\Rightarrow AB^2 = 50 \Rightarrow AB = \sqrt{50} = 5\sqrt{2} = 7.07$ units

EXERCISE 14A

GRADE A

1. Write down the equation of each circle.

a) Centre (0, 0), radius 7 **b)** Centre (0, 0), radius 9

c) Centre (0, 0), radius 6 **d)** Centre (0, 0), radius $\sqrt{2}$

e) Centre (0, 0), radius $\sqrt{5}$ **f)** Centre (0, 0), radius $2\sqrt{3}$

2. Write down the exact value of the radius for each of these circles.

a) $x^2 + y^2 = 100$ **b)** $x^2 + y^2 = 144$

c) $x^2 + y^2 = 225$ **d)** $x^2 + y^2 = 5$

e) $x^2 + y^2 = 28$ **f)** $x^2 + y^2 = 45$

3. The circle $x^2 + y^2 = 20$ intersects the line $y = x - 2$ at the points A and B.

a) Work out the coordinates of A and B.

b) Work out the length of the chord AB.

4. The circle $x^2 + y^2 = 25$ intersects the line $y = x$ at the points P and Q. Work out the length of the chord PQ.

HINTS AND TIPS

Drawing a sketch can help when you are answering questions like these.

5. Use graphical methods to find solutions to these pairs of simultaneous equations. In this question, suitable ranges for the axes are given. In an examination, a grid will be supplied.

a) $x^2 + y^2 = 25$ and $x + y = 1$ $(-6 \leqslant x \leqslant 6, -6 \leqslant y \leqslant 6)$

b) $x^2 + y^2 = 4$ and $y = x + 1$ $(-5 \leqslant x \leqslant 5, -5 \leqslant y \leqslant 5)$

c) $x^2 + y^2 = 9$ and $y = x - 1$ $(-5 \leqslant x \leqslant 5, -5 \leqslant y \leqslant 5)$

Lesson 14.2 The equation of a circle centred on any point (a, b)

THIS SECTION WILL SHOW YOU HOW TO ...

- ✓ recognise the equation of a circle, centre (a, b), radius r
- ✓ write down the equation of a circle, given the centre (a, b) and radius
- ✓ work out coordinates of points of intersection of any given circle and a given straight line
- ✓ calculate the length of a chord formed by the intersection of any given circle and a given straight line
- ✓ recognise that the circle $(x - a)^2 + (y - b)^2 = r^2$ is a translation of the circle $x^2 + y^2 = r^2$ by the vector $\begin{pmatrix} a \\ b \end{pmatrix}$

KEY WORDS

- ✓ general equation
- ✓ translate
- ✓ vector
- ✓ image

The **general equation** of a circle with centre (a, b) and radius r is $(x - a)^2 + (y - b)^2 = r^2$.

For example, the equation of a circle with centre (2, 6) and radius 13 is $(x - 2)^2 + (y - 6)^2 = 169$.

The point (−3, 18) lies on the circle as the values of x and y satisfy the equation of the circle, $(-3 - 2)^2 + (18 - 6)^2 = 25 + 144 = 169$.

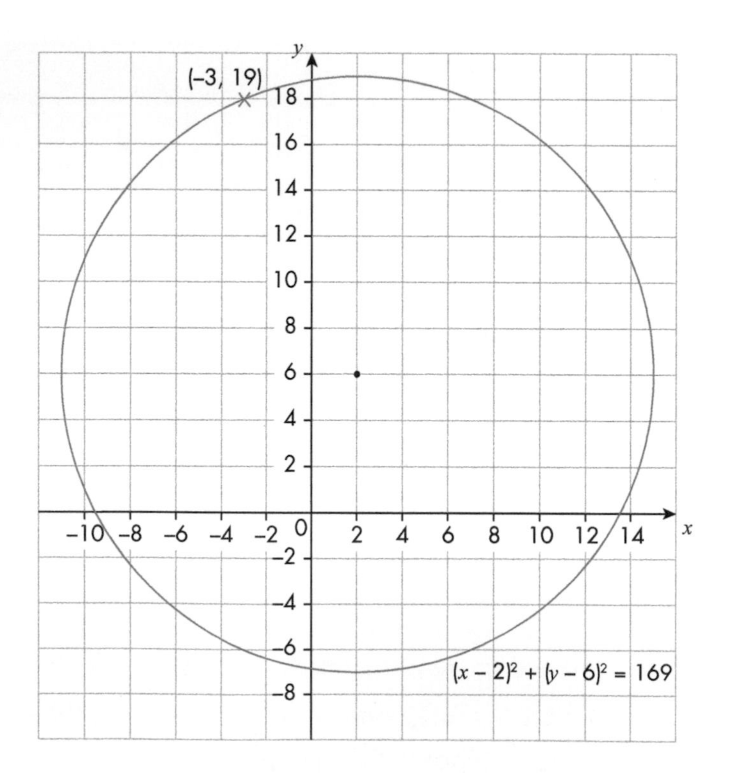

EXAMPLE 4

Write down the equation of a circle with centre (1, –4) and radius 7.

The equation is $(x-1)^2+(y+4)^2=49$.

EXAMPLE 5

Show that the point (–3, 2) lies on the circle with centre (5, 1) and radius $\sqrt{65}$.

The equation of a circle centre (5, 1) and radius $\sqrt{65}$ is $(x-5)^2+(y-1)^2=65$.

At the point (–3, 2), $x=-3$ and $y=2$.

$(-3-5)^2+(2-1)^2=64+1=65$

So (–3, 2) lies on the circle.

EXAMPLE 6

The circle $(x-2)^2+(y+3)^2=50$ intersects the line $y=x-5$ at the points P and Q.

a) Work out the coordinates of P and Q.

b) Work out the length of the chord PQ.

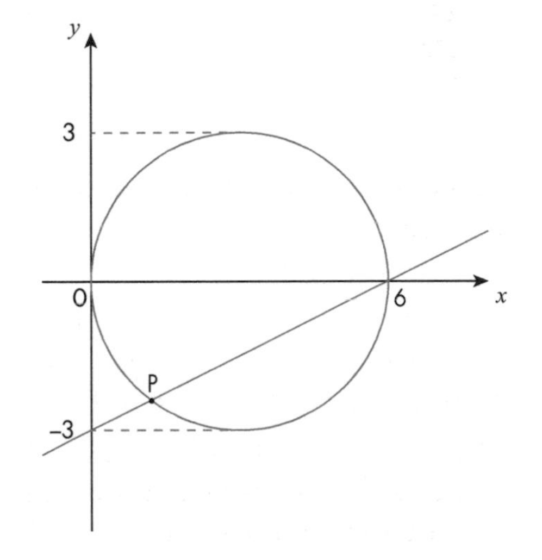

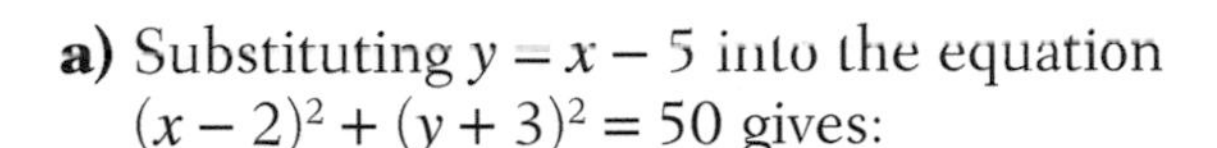

a) Substituting $y=x-5$ into the equation $(x-2)^2+(y+3)^2=50$ gives:

$(x-2)^2+(x-5+3)^2=50$

$\Rightarrow (x-2)^2+(x-2)^2=50$

$\Rightarrow x^2-4x+4+x^2-4x+4=50$

$\Rightarrow 2x^2-8x+8=50$

$\Rightarrow 2x^2-8x+8-50=0$

$\Rightarrow 2x^2-8x-42=0$

$\Rightarrow x^2-4x-21=0 \Rightarrow (x+3)(x-7)=0 \Rightarrow x=-3$ or $x=7$

When $x=-3$, $y=-8$ and when $x=7$, $y=2$.

So from the diagram P has coordinates (–3, –8) and Q has coordinates (7, 2).

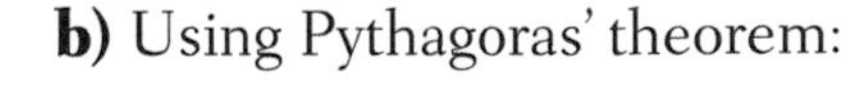

b) Using Pythagoras' theorem:

$PQ^2=(7--3)^2+(2--8)^2 \Rightarrow PQ^2=10^2+10^2$

$\Rightarrow PQ^2=100+100$

$\Rightarrow PQ^2=200 \Rightarrow PQ=\sqrt{200}=10\sqrt{2}$ units or 14.1 units.

Translating a circle

If a circle $x^2 + y^2 = r^2$ is **translated** by the **vector** $\begin{pmatrix} a \\ b \end{pmatrix}$ then the equation of the **image** will be $(x - a)^2 + (y - b)^2 = r^2$.

EXERCISE 14B

GRADE A

1. Write down the equation of each circle.

a) Centre (1, 2), radius 3
b) Centre (–1, 3), radius 4
c) Centre (4, –2), radius 5
d) Centre (3, 3), radius $\sqrt{5}$
e) Centre (8, –3), radius $3\sqrt{2}$
f) Centre (4, –4), radius $2\sqrt{3}$

2. Write down the centre and the exact value of the radius for each of these circles.

a) $(x - 1)^2 + (y - 4)^2 = 25$
b) $(x + 2)^2 + (y - 3)^2 = 9$
c) $(x - 5)^2 + (y + 5)^2 = 36$
d) $(x - 1)^2 + (y + 3)^2 = 6$
e) $(x - 2)^2 + (y + 3)^2 = 20$
f) $(x + 8)^2 + (y - 1)^2 = 24$

3. The line segment AB is the diameter of a circle.

A is the point (1, –4) and B is the point (5, 2).

Work out the equation of the circle.

HINTS AND TIPS

Drawing a sketch can help when you are answering questions like these.

4. The circle $(x - 2)^2 + (y + 3)^2 = 13$ intersects the line $y = -x + 4$ at the points A and B.

a) Work out the coordinates of A and B.
b) Work out the length of the chord AB.

5. The circle $(x + 4)^2 + (y - 5)^2 = 41$ intersects the line $y = -x + 8$ at the points P and Q.

Work out the length of the chord PQ.

6. The circle $x^2 + y^2 = 16$ is translated to the circle $(x - 4)^2 + (y + 7)^2 = 16$.

Write down the translation vector.

7. The circle $x^2 + y^2 = 25$ is translated by the vector $\begin{pmatrix} -1 \\ 2 \end{pmatrix}$.

Write down the equation of the circle after the translation.

Exam-style questions

1. The equation of a circle is $x^2 + y^2 = 25$. GRADE A

   a) Write down the coordinates of the centre of the circle.

   b) Work out the radius of the circle.

   c) Where does the line $y = \frac{3}{4}x$ cross the circle?

2. **a)** A circle has its centre at the origin and passes through the point $(6, -4)$.
   Work out the equation of the circle.

   b) Is the point (5, 5) inside or outside the circle?
   Give a reason for your answer. GRADE A

3. The equation of a circle is $(x - 2)^2 + (y + 3)^2 = c$.
   The line $x = 5$ is a tangent to the circle.
   Work out the value of c. GRADE A

4. The diagram shows a circle and a straight line. GRADE A

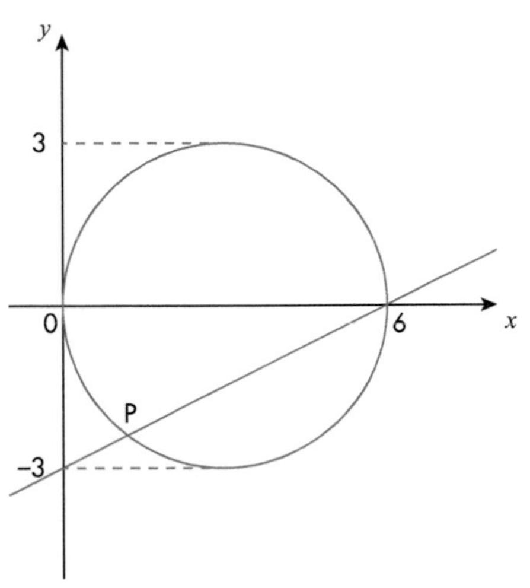

   a) Work out the equation of the circle.

   b) Work out the equation of the line.

   c) Find the coordinates of the point P, where the line and the circle cross.

5. The circle $x^2 + y^2 = 50$ is translated with the vector $\begin{pmatrix} 8 \\ -7 \end{pmatrix}$. GRADE A

   a) Find the equation of the new circle.

   b) Show that the new circle crosses one axis but not the other one.

6. A circle passes through the points (6, 0), (6, 10) and (0, 10).
   Work out the equation of the circle. GRADE A

15 Indices

15.1 Using indices

THIS SECTION WILL SHOW YOU HOW TO ...

- ✓ use indices to write down repeated multiplications of the same number or letter
- ✓ work out powers on your calculator
- ✓ recognise the special properties of the powers 0 and 1

KEY WORDS

✓ index ✓ indices ✓ power

An **index** is a convenient way of writing repetitive multiplications. The plural of index is **indices**.

The index tells you the number of times a number is multiplied by itself. For example:

$4^6 = 4 \times 4 \times 4 \times 4 \times 4 \times 4$ six lots of 4 multiplied together (we call this '4 to the **power** 6')

$6^4 = 6 \times 6 \times 6 \times 6$ four lots of 6 multiplied together (we call this '6 to the power 4')

$7^3 = 7 \times 7 \times 7$ $12^2 = 12 \times 12$

EXAMPLE 1

a) Write each of these numbers out in full.

i) 4^3 **ii)** 6^2 **iii)** 7^5 **iv)** 12^4

b) Write these multiplications as powers.

i) $3 \times 3 \times 3 \times 3 \times 3 \times 3 \times 3 \times 3$ **ii)** $13 \times 13 \times 13 \times 13 \times 13$

iii) $7 \times 7 \times 7 \times 7$ **iv)** $5 \times 5 \times 5 \times 5 \times 5 \times 5 \times 5$

a) i) $4^3 = 4 \times 4 \times 4$ **ii)** $6^2 = 6 \times 6$

iii) $7^5 = 7 \times 7 \times 7 \times 7 \times 7$ **iv)** $12^4 = 12 \times 12 \times 12 \times 12$

b) i) $3 \times 3 \times 3 \times 3 \times 3 \times 3 \times 3 \times 3 = 3^8$ **ii)** $13 \times 13 \times 13 \times 13 \times 13 = 13^5$

iii) $7 \times 7 \times 7 \times 7 = 7^4$ **iv)** $5 \times 5 \times 5 \times 5 \times 5 \times 5 \times 5 = 5^7$

Using indices on your calculator

The power button on your calculator will probably look like this $x^{\square}$.

To work out 5^7 on your calculator use the power key.

$5^7 = 5$ $x^{\square}$ $7 = 78\ 125$

Two special powers

Power 1

Any number to the power 1 is the same as the number itself. This is always true so normally you do not write the power 1.

For example: $5^1 = 5$ $32^1 = 32$ $(-8)^1 = -8$

Power zero

Any number to the power 0 is equal to 1.

For example: $5^0 = 1$ $32^0 = 1$ $(-8)^0 = 1$

You can check these results on your calculator.

EXERCISE 15A

GRADE C

1. Write these expressions in index notation. Do not work them out yet.

a) $2 \times 2 \times 2 \times 2$ **b)** $3 \times 3 \times 3 \times 3 \times 3$

c) 7×7 **d)** $5 \times 5 \times 5$

e) $10 \times 10 \times 10 \times 10 \times 10 \times 10 \times 10$ **f)** $6 \times 6 \times 6 \times 6$

g) 4 **h)** $1 \times 1 \times 1 \times 1 \times 1 \times 1 \times 1$

i) $0.5 \times 0.5 \times 0.5 \times 0.5$ **j)** $100 \times 100 \times 100$

2. Write these power terms out in full. Do not work them out yet.

a) 3^4 **b)** 9^3 **c)** 6^2 **d)** 10^5 **e)** 2^{10}

f) 8^1 **g)** 0.1^3 **h)** 2.5^2 **i)** 0.7^3 **j)** 1000^2

3. Using the power key on your calculator (or another method), work out the values of the power terms in question **1**.

4. Using the power key on your calculator (or another method), work out the values of the power terms in question **2**.

5. A storage container is in the shape of a cube. The length of the container is 5 m.

Work out the total storage space in the container.

> **HINTS AND TIPS**
>
> To work out the volume of a cube, use the formula: volume = (length of edge)^3

6. Write each number as a power of a different number.

The first one has been done for you.

a) $32 = 2^5$ **b)** 100 **c)** 8 **d)** 25

7. Without using a calculator, work out the values of these power terms.

a) 2^0 **b)** 4^1 **c)** 5^0 **d)** 1^9 **e)** 1^{235}

8. The answers to question **7**, parts **d** and **e**, should tell you something special about powers of 1. What is it?

9. Write the answer to question **1**, part **j** as a power of 10.

GRADE B

10. You are given that $x = 2^p$ and $y = 2^q$.

a) Write 2^{p+1} in terms of x. **b)** Write 2^{p+q} in terms of x and y.

c) Write 2^{p-q} in terms of x and y. **d)** Write 2^{5p} in terms of x.

e) Write 2^{2p+2q} in terms of x and y. **f)** Write 2^{2p-2q} in terms of x.

15.2 Rules of indices

THIS SECTION WILL SHOW YOU HOW TO ...

✓ multiply and divide expressions that include indices
✓ work out powers on your calculator
✓ recognise the special properties of the powers 0 and 1

To *multiply* powers of the same number or variable, *add* the indices.

$3^4 \times 3^5 = 3^{(4+5)} = 3^9$ $\quad 2^3 \times 2^4 \times 2^5 = 2^{12}$ $\quad a^x \times a^y = a^{(x+y)}$

To *divide* powers of the same number or variable, *subtract* the indices.

$a^4 \div a^3 = a^{(4-3)} = a^1 = a$ $\quad \frac{b^4}{b^7} = \frac{1}{b^4}$ $\quad a^x \div a^y = a^{(x-y)}$

When you *raise* a power to a further power, you *multiply* the indices.

$(4^2)^3 = 1^{2\times3} = 4^6$ $\quad (5^3)^4 = 5^{12}$ $\quad (a^x)^y = a^{xy}$

EXERCISE 15B

GRADE C

1. Write these as single powers.

a) $5^2 \times 5^2$ **b)** 5×5^2 **c)** $5^2 \times 5^4$ **d)** $5^6 \times 5^3$ **e)** $5^2 \times 5^3$

f) $6^5 \div 6^2$ **g)** $6^4 \div 6$ **h)** $6^4 \div 6^2$ **i)** $6^5 \div 6^2$ **j)** $6^5 \div 6^4$

2. Simplify these and write them as single powers of a.

a) $a^2 \times a$ **b)** $a^3 \times a^2$ **c)** $a^4 \times a^3$ **d)** $a^6 \div a^2$ **e)** $a^3 \div a$ **f)** $a^5 \div a^4$

3. Write down a possible pair of values of x and y.

a) $a^x \times a^y = a^{10}$ **b)** $a^x \div a^y = a^{10}$

GRADE B

4. Write these as single powers of 4.

a) $(4^2)^3$ **b)** $(4^3)^5$ **c)** $(4^1)^6$ **d)** $(4^3)^2$ **e)** $(4^4)^2$ **f)** $(4^7)^0$

5. Simplify these expressions.

a) $\frac{a^2}{a^3}$ **b)** $\frac{a^4}{a^6}$ **c)** $\frac{a^4}{a}$ **d)** $\frac{a}{a^4}$ **e)** $\frac{a^5}{a^4}$

6. Find the value of n. **a)** $a^2 \times a^5 = a^n$ **b)** $\frac{a^3 \times a^2}{a^n} = a.$

7. Simplify: **a)** $(a^2)^2$ **b)** $(a^4)^2$ **c)** $\sqrt{a^6}$ **d)** $\sqrt{a^{10}}$

8. Simplify these expressions.

a) $6a^3 \div 2a^2$ **b)** $12a^5 \div 3a^2$ **c)** $15a^5 \div 5a$

d) $\frac{18a}{3a^2}$ **e)** $\frac{24a^5}{6a^2}$ **f)** $\frac{30a}{6a^5}$

9. a, b and c are three different positive integers.
What is the smallest possible value of a^2b^3c?

15.3 Negative indices

THIS SECTION WILL SHOW YOU HOW TO ...

✓ simplify expressions involving negative indices, which may be written in a variety of forms

✓ recognise that, for example, x^{-n} is equivalent to $\frac{1}{x^n}$

KEY WORDS

✓ negative index ✓ reciprocal

A **negative index** is a convenient way of writing the **reciprocal** of a number or term.

(The reciprocal of a number or term is one divided by that number or term) For example:

$x^{-a} = \frac{1}{x^a}$

Here are some other examples.

$5^{-2} = \frac{1}{5^2}$ $3^{-1} = \frac{1}{3}$ $5x^{-2} = \frac{5}{x^2}$

EXAMPLE 2

Rewrite the following in the form 2^n.

a) 8 **b)** $\frac{1}{4}$ **c)** -32 **d)** $-\frac{1}{64}$

a) $8 = 2 \times 2 \times 2 = 2^3$ **b)** $\frac{1}{4} = \frac{1}{2^2} = 2^{-2}$

c) $-32 = -2^5$ **d)** $-\frac{1}{64} = -\frac{1}{2^6} = -2^{-6}$

What about x^0?

In fact $x^0 = 1$ for any value of x. This can seem surprising but it fits the pattern in this table.

n	3	2	1	0	−1	−2
2^n	8	4	2	1	$\frac{1}{2}$	$\frac{1}{4}$
3^n	27	9	3	1	$\frac{1}{3}$	$\frac{1}{9}$

The numbers in the second row are divided by 2 as you move from left to right.

The numbers in the third row are divided by 3 as you move from left to right.

2^0 and 3^0 are both equal to 1.

The rules for multiplying and dividing with indices still apply, whether the indices are positive, negative or zero.

EXERCISE 15C

GRADE B

1. Write each of these in fraction form, using indices.

a) 5^{-3} **b)** 6^{-1} **c)** 10^{-5} **d)** 3^{-2} **e)** 8^{-2}
f) 9^{-1} **g)** w^{-2} **h)** t^{-1} **i)** x^{-m} **j)** $4m^{-3}$

2. Write each of these in negative index form.

a) $\frac{1}{3^2}$ **b)** $\frac{1}{5}$ **c)** $\frac{1}{10^3}$ **d)** $\frac{1}{m}$ **e)** $\frac{1}{t^n}$

GRADE A

3. Change each expression into an index form of the type shown.

a) All of the form 2^n

i) 16 **ii)** $\frac{1}{2}$ **iii)** $\frac{1}{16}$ **iv)** -8

b) All of the form 10^n

i) 1000 **ii)** $\frac{1}{10}$ **iii)** $\frac{1}{100}$ **iv)** 1 million

c) All of the form 5^n

i) 125 **ii)** $\frac{1}{5}$ **iii)** $\frac{1}{25}$ **iv)** 1

d) All of the form 3^n

i) 9 **ii)** $\frac{1}{27}$ **iii)** 1 **iv)** -243

HINTS AND TIPS

If you move a power from top to bottom, or vice versa, the sign changes. Negative power means the reciprocal: it does not mean the answer is negative.

4. Rewrite each expression in fraction form.

a) $5x^{-3}$ **b)** $6t^{-1}$ **c)** $7m^{-2}$ **d)** $4q^{-4}$ **e)** $10y^{-5}$

f) $\frac{1}{2}x^{-3}$ **g)** $\frac{1}{2}m^{-1}$ **h)** $\frac{3}{4}t^{-4}$ **i)** $\frac{4}{5}y^{-3}$ **j)** $\frac{7}{8}x^{-5}$

5. Write each fraction in index form.

a) $\frac{7}{x^3}$ **b)** $\frac{10}{p}$ **c)** $\frac{5}{t^2}$ **d)** $\frac{8}{m^5}$ **e)** $\frac{3}{y}$

6. Find the value of each expression.

a) $x = 5$

i) x^2 **ii)** x^{-3} **iii)** $4x^{-1}$

b) $t = 4$

i) t^3 **ii)** t^{-2} **iii)** $5t^{-4}$

c) $m = 2$

i) m^3 **ii)** m^{-5} **iii)** $9m^{-1}$

d) $w = 10$

i) w^6 **ii)** w^{-3} **iii)** $25w^{-2}$

7. Write each of these in index form.

a) $a^{-3} \times a^{-4}$ **b)** $a^{-2} \times a^4$ **c)** $a^2 \div a^{-2}$ **d)** $a^{-3} \div a^2$ **e)** $(a^{-3})^2$ **f)** $(a^{-2})^{-3}$

15.4 Fractional indices

THIS SECTION WILL SHOW YOU HOW TO ...

- ✓ simplify expressions involving fractional and negative indices, which may be written in a variety of forms
- ✓ recognise, for example, that $x^{\frac{1}{n}}$ is equivalent to the nth root of x

Indices of the form $\frac{1}{n}$

Consider the problem $7^x \times 7^x = 7$. This can be written as:

$7^{(x + x)} = 7$

$7^{2x} = 7^1 \Rightarrow 2x = 1 \Rightarrow x = \frac{1}{2}$

If you now substitute $x = \frac{1}{2}$ back into the original equation, you see that:

$7^{\frac{1}{2}} \times 7^{\frac{1}{2}} = 7$

So $7^{\frac{1}{2}}$ is the same as $\sqrt{7}$.

You can similarly show that $7^{\frac{1}{3}}$ is the same as $\sqrt[3]{7}$. And that, generally:

$x^{\frac{1}{n}} = \sqrt[n]{x}$ (nth root of x)

So in summary:

power $\frac{1}{2}$ is the same as positive square root

power $\frac{1}{3}$ is the same as cube root

power $\frac{1}{n}$ is the same as nth root.

For example:

$49^{\frac{1}{2}} = \sqrt{49} = 7 \quad 8^{\frac{1}{3}} = \sqrt[3]{8} = 2 \quad 10000^{\frac{1}{4}} = \sqrt[4]{10000} = 10 \quad 36^{-\frac{1}{2}} = \frac{1}{\sqrt{36}} = \frac{1}{6}$

If you have an expression in the form $\left(\frac{a}{b}\right)^n$ it can be calculated as $\frac{a^n}{b^n}$ and then written as a fraction.

EXAMPLE 3

Write $\left(\frac{16}{25}\right)^{\frac{1}{2}}$ as a fraction.

Find the power of the numerator and denominator separately.

$$\left(\frac{16}{25}\right)^{\frac{1}{2}} = \frac{16^{\frac{1}{2}}}{25^{\frac{1}{2}}} = \frac{4}{5}$$

EXERCISE 15D

GRADE A

1. Evaluate each expression.

a) $25^{\frac{1}{2}}$ **b)** $100^{\frac{1}{2}}$ **c)** $64^{\frac{1}{2}}$ **d)** $81^{\frac{1}{2}}$ **e)** $625^{\frac{1}{2}}$

f) $27^{\frac{1}{3}}$ **g)** $64^{\frac{1}{3}}$ **h)** $1000^{\frac{1}{3}}$ **i)** $125^{\frac{1}{3}}$ **j)** $512^{\frac{1}{3}}$

k) $144^{\frac{1}{2}}$ **l)** $400^{\frac{1}{2}}$ **m)** $625^{\frac{1}{4}}$ **n)** $81^{\frac{1}{4}}$ **o)** $100\,000^{\frac{1}{5}}$

p) $729^{\frac{1}{6}}$ **q)** $32^{\frac{1}{5}}$ **r)** $1024^{\frac{1}{10}}$ **s)** $1296^{\frac{1}{4}}$ **t)** $216^{\frac{1}{3}}$

u) $16^{-\frac{1}{2}}$ **v)** $8^{-\frac{1}{3}}$ **w)** $81^{-\frac{1}{4}}$ **x)** $3125^{-\frac{1}{5}}$ **y)** $1\,000\,000^{-\frac{1}{6}}$

GRADE A*

2. Evaluate each expression.

a) $\left(\frac{25}{36}\right)^{\frac{1}{2}}$ **b)** $\left(\frac{100}{36}\right)^{\frac{1}{2}}$ **c)** $\left(\frac{64}{81}\right)^{\frac{1}{2}}$ **d)** $\left(\frac{81}{25}\right)^{\frac{1}{2}}$ **e)** $\left(\frac{25}{64}\right)^{\frac{1}{2}}$

f) $\left(\frac{27}{125}\right)^{\frac{1}{3}}$ **g)** $\left(\frac{8}{512}\right)^{\frac{1}{3}}$ **h)** $\left(\frac{1000}{64}\right)^{\frac{1}{3}}$ **i)** $\left(\frac{64}{125}\right)^{\frac{1}{3}}$ **j)** $\left(\frac{512}{343}\right)^{\frac{1}{3}}$

3. Use the general rule for raising a power to another power to prove that $x^{\frac{1}{n}}$ is equivalent to $\sqrt[n]{x}$.

4. Which of these is the odd one out?

$16^{-\frac{1}{4}}$ $64^{-\frac{1}{2}}$ $8^{-\frac{1}{3}}$

Show how you decided.

5. Imagine that you are the teacher.

Write down how you would teach the class that $27^{-\frac{1}{3}}$ is equal to $\frac{1}{3}$.

6. $x^{-\frac{2}{3}} = y^{\frac{1}{3}}$

Find values for x and y that make this equation work.

7. Solve these equations.

a) $2^x = 8$ **b)** $8^x = 2$ **c)** $4^x = 1$ **d)** $16^x = 4$ **e)** $100^x = 10$

f) $81^x = 3$ **g)** $16^x = 2$ **h)** $125^x = 5$ **i)** $1000^x = 10$ **j)** $400^x = 20$

k) $512^x = 8$ **l)** $128^x = 2$

Indices of the form $\frac{a}{b}$

Here are two examples of this form.

$t^{\frac{2}{3}} = t^{\frac{1}{3}} \times t^{\frac{1}{3}} = \left(\sqrt[3]{t}\right)^2$ $\quad$ $81^{\frac{3}{4}} = \left(\sqrt[4]{81}\right)^3 = 3^3 = 27$

If you have an expression of the form $\left(\frac{a}{b}\right)^{-n}$ you can invert it to calculate it as a fraction:

$$\left(\frac{a}{b}\right)^{-n} = \left(\frac{b}{a}\right)^{n}$$

EXAMPLE 4

Evaluate each expression. **a)** $16^{-\frac{1}{4}}$ **b)** $32^{-\frac{4}{5}}$

When dealing with the negative index remember that it means reciprocal.

Do problems like these one step at a time.

Step 1: Rewrite the calculation as a fraction by dealing with the negative power.

Step 2: Take the root of the base number given by the denominator of the fraction.

Step 3: Raise the result to the power given by the numerator of the fraction.

Step 4: Write out the answer as a fraction.

a) Step 1: $16^{-\frac{1}{4}} = \left(\frac{1}{16}\right)^{\frac{1}{4}}$ **Step 2:** $16^{\frac{1}{4}} = \sqrt[4]{16} = 2$

Step 3: $2^1 = 2$ **Step 4:** $16^{-\frac{1}{4}} = \frac{1}{2}$

b) Step 1: $32^{-\frac{4}{5}} = \left(\frac{1}{32}\right)^{\frac{4}{5}}$ **Step 2:** $32^{\frac{1}{5}} = \sqrt[5]{32} = 2$

Step 3: $2^4 = 16$ **Step 4:** $32^{-\frac{4}{5}} = \frac{1}{16}$

EXAMPLE 5

Write $\left(\frac{8}{27}\right)^{-\frac{2}{3}}$ as a fraction.

$$\left(\frac{8}{27}\right)^{-\frac{2}{3}} = \left(\frac{27}{8}\right)^{\frac{2}{3}} = \frac{27^{\frac{2}{3}}}{8^{\frac{2}{3}}} = \frac{\left(\sqrt[3]{27}\right)^2}{\left(\sqrt[3]{8}\right)^2} = \frac{3^2}{2^2} = \frac{9}{4}$$

The rules for multiplying and dividing with indices still apply for fractional indices. For example:

$a^{\frac{1}{2}} \times a^{-2} = a^{-\frac{3}{2}}$ $a^{\frac{1}{2}} \div a^{-2} = a^{\frac{5}{2}}$

EXERCISE 15E

GRADE A

1. Evaluate each expression.

a) $32^{\frac{4}{5}}$ **b)** $125^{\frac{2}{3}}$ **c)** $1296^{\frac{3}{4}}$ **d)** $243^{\frac{4}{5}}$

2. Rewrite the following in index form.

a) $\sqrt[3]{t^2}$ **b)** $\sqrt[4]{m^3}$ **c)** $\sqrt[5]{k^2}$ **d)** $\sqrt{x^3}$

3. Evaluate each expression.

a) $25^{-\frac{1}{2}}$ **b)** $36^{-\frac{1}{2}}$ **c)** $16^{-\frac{1}{4}}$ **d)** $81^{-\frac{1}{4}}$

e) $16^{-\frac{1}{2}}$ **f)** $8^{-\frac{1}{3}}$ **g)** $32^{-\frac{1}{5}}$ **h)** $27^{-\frac{1}{3}}$

GRADE A*

4. Evaluate each expression.

a) $25^{-\frac{3}{2}}$ **b)** $36^{-\frac{3}{2}}$ **c)** $16^{-\frac{3}{4}}$ **d)** $81^{-\frac{3}{4}}$

e) $64^{-\frac{4}{3}}$ **f)** $8^{-\frac{2}{3}}$ **g)** $32^{-\frac{2}{5}}$ **h)** $27^{-\frac{2}{3}}$

5. Evaluate each expression.

a) $100^{-\frac{5}{2}}$ **b)** $144^{-\frac{1}{2}}$ **c)** $125^{-\frac{2}{3}}$ **d)** $9^{-\frac{3}{2}}$

e) $4^{-\frac{5}{2}}$ **f)** $64^{-\frac{5}{6}}$ **g)** $27^{-\frac{4}{3}}$ **h)** $169^{-\frac{1}{2}}$

6. Which of these is the odd one out?

$16^{-\frac{3}{4}}$ $64^{-\frac{1}{2}}$ $8^{-\frac{2}{3}}$

Show how you decided.

7. Write each of these as a fraction.

a) $\left(\frac{9}{4}\right)^{\frac{3}{2}}$ **b)** $\left(\frac{27}{125}\right)^{\frac{2}{3}}$ **c)** $\left(\frac{16}{9}\right)^{\frac{5}{2}}$ **d)** $\left(\frac{4}{49}\right)^{\frac{3}{2}}$

e) $\left(\frac{64}{27}\right)^{\frac{2}{3}}$ **f)** $\left(\frac{16}{81}\right)^{\frac{3}{4}}$ **g)** $\left(\frac{125}{64}\right)^{\frac{4}{3}}$ **h)** $\left(\frac{64}{729}\right)^{\frac{5}{6}}$

8. Write these as fractions.

a) $\left(\frac{3}{5}\right)^{-2}$ **b)** $\left(\frac{4}{3}\right)^{-3}$ **c)** $\left(\frac{9}{5}\right)^{-3}$ **d)** $\left(\frac{2}{3}\right)^{-5}$

e) $\left(\frac{125}{64}\right)^{-\frac{2}{3}}$ **f)** $\left(\frac{25}{64}\right)^{-\frac{3}{2}}$ **g)** $\left(\frac{16}{81}\right)^{-\frac{5}{4}}$ **h)** $\left(\frac{2187}{128}\right)^{-\frac{5}{7}}$

GRADE A**

9. Simplify each expression.

a) $x^{\frac{3}{2}} \times x^{\frac{5}{2}}$ **b)** $x^{\frac{1}{2}} \times x^{-\frac{3}{2}}$ **c)** $(8y^3)^{\frac{2}{3}}$

d) $5x^{\frac{3}{2}} \div \frac{1}{2}x^{-\frac{1}{2}}$ **e)** $4x^{\frac{1}{2}} \times 5x^{-\frac{3}{2}}$ **f)** $\left(\frac{27}{y^3}\right)^{-\frac{1}{3}}$

10. Simplify each expression.

a) $x^{\frac{1}{2}} \times x^{\frac{1}{2}}$ **b)** $d^{-\frac{1}{2}} \times d^{-\frac{1}{2}}$ **c)** $t^{\frac{1}{2}} \times t$

d) $(x^{\frac{1}{2}})^4$ **e)** $(y^2)^{\frac{1}{4}}$ **f)** $a^{\frac{1}{2}} \times a^{\frac{3}{2}} \times a^2$

11. Simplify each expression.

a) $x \div x^{\frac{1}{2}}$ **b)** $y^{\frac{1}{2}} \div y^{1\frac{1}{2}}$ **c)** $a^{\frac{1}{3}} \times a^{\frac{4}{3}}$

d) $t^{-\frac{1}{2}} \times t^{-\frac{3}{2}}$ **e)** $\frac{1}{d^{-2}}$ **f)** $\frac{k^{\frac{1}{2}} \times k^{\frac{3}{2}}}{k^2}$

12. Simplify each expression.

a) $\sqrt{x^{\frac{3}{2}} \times x^{\frac{9}{2}}}$ **b)** $\sqrt{x^{\frac{1}{2}} \times x^{-\frac{5}{2}}}$ **c)** $\sqrt{x^{\frac{4}{3}} \times x^{\frac{2}{3}}}$

d) $\sqrt{\frac{x^{\frac{5}{2}}}{x^{\frac{1}{2}}}}$ **e)** $\sqrt{\frac{x^{-\frac{1}{2}} \times x^5}{x^{\frac{3}{2}}}}$ **f)** $\sqrt{\frac{x^{\frac{1}{3}} \times x^{\frac{7}{3}}}{x^{\frac{2}{3}}}}$

15.5 Solving equations with indices

THIS SECTION WILL SHOW YOU HOW TO …

✓ solve equations involving expressions that include fractional and negative indices

KEY WORDS

✓ index ✓ reciprocal
✓ indices ✓ root

If you want to solve equations involving fractional or negative **indices**, it is important that you understand what the **index** means.

Remember, a negative index means a **reciproca**l and a fractional index is a **root**.

$x^{-n} = \frac{1}{x^n}$ and $x^{\frac{1}{n}} = \sqrt[n]{x}$

For example: $x^{-3} = \frac{1}{x^3}$, $x^{\frac{1}{4}} = \sqrt[4]{x}$ and $x^{-\frac{2}{3}} = \frac{1}{x^{\frac{2}{3}}} = \frac{1}{\sqrt[3]{x^2}}$

To solve equation of this type, you need to rewrite the equation, using reciprocals and roots, and then work backwards.

EXAMPLE 6

Solve the equation $x^{-\frac{1}{2}} = 5$.

Rewrite the expression as a reciprocal. $\frac{1}{x^{\frac{1}{2}}} = 5$

Now write it as a root. $\frac{1}{\sqrt{x}} = 5$

Now rearranging gives: $\sqrt{x} = \frac{1}{5}$

Squaring both sides gives: $x = \frac{1}{25}$

EXAMPLE 7

Solve the equation $x^{\frac{2}{3}} = 4$.

Rewrite, using roots and powers: $\sqrt[3]{x^2} = 4$

Cube both sides: $x^2 = 4^3 \Rightarrow x^2 = 64$

Take the positive square root: $x = 8$

EXERCISE 15F

GRADE B

1. Solve these equations.

a) $\sqrt{x} = 8$ **b)** $\sqrt{x} = 5$ **c)** $\sqrt[3]{x} = 2$

d) $\sqrt[3]{x} = 10$ **e)** $\frac{1}{x} = 8$ **f)** $\frac{4}{x} = 5$

GRADE A

2. Solve these equations, giving your answers in surd form.

a) $x^2 = 8$ **b)** $x^2 = 20$ **c)** $5x^2 = 135$

d) $x^3 = 2\sqrt{2}$ **e)** $x^3 = 24\sqrt{3}$ **f)** $\frac{1}{x^2} = 2$

3. Solve these equations.

a) $x^{\frac{1}{2}} = 9$ **b)** $x^{\frac{1}{2}} = 7$ **c)** $x^{\frac{1}{2}} = \sqrt{3}$

d) $x^{\frac{1}{2}} = 3\sqrt{2}$ **e)** $x^{\frac{1}{2}} = 2\sqrt{5}$ **f)** $x^{\frac{1}{2}} = 5\sqrt{5}$

4. Solve these equations.

a) $x^{\frac{1}{3}} = 2$ **b)** $x^{\frac{1}{3}} = \sqrt{5}$ **c)** $x^{\frac{1}{4}} = 3$

d) $x^{\frac{1}{3}} = 5\sqrt{3}$ **e)** $x^{\frac{1}{5}} = \sqrt{2}$ **f)** $x^{\frac{1}{4}} = 2\sqrt{2}$

GRADE A*

5. Solve these equations.

a) $x^{\frac{2}{3}} = 9$ **b)** $x^{\frac{4}{3}} = 16$ **c)** $x^{\frac{3}{4}} = 8$

d) $x^{\frac{4}{3}} = 1$ **e)** $x^{\frac{2}{5}} = 4$ **f)** $x^{\frac{3}{10}} = 10\sqrt{10}$

6. Solve these equations.

a) $x^{-1} = 4$ **b)** $x^{-2} = 9$ **c)** $x^{-2} = 2$

d) $x^{-1} = 5$ **e)** $x^{-3} = 27$ **f)** $x^{-3} = \frac{1}{8}$

7. Solve these equations.

a) $x^{-\frac{1}{2}} = 7$ **b)** $x^{-\frac{1}{3}} = 4$ **c)** $x^{-\frac{1}{4}} = 3$

d) $x^{-\frac{1}{5}} = 2$ **e)** $x^{-\frac{1}{2}} = \sqrt{6}$ **f)** $x^{-\frac{1}{3}} = \sqrt{2}$

8. Solve these equations.

a) $x^{-\frac{3}{2}} = 8$ **b)** $x^{-\frac{2}{3}} = 9$ **c)** $x^{-\frac{3}{4}} = 27$

d) $x^{-\frac{3}{5}} = 125$ **e)** $x^{-\frac{3}{2}} = \frac{1}{27}$ **f)** $x^{-\frac{4}{3}} = \frac{1}{9}$

Exam-style questions

1. Write down **two** possible:

a) multiplication questions with an answer of $12x^2y^5$

b) division questions with an answer of $12x^2y^5$.

2. a) The cube root of $\frac{1}{y}$ is 6. GRADE A*

Find the exact value of y.

b) The cube of $\sqrt{x}$ is $\frac{1}{8}$.

Work out the reciprocal of x.

3. Simplify $\frac{x^{\frac{1}{2}} \times x^{1\frac{1}{2}}}{x^2 \times x^{-1}}$ as much as possible. GRADE A*

4. Write $\frac{\sqrt{x}}{\sqrt[3]{x}}$ as a power of x. GRADE A*

5. Solve these equations.

a) $x^{\frac{1}{3}} = 8$ **b)** $x^{-\frac{1}{2}} = 3$ GRADE A*

6. a) Simplify $\frac{z^{-\frac{3}{2}}}{\sqrt{z}}$.

b) Write $\sqrt{\frac{1}{z}}$ as a power of z.

c) Solve the equation $5z^{-2} = z^{-3}$. GRADE A**

16 Calculus

16.1 The gradient of a curve

THIS SECTION WILL SHOW YOU HOW TO …

- ✓ understand and use the notation $\frac{dy}{dx}$
- ✓ understand the concept of the gradient of a curve

KEY WORDS

✓ gradient ✓ $\frac{dy}{dx}$ ('dee y by dee x') ✓ gradient function

This is a graph of $y = x^2 - 3x + 4$.

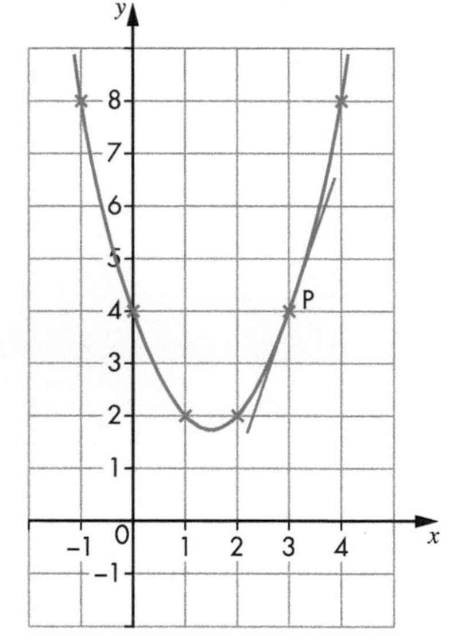

What is the **gradient** at the point P(3, 4)?

In an earlier chapter you learnt how to find the gradient by drawing a tangent to the curve at P. It is hard to do this accurately and you need to draw the graph first.

Now look at another method.

You need to choose two points on the curve that are close to P, one each side.

Choose two values for x, 2.9 and 3.1 are both close to P.

Now substitute these values for x in the equation $y = x^2 - 3x + 4$ to get the corresponding y values of 3.71 and 4.31.

This gives the two points A(2.9, 3.71) and B(3.1, 4.31).

Imagine drawing a straight line through A and B. You can assume this will have a similar gradient to the tangent at P.

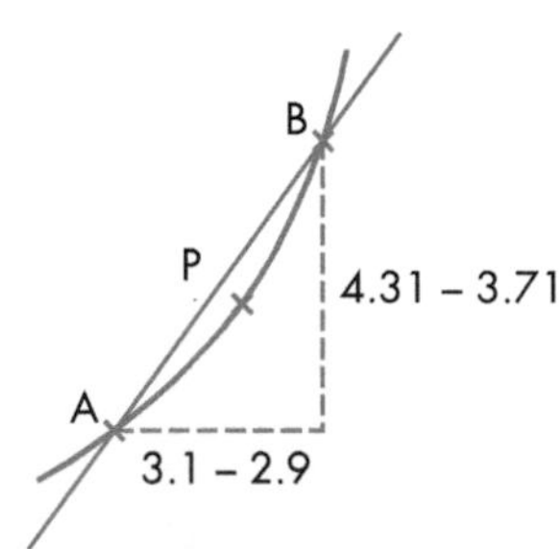

Gradient of AB $= \frac{4.31-3.71}{3.1-2.9} = 3$

If you choose any two points close to P you will get a similar result.

In fact the gradient of the curve at P (3, 4) is 3.

You could use this method at any point on the curve. It will show the following fact.

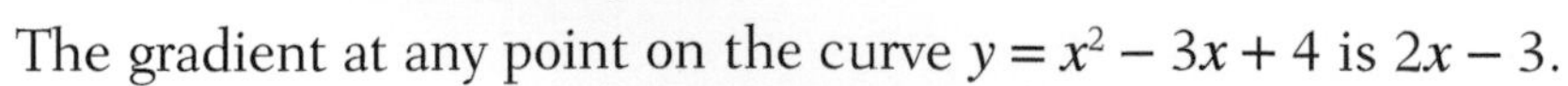

The gradient at any point on the curve $y = x^2 - 3x + 4$ is $2x - 3$.

For example, at (3, 4) the gradient is $2 \times 3 - 3 = 3$

at (4, 8) the gradient is $2 \times 4 - 3 = 5$

at (1, 2) the gradient is $2 \times 1 - 3 = -1$.

Looking at the graph should convince you that these seem to be reasonable values.

The notation to represent the gradient of a curve is $\frac{dy}{dx}$ (read it as 'dee y by dee x').

This may also be called the **gradient function**.

So the tangent to the curve on the graph shows that:

if $y = x^2 - 3x + 4$ then $\frac{dy}{dx} = 2x - 3$.

This is the case for all equations of this form. So the general result is:

If $y = ax^2 + bx + c$ then $\frac{dy}{dx} = 2ax + b$.

You can use this to calculate the gradient at any point on a quadratic curve. You no longer need to draw the graph and you can use this method for any point on the curve.

EXAMPLE 1

A curve has the equation $y = 0.5x^2 + 4x - 3$.

a) Find the gradient at $(0, -3)$ and at $(2, 7)$.

b) Find the coordinates of the point where the gradient is 0.

a) Using the general result above, if $y = 0.5x^2 + 4x - 3$:

then $\frac{dy}{dx} = 2 \times 0.5x + 4 \Rightarrow \frac{dy}{dx} = x + 4$

If $x = 0$, $\frac{dy}{dx} = 0 + 4 = 4$. The gradient at $(0, -3)$ is 4.

If $x = 2$, $\frac{dy}{dx} = 2 + 4 = 6$. The gradient at $(2, 7)$ is 6.

b) If the gradient is 0, then $\frac{dy}{dx} = 0 \Rightarrow x + 4 = 0 \Rightarrow x = -4$

If $x = -4$, $y = 0.5 \times (-4)^2 + 4 \times (-4) - 3 = -11 \Rightarrow$ the gradient is 0 at $(-4, -11)$.

EXERCISE 16A

GRADE C

1. A curve has the equation $y = x^2 - 2x$.

a) Copy and complete this table of values.

x	−2	−1	0	1	2	3	4
$x^2 - 2x$	8	3				3	8

b) Sketch the graph of $y = x^2 - 2x$.

c) Work out $\frac{dy}{dx}$.

d) Find the gradient of the curve at (3, 3).

e) Find the gradient of the curve at (4, 8).

f) Find the gradient at two more points on the curve.

g) At what point on the graph is the gradient 0?

h) By looking at your graph, check that your answers to parts **d**, **e**, **f** and **g** seem sensible.

2. $y = x^2 - 6x + 15$

a) Work out $\frac{dy}{dx}$.

b) Find the gradient at (0, 15).

c) Find the gradient at (5, 10).

d) Find the coordinates of the point where the gradient is 2.

3. $y = 2x^2 - 10$

a) Work out $\frac{dy}{dx}$.

b) Find the gradient at (2, –2).

c) Find the gradient at (–1, –9).

d) Find the point where the gradient is 12.

4. This is the graph of $y = 4x - x^2$.

a) Work out $\frac{dy}{dx}$.

b) Find the gradient at each point where the curve crosses the x-axis.

c) Where is the gradient equal to 2?

d) Where is the gradient equal to 1?

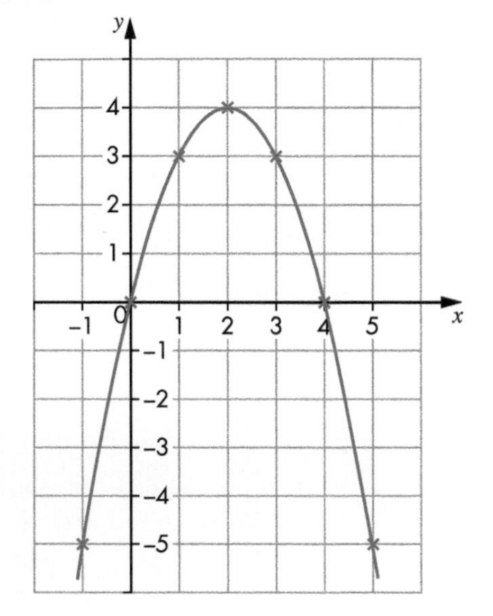

5. Work out $\frac{dy}{dx}$ for each equation.

a) $y = x^2 + x + 1$ **b)** $y = x^2 - 7x + 3$

c) $y = 4x^2 - x + 6$ **d)** $y = 0.3x^2 - 1.5x + 7.2$

e) $y = 6 - 2x + x^2$ **f)** $y = 10 + 3x - x^2$

g) $y = 2x + 5$ **h)** $y = 4$

GRADE B

6. If $y = (x + 4)(x - 2)$ what is $\frac{dy}{dx}$?

7. Work out $\frac{dy}{dx}$ for each equation.

a) $y = 2x(x + 1)$ **b)** $y = (x + 2)(x + 5)$ **c)** $y = (x + 3)(x - 3)$

HINTS AND TIPS

First multiply out the brackets.

8. A curve has the equation $y = x^2 + 2x - 5$.

a) Where does the curve cross the y-axis?

b) What is the gradient of the curve at that point?

16.2 More complex curves

THIS SECTION WILL SHOW YOU HOW TO …

✓ understand the concept of a rate of change
✓ know that the gradient of a function is the gradient of the tangent at that point

KEY WORDS

✓ rate of change ✓ differentiation ✓ differentiate

You have seen that if $y = x^2$ then $\frac{dy}{dx} = 2x$.

This table shows the value of $\frac{dy}{dx}$ for some other curves.

y	$\frac{dy}{dx}$
1	0
x	1
x^2	$2x$
x^3	$3x^2$
x^4	$4x^3$

→ The line $y = 1$ has gradient 0

→ The line $y = x$ has gradient 1

There is a general pattern here:

if $y = x^n$ then $\frac{dy}{dx} = nx^{n-1}$.

You say that $\frac{dy}{dx}$ is the **rate of change** of y with respect to x.

If a is a constant and $y = ax^n$ then $\frac{dy}{dx} = anx^{n-1}$.

For example:

if $y = 5x^2$ then $\frac{dy}{dx} = 5 \times 2x = 10x$

if $y = 4x^3$ then $\frac{dy}{dx} = 12x^2$

if $y = 6$ then $\frac{dy}{dx} = 0$.

So the line with equation $y = 6$ is horizontal and has gradient 0.

EXAMPLE 2

What is the gradient of the curve with equation $y = x^3 - 3x^2 + 4x + 7$ at the point $(2, -5)$?

$$\frac{dy}{dx} = 3x^2 - 6x + 4$$

If $x = 2$, $\frac{dy}{dx} = 3 \times 2^2 - 6 \times 2 + 4 = 4$

The gradient at $(2, -5)$ is 4.

The process of finding $\frac{dy}{dx}$ is called **differentiation**.

In example 2 you can see that each term was **differentiated** in turn.

Differentiate x^3 to get $3x^2$

Differentiate $-3x^2$ to get $-6x$

Differentiate $4x$ to get 4

Differentiate 7 to get 0

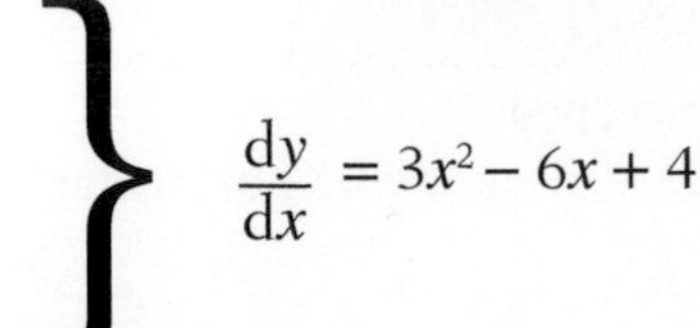

$$\frac{dy}{dx} = 3x^2 - 6x + 4$$

EXERCISE 16B

GRADE B

1. The equation of a curve is $y = 2x^3$.

a) Work out $\frac{dy}{dx}$.

b) Find the gradient of the curve at (1, 2) and (2, 16).

2. The equation of a curve is $y = x^3 - 6x^2 + 8x$.

a) Work out $\frac{dy}{dx}$.

b) Show that (0, 0), (2, 0) and (4, 0) are all on this curve.

c) Find the rate of change of y with respect to x at each point in part **b**.

3. Work out $\frac{dy}{dx}$ for each equation.

a) $y = 2x^4$ **b)** $y = 5x^3 - 2x + 4$ **c)** $y = 3x^3 + 5x - 7$

d) $y = 10 - x^3$ **e)** $y = x(x^3 - 1)$

GRADE A

4. This is a sketch of the curve $y = x^4 - 4x^2$.

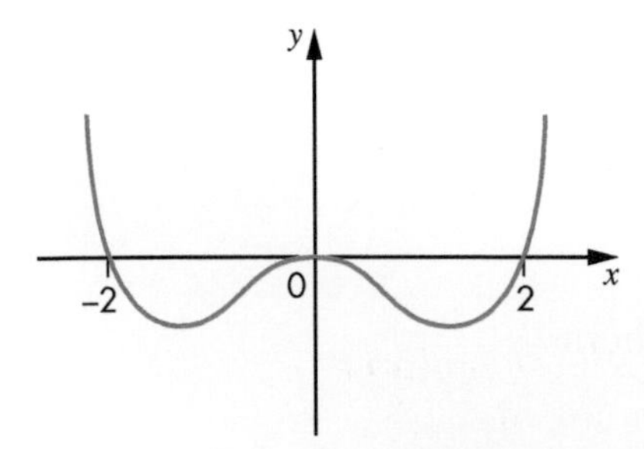

Find the gradient at the points where the curve meets the x-axis.

GRADE A*

5. A curve has the equation $y = \frac{1}{3}x^3 - 5x + 4$.

Show that there are two points on the curve where the rate of change of y with respect to x is 4. Find the coordinates of the two points.

16.3 Stationary points and curve sketching

THIS SECTION WILL SHOW YOU HOW TO …

✓ use differentiation to find stationary points on a curve: maxima, minima and points of inflection
✓ understand the meaning of maximum points, minimum points and points of inflection
✓ prove whether a stationary point is a maximum, minimum or point of inflection

KEY WORDS

✓ stationary point
✓ turning point
✓ maximum point
✓ minimum point
✓ maxima
✓ minima
✓ point of inflection
✓ increasing function
✓ decreasing function

A point where the gradient is zero is called a **stationary point** or **turning point**.

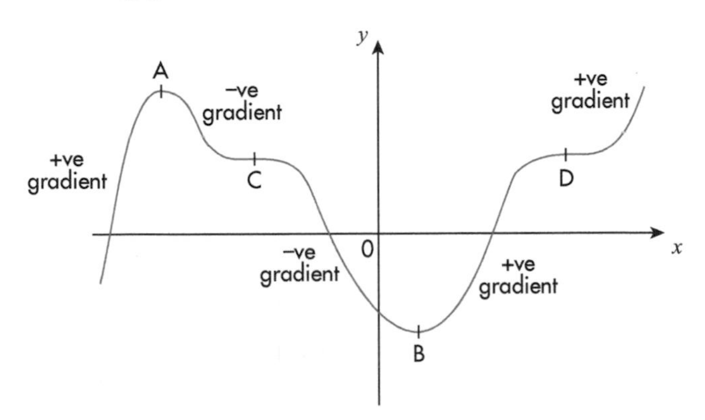

A and B are turning points.

At any turning point $\frac{dy}{dx} = 0$.

A is called a **maximum point** because it is higher than the points near it.

B is called a **minimum point**.

C and D are called **points of inflection** because the gradient is either positive on both sides of the point (as at point D) or negative on both sides of the point (as at point C).

The plural of maximum is **maxima** and the plural of minimum is **minima**. If a curve has two or more maximum points, they may be called its maxima. Similarly, if a curve has two or more minimum points, they may be called its minima.

EXAMPLE 3

Find the turning points of $y = x^3 - 12x + 4$ and state whether each is a maximum or a minimum point.

If $y = x^3 - 12x + 4$,

$$\frac{dy}{dx} = 3x^2 - 12$$

At a turning point $\frac{dy}{dx} = 0$,

$\Rightarrow 3x^2 - 12 = 0$

$\Rightarrow 3x^2 = 12$

$\Rightarrow x^2 = 4$

$\Rightarrow x = 2$ *or* -2

If $x = 2$, $y = 8 - 24 + 4 = -12 \Rightarrow (2, -12)$ is a turning point

If $x = -2$, $y = -8 + 24 + 4 = 20 \Rightarrow (-2, 20)$ is a turning point

Plotting the points on a grid makes it look likely that $(-2, 20)$ is a maximum point and $(2, -12)$ is a minimum point.

If you are not sure, check the gradient on each side of the point.

For $(2, -12)$:

x	1.9	2	2.1
$\frac{dy}{dx}$	−1.17	0	1.23
gradient	negative	0	positive

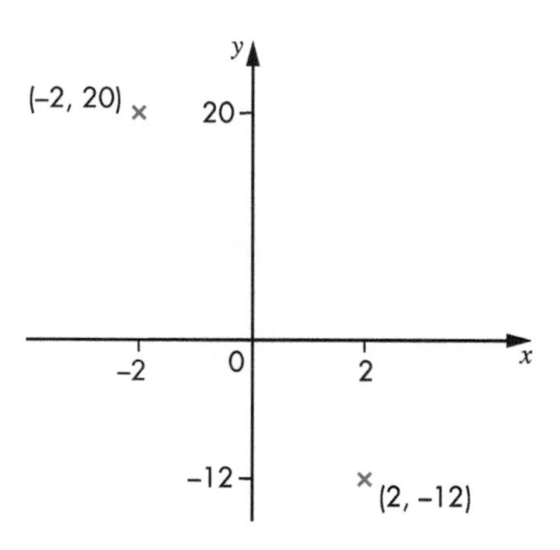

This is $3 \times 2.1^2 - 12$

$(2, -12)$ is a minimum point.

Because the gradient changes from negative to zero to positive as you move from left to right, $(2, -12)$ must be a minimum point.

For $(-2, 20)$:

x	−2.1	−2	−1.9
$\frac{dy}{dx}$	1.23	0	−1.17
gradient	positive	0	negative

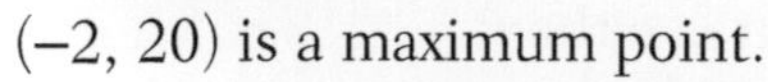

$(-2, 20)$ is a maximum point.

Because the gradient changes from positive to zero to negative as you move from left to right, $(-2, 20)$ must be a maximum point.

A sketch of the curve is shown on the right.

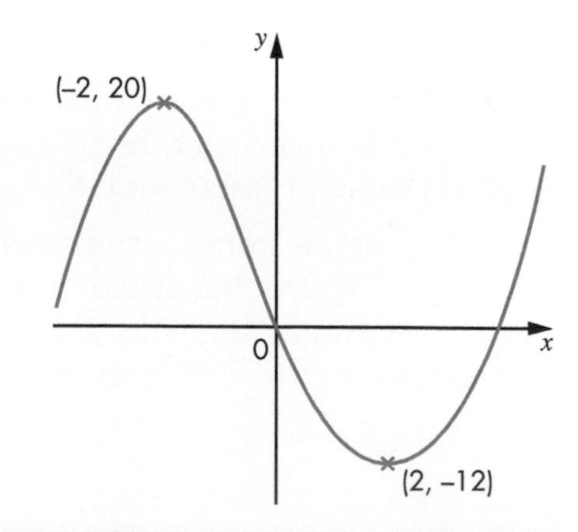

EXAMPLE 4

Find any stationary points of $y = x^3 - 3x^2 + 3x - 1$ and state whether each is a maximum, minimum or point of inflection.

$y = x^3 - 3x^2 + 3x - 1 \Rightarrow \frac{dy}{dx} = 3x^2 - 6x + 3$

At stationary points $\frac{dy}{dx} = 0 \quad \Rightarrow 3x^2 - 6x + 3 = 0 \Rightarrow 3(x-1)^2 = 0$
$\Rightarrow x = 1$

If $x = 1$, $y = 1 - 3 + 3 - 1 = 0$ so $(1, 0)$ is a stationary point.

Check the gradient on each side of the point $(1, 0)$.

x	0.9	1	1.1
$\frac{dy}{dx}$	0.03	0	0.03
gradient	positive	0	positive

Because, moving from left to right, the gradient changes from positive to zero to positive, $(1, 0)$ must be a point of inflection.

It is useful to draw a sketch of the curve.

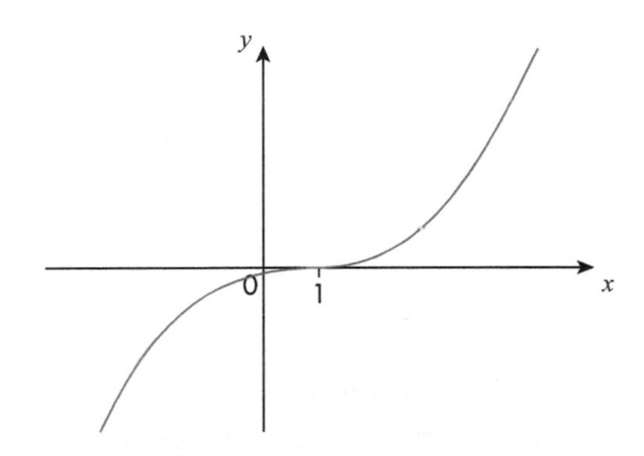

Increasing and decreasing functions

In an **increasing function**, as the x-value increases the y-value increases or stays the same.

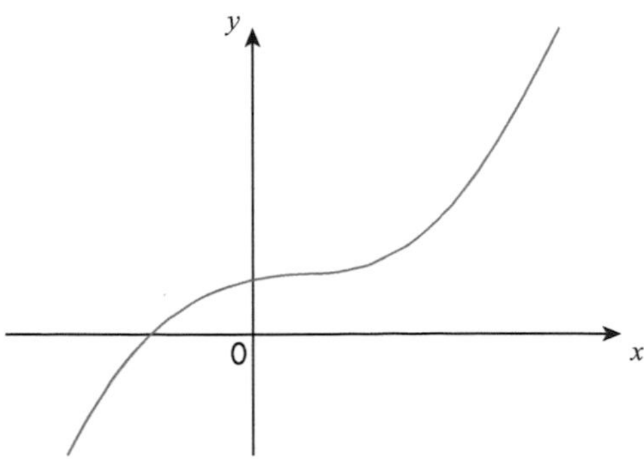

For an increasing function $\frac{dy}{dx} \geqslant 0$.

In a **decreasing function**, as the x-value increases the y-value decreases or stays the same.

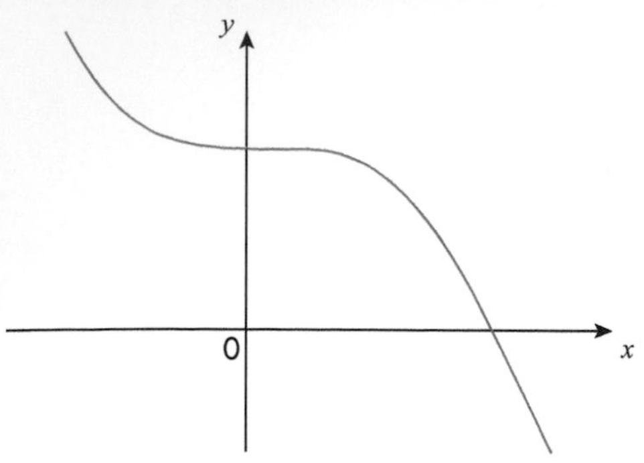

For a decreasing function $\frac{dy}{dx} \leqslant 0$.

EXAMPLE 5

Show that $y = 4x^3 + x$ is an increasing function for all values of x.

$y = 4x^3 + x \Rightarrow \frac{dy}{dx} = 12x^2 + 1$

$12x^2$ is positive for all values of x.

Therefore $12x^2 + 1 \geqslant 1$ for all values of x so it is always positive.

So $y = 4x^3 + x$ is an increasing function for all values of x.

EXAMPLE 6

For what values of x is $y = x^2 - 2x + 5$ a decreasing function?

$y = x^2 - 2x + 5 \Rightarrow \frac{dy}{dx} = 2x - 2$

At a turning point $\frac{dy}{dx} = 0 \Rightarrow 2x - 2 = 0$

$\Rightarrow x = 1$ and $y = 1 - 2 + 5 = 4$.

So $(1, 4)$ is a turning point.

When $x = 0.9$, $\frac{dy}{dx} = -0.2$

When $x = 1.1$, $\frac{dy}{dx} = 0.2$

So $y = x^2 - 2x + 5$ is a decreasing function when $x \geqslant 1$.

EXERCISE 16C

GRADE A*

1. $y = x^2 - 4x + 3$

a) Work out $\frac{dy}{dx}$.

b) Show that the curve has one turning point and find its coordinates.

c) State whether it is a maximum or minimum point.

d) State the values of x for which $y = x^2 - 4x + 3$ is an increasing function.

2. **a)** Work out the turning point of the curve $y = x^2 + 6x - 3$.
   b) Is it a maximum or a minimum point?

3. $y = 1 + 5x - x^2$
   a) Work out $\frac{dy}{dx}$.
   b) Find the turning point of the curve.
   c) Is it a maximum or a minimum point?
   d) State the values of x for which $y = 1 + 5x - x^2$ is a decreasing function.

4. This is a sketch of $y = x^3 - 3x^2$
   a) Work out $\frac{dy}{dx}$.
   b) Solve the equation $\frac{dy}{dx} = 0$.
   c) Find the coordinates of the two turning points shown on the graph.
   d) Write down the range of values of x for which $y = x^3 - 3x$ is a decreasing function.

5. $y = x^2 - 3x - 10$

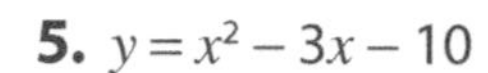

   a) Show that the graph crosses the x-axis at $(-2, 0)$ and $(5, 0)$.
   b) Work out $\frac{dy}{dx}$.
   c) Find the turning point of the curve.
   Is it a maximum or a minimum?
   d) Sketch the graph of $y = x^2 - 3x - 10$.
   e) The curve has a line of symmetry.
   What is the equation of this line?

6. The equation of a curve is $y = 2x^3 - 6x + 4$.
   a) Work out $\frac{dy}{dx}$.
   b) Find the turning points of the curve and state whether each one is a maximum or a minimum point.
   c) Sketch the graph of the curve.

GRADE A**

7. The equation of a curve is $y = x^3 + 6x^2 + 12x$.
   a) Work out $\frac{dy}{dx}$.
   b) Find any stationary points and state whether each is a maximum, minimum or point of inflection.

8. **a)** Prove that the curve $y = x^3 - 15x^2 + 75x + 6$ has only one stationary point.
   b) Show that the stationary point is a point of inflection.

16.4 The equation of a tangent and normal at any point on a curve

THIS SECTION WILL SHOW YOU HOW TO …

- ✓ use differentiation to find the stationary points on a curve – the maxima, minima and points of inflection
- ✓ understand the meaning of maximum points, minimum points and points of inflection
- ✓ establish whether a stationary point is a maximum, minimum or point of inflection

KEY WORDS

✓ tangent ✓ normal ✓ negative reciprocal

The equation of a tangent

You know that $\frac{dy}{dx}$ is the gradient of the **tangent** to a curve.

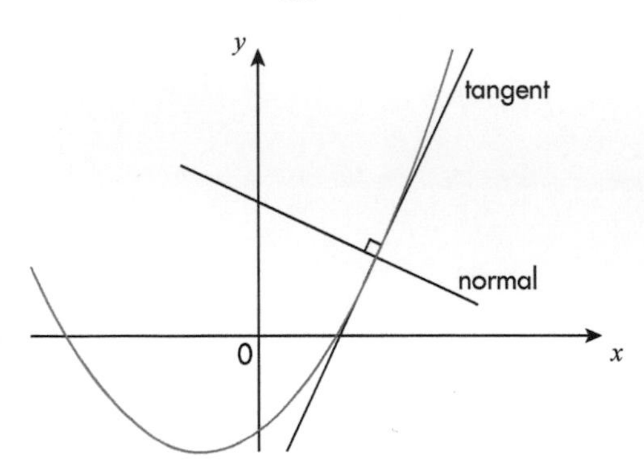

If you know the gradient of the tangent and the coordinates of the point on the curve where the tangent touches, you can work out the equation of the tangent.

The equation of a normal

A **normal** is a line drawn at right angles (perpendicular) to a tangent.

You have already learnt about perpendicular lines.

If two lines are perpendicular, their gradients are the **negative reciprocals** of each other.

> **HINTS AND TIPS**
>
> This means that the product of the gradients is −1.

EXAMPLE 7

Work out the equation of the tangent and normal to the curve $y = x^2 + 5x - 8$ at the point (2, 6).

$y = x^2 + 5x - 8 \Rightarrow \frac{dy}{dx} = 2x + 5$

At the point (2, 6), $\frac{dy}{dx} = 2 \times 2 + 5 = 9$

So the gradient of the tangent = 9.

- **Using the general form of the equation of a straight line $y = mx + c$**

  The equation of the tangent is of the form $y = 9x + c$.

  When $x = 2$, $y = 6 \Rightarrow 6 = 9 \times 2 + c \Rightarrow c = -12$

  The equation of the tangent is $y = 9x - 12$.

- **Using $y - y_1 = m(x - x_1)$**

  Then the equation of the tangent is $y - 6 = 9(x - 2) \Rightarrow y - 6 = 9x - 18$
  $\Rightarrow y = 9x - 12$

  As the gradient of the tangent = 9, the gradient of the normal $= -\frac{1}{9}$.

- **Using $y = mx + c$**

  The equation of the normal is of the form $y = -\frac{1}{9}x + c$.

  When $x = 2$, $y = 6 \Rightarrow 6 = -\frac{1}{9} \times 2 + c \Rightarrow c = 6\frac{2}{9}$ or $c = \frac{56}{9}$

  So the equation of the normal is $y = -\frac{1}{9}x + \frac{56}{9}$ or $9y = -x + 56$.

- **Using $y - y_1 = m(x - x_1)$**

  The equation of the normal is $y - 6 = -\frac{1}{9}(x - 2) \Rightarrow y - 6 = -\frac{1}{9}x + \frac{2}{9}$

  $\Rightarrow 9y - 54 = -x + 2 \Rightarrow 9y = -x + 56$.

EXERCISE 16D

GRADE A

1. The equation of a curve is $y = 3x^2$.

a) Work out $\frac{dy}{dx}$.

b) Work out the equation of the tangent at the point where $x = 2$.

c) Work out the equation of the normal at the point where $x = 2$.

GRADE A*

2. The equation of a curve is $y = x^2 - 4x + 3$.

a) Work out $\frac{dy}{dx}$.

b) Work out the equation of the tangent at the point (1, 0) .

c) Work out the equation of the normal at the point (4, 3).

3. The equation of a curve is $y = x^3 + x^2 - x + 5$.

a) Work out $\frac{dy}{dx}$.

b) Work out the equation of the tangent at the point (1, 6).
Give your answer in the form $y = mx + c$.

c) Work out the equation of the normal at the point (–2, 3).
Give your answer in the form $ax + by + c = 0$.

4. The equation of a curve is $y = x^2 - x - 6$.

Work out the equation of the tangents at the points where the curve intersects the x-axis.

5. a) Work out the equation of the normal to the curve $y = 2x^2 - 3x + 5$ at the point (1, 4).

b) The normal intersects the curve again at the point P.
Work out the coordinates of P.

6. a) Work out the equation of the tangent to the curve $y = 5 - x^2$ at the point (2, 1).

b) Work out the equation of the normal to the curve $y = 5 - x^2$ at the point (–1, 4).

c) Show that these lines meet at a point with x-coordinate $\frac{11}{7}$.

Exam-style questions

1. Work out $\frac{dy}{dx}$ for each equation. GRADE A

a) $y = 2(x^4 - 3x^2 + 17)$ **b)** $y = x^2(5 - x)$ **c)** $y = (x - 3)(x - 4)$

2. A curve has the gradient function $\frac{dy}{dx} = 9 - 2x^2$. GRADE A*

a) Find the rate of change of y with respect to x when $x = 3$.

b) Find the x-coordinates of the two points where the gradient of the curve is 1.

3. The equation of a curve is $y = 3x^2 - 2x + 1$. GRADE A*

a) Work out the gradient of the tangent to the curve at the point (2, 9).

b) Find the coordinates of the point where the gradient of the tangent is –8.

c) Find the x-coordinate of the point where the gradient of the normal is 2.

4. The equation of a curve is $y = x^3 - 4x + 5$. GRADE A*

a) Show that the point (2, 5) is on the curve.

b) Find the gradient of the tangent at the point (2, 5).

c) Find the coordinates of the point where the tangent crosses the y-axis.

5. A curve has the equation $y = x^3 - 3x^2 + 4$. GRADE A*

a) Show that the curve has a turning point at (0, 4).

b) The curve has another turning point. Find its coordinates.

c) Sketch the curve.

6. A curve has the equation $y = x^4 - 4x^2 + 5$. GRADE A*

a) Show that the point (2, 5) is on the curve.

b) Find the equation of the tangent at the point (2, 5).

c) Show that the curve has three turning points.

7. The sketch shows the curve $y = x^2 - 5x$ and the normal at the point (2, –6).

a) Find the equation of the normal at the point A(2, –6).

b) Find the coordinate of the point B where the normal meets the curve again.

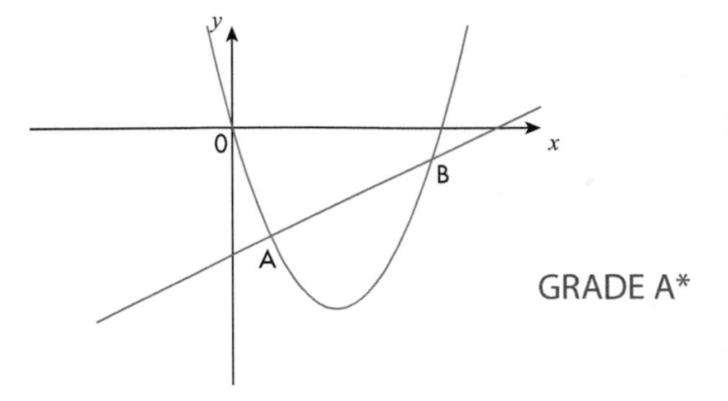

GRADE A*

8. The sketch shows the curve $y = 0.5x^2 - 2x + 4$ and the tangent and normal at the point P(4, 4).

The tangent and the normal cross the x-axis at points A and B respectively.

Find the length of AB.

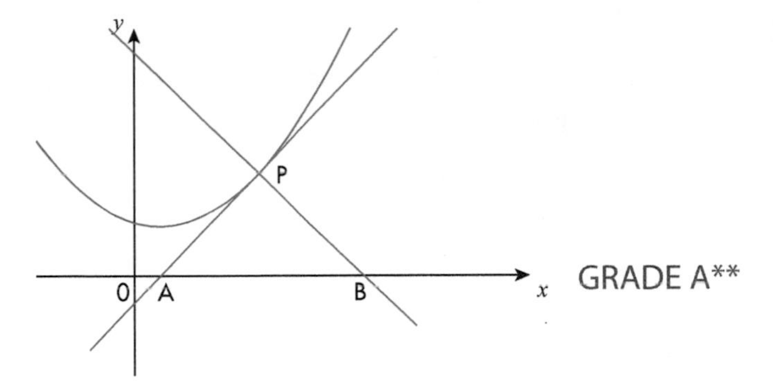

GRADE A**

17 Ratios of angles and their graphs

17.1 Trigonometric ratios of angles between 90° and 360°

THIS SECTION WILL SHOW YOU HOW TO ...

✓ find the sine, cosine and tangent of any angle from 0° to 360°

KEY WORDS

✓ cosine ✓ sine ✓ tangent

ACTIVITY 1

a) Copy and complete this table, using your calculator and rounding to three decimal places.

x	$\sin x$	x	$\sin x$	x	$\sin x$	x	$\sin x$
0°		180°		180°		360°	
15°		165°		195°		345°	
30°		150°		210°		330°	
45°		135°		225°		315°	
60°		120°		240°		300°	
75°		105°		255°		285°	
90°		90°		270°		270°	

b) Comment on what you notice about the **sine** of each acute angle, and the sines of its corresponding non-acute angles.

c) Draw a graph of $\sin x$ against x. Take x from 0° to 360° and $\sin x$ from –1 to 1.

d) Comment on any symmetries your graph has.

You should have discovered these three facts.

- When $90° < x$, $180°$, $\sin x = \sin(180° - x)$, for example, $\sin 153° = \sin 27°$
- When $180° < x < 270°$, $\sin x = -\sin(x - 180°)$, for example, $\sin 214° = -\sin 34° = -0.559$
- When $270° < x < 360°$, $\sin x = -\sin(360° - x)$, for example, $\sin 287° = -\sin 73°$

Look at the graph to check when $\sin x$ is positive and when $\sin x$ is negative.

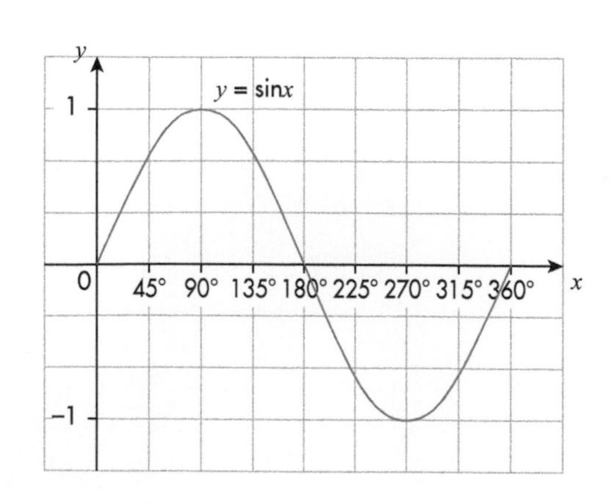

EXAMPLE 1

Find the angles with a sine of 0.56.

You know that both angles are between 0° and 180°.

Using your calculator to find $\sin^{-1}$ 0.56, you obtain 34.1°.

The other angle is, therefore, 180° − 34.1° = 145.9°.

So, the angles are 34.1° and 145.9°.

EXAMPLE 2

Find the angles with a sine of −0.197.

You know that both angles are between 180° and 360°.

Using your calculator to find $\sin^{-1}$ 0.197, you obtain 11.4°.

So the angles are 180° + 11.4° and 360° − 11.4°, which give 191.4° and 348.6°.

HINTS AND TIPS

You can always use your calculator to check your answer to this type of problem.

EXERCISE 17A

GRADE A*

1. State the two angles between 0° and 360° for each of these sine values.

a) 0.6 **b)** 0.8 **c)** 0.75 **d)** −0.7 **e)** −0.25
f) −0.32 **g)** −0.175 **h)** −0.814 **i)** 0.471 **j)** −0.097

2. Which of these values is the odd one out and why?

sin 36° sin 144° sin 234° sin 324°

3. The graph of sine x is cyclic, which means that it repeats forever in each direction.

a) Write down one value of x greater than 360° for which the sine value is 0.978 147 600 73.

b) Write down one value of x less than 0° for which the sine value is 0.978 147 600 73.

c) Describe any symmetries of the graph of $y = \sin x$.

ACTIVITY 2

a) Copy and complete this table, using your calculator and rounding to 3 decimal places. Use the same values of x as for Activity 1.

x	$\cos x$	x	$\cos x$	x	$\cos x$	x	$\cos x$
0°		180°		180°		360°	

b) Comment on what you notice about the cosines of the angles.

c) Draw a graph of $\cos x$ against x. Take x from 0° to 360° and $\cos x$ from −1 to 1.

d) Comment on the symmetry of the graph.

You should have discovered these three facts.

- When $90° < x < 180°$, $\cos x = -\cos(180° - x)$, for example, $\cos 161° = -\cos 19°$
- When $180° < x < 270°$, $\cos x = -\cos(x - 180°)$, for example, $\cos 245° = -\cos 65°$
- When $270° < x < 360°$, $\cos x = \cos(360° - x)$, for example, $\cos 310° = \cos 50°$

Look at the graph to check when $\cos x$ is positive and when $\cos x$ is negative.

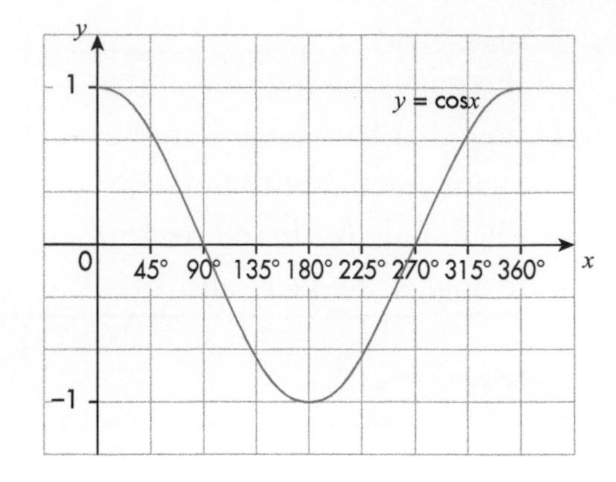

EXAMPLE 3

Find the angles with a cosine of 0.75.

One angle is between 0° and 90°, and the other is between 270° and 360°.

Using your calculator to find $\cos^{-1} 0.75$, you obtain 41.4°.

The other angle is, therefore, $360° - 41.4° = 318.6°$.

So, the angles are 41.4° and 318.6°.

EXAMPLE 4

Find the angles with a cosine of –0.285.

You know that both angles are between 90° and 270°.

Using your calculator to find $\cos^{-1} 0.285$, you obtain 73.4°.

The two angles are, therefore, $180° - 73.4°$ and $180° + 73.4°$, which give 106.6° and 253.4°.

Here again, you can use your calculator to check your answer, by keying in cosine.

EXERCISE 17B

GRADE A*

1. State the two angles between 0° and 360° for each of these cosine values.

a) 0.6 **b)** 0.58 **c)** 0.458 **d)** 0.575 **e)** 0.185
f) –0.8 **g)** –0.25 **h)** –0.175 **i)** –0.361 **j)** –0.974

2. Which of these values is the odd one out and why?

cos 58° cos 118° cos 238° cos 262°

3. The graph of cosine x is cyclic, which means that it repeats forever in each direction.

   a) Write down one value of x greater than 360° for which the cosine value is –0.66913060636.
   b) Write down one value of x less than 0° for which the cosine value is –0.66913060636.
   c) Describe any symmetries of the graph of $y = \cos x$.

EXERCISE 17C

GRADE A*

1. Write down the sine of each of these angles.

   a) 135° **b)** 269° **c)** 305° **d)** 133°

2. Write down the cosine of each of these angles.

   a) 129° **b)** 209° **c)** 95° **d)** 357°

3. Write down the two possible values of x ($0° < x < 360°$) for each equation. Give your answers to 1 decimal place.

   a) $\sin x = 0.361$ **b)** $\sin x = -0.486$ **c)** $\cos x = 0.641$
   d) $\cos x = -0.866$ **e)** $\sin x = 0.874$ **f)** $\cos x = 0.874$

4. Find two angles such that the sine of each is 0.5.

5. $\cos 41° = 0.755$. What is $\cos 139°$?

GRADE A**

6. Write down the value of each of the following, correct to 3 significant figures.

   a) $\sin 50° + \cos 50°$ **b)** $\cos 120° - \sin 120°$ **c)** $\sin 136° + \cos 223°$
   d) $\sin 175° + \cos 257°$ **e)** $\sin 114° - \sin 210°$ **f)** $\cos 123° + \sin 177°$

7. It is suggested that $(\sin x)^2 + (\cos x)^2 \equiv 1$ is true for all values of x. Test out this suggestion to see if you agree.

8. Suppose the sine key on your calculator is broken, but not the cosine key. Show how you could calculate these.

   a) $\sin 25°$ **b)** $\sin 130°$

9. Find a solution to each of these equations.

   a) $\sin (x + 20°) = 0.5$ **b)** $\cos (5x) = 0.45$

10. Use any suitable method to find the solution to the equation $\sin x = (\cos x)^2$.

The third ratio is the **tangent** of the angle.

Here is the graph of $y = \tan x$.

- When $90° < x < 180°$, $\tan x = -\tan(180° - x)$, for example, $\tan 147° = -\tan 33°$
- When $180° < x < 270°$, $\tan x = \tan(x - 180°)$, for example, $\tan 197° = \tan 17°$
- When $270° < x < 360°$, $\tan x = -\tan(360 - x°)$, for example, $\tan 302° = -\tan 58°$

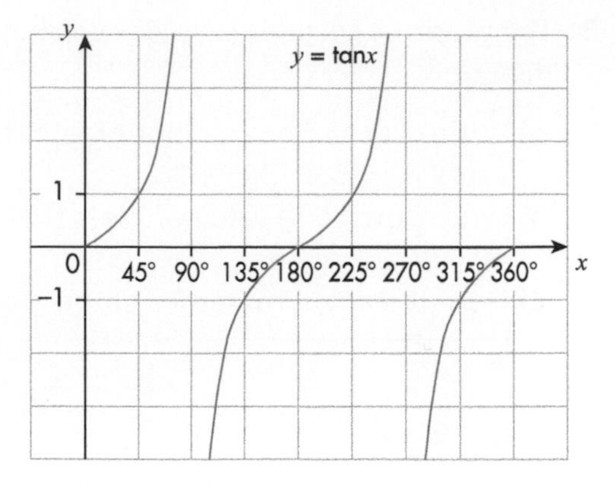

EXAMPLE 5

Find the angles between 0° and 360° with a tangent of 0.875.

One angle is between 0° and 90°, and the other is between 180° and 270°.

Using your calculator to find $\tan^{-1} 0.875$, you obtain 41.2°.

The other angle is, therefore, $180° + 41.2° = 221.2°$

So, the angles are 41.2° and 221.2°.

EXAMPLE 6

Find the angles between 0° and 360° with a tangent of –1.5.

You know that one angle is between 90° and 180°, and that the other is between 270° and 360°.

Using your calculator to find $\tan^{-1} 1.5$, you obtain 56.3°.

The angles are, therefore, $180° - 56.3°$ and $360° - 56.3°$,which give 123.7° and 303.7°.

EXERCISE 17D

GRADE A*

1. State the angles between 0° and 360° which have each of these tangent values.

a) 0.258 **b)** 0.785 **c)** 1.19 **d)** 1.875 **e)** 2.55
f) –0.358 **g)** –0.634 **h)** –0.987 **i)** –1.67 **j)** –3.68

2. Which of these values is the odd one out and why?

tan 45° tan 135° tan 235° tan 315°

3. The graph of $\tan x$ is cyclic, which means that it repeats forever in each direction.

a) Write down one value of x greater than 360° for which the tangent value is 2.144 506 920 51.

b) Write down one value of x less than 0° for which the tangent value is 2.144 506 920 51.

c) Describe any symmetries of the graph of $y = \tan x$.

17.2 The circular function graphs

THIS SECTION WILL SHOW YOU HOW TO …

✓ use the symmetry of the graphs $y = \sin x$, $y = \cos x$ and $y = \tan x$ in answering questions

KEY WORDS

✓ sine
✓ cosine
✓ tangent
✓ circular functions
✓ cyclic
✓ inverse cosine
✓ inverse sine
✓ line symmetry
✓ rotational symmetry

You have met **sine, cosine** and **tangent** graphs.

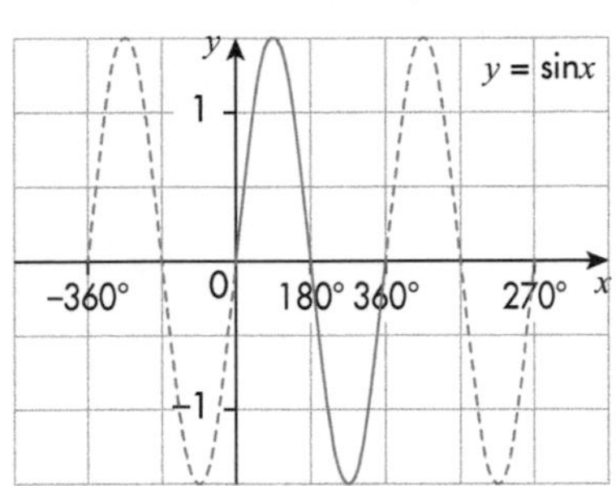

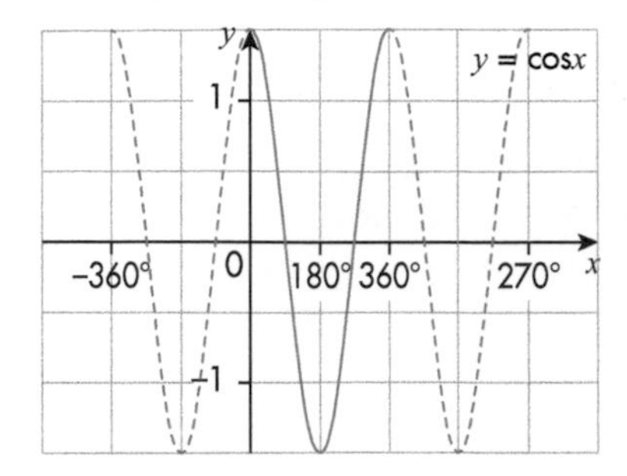

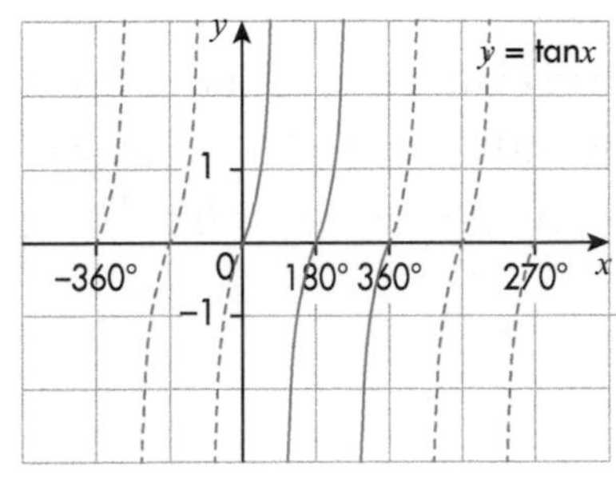

The trigonometric functions sine, cosine and tangent are referred to as **circular functions**.

Their graphs have some special properties.

- They are **cyclic**. This means that they repeat indefinitely in both directions.
- The sine graph has **rotational symmetry** about (180°, 0) and has **line symmetry** between 0° and 180° about $x = 90°$, and between 180° and 360° about $x = 270°$.
- The cosine graph has line symmetry about $x = 180°$, and has rotational symmetry between 0° and 180° about (90°, 0) and between 180° and 360° about (270°, 0).
- The tangent graph has rotational symmetry about (180°, 0).

The graphs can be used to find angles with certain values of sine, cosine and tangent.

EXAMPLE 7

Given that sin 42° = 0.669, find another angle between 0° and 360° that also has a sine of 0.669.

Plot the approximate value 0.669 on the sine graph and use the symmetry to work out the other value.

The other value is 180° – 42° = 138°.

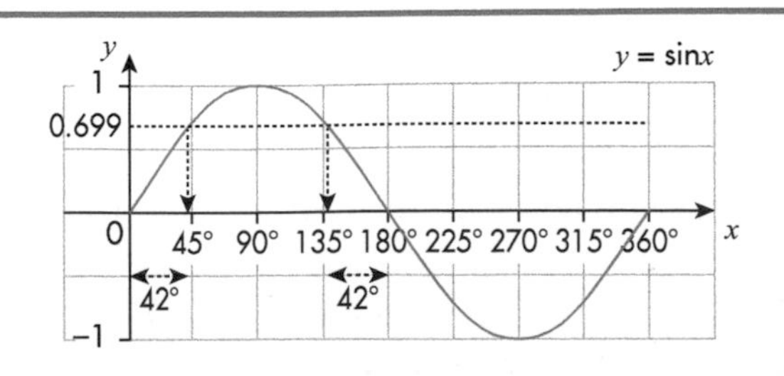

EXAMPLE 8

Given that cos 110° = –0.342, find two angles between 0° and 360° that have a cosine of +0.342.

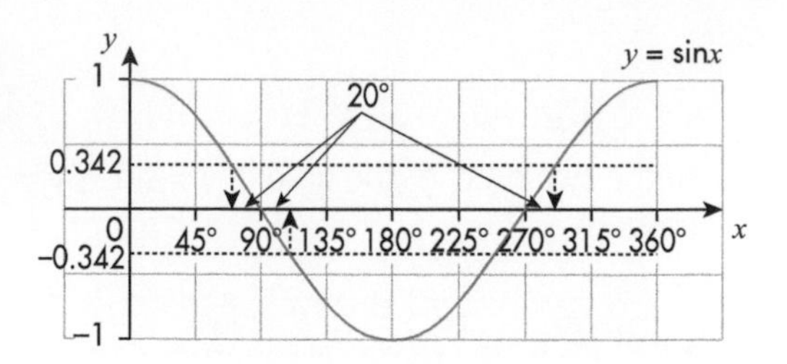

Plot the approximate values –0.342 and 0.342 on the cosine graph and use the symmetry to work out the values.

The required values are 90° – 20° = 70° and 270° + 20° = 290°.

EXAMPLE 9

Given that tan 48° = 1.111, find two angles between 0° and 360° that have a tangent of –1.111.

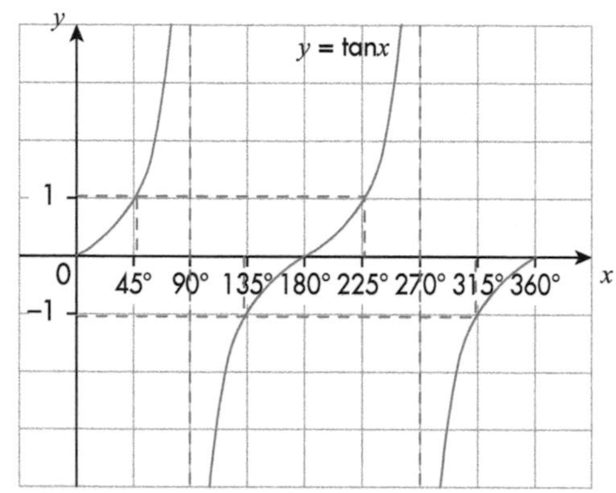

Plot the approximate values 1.111 and –1.111 on the tangent graph and use symmetry to work out the other values.

The required values are 180° – 48°= 132° and 360° – 48° = 312°.

EXERCISE 17E

GRADE A**

1. Given that sin 65° = 0.906, find another angle between 0° and 360° that also has a sine of 0.906.

2. Given that sin 213° = –0.545, find another angle between 0° and 360° that also has a sine of –0.545.

3. Given that cos 36° = 0.809, find another angle between 0° and 360° that also has a cosine of 0.809.

4. Given that cos 165° = –0.966, find another angle between 0° and 360° that also has a cosine of –0.966.

5. Given that tan 36° = 0.727, find another angle between 0° and 360° that also has a tangent of 0.727.

6. Given that tan 151° = –0.554, find another angle between 0° and 360° that also has a tangent of –0.554.

7. Given that $\sin 30° = 0.5$, find two angles between 0° and 360° that have a sine of –0.5.

8. Given that $\cos 45° = 0.707$, find two angles between 0° and 360° that have a cosine of –0.707.

9. Given that $\tan 60° = 1.732$, find two angles between 0° and 360° that have a tangent of –1.732.

10. **a)** Choose an acute angle a. Write down the values of:

    i) $\sin a$ **ii)** $\cos (90° - a)$.

    b) Repeat with another acute angle b.

    c) Write down a rule connecting the sine of an acute angle x and the cosine of the complementary angle (i.e. the difference with 90°).

    d) Find a similar rule for the cosine of x and the sine of its complementary angle.

11. Given that $\sin 26° = 0.438$:

    a) write down an angle between 0° and 90° that has a cosine of 0.438

    b) find two angles between 0° and 360° that have a sine of –0.438

    c) find two angles between 0° and 360° that have a cosine of –0.438.

12. Use the cosine rule to work out the size of angle A for a triangle where $a = 20$, $b = 11$ and $c = 13$.

13. Use the sine rule to work out the size of obtuse angle A for this triangle, where $a = 16$, $b = 14$ and $B = 46°$. What do you notice?

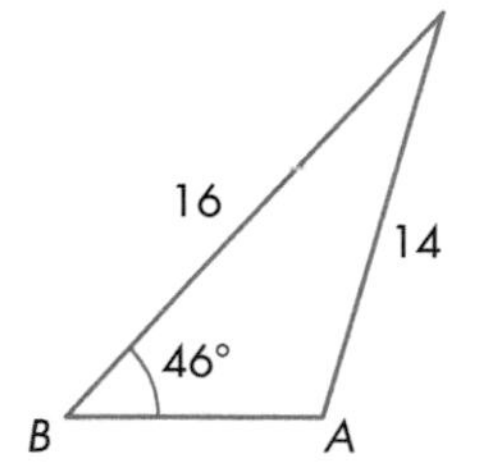

14. Use the sine rule to work out the size of the obtuse A in this triangle.

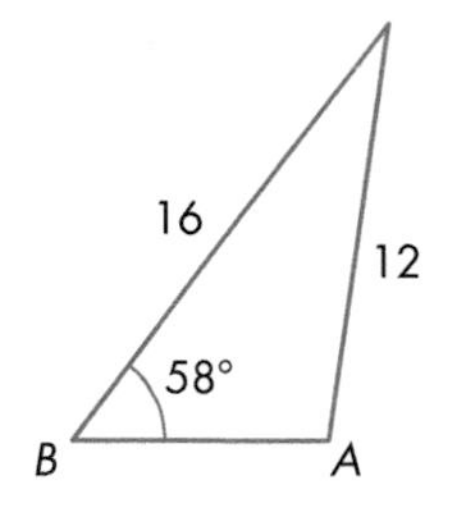

15. State if the following rules are true or false.

    a) $\sin x = \sin (180° - x)$

    b) $\sin x = -\sin (360° - x)$

    c) $\cos x = \cos (360° - x)$

    d) $\sin x = -\sin (180° + x)$

    e) $\cos (180° - x) = \cos (180° + x)$

    f) $\tan x = \tan (180° + x)$

    g) $\tan (180° - x) = \tan (180° + x)$

17.3 Special right-angled triangles

THIS SECTION WILL SHOW YOU HOW TO …

✓ work out the exact values of trigonometrical ratios for angles 30°, 45° and 60°

An equilateral triangle can be divided into two right-angled triangles.

Let the sides of the equilateral triangle have lengths 2 units, as shown.

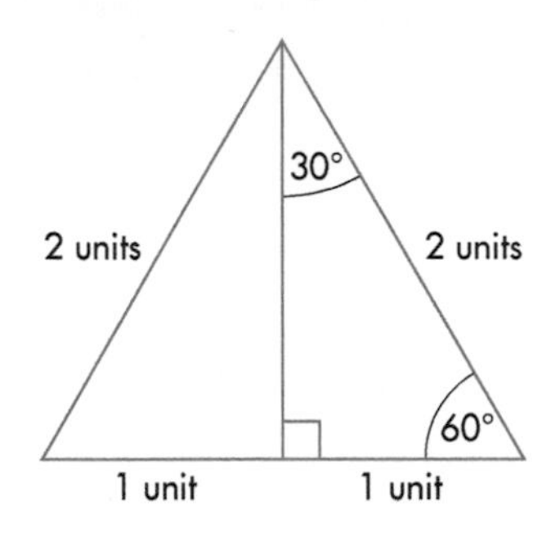

You can use Pythagoras' theorem to work out the height of the triangle.

$h^2 = 2^2 - 1^2 = 3 \Rightarrow h = \sqrt{3}$ units

Then one of the right-angled triangles looks like this.

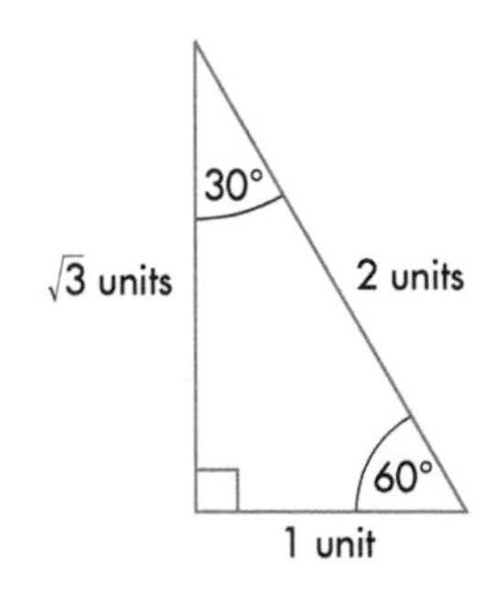

Using trigonometrical ratios:

$$\sin x = \frac{\text{opposite}}{\text{hypotenuse}}, \cos x = \frac{\text{adjacent}}{\text{hypotenuse}}$$

and $\tan x = \dfrac{\text{opposite}}{\text{adjacent}}$ giving:

$$\sin 30° = \frac{1}{2}, \cos 30° = \frac{\sqrt{3}}{2}, \tan 30° = \sqrt{3}$$

$$\sin 60° = \frac{\sqrt{3}}{2}, \cos 60° = \frac{1}{2}, \tan 60° = \frac{1}{\sqrt{3}}$$

An isosceles right-angled triangle has angles of 45°, 45° and 90°.

Let the sides forming the right angle have lengths 1 unit, as shown.

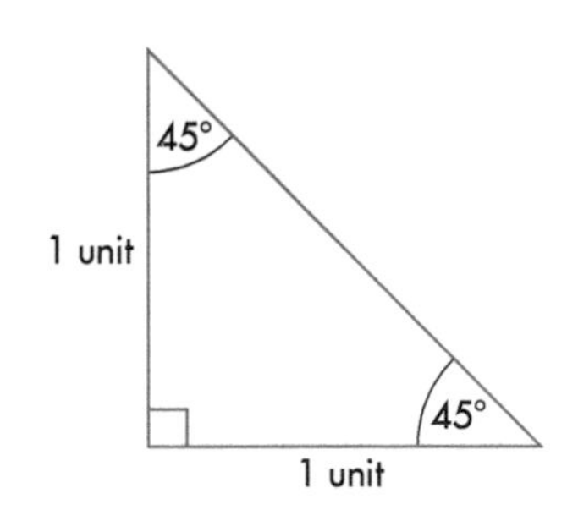

You can use Pythagoras' theorem to work out the length of the hypotenuse.

$x^2 = 1^2 + 1^2 = 2$

$x = \sqrt{2}$ units

Using trigonometrical ratios:

$$\sin x = \frac{\text{opposite}}{\text{hypotenuse}}, \cos x = \frac{\text{adjacent}}{\text{hypotenuse}} \text{ and } \tan x = \frac{\text{opposite}}{\text{adjacent}} \text{ giving:}$$

$$\sin 45° = \frac{1}{\sqrt{2}}, \cos 45° = \frac{1}{\sqrt{2}}, \tan 45° = 1$$

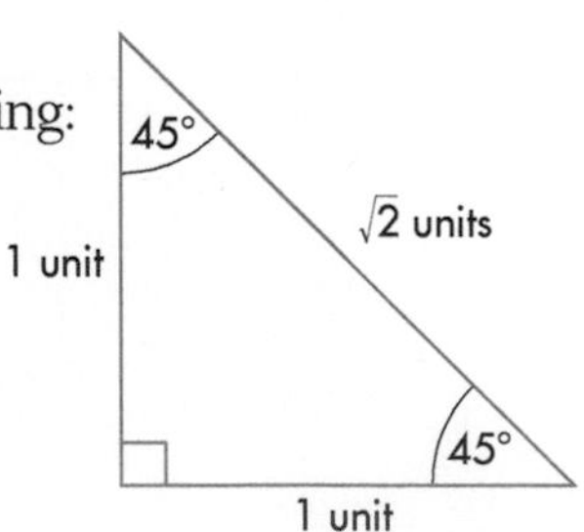

You will need to learn these ratios. Use this table to help you.

	sin	*cos*	*tan*
30°	$\frac{1}{2}$	$\frac{\sqrt{3}}{2}$	$\frac{1}{\sqrt{3}}$
45°	$\frac{1}{\sqrt{2}}$	$\frac{1}{\sqrt{2}}$	1
60°	$\frac{\sqrt{3}}{2}$	$\frac{1}{2}$	$\sqrt{3}$

EXAMPLE 10

Work out the value of x in this triangle.

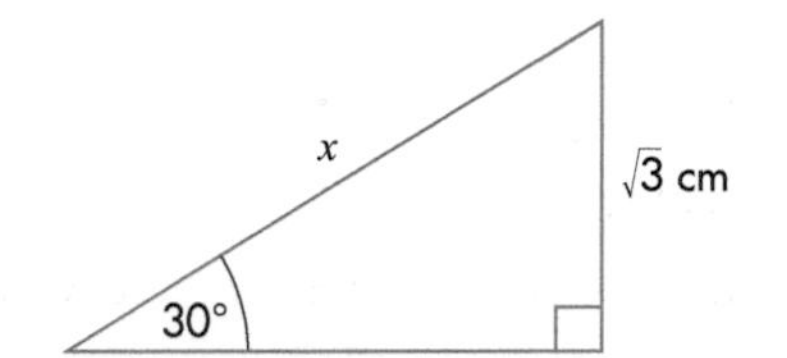

Using $\sin 30° = \frac{\sqrt{3}}{x}$ gives $\frac{1}{2} = \frac{\sqrt{3}}{x}$.

Rearranging gives $x = 2\sqrt{3}$ cm.

EXERCISE 17F

GRADE A* non-calculator

In this exercise, leave your answers in surd form.

1. Work out the value of x in each triangle.

a)

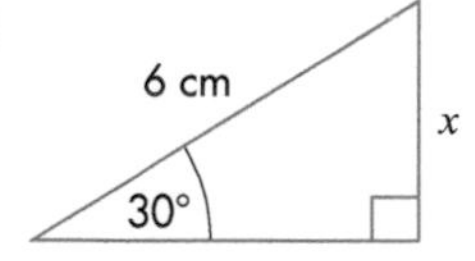

b)

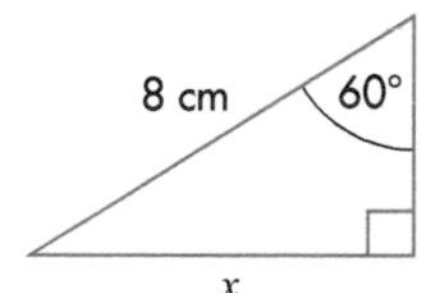

c)

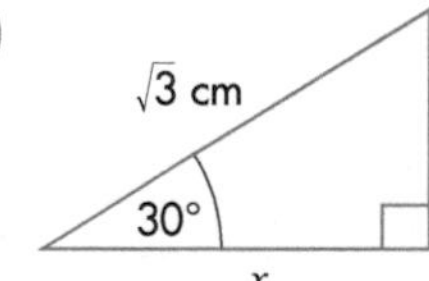

d)

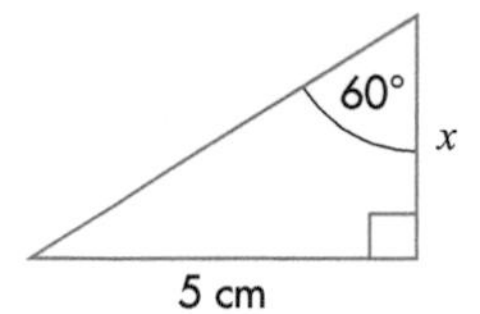

e)

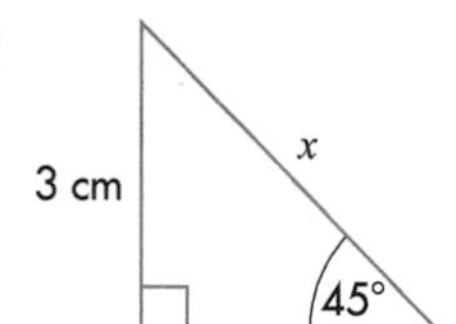

f)

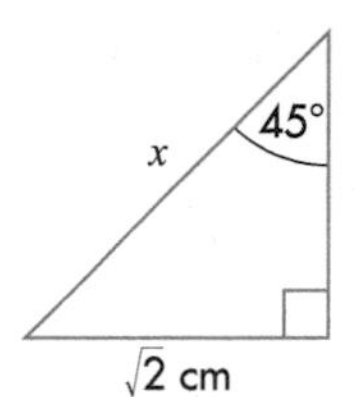

GRADE A**

2. Work out the value of x in each diagram.

a)

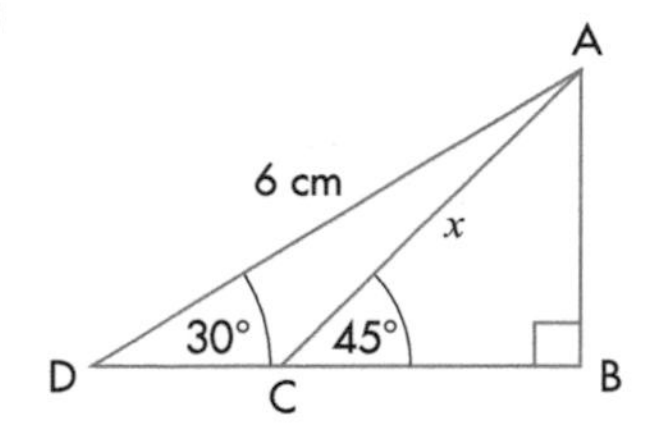

b)

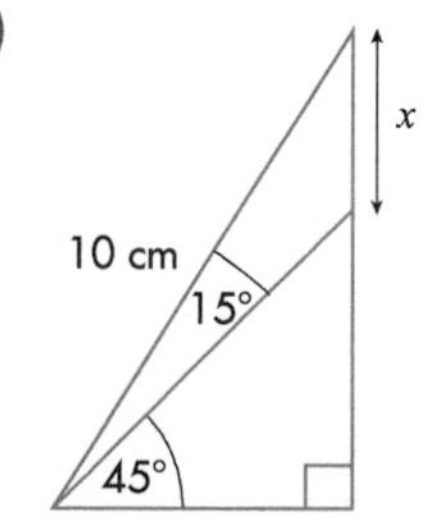

c)

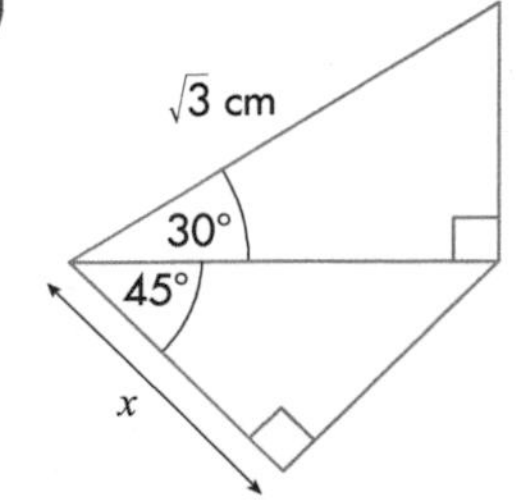

17.4 Trigonometrical expressions and equations

THIS SECTION WILL SHOW YOU HOW TO …

✓ solve trigonometrical equation in given intervals

✓ use the identities tan $\theta \equiv \dfrac{\sin\theta}{\cos\theta}$ and $\sin^2\theta + \cos^2\theta \equiv 1$ to simplify expression and to solve equations

You have already seen the symmetries in trigonometrical graphs. This means that trigonometrical equations can have more than one solution.

EXAMPLE 11

Solve the equation $\sin x = 0.7$ for values of x between 0° and 360°.

Using a calculator to work out the acute angle gives $x = 44.4°$ (to 1 decimal place).

Using symmetries, note that x can also be obtuse so $x = 44.4°$ or $x = 180° - 44.4° = 135.6°$.

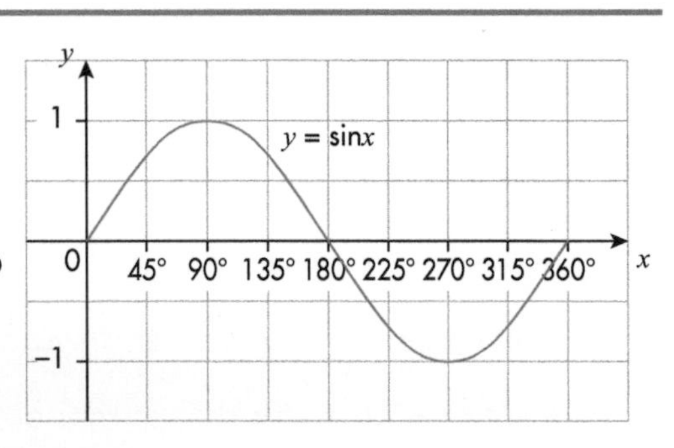

EXAMPLE 12

Solve the equation $5 \cos x + 2 = 0$, where x is obtuse.

$\cos x = -\frac{2}{5}$

When $\cos x = \frac{2}{5} = 0.4$, the acute angle $x = 66.4°$.

But as x is obtuse, $x = 180° - 66.4° = 113.6°$.

EXAMPLE 13

Solve the equation $2 \sin^2 x + \sin x - 1 = 0$ for $0° \leq x \leq 360°$.

Factorising gives $(2 \sin x - 1)(\sin x + 1) = 0$.

So $2 \sin x - 1 = 0$ or $\sin x + 1 = 0 \Rightarrow \sin x = \frac{1}{2}$ or $\sin x = -1$.

Looking again at the graph of $y = \sin x$ you can see that $x = 30°$, $x = 150°$ or $x = 270°$.

You have used these trigonometrical ratios in right-angled triangles.

$\sin\,\theta = \dfrac{\text{opposite}}{\text{hypotenuse}}$, $\cos\,\theta = \dfrac{\text{adjacent}}{\text{hypotenuse}}$ and $\tan\,\theta = \dfrac{\text{opposite}}{\text{adjacent}}$

Dividing $\sin\,\theta$ by $\cos\,\theta$ gives $\dfrac{\sin\theta}{\cos\theta} = \dfrac{\text{opposite}}{\text{hypotenuse}} \div \dfrac{\text{adjacent}}{\text{hypotenuse}}$.

So $\dfrac{\sin\theta}{\cos\theta} = \dfrac{\text{opposite}}{\text{adjacent}}$ which is the same as $\tan\,\theta$.

opposite
hypotenuse
adjacent

This shows that $\tan\,\theta = \dfrac{\sin\theta}{\cos\theta}$ for acute angles.

However, it is actually true for all angles.

You can use your calculator to check that this is true.

Similarly, using Pythagoras' theorem on a right-angled triangle gives:

$\text{opposite}^2 + \text{adjacent}^2 = \text{hypotenuse}^2$

Dividing both sides by (hypotenuse^2) gives:

$$\frac{\text{opposite}^2}{\text{hypotenuse}^2} + \frac{\text{adjacent}^2}{\text{hypotenuse}^2} = \frac{\text{hypotenuse}^2}{\text{hypotenuse}^2}$$

This can be written as: $\sin^2\,\theta + \cos^2\,\theta = 1$

This is actually true for all angles.

You can use your calculator to check that this is true.

Note:

You always write $(\sin\,\theta)^2$ as $\sin^2\,\theta$ and $(\cos\,\theta)^2$ as $\cos^2\,\theta$.

So now you have these two important identities:

$\tan\,\theta \equiv \dfrac{\sin\theta}{\cos\theta}$ and $\sin^2\,\theta + \cos^2\,\theta \equiv 1$

You can use these to simplify expressions and equations.v

EXAMPLE 14

Prove that $\tan\,\theta + \dfrac{\cos\theta}{\sin\theta} \equiv \dfrac{1}{\sin\theta\cos\theta}$.

When proving identities, start with the left-hand side and work to the right-hand side.

$\tan\,\theta + \dfrac{\cos\theta}{\sin\theta} \equiv \dfrac{\sin\theta}{\cos\theta} + \dfrac{\cos\theta}{\sin\theta}$ As $\tan\,\theta \equiv \dfrac{\sin\theta}{\cos\theta}$.

$= \dfrac{\sin^2\,\theta + \cos^2\,\theta}{\sin\theta\cos\theta}$ Finding a common denominator.

$= \dfrac{1}{\sin\theta\cos\theta}$ Using $\sin^2\,\theta + \cos^2\,\theta \equiv 1$.

EXAMPLE 15

Solve the equation $\tan x \sin x + \cos x = 2$ for $0° \leqslant x \leqslant 360°$.

As $\tan x \equiv \dfrac{\sin x}{\cos x} \Rightarrow \dfrac{\sin x}{\cos x} \sin x + \cos x = 2$

$$\Rightarrow \frac{\sin^2 x + \cos^2 x}{\cos x} = 2 \Rightarrow \frac{1}{\cos x} = 2 \Rightarrow \cos x = \frac{1}{2}$$

Looking at the symmetries on the cosine graph, the acute angle is $x = 60°$ and the other angle is $360° - 60° = 300°$.

EXERCISE 17G

GRADE A**

1. Solve each equation for $0° \leqslant x \leqslant 360°$.

a) $\sin x = 0.3$ **b)** $\cos x = 0.2$ **c)** $\tan x = 1.2$
d) $\sin x = 0.6$ **e)** $\cos x = 1$ **f)** $\tan x = 0.7$

2. Solve each equation for $0° \leqslant x \leqslant 360°$.

a) $\sin x = -0.3$ **b)** $\cos x = -0.2$ **c)** $\tan x = -1.2$
d) $\sin x = -0.6$ **e)** $\cos x = -1$ **f)** $\tan x = -0.7$

3. Solve each of equation for $0° \leqslant x \leqslant 180°$.

a) $4 \sin x + 1 = 0$ **b)** $5 \cos x + 2 = 0$ **c)** $\tan x - 1 = 0$
d) $-2 \sin x + \sqrt{2} = 0$ **e)** $3 \cos x - 1 = 0$ **f)** $5 \tan x = 0$

4. Solve each equation for $0° \leqslant x \leqslant 180°$.

a) $\sin x = 2 \cos x$ **b)** $\sin x = 4 \cos x$ **c)** $-3 \sin x = \cos x$
d) $5 \sin x = -2 \cos x$ **e)** $-\sin x = \cos x$ **f)** $\sin x + 4 \cos x = 0$

5. Prove each of these identities.

a) $\sin^3 x + \sin x \cos^2 x \equiv \sin x$ **b)** $\cos x \tan x \equiv \sin x$ **c)** $\dfrac{1 - \cos^2 x}{\cos^2 x} \equiv \tan^2 x$

d) $\dfrac{(\sin^2 x + \cos^2 x)\sin x}{\tan x} \equiv \cos x$ **e)** $\dfrac{1}{1 + \tan^2 x} \equiv \cos^2 x$

f) $\sin^4 x - \cos^4 x \equiv 2 \sin^2 x - 1$

6. Solve each equation for $0° \leqslant x \leqslant 360°$.

a) $\sin^2 x - 1 = 0$ **b)** $2\cos^2 x - 1 = 0$
c) $\tan^2 x - \sqrt{3} \tan x = 0$ **d)** $(2\cos x + 1)(\cos x + 1) = 0$
e) $\tan^2 x - \tan x - 2 = 0$ **f)** $2\sin^2 x - 5 \sin x + 2 = 0$

Exam-style questions

Do not use a calculator for questions 1 to 6.

1. a) Sketch a graph of $y = \tan x$ for $0° \leqslant x \leqslant 360°$.

b) One solution of the equation $\tan x = 4$ is $x = 76°$. Use your graph to find another solution. GRADE A*

2. Sin 27° = 0.454 to 3 decimal places. GRADE A**

a) What other angle between 0° and 360° has the same sine?

b) Find all the angles between 0° and 360° that have a sine of –0.454.

3. This table shows the cosines of some angles, correct to 2 decimal places. GRADE A**

θ	10°	20°	70°	80°
$\cos\theta$	0.98	0.94	0.34	0.17

Use the values in the table to find:

a) cos 100° **b)** cos 200° **c)** sin 10°.

4. BCD is a straight line. GRADE A**

Find the exact length of BD in the form $a + b\sqrt{c}$.

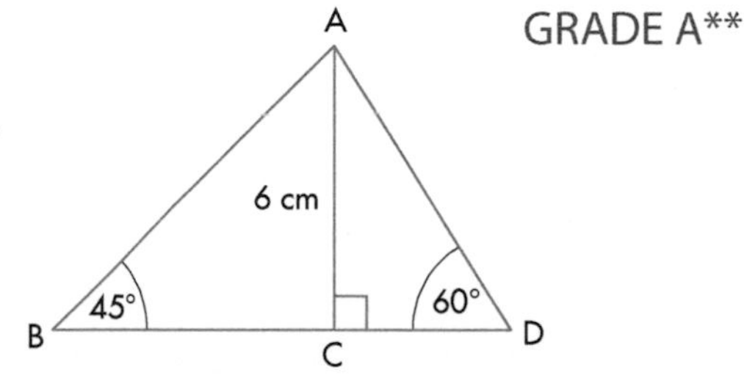

5. Show that the area of this triangle is $\frac{\sqrt{3}a^2}{8}$. GRADE A**

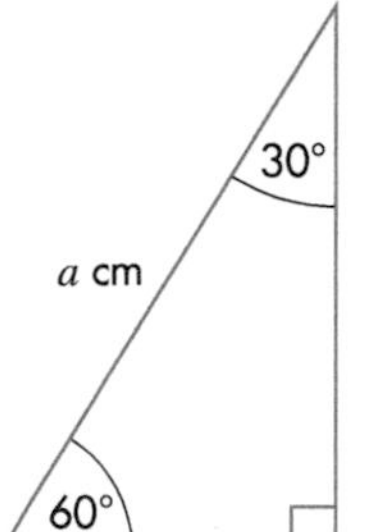

6. You are given that θ is an acute angle and $\sin\theta = \frac{3}{5}$. GRADE A**

Find the exact value of:

a) $\cos\theta$ **b)** $\tan\theta$ **c)** $\sin(360° - \theta)$.

7. Show that $(1 + \cos\theta)(1 - \cos\theta) \equiv \sin\theta\cos\theta\tan\theta$. GRADE A**

8. Solve the equation $3\sin x = \cos x$ for $0° < x < 90°$. GRADE A**

9. Solve the equation $4\sin^2 x = 1$ for $0° < x < 360°$. GRADE A**

18 Proof

18.1 Algebraic proof

THIS SECTION WILL SHOW YOU HOW TO …

✓ recognise and continue some special number sequences

KEY WORDS

✓ proof ✓ show ✓ verify

You will have met the fact that the sum of any two odd numbers is always even but this is how to prove it algebraically.

You can take any two odd numbers, add them together and get a number that divides exactly by 2. This does not prove the result, even if everyone in your class, or your school, or the whole of Britain, did this for a different pair of starting odd numbers. Unless you tried every pair of odd numbers (and there is an infinite number of them) you cannot be 100% certain this result is always true.

This is how to prove the result.

Let n be any whole number.

Whatever whole number is represented by n, $2n$ will always be even. So, $2n + 1$ represents any odd number.

Let one odd number be $2n + 1$, and let the other odd number be $2m + 1$.

The sum of these is:

$(2n + 1) + (2m + 1) = 2n + 2m + 1 + 1 = 2n + 2m + 2 = 2(n + m + 1)$

which must be even. This proves the result, as n and m can be any numbers.

In an algebraic proof, you must show every step clearly and the algebra must be done properly.

There are three levels of 'proof': **Verify** that ..., **Show** that ... and **Prove** that ...

- At the lowest level (verification), all you have to do is substitute numbers into the result to show that it works.
- At the middle level, you have to show that both sides of the result are the same algebraically.
- At the highest level (proof), you have to manipulate the left-hand side of the result to become its right-hand side. The following example demonstrates these three different procedures.

EXAMPLE 1

You are given that $n^2 + (n + 1)^2 - (n + 2)^2 = (n - 3)(n + 1)$.

a) Verify that this result is true.

b) Show that this result is true.

c) Prove that this result is true.

a) Choose a number for n, say $n = 5$. Put this value into both sides of the expression, which gives:

$5^2 + (5 + 1)^2 - (5 + 2)^2 = (5 + 3)(5 + 1) \Rightarrow 25 + 36 - 49 = 2 \times 6 \Rightarrow 12 = 12$

Hence, the result is true.

b) Expand the LHS and RHS of the expression to get:

$n^2 + n^2 + 2n + 1 - (n^2 + 4n + 4) = n^2 - 2n - 3 \Rightarrow n^2 - 2n - 3 = n^2 - 2n - 3$

That is, both sides are algebraically the same.

c) Expand the LHS of the expression to get: $n^2 + n^2 + 2n + 1 - (n^2 + 4n + 4)$

Collect like terms, which gives: $n^2 + n^2 - n^2 + 2n - 4n + 1 - 4 = n^2 - 2n - 3$

Factorise the collected result: $n^2 - 2n - 3 = (n - 3)(n + 1)$, which is the RHS of the original expression.

EXERCISE 18A

GRADE A**

1. a) Choose any odd number and any even number. Add these together. Is the result odd or even? Does this always work for any odd number and even number you choose?

b) Let any odd number be represented by $2n + 1$. Let any even number be represented by $2m$, where m and n are integers. Prove that the sum of an odd number and an even number is always an odd number.

2. Prove the following results.

a) The sum of two even numbers is even.

b) The product of two even numbers is even.

c) The product of an odd number and an even number is even.

d) The product of two odd numbers is odd.

e) The sum of four consecutive numbers is always even.

f) Half the sum of four consecutive numbers is always odd.

3. A Fibonacci sequence is formed by adding the previous two terms to get the next term. For example, starting with 3 and 4, the series is:

3, 4, 7, 11, 18, 29, 47, 76, 123, 199, ...

a) Continue the Fibonacci sequence 1, 1, 2, ... up to 10 terms.

b) Continue the Fibonacci sequence $a, b, a + b, a + 2b, 2a + 3b, \ldots$ up to 10 terms.

c) Prove that the difference between the 8th term and the 5th term of any Fibonacci sequence is twice the 6th term.

4. The nth term in the sequence of triangular numbers 1, 3, 6, 10, 15, 21, 28, ... is given by $\frac{1}{2}n(n + 1)$.

a) Show that the sum of the 11th and 12th terms is a perfect square.

b) Explain why the $(n+1)$th term of the triangular number sequence is given by $\frac{1}{2}(n+1)(n+2)$.

c) Prove that the sum of any two consecutive triangular numbers is always a square number.

5. The sum of the series $1+2+3+4+...+(n-2)+(n-1)+n$ is given by $\frac{1}{2}n(n+1)$.

a) Verify that this result is true for $n=6$.

b) Write down a simplified value, in terms of n, for the sum of these two series.

$1+2+3+...+(n-2)+(n-1)+n$ and $n+(n-1)+(n-2)+...+3+2+1$

c) Prove that the sum of the first n integers is $\frac{1}{2}n(n+1)$.

6. T represents any triangular number. Prove that:

a) $8T+1$ is always a square number

b) $9T+1$ is always another triangular number.

7. Lewis Carroll, who wrote *Alice in Wonderland*, was also a mathematician. In 1890, he suggested the following results.

a) For any pair of numbers, x and y, if x^2+y^2 is even, then $\frac{1}{2}(x^2+y^2)$ is the sum of two squares.

b) For any pair of numbers, x and y, $2(x^2+y^2)$ is always the sum of two squares.

c) Any number of which the square is the sum of two squares is itself the sum of two squares.

Can you prove these statements to be true or false?

8. Pythagoras' theorem says that for a right-angled triangle with two short sides a and b and a long side c, $a^2+b^2=c^2$. For any integer n, $2n$, n^2-1 and n^2+1 form three numbers that obey Pythagoras' theorem. Can you prove this?

9. Waring's theorem states that: 'Any whole number can be written as the sum of not more than four square numbers.'

For example, $27=3^2+3^2+3^2$ and $23=3^2+3^2+2^2+1^2$. Is this always true?

10. The difference of two squares is an identity: $a^2-b^2\equiv(a+b)(a-b)$, which means that it is true for all values of a and b, whether they are numeric or algebraic. Prove that $a^2-b^2\equiv(a+b)(a-b)$ is true when $a=2x+1$ and $b=x-1$.

18.2 Geometric proof

THIS SECTION WILL SHOW YOU HOW TO ...

✓ understand the difference between a proof and a demonstration

KEY WORDS

✓ demonstration ✓ proof ✓ prove

You should already know:

- the angle sum of the interior angles in a triangle (180°)
- the circle theorems
- Pythagoras' theorem.

Can you prove them?

For a mathematical proof, you must proceed in logical steps, establishing a series of mathematical statements by using facts that are already known to be true.

Proof that the sum of the interior angles of a triangle is 180°

Look at the following proof.

Start with triangle ABC with angles α, β and γ (figure i).

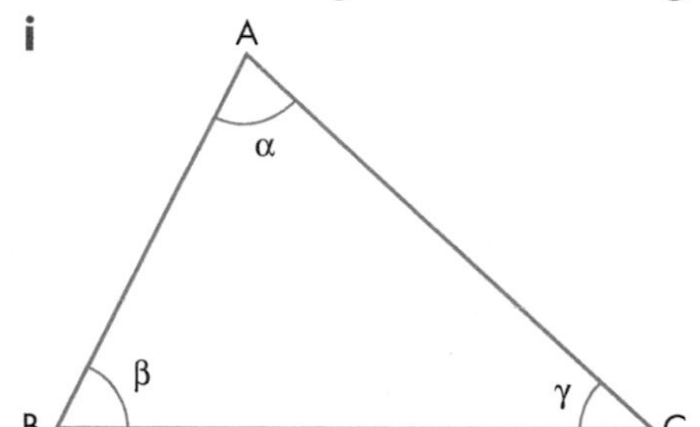

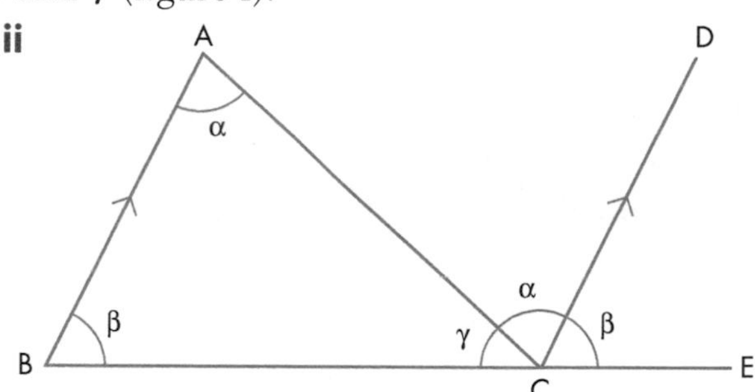

On figure i, draw a line CD parallel to side AB and extend BC to E, to give figure ii.

Since AB is parallel to CD:

$\angle ACD = \angle BAC = \alpha$ (alternate angles)

$\angle DCE = \angle ABC = \beta$ (corresponding angles)

BCE is a straight line, so $\gamma + \alpha + \beta = 180°$. Therefore the sum of the interior angles of a triangle is 180°.

This proof assumes that alternate angles are equal and that corresponding angles are equal. Strictly, you should prove these results, but you have to accept certain results as true. These are based on Euclid's axioms from which all geometric proofs are derived.

Proof of Pythagoras' theorem

Draw a square of side c inside a square of side $(a + b)$, as shown.

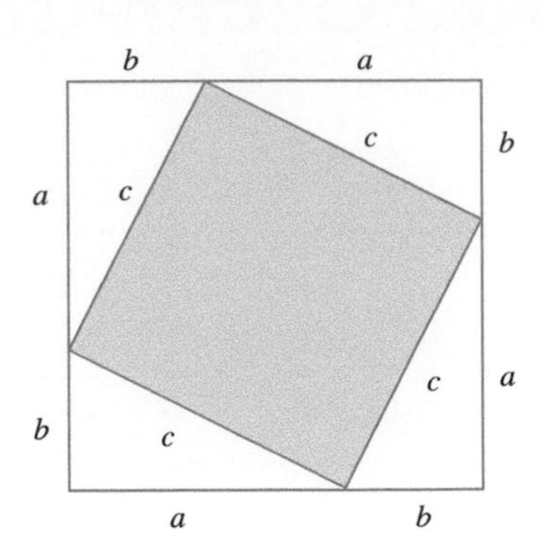

The area of the exterior square is $(a + b)^2 = a^2 + 2ab + b^2$.

The area of each small triangle around the shaded square is $\frac{1}{2}ab$.

The total area of all four triangles is $4 \times \frac{1}{2}ab = 2ab$.

Subtracting the total area of the four triangles from the area of the large square gives the area of the shaded square:

$a^2 + 2ab + b^2 - 2ab = a^2 + b^2$

But the area of the shaded square is c^2, so $c^2 = a^2 + b^2$, which is Pythagoras' theorem.

Congruency proof

There are four conditions to prove congruency. These are commonly known as SSS (three sides the same), SAS (two sides and the included angle the same), ASA (or AAS) (two angles and one side the same) and RHS (right-angled triangle, hypotenuse, and one short side the same).
Note: AAA (three angles the same) is *not* a condition for congruency.

When you prove a result, you must explain or justify every statement or line. Proofs have to be rigorous and logical.

EXAMPLE 2

ABCD is a parallelogram. X is the point where the diagonals meet.

Prove that triangles AXB and CXD are congruent.

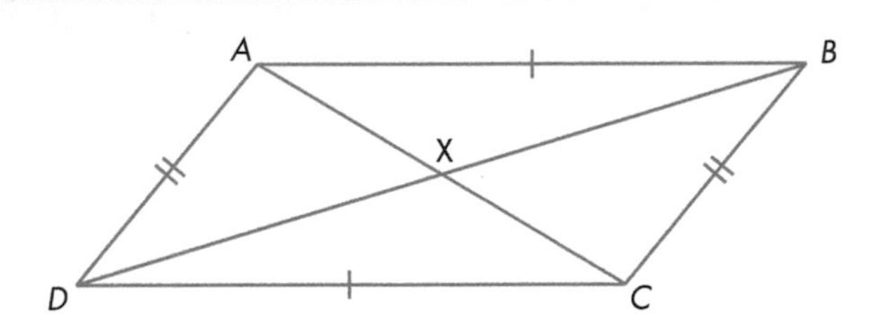

$\angle BAX = \angle DCX$ (alternate angles)

$\angle ABX = \angle CDX$ (alternate angles)

AB = CD (opposite sides in a parallelogram)

Hence ΔAXB is congruent to ΔCXD (ASA).

Note: You could have used $\angle AXB = \angle CXD$ (vertically opposite angles) as the second line but whichever approach is used you must give a reason for each statement.

EXERCISE 18B

GRADE A**

1. Prove that the triangle DEF with one angle of $x°$ and an exterior angle of $90° + \frac{x°}{2}$ is isosceles.

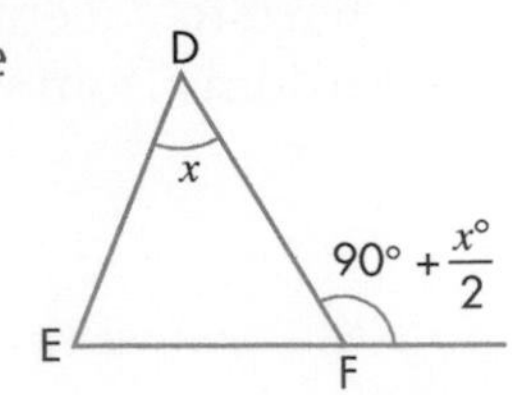

2. Prove that a triangle with an interior angle of $\frac{1}{2}x°$ and an exterior angle of $x°$ is isosceles.

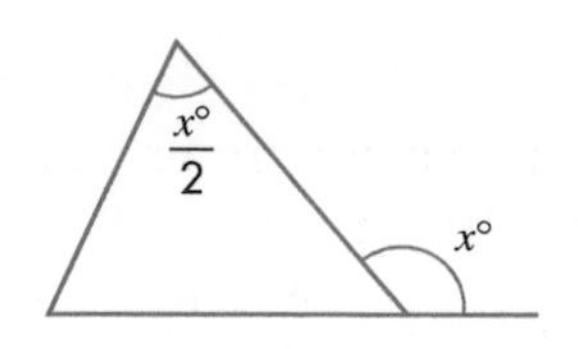

3. Prove that the sum of the opposite angles of a cyclic quadrilateral is 180°. (You may find the diagram useful.)

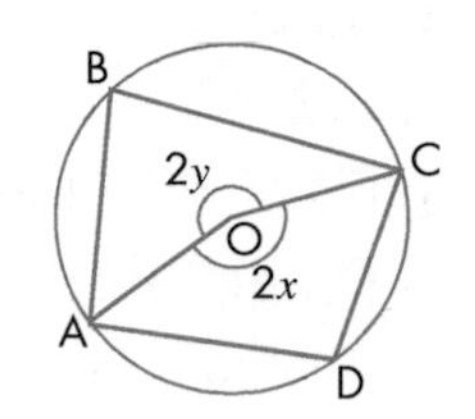

4. Prove that angle ACB = angle CED.

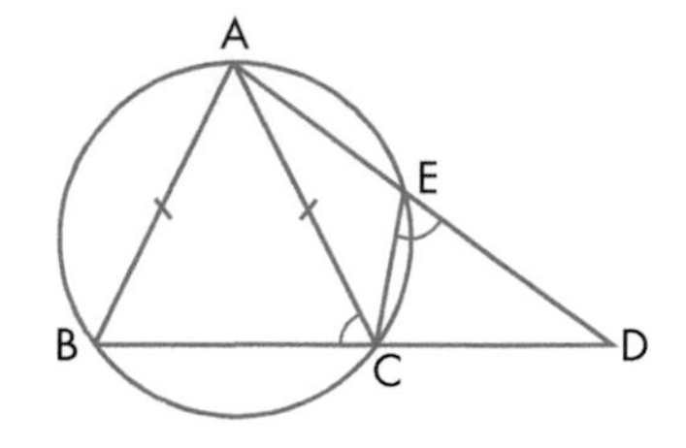

5. PQRS is a parallelogram. Prove that triangles PQS and RQS are congruent.

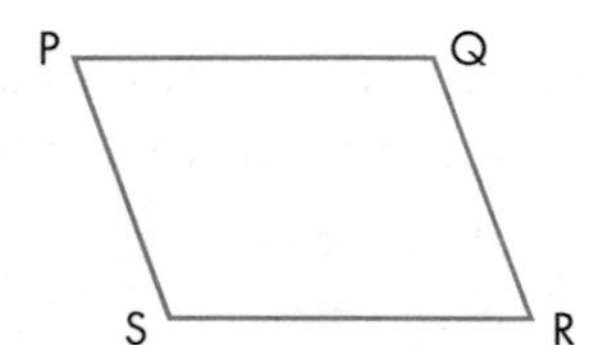

6. a) Prove the alternate segment theorem.

b) Two circles touch internally at T. The common tangent at T is drawn. Two lines TAB and TXY are drawn from T. Prove that AX is parallel to BY.

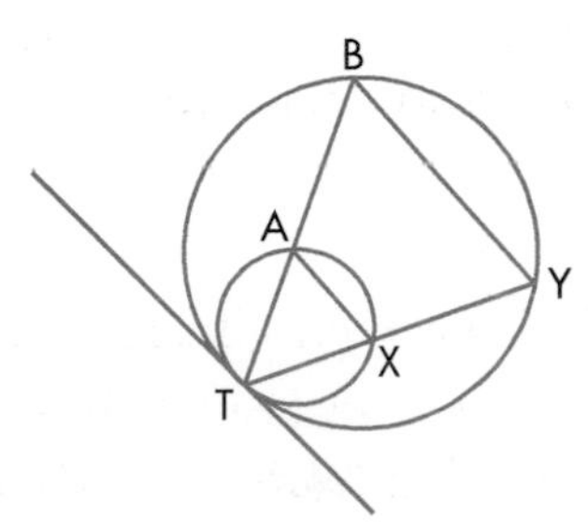

7. $\mathbf{a}$ and $\mathbf{b}$ are vectors.

$\overrightarrow{XY} = \mathbf{a} + \mathbf{b}$ $\quad$ $\overrightarrow{YZ} = 2\mathbf{a} + \mathbf{b}$ $\quad$ $\overrightarrow{ZW} = \mathbf{a} + 2\mathbf{b}$

a) Show that YW is parallel to XY.

b) Write down the ratio YW : XY.

c) What do your answers to **a** and **b** tell you about the points X, Y and W?

d) O is the origin. A, B and C are three points such that:

$\overrightarrow{OA} = \begin{pmatrix} 6 \\ 2 \end{pmatrix}$ $\quad$ $\overrightarrow{OB} = \begin{pmatrix} 1 \\ 1 \end{pmatrix}$ $\quad$ $\overrightarrow{OC} = \begin{pmatrix} 2 \\ -4 \end{pmatrix}$

Prove that angle ABC is a right angle.

Exam-style questions

1. Prove that if you add any two-digit number from the 9 times table to the reverse of itself (that is, swap the tens digit and units digit), the result will always be 99. GRADE A**

2. You are given that: $(a+b)^2+(a-b)^2=2(a^2+b^2)$ GRADE A**

a) Verify that this result is true for $a=3$ and $b=4$.

b) Show that the LHS is the same as the RHS.

c) Prove that the LHS can be simplified to the RHS.

3. Prove that: $(a+b)^2-(a-b)^2=4ab$. GRADE A**

4. 'When two numbers have a difference of 2, the difference of their squares is twice the sum of the two numbers.' GRADE A**

a) Verify that this is true for 5 and 7.

b) Prove that the result is true.

c) Prove that when two numbers have a difference of n, the difference of their squares is n times the sum of the two numbers.

5. Four consecutive numbers are 4, 5, 6 and 7. GRADE A**

a) Verify that their product plus 1 is a perfect square.

b) Use a suitable method to show that:
$(n^2-n-1)^2=n^4-2n^3-n^2+2n+1$.

c) Let four consecutive numbers be $(n-2)$, $(n-1)$, n, $(n+1)$. Prove that the product of four consecutive numbers plus 1 is a perfect square.

6. Prove that the sum of the squares of two consecutive integers is an odd number. GRADE A**

7. The square of the sum of the first n consecutive whole numbers is equal to the sum of the cubes of the first n consecutive whole numbers. GRADE A**

a) Verify that $(1+2+3+4)^2=1^3+2^3+3^3+4^3$.

b) The sum of the first n consecutive whole numbers is $\frac{1}{2}n(n+1)$. Write down a formula for the sum of the cubes of the first n whole numbers.

c) Test your formula for $n=6$.

8. Two circles touch externally at T. A line ATB is drawn through T. The common tangent at T and the tangents at A and B meet at P and Q. Prove that PB is parallel to AQ. GRADE A**

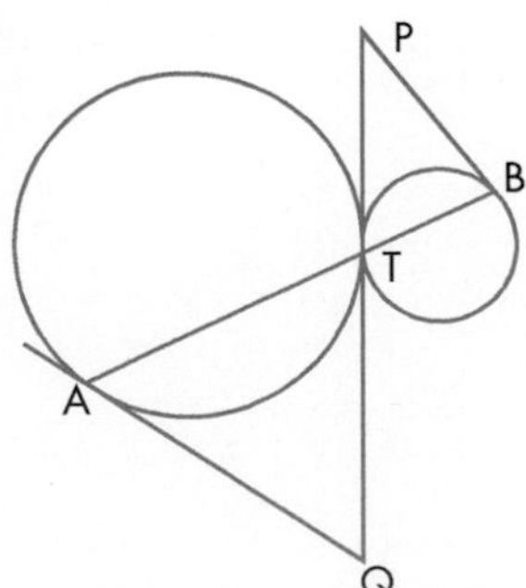

Answers

CHAPTER 1 NUMBER RECALL

1.1 Number: Recall and extension

Exercise 1A page 5

1. a) $6\frac{5}{6}$ b) $8\frac{4}{15}$ c) $3\frac{19}{42}$
   d) $4\frac{7}{8}$ e) $3\frac{13}{120}$ f) $2\frac{25}{36}$
2. a) $10\frac{1}{2}$ b) $2\frac{68}{77}$ c) $3\frac{1}{19}$
   d) $3\frac{37}{120}$ e) $6\frac{7}{90}$ f) $2\frac{1}{6}$
   g) $41\frac{7}{10}$ h) 2 i) $\frac{4}{27}$
3. 15 : 28
4. 32%
5. 36%
6. 4.464×10^4 minutes
7. 480
8. 5.12×10^{21}
9. 80 litres
10. More money, e.g. 100 loaves at £1 would change to 88 loaves at £1.14 = £100.32
11. a) 4.617×10^{12} b) 1.311×10^{14}
12. 70 weeks

1.2 Manipulation of surds

Exercise 1B

1. a) $\sqrt{6}$ b) $\sqrt{15}$ c) 2 d) 4
   e) $\sqrt{14}$ f) 6 g) 6 h) $\sqrt{30}$
2. a) 2 b) $\sqrt{5}$ c) $\sqrt{6}$ d) $\sqrt{3}$
   e) 2 f) $\sqrt{6}$ g) 1 h) 3
3. a) $2\sqrt{3}$ b) $4\sqrt{2}$ c) $8\sqrt{5}$ d) 24
   e) $2\sqrt{7}$ f) $6\sqrt{3}$
4. a) $\sqrt{3}$ b) 1 c) $2\sqrt{2}$ d) $\sqrt{5}$ e) $\sqrt{3}$
   f) $\sqrt{2}$ g) $\sqrt{7}$ h) $2\sqrt{3}$ i 1
5. a) a b) 1 c) $\sqrt{a}$
6. a) $3\sqrt{2}$ b) $2\sqrt{6}$ c) $2\sqrt{3}$ d) $5\sqrt{2}$
   e) $2\sqrt{2}$ f) $3\sqrt{3}$ g) $4\sqrt{2}$ h) $10\sqrt{2}$
   i) $10\sqrt{10}$ j) $5\sqrt{10}$ k) $7\sqrt{2}$ l $9\sqrt{3}$
7. a) 36 b) $16\sqrt{30}$ c) 54 d) 32
   e) $48\sqrt{6}$ f) $48\sqrt{6}$ g) $18\sqrt{15}$ h) 84
8. a) $20\sqrt{6}$ b) 24 c) 16 d) 18
   e) $10\sqrt{21}$ f) $6\sqrt{14}$ g) 36 h) $12\sqrt{30}$
9. a) 6 b) $3\sqrt{5}$ c) $6\sqrt{6}$ d) $2\sqrt{3}$
   e) 6 f) $2\sqrt{7}$ g) 5 h) 24
10. a) $2\sqrt{3}$ b) 4 c) $6\sqrt{2}$ d) $4\sqrt{2}$ e) $3\sqrt{2}$
    f) $\sqrt{7}$ g) $8\sqrt{3}$ h) $10\sqrt{3}$ i) 6
11. a) abc b) $\frac{a}{c}$ c) $c\sqrt{b}$
12. a) 20 b) 24 c) 10 d) 24
    e) 3 f) 6
13. a) $\frac{3}{4}$ b) $8\frac{1}{3}$ c) $\frac{5}{16}$ d) 12 e) 2
14. a) False b) False
15. For example, $\sqrt{2} \times \sqrt{8} = 4$
16. a) $5\sqrt{2}$ b) $2\sqrt{3}$ c) $14\sqrt{2}$ d) $6\sqrt{3}$
    e) $3\sqrt{5}$ f) $23\sqrt{2}$

Exercise 1C

1. Expand the brackets each time.
2. a) $2\sqrt{3} - 3$ b) $3\sqrt{2} - 8$ c) $10 + 4\sqrt{5}$
   d) $12\sqrt{7} - 42$ e) $15\sqrt{2} - 24$ f) $9 - \sqrt{3}$
3. a) $2\sqrt{3}$ b) $1 + \sqrt{5}$ c) $-1 - \sqrt{2}$
   d) $\sqrt{7} - 30$ e) -41 f) $7 + 3\sqrt{6}$
   g) $9 + 4\sqrt{5}$ h) $3 - 2\sqrt{2}$ i) $11 + 6\sqrt{2}$
4. a) $3\sqrt{2}$ cm b) $2\sqrt{3}$ cm c) $2\sqrt{10}$ cm
5. a) $\sqrt{3} - 1$ cm² b) $2\sqrt{5} + 5\sqrt{2}$ cm² c) $2\sqrt{3} + 18$ cm²
6. a) $\frac{\sqrt{3}}{3}$ b) $\frac{\sqrt{2}}{2}$ c) $\frac{\sqrt{5}}{5}$
   d) $\frac{\sqrt{3}}{6}$ e) $\sqrt{3}$ f) $\frac{5\sqrt{2}}{2}$
   g) $\frac{3}{2}$ h) $\frac{5\sqrt{2}}{2}$ i) $\frac{\sqrt{21}}{3}$
   j) $\frac{\sqrt{2}+2}{2}$ k) $\frac{2\sqrt{3}-3}{3}$ l) $\frac{5\sqrt{3}+6}{3}$
7. a) i 1 ii −4 iii 2 iv 17 v −44
   b) They become whole numbers. Difference of two squares makes the 'middle terms' (and surds) disappear.
8. Possible answer: $80^2 = 6400$, so $80 = \sqrt{6400}$ and $10\sqrt{70} = \sqrt{7000}$
   Since 6400 < 7000, there is not enough cable.

9. $9 + 6\sqrt{2} + 2 - (1 - 2\sqrt{8} + 8) = 11 - 9 + 6\sqrt{2} + 4\sqrt{2} = 2 + 10\sqrt{2}$

10. $x^2 - y^2 = (1 + \sqrt{2})^2 - (1 - \sqrt{8})^2 = 1 + 2\sqrt{2} + 2 - (1 - 2\sqrt{8} + 8) = 3 - 9 + 2\sqrt{2} + 4\sqrt{2} = -6 + 6\sqrt{2}$
$(x + y)(x - y) = (2 - \sqrt{2})(3\sqrt{2}) = 6\sqrt{2} - 6$

11. a) $\frac{3\sqrt{2}-1}{17}$ **b)** $\frac{5(3+\sqrt{3})}{6}$ **c)** $\frac{7\sqrt{2}-4}{41}$
d) $\frac{11+6\sqrt{2}}{7}$ **e)** $13-5\sqrt{5}$ **f)** $\frac{13+7\sqrt{3}}{22}$
g) $-5-2\sqrt{6}$ **h)** $\frac{32+7\sqrt{6}}{10}$

12. a) $x = 23, y = 9$ **b)** $x = 128, y = 648$
c) $x = \frac{1}{2}, y = -\frac{1}{2}$ **b)** $x = 2, y = -6$ or $x = -2, y = 6$

EXAM-STYLE QUESTIONS

1. $1\frac{1}{9}$

2. a) 1.08×10^{12} **b)** 3.33×10^{-9} seconds

3. $1.35b = 0.9g$, $b = \frac{2}{3}g$, the ratio of boys to girls was 2 : 3

4. 139.5p

5. a) $13 - 6\sqrt{5}$ **b)** $\frac{3}{11} - \frac{2}{11}\sqrt{5}$

2 ALGEBRA RECALL

2.1 Recall of basic algebra

Exercise 2A

1. a) All of them **b)** $\frac{1}{2}$

2. a) $15 - 5m$ **b)** $6x + 21$ **c)** $x^2 + 2x$
d) $10m - 2m^2$ **e)** $3nm - 3np$

3. a) $3(6 - m)$ **b)** $x(x + 5)$ **c)** $m(10 - m)$
d) $3(5s^2 + 1)$ **e)** $n(3 - p)$

4. a) $-3x - 8y$ **b)** $-2a + 4b$

5. a) Side AF – side DE $= 4x - 1 - x = 3x - 1$
b) $14x$ **c)** 84 cm

6. Darren has added 2 and 3 instead of multiplying, and has added 2 and –5 instead of multiplying. The correct answer is $6x - 10$.

7. a) $4(2y + 4)$ **b)** $3(2z + 1)$

8. 4 cm × 12 cm

9. a) 4.1 **b)** 8 **c)** 4.525

10. Any values that work, e.g. $x = 8, b = 4, h = 32$

11. a) x must be 2, y can be any other prime number
b) x must be an odd prime, y can be any other prime number

12. a) $6 + 3 \times 9 - 5 \times 3 = 18$ **b)** $2 \times 6 - 9 + 3 \times 3 = 12$

13. a) $\frac{450}{n}$ **b)** £390

14. $6(3x + 5) - 2(x - 2) = 18x + 30 - 2x + 4 = 16x + 34$

15. a) Both calculations give the cost of 5 meals and 5 desserts
b) It is easier to work out as the bracketed term evaluates to 10
c) £50

16. The terms have no common factors

17. a) $3 \times (5 + 1) = 3 \times 6 = 18$; $3 \times 5 + 3 = 15 + 3 = 18$
b) $3 \times (n + 2 + 1) = 3 \times (n + 3) = 3n + 9$; $3 \times (n + 2) + 3 = 3n + 6 + 3 = 3n + 9$

18. a) $12p^3 - 4p^2q$ **b)** $10t^4 + 35t^2$ **c)** $10x^2 + 35xy$
d) $10m^2 - 2m^5$ **e)** $8s^4 + 24s^3t$ **f)** $6nm^3 - 6n^2m^2$

19. a) $23x + 11$ **b)** $9y + 7$ **c)** $2x - 8$
d) $22x + 9$ **e)** $14x^2 - 10x$ **f)** $2x^3 + 17x^2 - 9$

20. a) $3p(3p + 2t)$ **b)** $4m(3p - 2m)$ **c)** $4ab(4a + 1)$
d) $2(2a^2 - 3a + 1) = 2(2a - 1)(a - 1)$
e) $5xy(4y + 2x + 1)$ **f)** $4mt(2t - m)$

2.2 Expanding brackets and collecting like terms

Exercise 2B

1. a) $5d + 4de$ **b)** $2t$ **c)** $4y^2$ **d)** $3a^2d$

2. a) $2x$ and $2y$ **b)** a and $7b$

3. a) The length of AF is the same as the lengths of BC and DE together. $3x - 1 - x$
b) $10x$ **c)** 25 cm

4. a) $22 + 5t$ **b)** $14 + 3g$

5. a) $2 + 2h$ **b)** $6e + 20$

6. a) $4m + 3p + 2mp$ **b)** $3k + 4h + 5hk$
c) $12r + 24p + 13pr$ **d)** $19km + 20k - 6m$

7. a) $9t^2 + 13t$ **b)** $13y^2 + 5y$
c) $10e^2 - 6e$ **d)** $14k^2 - 3kp$

8. a) $17ab + 12ac + 6bc$ **b)** $18wy + 6ty - 8tw$
c) $14mn - 15mp - 6np$ **d)** $8r^3 - 6r^2$

9. a) $5(f + 2s) + 2(2f + 3s) = 9f + 16s$
b) £$(270f + 480s)$
c) £42 450 – £30 000 = £12 450

10. For x-coefficients, 3 and 1 or 1 and 4; for y-coefficients, 5 and 1 or 3 and 4 or 1 and 7

11. $5(3x + 2) - 3(2x - 1) = 9x + 13$

Exercise 2C

1. $x^2 + 5x + 6$

2. $w^2 + 4w + 3$

3. $a^2 + 5a + 4$

4. $x^2 + 2x - 8$

5. $w^2 + 2w - 3$

6. $f^2 - f - 6$

7. $y^2 + y - 12$

8. $x^2 + x - 12$

9. $p^2 - p - 2$

10. $a^2 + 2a - 3$

11. $x^2 - 9$

12. $t^2 - 25$

13. $t^2 - 4$

14. $y^2 - 64$

15. $25 - x^2$

16. $x^2 - 36$

17. $(x + 2)$ and $(x + 3)$

18. **a)** B: $1 \times (x - 2)$ **b)** $(x - 2) + 2 + 2(x - 1)$
C: 1×2 $= 3x - 2$
D: $(x - 1) \times 2$
c) Area A $= (x - 1)(x - 2)$
= area of square minus areas (B + C + D)
$= x^2 - (3x - 2)$
$= x^2 - 3x + 2$

19. **a)** $x^2 - 9$ **b) i** 9991 **ii** 39991

Exercise 2D

1. $6x^2 + 11x + 3$
2. $8t^2 + 2t - 3$
3. $12k^2 - 11k - 15$
4. $6a^2 - 7a - 3$
5. $15g^2 - 16g + 4$
6. $12d^2 + 5d - 2$
7. $8p^2 + 26p + 15$
8. $-10t^2 - 7t + 6 = 6 - 7t - 10t^2$
9. $6f^2 - 5f - 6$
10. $-6t^2 + 10t + 4 = 4 + 10t - 6t^2$
11. **a)** $x^2 + 2x + 1$ **b)** $x^2 - 2x + 1$ **c)** $x^2 - 1$
d) $p + q = (x + 1 + x - 1) = 2x$
$(p + q)^2 = (2x)^2 = 4x^2$
$p^2 + 2pq + q^2 = x^2 + 2x + 1 + 2(x^2 - 1) + x^2 - 2x + 1$
$= 4x^2 + 2x - 2x + 2 - 2 = 4x^2$
12. **a)** $(3x - 2)(2x + 1) = 6x^2 - x - 2$, $(2x - 1)(2x - 1) = 4x^2 - 4x + 1$, $(6x - 3)(x + 1) = 6x^2 + 3x - 3$, $(3x + 2)(2x + 1) = 6x^2 + 7x + 2$
b) Multiply the x terms to match the x^2 term and/or multiply the constant terms to get the constant term in the answer.

Exercise 2E

1. $4x^2 - 1$
2. $25y^2 - 9$
3. $16m^2 - 9$
4. $16h^2 - 1$
5. $4 - 9x^2$
6. $36 - 25y^2$
7. $a^2 - b^2$
8. $4m^2 - 9p^2$
9. $a^2b^2 - c^2d^2$
10. $a^4 - b^4$
11. **a)** $a^2 - b^2$
c) Dimensions: $a + b$ by $a - b$; Area: $a^2 - b^2$
d) Areas are the same, so $a^2 - b^2 = (a + b) \times (a - b)$
12. First shaded area is $(2k)^2 - 1^2 = 4k^2 - 1$. Second shaded area is $(2k + 1)(2k - 1) = 4k^2 - 1$

Exercise 2F

1. $x^2 + 10x + 25$
2. $t^2 + 12t + 36$
3. $m^2 - 6m + 9$
4. $k^2 - 14k + 49$
5. $9x^2 + 6x + 1$
6. $25y^2 + 20y + 4$
7. $9x^2 - 12x + 4$
8. $x^2 + 2xy + y^2$
9. $m^2 - 2mn + n^2$
10. $m^2 - 6mn + 9n^2$
11. $x^2 - 10x$
12. $x^2 - 4x$
13. **a)** Bernice has just squared the first term and the second term.
She hasn't written down the brackets twice.
b) Pete has written down the brackets twice but has worked out $(3x)^2$ as $3x^2$ and not $9x^2$.
c) $9x^2 + 6x + 1$
14. The area of the whole square is $(2x)^2 = 4x^2$.
The area of the top left square is $(2x - 1)^2 = 4x^2 - 4x + 1$
The areas of the three unmarked rectangles are $2x - 1$, $2x - 1$ and 1.
$4x^2 - (2x - 1 + 2x - 1 + 1) = 4x^2 - (4x - 1) = 4x^2 - 4x + 1$

Exercise 2G

1. **a)** $x^3 + 4x^2 + 2x - 1$ **b)** $8x^3 + 22x^2 + 9x + 1$
c) $2x^4 - 9x^3 - 14x^2 + 9$ **d)** $x^6 + x^5 - 6x^4 + 4x^3 - x$
2. **a)** $x^3 + 3x^2 + 3x + 1$ **b)** $8x^3 - 12x^2 + 6x - 1$
c) $27x^3 + 54x^2 + 36x + 8$ **d)** $64x^3 - 144x^2 + 108x - 27$
3. **a)** $2x - 2$ **b)** $x - 1$ **c)** $6x - 13 + \frac{6}{x}$ **d)** $16x - x^{-\frac{2}{3}}$

2.3 Factorising

Exercise 2H

1. **a)** $4(2m + 3k)$ **b)** $m(n + 3)$ **c)** $g(5g + 3)$
d) $y(3y + 2)$ **e)** $t(4t - 3)$ **f)** $3m(m - p)$
g) $3p(2p + 3t)$ **h)** $2p(4t + 3m)$ **i)** $2(2a^2 + 3a + 4)$
j) $3b(2a + 3c + d)$ **k)** $t(5t + 4 + a)$ **l)** $3mt(2t - 1 + 3m)$
m) $2ab(4b + 1 - 2a)$ **n)** $5pt(2t + 3 + p)$

2. a) $m(5 + 2p)$ b) $t(t - 7)$ c) does not factorise
d) does not factorise e) $a(4a - 5b)$ f $b(5a - 3bc)$

3. a) Bernice
b) Aidan has not taken out the largest possible common factor. Craig has taken m out of both terms but there isn't an m in the second term.

4. There are no common factors.

5. $4x^2 - 12x$, $2x - 6$; $\frac{4x^2 - 12x}{2x - 6}$

Exercise 2I

1. $(x + 2)(x + 3)$
2. $(m + 2)(m + 5)$
3. $(p + 2)(p + 12)$
4. $(w + 2)(w + 9)$
5. $(a + 2)(a + 6)$
6. $(b + 8)(b + 12)$
7. $(t - 2)(t - 3)$
8. $(d - 4)(d - 1)$
9. $(x - 3)(x - 12)$
10. $(t - 4)(t - 9)$
11. $(y - 4)(y - 12)$
12. $(j - 6)(j - 8)$
13. $(y + 6)(y - 1)$
14. $(m + 2)(m - 6)$
15. $(n + 3)(n - 6)$
16. $(m + 4)(m - 11)$
17. $(t + 9)(t - 10)$
18. $(h + 8)(h - 9)$
19. $(t + 7)(t - 9)$
20. $(y + 10)^2$
21. $(m - 9)^2$
22. $(x - 12)^2$
23. $(d + 3)(d - 4)$
24. $(q + 7)(q - 8)$
25. $(x + 2)(x + 3)$, giving areas of $2x$ and $3x$, or $(x + 1)(x + 6)$, giving areas of x and $6x$.
26. a) $x^2 + (a + b)x + ab$ b) i $p + q = 7$ ii $pq = 12$
c) 7 can only be 1×7 and $1 + 7 \neq 12$

Exercise 2J

1. $(x + 3)(x - 3)$
2. $(t + 5)(t - 5)$
3. $(m + 4)(m - 4)$
4. $(3 + x)(3 - x)$
5. $(7 + t)(7 - t)$
6. $(k + 10)(k - 10)$
7. $(2 + y)(2 - y)$
8. $(x + 8)(x - 8)$
9. $(t + 9)(t - 9)$
10. a) $x^2 + 4x + 4 - (x^2 + 2x + 1) = 2x + 3$
b) $(a + b)(a - b)$
c) $(x + 2 + x + 1)(x + 2 - x - 1) = (2x + 3)(1) = 2x + 3$
d) The answers are the same.
e) $(x + 1 + x - 1)(x + 1 - x + 1) = (2x)(2) = 4x$
11. $(x + y)(x - y)$
12. $(x + 2y)(x - 2y)$
13. $(x + 3y)(x - 3y)$
14. $(3x + 1)(3x - 1)$
15. $(4x + 3)(4x - 3)$
16. $(5x + 8)(5x - 8)$
17. $(2x + 3y)(2x - 3y)$
18. $(3t + 2w)(3t - 2w)$
19. $(4y + 5x)(4y - 5x)$

Exercise 2K

1. $(2x + 1)(x + 2)$
2. $(7x + 1)(x + 1)$
3. $(4x + 7)(x - 1)$
4. $(3t + 2)(8t + 1)$
5. $(3t + 1)(5t - 1)$
6. $(4x - 1)^2$
7. $3(y + 7)(2y - 3)$
8. $4(y + 6)(y - 4)$
9. $(2x + 3)(4x - 1)$
10. $(2t + 1)(3t + 5)$
11. $(x - 6)(3x + 2)$
12. $(x - 5)(7x - 2)$
13. $4x + 1$ and $3x + 2$
14. a) All the terms in the quadratic have a common factor of 6, but this has not been taken out, so these are not complete factorisations. 36 has several different pairs of factors which all work.
b) $6(x + 2)(x + 3)$. This has the highest common factor taken out.

Exercise 2L

1. $(x + 2y)(x + 3y)$
2. $(x + 7y)(x + 3y)$
3. $(x - y)(x - 4y)$
4. $(x + y)(x - 7y)$
5. $(x - 9y)(x + 8y)$
6. $(2x + y)(x + 2y)$
7. $(3x + 2y)(x - 6y)$
8. $(5x + 2y)(x - 3y)$
9. $(3x + 5y)(2x + y)$
10. $(5x - y)(3x + y)$
11. $(x^2 + 5y^2)(x^2 - 5y^2)$

12. $2x(2x + 5)(2x - 5)$

13. $(4x^2 + 5y^2)(4x^2 - 5y^2)$

14. $16(x + 1)$

15. $8x$

16. $5(x + 1)(x - 1)$

17. $-(4x + 1)$ or $-4x - 1$

18. $(8x + 1)(2x + 1)$

Exam-style questions

1. $a = 0.1$
2. $4x^2 - 3$
3. $26y$
4. **a)** $(3x + 2)(3x - 2)$ **b)** $(3x - 2)(2x + 1)$ **c)** $\frac{2x+1}{3x+2}$
5. $7(d - 2)$
6. **a)** $x^3 - 6x^2 + 3x + 18$ **b)** $64 - 48x + 12x^2 - x^3$ **c)** $x + 1$
7. $16(x + 1)(x - 1)$
8. $6x^2(2x - 3)(x + 4)$

CHAPTER 3 GEOMETRY RECALL 1

3.1 Perimeter of compound shapes

Exercise 3A

1. 10 cm 2. 12 cm 3. 12 cm
4. 14 cm 5. 12 cm 6. 12 cm

3.2 Area of basic shapes

Exercise 3B

1. **a)** 21 cm^2 **b)** 12 cm^2 **c)** 140 cm^2 **d)** 40 cm^2 **e)** 65 m^2 **f)** 80 cm^2
2. **a)** 65 cm^2 **b)** 50 m^2
3. **a)** 96 cm^2 **b)** 70 cm^2 **c)** 10 cm^2
4. **a)** 27.5 cm, 36.25 cm^2 **b)** 33.4 cm, 61.2 cm^2 **c)** 38.5 m, 90 m^2
5. **a)** 57 m^2 **b)** 702.5 cm^2 **c)** 84 m^2

3.3 Circumference and area of a circle

Exercise 3C

1. **a)** 8 cm, 25.1 cm, 50.3 cm^2 **b)** 5.2 m, 16.3 m, 21.2 m^2 **c)** 6 cm, 37.7 cm, 113 cm^2 **d)** 1.6 m, 10.1 m, 8.04 m^2
2. **a)** 5π cm **b)** 8π cm **c)** 18π m **d)** 12π cm
3. **a)** 25π cm^2 **b)** 36π cm^2 **c)** 100π cm^2 **d)** 0.25π m^2
4. 8.80 m
5. 4 complete revolutions
6. 1p : 3.1 cm^2, 2p : 5.3 cm^2, 5p : 2.3 cm^2, 10p : 4.5 cm^2
7. 0.83 m
8. 38.6 cm
9. The claim is correct (ratio of the areas is just over 1.5 : 1).
10. **a)** 18π cm^2 **b)** 4π cm^2
11. 9π cm^2
12. 28.3 m^2
13. Diameter of tree is 9.96 m – is your classroom this wide in all directions?
14. 45 complete revolutions

3.4 Volume of a cube, cuboid, prism and pyramid

Exercise 3D

1. **a) i** 21 cm^2 **ii** 63 cm^3 **b) i** 48 cm^2 **ii** 432 cm^3 **c) i** 36 m^2 **ii** 324 m^3
2. **a)** A cross-section parallel to the side of the pool always has the same shape
   b) About 3.5 hours (3 hours 32.5 minutes)
3. **a)** 21 cm^3, 210 cm^3 **b)** 54 cm^2, 270 cm^2
4. 146 cm^3
5. 327 litres
6. 1.02 tonnes

Exercise 3E

1. **a) i** 226 cm^3 **ii** 207 cm^2 **b) i** 14.9 cm^3 **ii** 61.3 cm^2 **c) i** 346 cm^3 **ii** 275 cm^2 **d) i** 1060 cm^3 **ii** 636 cm^2
2. **a) i** 72π cm^3 **ii** 48π cm^2 **b) i** 112π cm^3 **ii** 56π cm^2
3. £80
4. 1.23 tonnes
5. 332 litres

Exercise 3F

1. **a)** 56 cm^3 **b)** 1040 cm^3 **c)** 160 cm^3
2. **a)** Put the apexes of the pyramids together. The 6 square bases will then form a cube.

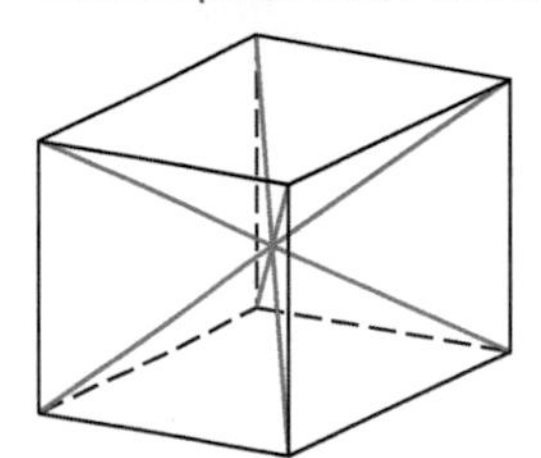

   b) If the side of the base of the cube is a then the height of each pyramid will be $\frac{1}{2}a$. Total volume of the 6 pyramids is equal to the volume of the cube, which is a^3. The volume of one pyramid is $\frac{1}{6}a^3 = \frac{1}{3} \times \frac{1}{2} \times a \times a^2 = \frac{1}{3}$ height $\times$ base area
3. **a)** 73.3 m^3 **b)** 45 m^3 **c)** 3250 cm^3
4. 5.95 cm
5. 260 cm^3

3.5 Volume of a cone and a sphere

Exercise 3G

1. a) i 3560 cm^3 ii 1430 cm^2
   b) i 314 cm^3 ii 283 cm^2
   c) i 1020 cm^3 ii 679 cm^2
2. 24π cm^2
3. a) 816π cm^3 b) 720π mm^3
4. a) 4 cm b) 6 cm
   c) Various answers, e.g. 60° gives 2 cm, 240° gives 8 cm
5. 24π cm^2
6. If radius of base is r, slant height is $2r$.
   Area of curved surface $= \pi r \times 2r = 2\pi r^2$, area of base $= \pi r^2$
7. 2.81 cm

Exercise 3H

1. a) 36π cm^3 b) 288π cm^3 c) 1333π cm^3
2. a) 36π cm^2 b) 100π cm^2 c) 196π cm^2
3. 65 400 cm^3, 7850 cm^2
4. a) 1960 cm^2 b) 8180 cm^3
5. 125 cm
6. 7.8 cm
7. 47.4%
8. Radius of sphere = base radius of cylinder = r, height of cylinder = $2r$
   Area of curved surface of cylinder = circumference × height $= 2\pi r \times 2r = 4\pi r^2$ = surface area of sphere

Exam-style questions

1. 7.7 cm
2. a) $24a^3$ cm^3 b) $52a^2$ cm^2
3. $r < 3$ (If they were equal, $4\pi r^2 = \frac{4}{3}\pi r^3$, so $r = 3$)
4. a) $x(x+1)$ b) $\sqrt{\frac{x(x+1)}{\pi}}$
5. 270 cm^3
6. 14.4 cm
7. 2122

CHAPTER 4 GEOMETRY RECALL 2

4.1 Special triangles and quadrilaterals

Exercise 4A

1. a) $a = b = 70°$ b) $e = 55°, f = 70°$ c) $a = 110°, b = 55°$
   d) $c = e = 105°, d = 75°$ e) $h = i = 94°$
   f) $m = o = 49°, n = 131°$
2. 40°, 40°, 100°
3. $a = b = 65°, c = d = 115°, e = f = 65°, g = 80°, h = 60°, i = 60°, j = 60°, k = 20°$
4. a) $x = 25°, y = 15°$ b) $x = 7°, y = 31°$
   c) $x = 60°, y = 30°$
5. a) $x = 50°$; 60°, 70°, 120°, 110° – possibly trapezium
   b) $x = 60°$; 50°, 130°, 50°, 130° – parallelogram or isosceles trapezium
   c) $x = 30°$; 20°, 60°, 140°, 140° – possibly kite
   d) $x = 20°$; 90°, 90°, 90°, 90° – square or rectangle
6. 52°
7. Both 129°
8. $y = 360° - 4x$
9. a) 65°
   b) Trapezium, angle A + angle D = 180° and angle B + angle C = 180°

4.2 Angles in polygons

Exercise 4B

1. a) 1440° b) 2340° c) 17 640° d) 7740°
2. a) 150° b) 162° c) 140° d) 174°
3. a) 9 b) 15 c) 102 d) 50
4. a) 15 b) 36 c) 24 d) 72
5. a) 12 b) 9 c) 20 d) 40
6. a) 130° b) 95° c) 130°
7. a) 50° b) 40° c) 59°
8. Hexagon
9. a) Octagon b) 89°
10. a) i 71° ii 109° iii Equal
    b) If S = sum of the two opposite interior angles, then $S + I = 180°$ (angles in a triangle), but $E + I = 180°$ (angles on a straight line), so $S + I = E + I$, therefore $S = E$
11. $a = 144°$
12. Three angles are 135° and two angles are 67.5°.
13. 88°; $\frac{1440° - 5 \times 200}{5}$
14. a) 36° b) 10

4.3 Circle theorems

Exercise 4C

1. a) 56° b) 62° c) 105° d) 45°
   e) 55° f) 52° g) 24° h) 80°
2. a) 41° b) 49° c) 41°
3. a) 72° b) 37° c) 72°
4. ∠AZY = 35° (angles in a triangle), $a = 50°$ (angle in a semicircle = 90°)
5. 68°
6. ∠ABC = $180° - x$ (angles on a line), ∠AOC = $360° - 2x$ (angle at centre is twice angle at circumference), reflex ∠AOC = $360° - (360° - 2x) = 2x$ (angles at a point)
7. a) x b) $2x$ c) ∠ABC = $(x + y)$ and ∠AOC = $2(x + y)$

4.4 Cyclic quadrilaterals

Exercise 4D

1. a) $a = 50°, b = 95°$ b) $d = 110°, e = 110°, f = 70°$
   c) $g = 105°, h = 99°$ d) $x = 40°, y = 34°$
2. a) $x = 48°, y = 78°$ b) $x = 36°, y = 144°$
   c) $x = 55°, y = 125°$ d) $x = 35°$
3. a) $x = 49°, y = 49°$ b) $x = 80°, y = 100°$
   c) $x = 100°, y = 75°$ d) $x = 92°, y = 88°$
   e) $x = 55°, y = 75°$ f) $x = 95°, y = 138°$
   g) $x = 32°, y = 48°$ h) $x = 52°$
4. a) 71° b) 125.5° c) 54.5°
5. a) $x + 2x - 30° = 180°$ (opposite angles in a cyclic quadrilateral), so $3x - 30° = 180°$
   b) $x = 70°$, so $2x - 30° = 110°$
   $\angle$DOB = 140° (angle at centre equals twice angle at circumference), $y = 80°$ (angles in a quadrilateral)
6. a) x
   b) $360° - 2x$
   c) $\angle ADC = \frac{1}{2}$ reflex $\angle AOC = 180° - x$, so $\angle ADC + \angle ABC = 180°$
7. Let $\angle AED = x$, then $\angle ABC = x$ (opposite angles are equal in a parallelogram), $\angle ADC = 180° - x$ (opposite angles in a cyclic quadrilateral), so $\angle ADE = x$ (angles on a line)

4.5 Tangents and chords

Exercise 4E

1. a) 38° b) 110° c) 45°
2. a) 6 cm b) 3.21 cm c) 8 cm
3. a) $x = 12°, y = 156°$ b) $x = 62°, y = 28°$
   c) $x = 30°, y = 60°$
4. a) 62° b) 66° c) 19° d) 20°
5. 19.5 cm
6. 5.77 cm
7. $\angle$OCD = 58° (triangle OCD is isosceles), $\angle$OCB = 90° (tangent/radius theorem), so $\angle$DCB = 32°, hence triangle BCD is isosceles (2 equal angles)
8. a) OAB and OAC are congruent by RHS: A and C are right angles, OB is a hypotenuse for both, and OA and OC are equal (radii). The results follow.
   b) As $\angle$AOB = $\angle$COB, so $\angle$ABO = $\angle$CBO, so OB bisects $\angle$ABC

4.6 Alternate segment theorem

Exercise 4F

1. a) $a = 65°, b = 75°, c = 40°$ b) $d = 79°, e = 58°, f = 43°$
   c) $g = 41°, h = 76°, i = 76°$ d) $k = 80°, m = 52°, n = 80°$
2. a) $a = 75°, b = 75°, c = 75°, d = 30°$
   b) $a = 47°, b = 86°, c = 86°, d = 47°$
   c) $a = 53°, b = 53°$ d) $a = 55°$
3. a) 36° b) 70°
4. a) $x = 25°$ b) $x = 46°, y = 69°, z = 65°$
   c) $x = 38°, y = 70°, z = 20°$ d) $x = 48°, y = 42°$
5. $\angle$ACB = 64° (angle in alternate segment), $\angle$ACX = 116° (angles on a line), $\angle$CAX = 32° (angles in a triangle), so triangle ACX is isosceles (two equal angles)
6. $\angle$AXY = 69° (tangents equal and so triangle AXY is isosceles), $\angle$XZY = 69° (alternate segment), $\angle$XYZ = 55° (angles in a triangle)

Exam-style questions

1. Angle AED = the angle of a regular pentagon = 108°. Triangle AEB is isosceles so angle AEB = half of (180° − 108°) = 36°. Angle DEB = 108° − 36° = 72°. The result follows.
2. $3x + 4x + 5x + 6x = 360 \Rightarrow 18x = 360 \Rightarrow x = 20$.
   So angle A = 80° and B = 100°. These add up to 180° which means that AD and BC are parallel.
3. 240°
4. 60 cm²
5. 65°

6.

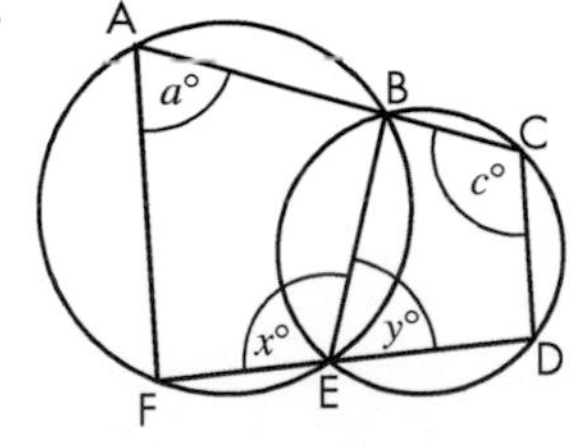

$a + x = 180$ and $y + c = 180$ (opposite angles of a cyclic quadrilateral). So $a + x + y + c = 360$. But $x + y = 180$ so $a + b = 180$ and the result follows.

CHAPTER 5 FUNCTIONS

5.1 Function notation

Exercise 5A

1. a) 12 b) 26 c) 7 d) −2 e) 3
2. a) 0.5 b) 5 c) 50.5 d) 2.5 e) 0.625 or $\frac{5}{8}$
3. a) 5 b) −3 c) 999 801 d) 1 e) $\frac{1}{8}$
4. a) 4 b) 32 c) 1 d) $\frac{1}{2}$ e) $\frac{1}{8}$
5. a) 3 b) 2 c) 0 d) −1 e) 5
6. a) 7.5 b) −2.5 c) −5
7. a) 6 b) 97 c) 3.25
8. a) 6 b) The functions intersect at (6, 4).

9. a) 3 b) $-\frac{1}{2}$ c) $x = 2$ d) $x = 0$

10. a) $4x + 7$ b) $12x - 1$ c) $12x + 2$ d) 4

11. a) 4 and −4 b) 2 and −2
    c) $\frac{4}{3}$ and $-\frac{4}{3}$. Their product is −1.

12. a) 0.19 and −5.19 b) 0.10 and −2.60

13. a) $-\frac{5}{2}$ and 4 b) $\frac{3}{2}$ and 0 c) $-\frac{20}{3}$

5.2 Domain and range of a function

Exercise 5B

1. a) $x < 0$ b) −1 c) $x \leqslant -1$
   d) $-\frac{1}{2}$ e) $x = 1$ and $x = 2$
2. a) {10, 17, 26} b) {1, 2, 5} c) $\{y : 2 \leqslant y \leqslant 5\}$
   d) $\{y : y \geqslant 101\}$ e) Same as **d**
3. a) {0, 1, 4} b) $\{1, \frac{1}{2}, \frac{1}{3}, \frac{1}{4}\}$ c) {5, 7, 9, 11}
   d) {5, 4, 3, 2} e) {0, −2}
4. −2 can be squared so it could be in the domain. $x^2 = -2$ has no solution so −2 cannot be in the range.
5. a) Yes b) No c) Yes
6. 5
7. a) $f(x) > 16$ b) Domain $x > 5$, range $f(x) > 61$
8. a) $f(x) > 3$ b) Domain $x > 0$, range $f(2x) > 3$
9. a) 7 b) 11
10. $a = 2$, $b = 8$
11. $-4 \leqslant f(x) \leqslant 5$

5.3 Sketching graphs of linear and quadratic functions

Exercise 5C

1. a) Values of y: 27, 12, 3, 0, 3, 12, 27
   b) 6.8
   c) 1.8 or −1.8
2. a) Values of y: 27, 18, 11, 6, 3, 2, 3, 6, 11, 18, 27
   b) 8.3
   c) 3.5 or −3.5
3. a) Values of y: 27, 16, 7, 0, −5, −8, −9, −8, −5, 0, 7
   b) −8.8
   c) 3.4 or −1.4
4. a) Values of y: 2, −1, −2, −1, 2, 7, 14
   b) 0.25
   c) 0.7 or −2.7
   d

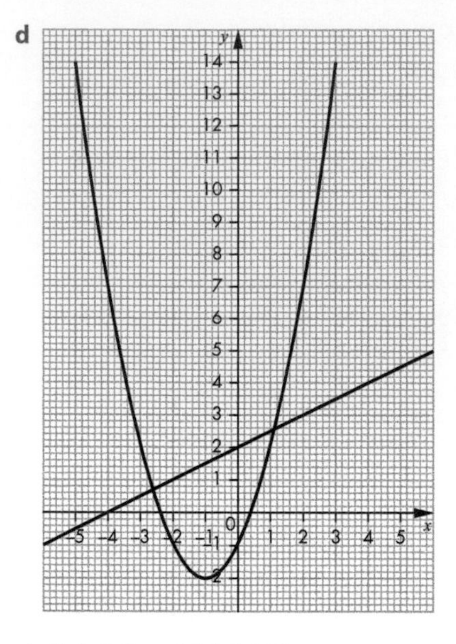

   e) (1.1, 2.6) and (−2.6, 0.7)
5. a) Values of y: 18, 12, 8, 6, 6, 8, 12
   b) 9.75 c) 2 or −1
   d) Values of y: 14, 9, 6, 5, 6, 9, 14 e) (1, 6)
6. a) Values of y: 4, 1, 0, 1, 4, 9, 16 b) 7.3
   c) 0.4 or −2.4
   d)

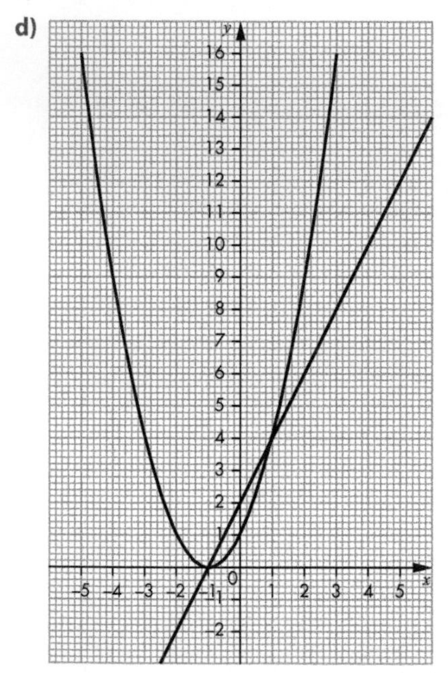

   e) (1, 4) and (−1, 0)
7. a) Values of y: 15, 9, 4, 0, −3, −5, −6, −6, −5, −3, 0, 4, 9
   b) −0.5 and 3

5.4 The significant points of a quadratic graph

Exercise 5D

1. a) Values of y: 12, 5, 0, –3, –4, –3, 0, 5, 12
   b) 2 and –2
2. a) The roots are positive and negative square roots of the constant term.
   b) Make predictions.
   c) Values of y: 15, 8, 3, 0, –1, 0, 3, 8, 15
   d) Values of y: 11, 4, –1, –4, –5, –4, –1, 4, 11
   e) 1 and –1, 2.2 and –2.2. Check predictions.
3. a) Values of y: 5, 0, –3, –4, –3, 0, 5, 12
   b) –4 and 0
4. a) Values of y: 10, 4, 0, –2, –2, 0, 4, 10, 18
   b) –3 and 0
5. a) The roots are 0 and the negative of the coefficient of x.
   b) Make predictions.
   c) Values of y: 10, 4, 0, –2, –2, 0, 4, 10
   d) Values of y: 6, 0, –4, –6, –6, –4, 0, 6, 14
   e) 0 and 3, –5 and 0. Check predictions.
6. a) Values of y: 9, 4, 1, 0, 1, 4, 9
   b) 2
   c) Only 1 root
7. a) Values of y: 10, 3, –2, –5, –6, –5, –2, 3, 10
   b) 0.6 and 5.4
8. a) Values of y: 19, 6, –3, –8, –9, –6, 1, 12
   b) 0.9 and –3.4
9. a) Q1:(0, –4); Q2: (0, –1), (0, –5); Q3: (0, 0); Q4: (0, 0); Q5: (0, 0), (0, 0)
   b) Q1: (0, –4); Q2: (0, –1), (0, –5); Q3: (–2, –4): Q4: (–1.5, –2.25); Q5: (1.5, –2.25), (–2.5, –6.25)
   c) The y-intercept; the point where the x-value is the mean of the roots.

Exercise 5E

1. a)

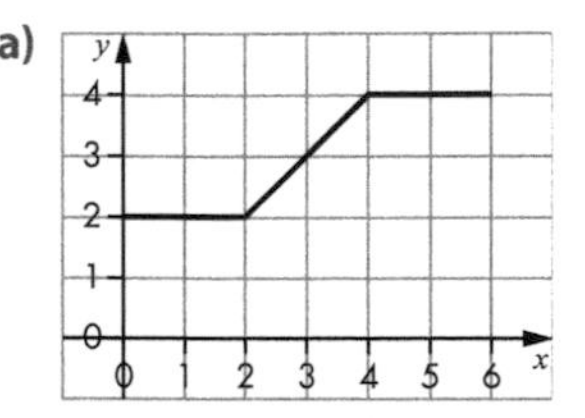

b)

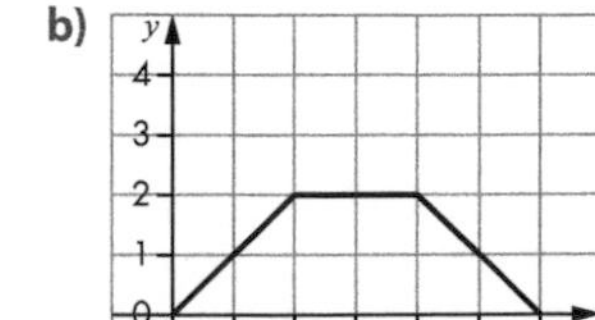

c)

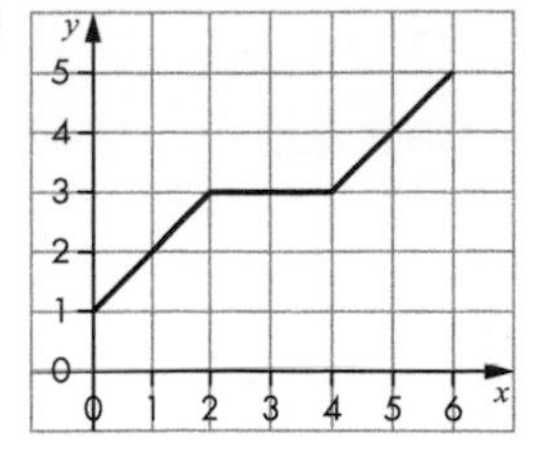

d)

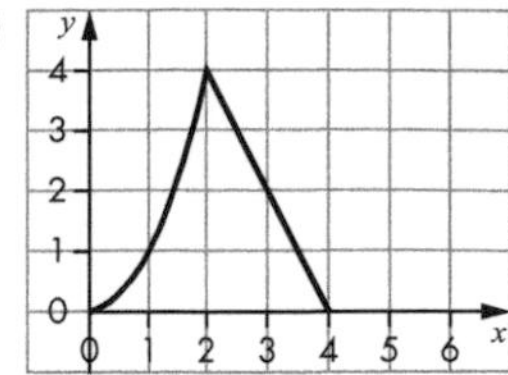

e)

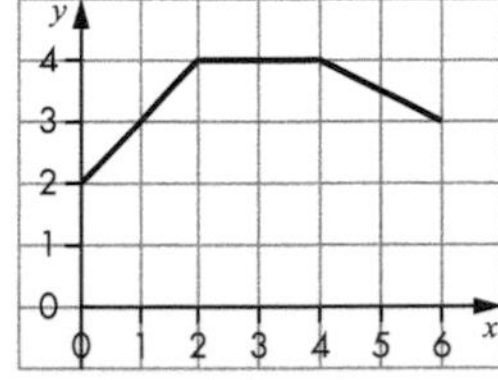

f)

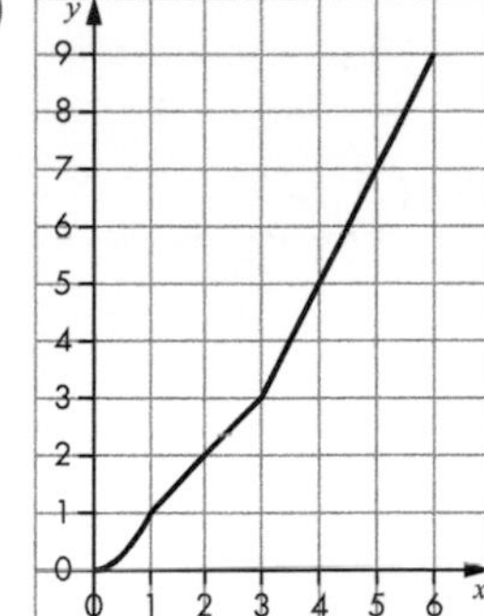

2. a) $2 \leqslant f(x) \leqslant 4$ b) $0 \leqslant f(x) \leqslant 2$ c) $1 \leqslant f(x) \leqslant 5$
   d) $0 \leqslant f(x) \leqslant 4$ e) $2 \leqslant f(x) \leqslant 4$ f) $0 \leqslant f(x) \leqslant 9$
3. a) $f(x) = 2 \quad -2 \leqslant x < 0$
   $= x + 2 \quad 0 \leqslant x < 2$
   $= 4 \quad 2 \leqslant x \leqslant 4$
   b) $f(x) = 6 \quad 0 \leqslant x < 2$
   $= 8 - x \quad 2 \leqslant x < 4$
   $= 4 \quad 4 \leqslant x \leqslant 6$
   c) $f(x) = x^2 \quad 0 \leqslant x < 3$
   $= 9 \quad 3 \leqslant x \leqslant 5$
   d) $f(x) = 1 \quad -4 \leqslant x < -1$
   $= -x \quad -1 \leqslant x < 1$
   $= -1 \quad 1 \leqslant x \leqslant 4$

EXAM-STYLE QUESTIONS

1. a) 110 and 90 b) $n = 0$ or -1
2. a) $x = 19.25$ b) $f(x) \geqslant 15$
3. a) $x = -4$ b) $x = -6$

CHAPTER 6 MATRICES

6.1 Introduction to matrices

Exercise 6A

1. **a)** $\begin{pmatrix} 15 \\ -30 \end{pmatrix}$ **b)** $\begin{pmatrix} -64 \\ 16 \end{pmatrix}$

c) $\begin{pmatrix} 20 & -8 \\ -18 & 14 \end{pmatrix}$ **d)** $\begin{pmatrix} 32 & 12 \\ -48 & -36 \end{pmatrix}$

e) $\begin{pmatrix} -9 \\ 18 \end{pmatrix}$ **f)** $\begin{pmatrix} -60 & 24 \\ 54 & -42 \end{pmatrix}$

g) $\begin{pmatrix} 1 \\ -2 \end{pmatrix}$ **h)** $\begin{pmatrix} -2 \\ 0.5 \end{pmatrix}$

2. **a)** $\begin{pmatrix} 10 & 5 \\ 5 & 5 \end{pmatrix}$ **b)** $\begin{pmatrix} 2 & 10 \\ 5 & 27 \end{pmatrix}$

c) $C^2 = \begin{pmatrix} 5 & 4 \\ 4 & 5 \end{pmatrix}$; $C^3 = \begin{pmatrix} 13 & 14 \\ 14 & 13 \end{pmatrix}$

3. **a)** $\begin{pmatrix} 7 & -7 \\ 8 & 5 \end{pmatrix}$ **b)** $\begin{pmatrix} 1 & -5 \\ 16 & 11 \end{pmatrix}$

c) $\begin{pmatrix} 19 \\ 5 \end{pmatrix}$ **d)** $\begin{pmatrix} 6 \\ 5 \end{pmatrix}$

e) $\begin{pmatrix} 27 & 8 \\ 16 & 11 \end{pmatrix}$ **f)** $\begin{pmatrix} -3 & -8 \\ 8 & 5 \end{pmatrix}$

4. **a)** $\begin{pmatrix} 2 & 4 \\ 13 & 5 \end{pmatrix}$ **b)** $\begin{pmatrix} 4 & 6 \\ 9 & 10 \end{pmatrix}$

5. **a)** **i** $AB = \begin{pmatrix} 7 & 3 \\ 6 & 2 \end{pmatrix}$ **ii** $BA = \begin{pmatrix} 4 & 4 \\ 6 & 5 \end{pmatrix}$

b) No

6. $x = -2$ and $y = 2$.

7. Sometimes true.

8. $x = 1.5, y = -0.5$

9. $x = 2, y = 1$

6.2 The zero matrix and the identity matrix

Exercise 6B

1. **a)** $\begin{pmatrix} -5 & 9 \\ 7 & -3 \end{pmatrix}$ **b)** $\begin{pmatrix} -12 & 0 \\ 10 & 4 \end{pmatrix}$

c) $\begin{pmatrix} 14 & 5 \\ -10 & -1 \end{pmatrix}$ **d)** $\begin{pmatrix} 10 & -2 \\ -3 & 7 \end{pmatrix}$

2. **a)** **Z** **b)** **Q** **c)** **Q** **d)** **Z**

3. **a)** **Z** **b)** **I**

4. **a)** $\begin{pmatrix} 8 & 2 \\ 2 & 4 \end{pmatrix}$ **b)** $\begin{pmatrix} 17 & 6 \\ 6 & 5 \end{pmatrix}$

5. **a)** $\begin{pmatrix} 1 & 0 \\ 0 & 1 \end{pmatrix}$ **b)** $\begin{pmatrix} 1 & 0 \\ 0 & 1 \end{pmatrix}$ **c)** Both equal I

6. **a)** $\begin{pmatrix} 0 & 0 \\ 0 & 0 \end{pmatrix}$ **b)** $\begin{pmatrix} 0 & 0 \\ 0 & 0 \end{pmatrix}$ **c)** Both equal Z

7. Both AB and BA equal I

8. $A \times A = I$

6.3 Transformations

Exercise 6C

1. (0, 1)

2. (–1, –1)

3. $x = 1, y = -3$

4. **a)**

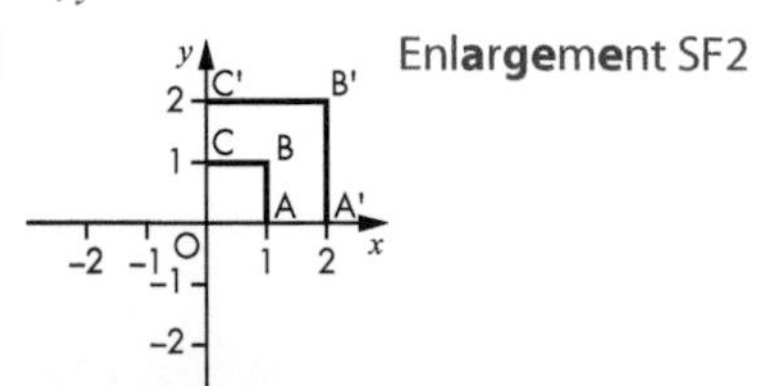

Enlargement SF2

b)

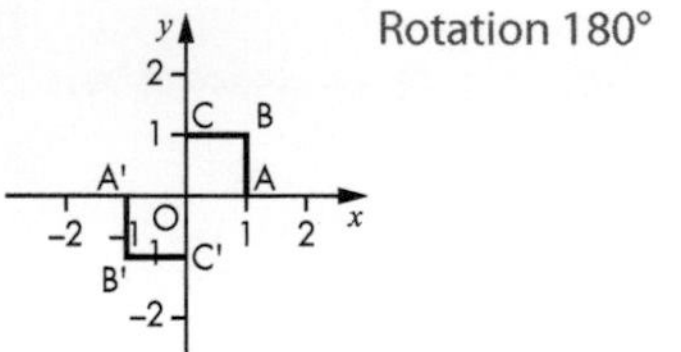

Rotation 180°

c)

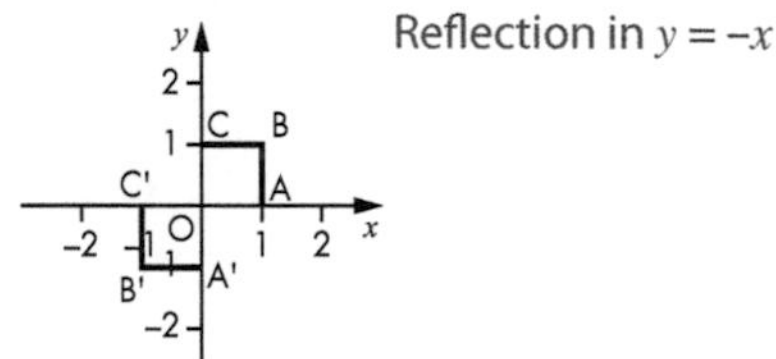

Reflection in $y = -x$

d)

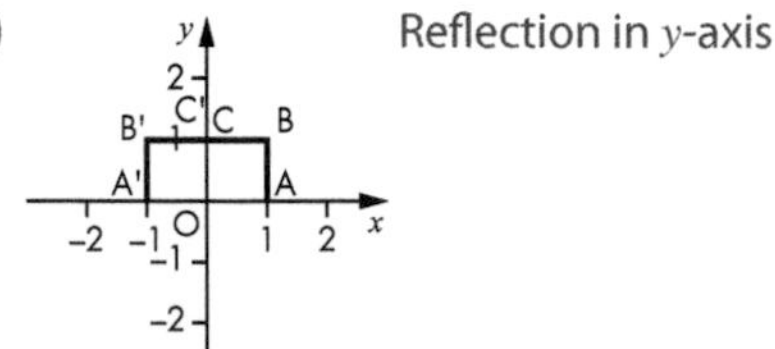

Reflection in y-axis

e)

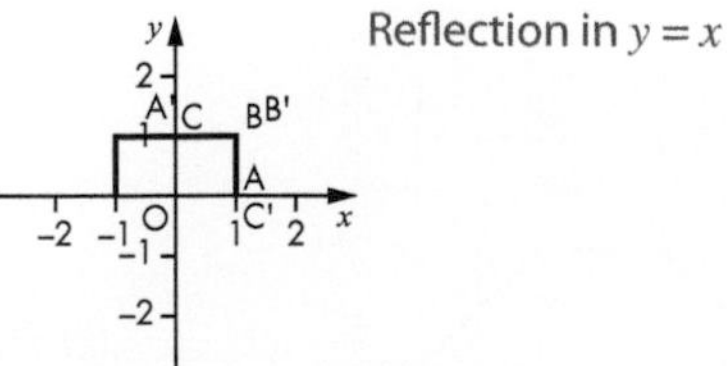

Reflection in $y = x$

f) 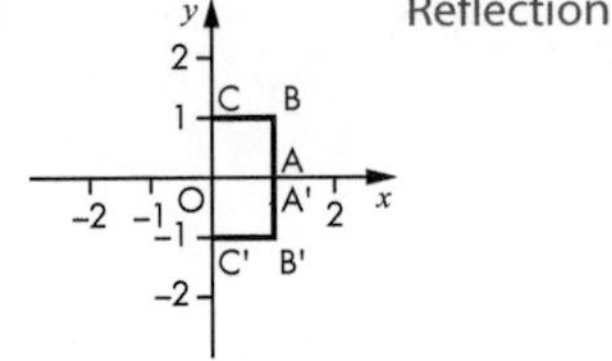

Reflection in x-axis

5. Enlargement scale factor 4, centre O.

6. $\begin{pmatrix} 0 & -1 \\ 1 & 0 \end{pmatrix}$

7. $4\sqrt{5}$

6.4 Combinations of transformations

Exercise 6D

1. **a)** Reflection in the x-axis
b) Reflection in the y-axis
c) Rotation 180° about O
d) Rotation 180° about O
e) Does not move (identity matrix)
f) Does not move (identity matrix)
g) Does not move (identity matrix)

2. **a)** **i** Reflection in the x-axis **ii** $\begin{pmatrix} 1 & 0 \\ 0 & -1 \end{pmatrix}$

b) **i** Reflection in the y-axis **ii** $\begin{pmatrix} -1 & 0 \\ 0 & 1 \end{pmatrix}$

c) **i** Reflection in the x-axis **ii** $\begin{pmatrix} 1 & 0 \\ 0 & -1 \end{pmatrix}$

d) **i** Reflection in the y-axis **ii** $\begin{pmatrix} -1 & 0 \\ 0 & 1 \end{pmatrix}$

3. Rotating 90° clockwise four times takes shape back to starting point (identity matrix).

4. $X=\begin{pmatrix} 0 & 1 \\ -1 & 0 \end{pmatrix}$, $Y=\begin{pmatrix} -1 & 0 \\ 0 & -1 \end{pmatrix}$, $Z=\begin{pmatrix} 0 & -1 \\ 1 & 0 \end{pmatrix}$

Show that $Y \times X = Z$

EXAM-STYLE QUESTIONS

1. **a)** $\begin{pmatrix} 8 & 12 \\ -8 & 4 \end{pmatrix}$ **b)** $\begin{pmatrix} 10 & 4 \\ 6 & -4 \end{pmatrix}$

c) $\begin{pmatrix} 9 & -2 \\ 4 & 8 \end{pmatrix}$

2. $x = 12$ and $y = 2$

3. (28, –17)

4. $MN = \begin{pmatrix} 4-7 & -14+14 \\ 2-2 & -7+4 \end{pmatrix} = \begin{pmatrix} -3 & 0 \\ 0 & -3 \end{pmatrix}$

$= -3I$ and $k = -3$

5. **a)** A′(0, 1), B′(–1, 1) and C′(–1, 0)
b) A 90° anticlockwise rotation about the origin
c) A 180° rotation about the origin

6. Enlargement, centre the origin, scale factor –2

7. **a)** R a reflection in the y-axis; S a reflection in the line $y = x$

b) $SR = \begin{pmatrix} 0 & 1 \\ -1 & 0 \end{pmatrix}$ which represents a rotation of 90° clockwise about the origin

c) SR represents first R, then S. RS represents first S, then R and is equivalent to a rotation of 90° anticlockwise about the origin. This is different to the SR transformation so the matrices must be different.

CHAPTER 7 ALGEBRA

7.1 Manipulation of rational expressions

Exercise 7A

1. **a)** $\frac{x^2y+8}{4x}$ **b)** $\frac{7x+3}{4}$ **c)** $\frac{13x+5}{15}$
d) $\frac{5x-10}{4}$ **e)** $\frac{xy^2-8}{4y}$ **f)** $\frac{x+1}{4}$
g) $\frac{-7x-5}{15}$ **h)** $\frac{2-3x}{4}$ **i)** $\frac{2xy}{3}$
j) $\frac{x^2-2x}{10}$ **k)** $\frac{1}{6}$ **l)** $\frac{1}{2x}$
m) 1 **n)** 3 **o)** $\frac{13x+9}{10}$
p) $\frac{x+3}{2}$

2. **a)** $\frac{7x+9}{(x+1)(x+2)}$ **b)** $\frac{11x-10}{(x-2)(x+1)}$
c) $\frac{2-13x}{(4x+1)(x+2)}$ **d)** $\frac{8-10x}{(2x-1)(x+1)}$
e) $\frac{x+1}{(2x-1)(3x-1)}$

3. $\frac{2x^2+x-3}{4x^2-9}$

4. **a)** $\frac{9x+13}{(x+1)(x+2)}$ **b** $\frac{14x+19}{(4x-1)(x+1)}$
c) $\frac{2x^2+x-13}{2(x+1)}$ **d** $\frac{x+1}{(2x-1)(3x-1)}$

5. **a)** $\frac{x-1}{2x+1}$ **b** $\frac{2x+1}{x+3}$ **c** $\frac{2x-1}{3x-2}$
d) $\frac{x+1}{x-1}$ **e)** $\frac{2x+5}{4x-1}$

6. **a)** $\frac{x(x-1)}{x+2}$ **b)** $\frac{x(x-5)}{x-4}$ **c)** $\frac{x(x-4)}{x-5}$
d) $\frac{x(x-6)}{x-7}$ **e)** $\frac{x(x-5)}{x+3}$ **f)** $\frac{x(x+3)}{x+8}$

7. a) $x+2$ b) $x+3$ c) $4x+5$

d $2x-7$ e $3(x+1)$ f $x+2$

8. a $\frac{4x(3x+1)}{2x-3}$ b $\frac{x(9x-8)}{3x-7}$ c $\frac{2x(x+4)}{4x-3}$

d $\frac{6(x-1)}{5x+6}$ e $\frac{6x(x+1)}{8x-3}$ d $\frac{x(4x-5)}{3x+4}$

7.2 Use and manipulation of formulae and expressions

Exercise 7B

1. $m = gv$ 2. $m=\sqrt{t}$ 3. $r=\frac{C}{2\pi}$

4. $b=\frac{A}{h}$ 5. $l=\frac{P-2w}{2}$ 6. $p=\sqrt{m-2}$

7. $a=\frac{v-u}{t}$ 8. $d=\sqrt{\frac{4A}{\pi}}$

9. a) $n=\frac{W-t}{3}$ b) $t=W-3n$

10. $p=\sqrt{\frac{k}{2}}$

11. a) $t=u^2-v$ b) $u=\sqrt{v+t}$

12. a) $m=k-n^2$ b) $n=\sqrt{k-m}$

13. $r=\sqrt{\frac{T}{5}}$

14. a) $w=K-5n^2$ b) $n=\sqrt{\frac{K-w}{5}}$

Exercise 7C

1. a) $a=\sqrt{c^2-b^2}$ b) 2.5

2. a) $a=\frac{2(s-ut)}{t^2}$ b) 60

3. a) $c=\frac{b+2}{a}$ b) $b=ac-2$

4. $t=\frac{r}{p}+3$

5. $e=\left(\frac{12}{d}-1\right)^2$

6. a) 5 b) $u=\sqrt{v^2-2as}$ c) $s=\frac{v^2-u^2}{2a}$

7. a) $L=\left(\frac{t}{2\pi}\right)^2 G$

b) $T^2=4\pi^2\left(\frac{L}{G}\right)\Rightarrow GT^2=4\pi^2L\Rightarrow G=L\frac{4\pi^2}{T^2}\Rightarrow$

$G=L\left(\frac{2\pi}{T}\right)^2$

8. a) $R=\sqrt{\frac{D+\pi r^2}{\pi}}$ b) $r=\sqrt{\frac{\pi R^2-D}{\pi}}$

c) $\pi=\frac{D}{R^2-r^2}$

9. a) $x=5$ or -5 b) $x=\sqrt{\frac{11+4y^2}{3}}$

c) $y=\sqrt{\frac{3x^2-11}{4}}$

10. a) $a=\left(\frac{T}{2}\right)^2(c+3)$ b) $c=a\left(\frac{2}{T}\right)^2-3$

11. $T=\frac{b^2+c^2-a^2}{2bc}$

12. a) $f=\frac{uv}{u+v}$ b) 12 c) $u=\frac{fv}{v-f}$ d) $v=\frac{fu}{u-f}$

13. $x=\frac{yz}{y+z}$

14. $s=\frac{2r}{t+1}$

15. $h=\frac{2gi}{1+5i}$

16. $k=\frac{j}{t}$

17. $c=\frac{81b}{4ad}$

18. $h=9gi\left(\frac{12e}{5}-\frac{1}{8f}\right)$

7.3 The factor theorem

Exercise 7D

1. a) $(-1)^3+6(-1)^2-9(-1)-14=-1+6+9-14=0$
   b) $(3)^3+3(3)^2-13(3)-14=27+27-39-14=0$
   c) $(4)^3-7(4)^2+2(4)+40=64-112+8+40=0$
   d) $(-6)^3+13(-6)^2+54(-6)+72=-216+468-324+72=0$
   e) $(-7)^3-37(-7)+84=-343+259+84=0$
   f) $2(1.5)^3-5(1.5)^2+(1.5)+6=6.75-11.25+1.5+6=0$
2. a) $(x+1)(x-3)(x+5)$ b) $(x+4)(x-1)(x+2)$
   c) $(x+1)(x+2)(x+3)$ d) $(x-2)(x+3)(x+5)$
   e) $(2x+1)(x-4)(x-3)$ f) $(3x-1)(2x+3)(x+2)$
3. a) $x=1, x=-2, x=5$ b) $x=1$
   c) $x=4, x=-4, x=-2$ d) $x=-3, x=-4, x=1$
   e) $x=-2, x=-6, x=-7$ f) $x=-4, x=10, x=5$
4. -2

5. a) $(-5)^3 + 3(-5)^2 - 13(-5) + c = 0 \Rightarrow -125 + 75 + 65 + c = 0 \Rightarrow c = -15$
   b) −1 and 3
6. $a = 3, b = 1, c = 1$
7. $x - 4, x + 4, x + 3$
8. $x + 2$ and $x - 4$

Exam-style questions

1. $\dfrac{3}{(x+1)(x+2)}$
2. $\dfrac{(2x-1)(x+3)}{2x+1}$
3. $\dfrac{x}{(x+2)^2}$
4. $x = \dfrac{4y-2}{9}$
5. $v = \dfrac{uf}{u-f}$
6. $C^2 = (2\pi r)^2 = 4\pi^2 r^2 \Rightarrow \dfrac{C^2}{4\pi} = \dfrac{4\pi^2 r^2}{4\pi} = \pi r^2 = A$
7. a) $f(-4) = -64 - 16 + 56 + 24 = 0$
   b) $(x - 2)(x - 3)(x + 4)$
8. a) $f(2) = 8 - 24 + 24 - 8 = 0$ b) $(x - 2)^3$
9. $(x - 4)(x + 5)(x - 5)$
10. $x = -1, -2$ or -3
11. a) $x = -1$ or 4 b) $x = 0, -1$ or 4 c) $x = -2, 2$ or 3

CHAPTER 8 SEQUENCES

8.1 Number sequences

Exercise 8A

1. a) 28, 34, 40; add 6 b) 23, 28, 33; add 5
   c) 20 000, 200 000, 2 000 000; multiply by 10
   d) 19, 22, 25; add 3 e) 46, 55, 64; add 9
   f) 405, 1215, 3645; multiply by 3
   g) 18, 22, 26; add 4
   h) 625, 3125, 15 625; multiply by 5
2. a) 16, 22 b) 26, 37 c) 31, 43 d) 46, 64
   e) 121, 169 f) 11, 13 g) 33, 65 h) 78, 108
3. a) 48, 96, 192; double b) 33, 39, 45; add 6
   c) 4, 2, 1; halve d) 38, 35, 32; subtract 3
   e) 26, 33, 41; add the next integer (1, 2, 3, 4…)
   f 19, 22, 25; add 3
   g 28, 36, 45; add the next integer, starting from 2
   h 0.0625, 0.031 25, 0.015 625; halve
4. a) 21, 34; add previous 2 terms
   b) 49, 64; next square number
   c) 47, 76; add previous 2 terms
   d) 216, 343; cube numbers

8.2 The *n*th term of a sequence

Exercise 8B

1. a) 4, 5, 6, 7, 8 b) 2, 5, 8, 11, 14
   c) 3, 8, 13, 18, 23 d) 9, 13, 17, 21, 25
2. a) £305 b) £600 c) 3 d) 5

8.3 The *n*th term of a linear sequence

Exercise 8C

1. a) 13, 15; $2n + 1$ b) 25, 29; $4n + 1$
   c) 33, 38; $5n + 3$ d) 32, 38; $6n - 4$
   e) 20, 23; $3n + 2$ f) 37, 44; $7n - 5$
   g) 21, 25; $4n - 3$ h) 23, 27; $4n - 1$
   i) 17, 20; $3n - 1$
2. a) $3n + 1$; 151 b) $2n + 5$; 105
   c) $5n - 2$; 248 d) $4n - 3$; 197
   e) $8n - 6$; 394 f) $n + 4$; 54
3. a) 33rd b) 30th c) 100th = 499
4. a) i $4n + 1$ ii 401 iii 101, 25th
   b) i $2n + 1$ ii 201 iii 99 or 101, 49th and 50th
   c) i $3n + 1$ ii 301 iii 100, 33rd
   d) i $2n + 6$ ii 206 iii 100, 47th
   e) i $4n + 5$ ii 405 iii 101, 24th
   f) i $5n + 1$ ii 501 iii 101, 20th
5. a) $\dfrac{2n+1}{3n+1}$ b) Getting closer to $\dfrac{2}{3}$ $(0.\dot{6})$
   c) i 0.667 774 (6 d.p.) ii 0.666 778 (6 d.p.)
   d) 0.666 678 (6 d.p.), 0.666 667 (6 d.p.)
6. a) $\dfrac{1}{2}, \dfrac{3}{5}, \dfrac{5}{8}$
   b) i 0.666 666 555 6 ii $\dfrac{2}{3}$
   c) For n, $\dfrac{2n-1}{3n-1} \approx \dfrac{2n}{3n} = \dfrac{2}{3}$
7. a) Sequence goes up in 2s; first term is 2 + 29
   b) $n + 108$ c) Because it approached $2n \div n$
   d) 79th

8.4 The *n*th term of a quadratic sequence

Exercise 8D

1. a) 36, 49 b) 38, 51 c) 42, 56 d) 74, 100
   e) 78, 105 f) 109, 148 g) 43, 57 h) 178, 243
   i) 114, 154 j) 66, 91
2. a) n^2 b) $n^2 + 2$ c) $n^2 + n$ d) $2n^2 + 2$
   e) $2n^2 + n$ f) $3n^2 + 1$ g) $n^2 + n + 1$ h) $5n^2 - 2$
   i) $3n^2 + n$ j) $2n^2 - n$
3. a) $0, \dfrac{1}{3}, \dfrac{8}{13}$ b) 6th term

4. $4n^2$ is positive since n is always positive and n^2 is positive. $3n - 2$ is always positive since when $n = 1$, $3n - 2 = 1$, so first term = 1 and terms are increasing by 3 each time, so $3n - 2$ is always positive. Positive divided by positive is always positive.

5. $\frac{n^2}{2n+1}$

6. **a)** $3n - 2$ **b)** $4n - 2$
   c) $(3n - 2)(4n - 2)$ or $12n^2 - 14n + 4$

8.5 The limiting value of a sequence as $n \to \infty$

Exercise 8E

1. **a)** 2 **b)** $\frac{1}{3}$ **c)** 4
   d) –1 **e)** 0 **f)** 2
2. **a)** 1 **b)** 0 **c)** 0
   d) 0 **e)** $\frac{1}{5}$ **f)** 0

Exam-style questions

1. 29, 41; the numbers go up in 12s.
2. No, one sequence the numbers are always a multiple of three ($3n$); in the other the numbers are always one less that a multiple of three ($3n - 1$). Hence no term in common.
3. **a)** 4, $5\frac{1}{3}$ and 6 **b)** The eighth **c)** 8
4. $n^2 + 2n$ or $n(n + 2)$
5. Four (5, 10, 50, 65)

CHAPTER 9 PYTHAGORAS' THEOREM AND TRIGONOMETRY

9.1 Pythagoras' theorem

Exercise 9A

1. 10.3 cm
2. 8.5 cm
3. 20.6 cm
4. The square in the first diagram has the same area as the two squares in the second diagram together. The four triangles are the same in both diagrams.
5. **a)** 15 cm **b)** 14.7 cm **c)** 6.3 cm **d)** 18.3 cm
6. **a)** 5 m **b)** 6 m **c)** 50 cm
7. There are infinite possibilities, e.g. any multiple of 3, 4, 5 such as 6, 8, 10; 9, 12, 15; 12, 16, 20; multiples of 5, 12, 13 and of 8, 15, 17.
8. 42.6 cm

Exercise 9B

1. No. The foot of the ladder is about 6.6 m from the wall.
2. About 17 minutes, assuming it travels at the same speed.
3. 13 units
4. **a)** 4.85 m
   b) 4.83 m (There is only a small difference.)
5. Yes, because $24^2 + 7^2 = 25^2$
6. About 6 cm (5.7 cm)

Exercise 9C

1. **a)** 32.2 cm² **b)** 2.83 cm² **c)** 50.0 cm²
2. **2** 22.2 cm²
3. **3** 15.6 cm²
4. **a)**

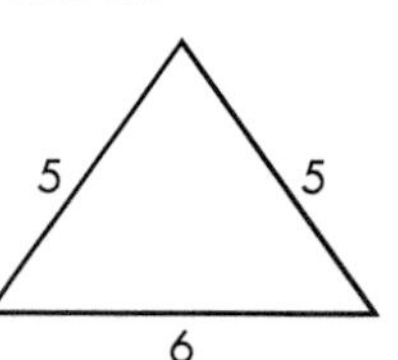

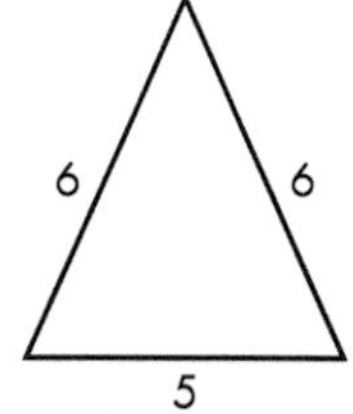

   b) The areas are 12 cm² and 13.6 cm² respectively, so the triangle with 6 cm, 6 cm, 5 cm sides has the greater area.
5. 259.8 cm²
6. 48 cm²
7. **a)** 10 cm **b)** 26 cm **c)** 9.6 cm

Exercise 9D

1. **a)** **i** 14.4 cm **ii** 13 cm **iii** 9.4 cm
   b) 15.0 cm (1 dp)
2. No, 6.55 m is longest length
3. **a)** 20.6 cm **b)** 15.0 cm
4. 21.3 cm
5. **a)** 8.49 m **b)** 9 m
6. $\sqrt{72}$ = 17.3 cm
7. 20.6 cm
8. **a)** 11.3 cm **b)** 7 cm **c)** 8.06 cm
9. **a)** 50.0 cm **b)** 54.8 cm **c)** 48.3 cm
   d) 27.0 cm

9.2 Trigonometry in right-angled triangles

Exercise 9E

1. **a)** 0.682 **b)** 0.829 **c)** 0.922 **d)** 1
2. **a)** **i** 0.574 **ii** 0.574 **b)** **i** 0.208 **ii** 0.208
   c) **i** 0.391 **ii** 0.391
   d) The ratios in each pair have the same value.

e) i sin 15° is the same as cos 75°.

ii cos 82° is the same as sin 8°.

iii sin x is the same as cos $(90° - x)$.

3. a) 0.933 b) 1.48 c) 2.38 d) infinite

4. It can take values > 1

5. a) 3.56 b) 8.96 c) 5.61 d) 7.08

6. a) $\frac{4}{5}, \frac{3}{5}, \frac{4}{3}$ b) $\frac{5}{13}, \frac{12}{13}, \frac{5}{12}$ c) $\frac{7}{25}, \frac{24}{25}, \frac{7}{24}$

7. a) 30° b) 51.7° c) 39.8° d) 61.3°
e) 87.4° f) 45.0°

8. a) 60° b) 50.2° c) 2.6° d) 45.0°
e) 78.5° f) 45.6°

9. a) 31.0° b) 20.8° c) 41.8° d) 46.4°
e) 69.5° f) 77.1°

10. Error message, largest value 1, smallest value –1

Exercise 9F

1. a) 17.5° b) 22.0° c) 32.2°
2. a) 5.29 cm b) 5.75 cm c) 13.2 cm
3. a) 4.57 cm b) 6.86 cm c) 100 cm
4. a) 5.12 cm b) 9.77 cm c) 11.7 cm d) 15.5 cm
5. a) 51.3° b) 75.5° c) 51.3°
6. a) 5.35 cm b) 14.8 cm c) 12.0 cm d) 8.62 cm
7. a) 5.59 cm b) 46.6° c) 9.91 cm d) 40.1°
8. a) 33.7° b) 36.9° c) 52.1°
9. a) 9.02 cm b) 7.51 cm c) 7.14 cm d) 8.90 cm
10. a) 13.7 cm b) 48.4° c) 7.03 cm d) 41.2°

Exercise 9G

1. a) 12.6 b) 59.6 c) 74.7 d) 16.0
e) 67.9 f) 20.1

2. a) 44.4° b) 39.8° c) 44.4° d) 49.5°
e) 58.7° f) 38.7°

3. a) 67.4° b) 11.3 c) 134 d) 28.1°
e) 39.7 f) 263 g) 50.2° h) 51.3°
i) 138 j) 22.8

4. a) Sides of right-hand triangle are sine θ and cosine θ
b) Pythagoras' theorem
c) Check the formulae

Exercise 9H

1. 65°
2. The safe limits are between 1.04 m and 2.05 m. The ladder will reach between 5.63 m and 5.90 m up the wall.
3. 31°
4. a) 338 km b) 725 km
5. 43.5 km
6. 170 km
7. One way is to stand directly opposite a feature, such as a tree, on the opposite bank, and record your position. Then move a measured distance, x, along your bank and measure the angle, θ, between your new position and the feature. Then the width of river is $x \tan \theta$. This, of course, requires measuring equipment! An alternative is to walk along the bank until the angle is 45° (if that is possible). This angle is easily found by folding a sheet of paper. This way you do not require an angle measurer.

Exercise 9I

1. 10.1 km
2. 22°
3. 429 m
4. a) 156 m
b) No. The new angle of depression is
$\tan^{-1} \frac{200}{312} = 33°$ and half of 52° is 26°.
5. a) 222 m b) 42°
6. a) 21.5 m b) 17.8 m
7. 13.4 m
8. 19°
9. The angle is 16° so Cara is not quite correct.

Exercise 9J

1. 25.1°
2. a) 25 cm b 58.6° c 20.5 cm
3. a) 3.46 m b 75.5° c 73.2°
4. a) 24.0° b 48.0° c 13.5 cm
d) 16.5° or 16.7°
5. a) It is 44.6°; use triangle XDM where M is the midpoint of BD; triangle DXB is isosceles, as X is over the point where the diagonals of the base cross; the length of DB is $\sqrt{656}$ and the cosine of the required angle is $0.5\sqrt{656} \div 18$.
b) 57.7°

9.3 The sine rule and the cosine rule

Exercise 9K

1. a) 3.64 m b) 8.05 cm c) 19.4 cm
2. a) 46.6° b) 112.0° c) 36.2°
3. 3.47 m
4. a) i 30° ii 40° b) 19.4 m
5. 36.5 m
6. 22.2 m
7. 64.6 km

Exercise 9L

1. a) 7.71 m b) 29.1 cm c) 27.4 cm
2. a) i 76.2° ii 125.1° iii 90°
b) Right-angled triangle

3. a) 10.7 cm b) 41.7° c) 38.3° d) 6.7 cm
4. 58.4 km at 092.5°
5. 21.8°; the smallest angle is opposite the shortest side.
6. 42.5 km
7. 111°; the largest angle is opposite the longest side.

Exercise 9M

1. a) 8.6 m b) 90° c) 27.2 cm d) 26.9°
   e 27.5° f) 62.4 cm
2. 7 cm
3. 11.1 km
4. a) $A = 90°$; this is Pythagoras' theorem
   b) A is acute c) A is obtuse
5. 142 m or 143 m

Exam-style questions

1. 84 cm²
2. 8 cm
3. $\frac{7}{32}$
4. 73°
5. a) 39° b) 11.1 cm or 11.2 cm
6. 5.16 cm
7. 65.5 cm
8. 17.3 cm
9. 58 km
10. 17.8 m
11. a) 12.3 cm b) 29° c) 59°

CHAPTER 10 SOLVING EQUATIONS

10.1 Solving linear equations

Exercise 10A

1. a) $x = 6$ b) $y = 3$ c) $t = 4$ d) $f = 2\frac{1}{2}$
   e) $k = 3\frac{1}{2}$ f) $x = 2\frac{1}{2}$ g) $m = 56$ h) $x = 0$
   i) $h = -7$ j) $w = -18$ k) $x = 36$ l) $y = 36$
   m) $x = 7$ n) $y = 1$ o) $x = 11.5$ p) $t = 0.2$
2. a) $x = 3$ b) $x = 7$ c) $t = 5$ d) $x = 1\frac{1}{2}$
   e) $y = 2\frac{1}{2}$ f) $k = \frac{1}{2}$ g) $x = -1$ h) $t = -2$
   i) $x = -2$
3. Any values that work, e.g. $a = 2$, $b = 3$ and $c = 30$.
4. a) $x = 2$ b) $y = 1$ c) $a = 7$ d) $k = -1$
   e) $m = 3$ f) $s = -2$
5. a) $d = 6$ b) $x = 11$ c) $y = 1$ d) $h = 4$
   e $b = 9$ f $c = 6$
6. $6x + 3 = 6x + 10$; $6x - 6x = 10 - 3$; $0 = 7$, which is obviously false. Both sides have $6x$, which cancels out.
7. When the brackets are expanded both sides of the equation are identical, so any value of x will be a solution.

10.2 SETTING UP EQUATIONS

Exercise 10B

1. a) 1.5 b) 2
2. a) 1.5 cm b) 6.75 cm²
3. Length is 5.5 m, width is 2.5 m and area is 13.75 m². Tiles cost 123.75 dollars
4. 3 years
5. 9 years
6. 3 cm
7. 5
8. a) $4x + 40 = 180$ b) $x = 35°$
9. a) $\frac{x+10}{5} = 9.50$ b) $37.50
10. No, as $x + (x + 2) + (x + 4) + (x + 6) = 360$ gives $x = 87°$ so the consecutive numbers (87, 89, 91, 93) are not even but odd.
11. $4x + 18 = 3x + 1 + 50$, $x = 33$. Large bottle 1.5 litres, small bottle 1 litre

Exercise 10C

1. a) $x = 7$ b) $x = 9$ c) $x = 14$ d) $x = 5$
   e) $x = 2.5$ f) $x = -2$
2. a) $x = 3$ b) $x = 6$ c) $x = 2$ d) $x = 5$
   e) $x = \frac{5}{16}$ f) $x = 3$
3. a) $x = 6$ b) $x = 14$ c) $x = 7$
4. 5, 6 and 7
5. 50, 55 and 75 degrees

10.3 *Solving quadratic equations by factorisation or the quadratic formula*

Exercise 10D

1. −2, −5
2. −3, −1
3. −6, −4
4. −3, 2
5. −1, 3
6. −4, 5
7. 1, −2
8. 2, −5
9. 7, −4
10. 3, 2
11. 1, 5
12. 4, 3
13. −4, −1
14. −9, −2
15. 2, 4
16. 3, 5
17. −2, 5
18. −3, 5

19. −6, 2

20. −6, 3

21. −1, 2

22. −2

23. −5

24. 4

25. −2, −6

26. 7

27. a) $x(x-3) = 550$, $x^2 - 3x - 550 = 0$
b) $(x-25)(x+22) = 0$, $x = 25$

28. $x(x+40) = 48\,000$, $x^2 + 40x - 48\,000 = 0$, $(x+240)(x-200) = 0$. Fence is $2 \times 200 + 2 \times 240 = 880$ m.

29. −6, −4

30. 2, 16

31. −6, 4

32. −9, 6

33. −10, 3

34. −4, 11

35. −8, 9

36. 8, 9

37. 1

Exercise 10E

1. a) $\frac{1}{3}, -3$ **b)** $1\frac{1}{3}, -\frac{1}{2}$ **c)** $-\frac{1}{5}, 2$ **d)** $-2\frac{1}{2}, 3\frac{1}{2}$
e) $-\frac{1}{6}, -\frac{1}{3}$ **f)** $\frac{2}{3}, 4$ **g)** $\frac{1}{2}, -3$ **h)** $\frac{5}{2}, -\frac{7}{6}$
i) $-1\frac{2}{3}, 1\frac{2}{5}$ **j)** $1\frac{3}{4}, 1\frac{2}{7}$ **k)** $\frac{2}{3}, \frac{1}{8}$ **l)** $\pm\frac{1}{4}$
m $-2\frac{1}{4}, 0$ **n)** $\pm 1\frac{2}{5}$ **o)** $-\frac{1}{3}, 3$

2. a) −3, 4 **b)** $-\frac{5}{2}, \frac{3}{2}$ **c)** −6, 7 **d)** $-1, \frac{11}{33}$
e) −2, 3 **f)** $-\frac{2}{5}, \frac{1}{2}$ **g)** $-\frac{1}{2}, -\frac{1}{3}$ **h)** $-2, \frac{1}{5}$
i) 4 **j)** $-2, \frac{1}{8}$ **k)** $-\frac{1}{3}, 0$ **l)** −5, 5
m) $-\frac{5}{3}$ **n)** $-\frac{7}{2}, \frac{7}{2}$ **o)** $-\frac{5}{2}, 3$

3. a) Both have only one solution: $x = 1$.
b B is a linear equation, but A and C are quadratic equations.

4. a) $(5x-1)^2 = (2x+3)^2 + (x+1)^2$, when expanded and collected into the general quadratics, gives the required equation.
b) $(10x+3)(2x-3)$, $x = 1.5$; area = 7.5 cm².

Exercise 10F

1. 1.77, −2.27

2. −0.23, −1.43

3. 3.70, −2.70

4. 0.29, −0.69

5. −0.19, −1.53

6. −1.23, −2.43

7. −0.41, −1.84

8. −1.39, −2.27

9. 1.37, −4.37

10. 2.18, 0.15

11. −0.39, −5.11

12. 0.44, −1.69

13. 1.64, 0.61

14. 0.36, −0.79

15. 1.89, 0.11

16. 13

17. $x^2 - 3x - 7 = 0$

18. $b^2 - 4ac < 0$ or $b^2 - 4ac = -8$ so no real solution as a negative number does not have a real square root.

19. a) 1 **b)** 0 **c)** 2

Exercise 10G

1. a) $2(x+2) + 5(x+1) = 3(x+1)(x+2)$
which simplifies to the required equation.
b) $4(x+1) + 7(x-2) = 3(x-2)(x+1)$
which simplifies to the required equation.
c) $3(x+2) - 4(4x+1) = 2(4x+1)(x+2)$
which simplifies to the required equation.
d) $2(x+1) - 6(2x-1) = 11(2x-1)(x+1)$
which simplifies to the required equation.
e) $3(3x-1) - 4(2x-1) = (2x-1)(3x-1)$
which simplifies to the required equation.

2. a) 3, −1.5 **b)** 4, −1.25 **c)** 3, −2.5 **d)** 0, 1

3. a) $x = 2$ and $x = 5$ **b)** $x = 4$ and $x = 25$

4. a) $x = -2$ and $x = 1$ **b)** $x = 4$ and $x = 1$

10.4 Solving quadratic equations by completing the square

Exercise 10H

1. a) $(x+2)^2 - 4$ **b)** $(x+7)^2 - 49$ **c)** $(x-3)^2 - 9$
d) $(x+3)^2 - 9$ **e)** $(x-1.5)^2 - 2.25$ **f)** $(x-4.5)^2 - 20.25$
g) $(x+6.5)^2 - 42.25$ **h)** $(x+5)^2 - 25$ **i)** $(x+4)^2 - 16$
j) $(x-1)^2 - 1$ **k)** $(x+1)^2 - 1$

2. a) $(x+2)^2 - 5$ **b)** $(x+7)^2 - 54$ **c)** $(x-3)^2 - 6$
d) $(x+3)^2 - 2$ **e)** $(x-1.5)^2 - 3.25$ **f)** $(x+3)^2 - 6$
g) $(x-4.5)^2 - 10.25$ **h)** $(x+6.5)^2 - 7.25$
i) $(x+4)^2 - 22$ **j)** $(x+1)^2 - 2$ **k)** $(x-1)^2 - 8$
l) $(x+1)^2 - 10$

3. a) $-2 \pm \sqrt{5}$ **b)** $-7 \pm 3\sqrt{6}$ **c)** $3 \pm \sqrt{6}$
d) $-3 \pm \sqrt{2}$ **e)** $1.5 \pm \sqrt{3.25}$ **f)** $3 \pm \sqrt{6}$
g) $4.5 \pm \sqrt{10.25}$ **h)** $-6.5 \pm \sqrt{7.25}$ **i)** $-4 \pm \sqrt{22}$
j) $-1 \pm \sqrt{2}$ **k)** $1 \pm 2\sqrt{2}$ **l)** $-1 \pm \sqrt{10}$

4. a) 1.45, −3.45 **b)** 5.32, −1.32 **c)** −4.16, 2.16

5. a) $2(x+1)^2 + 5$ **b)** $3(x+2)^2 - 9$
c) $6(x+1)^2 - 2$ **d)** $5(x-3)^2 - 33$
e) $8(x-2)^2 - 22$ **f)** $9(x+0.5)^2 + 6.75$

g) $12(x - 1.5)^2 - 13$ h) $5(x + 1)^2 + 1$
i) $7(x + 1)^2 - 2$ j) $7(x + 0.5)^2 + 0.25$
k) $10(x - 1)^2 - 5$ l) $11(x + 1)^2 - 5$

6. $a = 24, b = 4, c = -41$
7. $a = 4, b = 4, c = 9$
8. $a = 9, b = 2$
9. a) $x = 1.5 \pm \sqrt{3.75}$ b) $x = 1 \pm \sqrt{0.75}$
   c $x = -1.25 \pm \sqrt{6.5625}$ d $x = 7.5 \pm \sqrt{40.25}$
10. $p = -14, q = -3$
11. third, first, fourth, second, in that order.

Exam-style questions

1. a) $(x + 1)(x + 2) = 110$
   b) $x = 9$ or -12 so the length of CD is 13 cm.
2. a) i $6x = 4x + 16$
   ii $x = 8$, the side of the square is 12 cm
   b) i $2x^2 = (x + 4)^2$
   ii $x = 4 \pm \sqrt{32}$, the side of the square is 13.7 cm
3. $a = 6, b = 0.5, c = -2.5$
4. a) $a = 3, b = -2, c = -8$ b) $x = 3.63$ or 0.37
5. a) $x = \frac{x+1}{x}$
   b The solution is $x = \frac{1}{2} \pm \frac{1}{2}\sqrt{5}$ and the length of AB is $\frac{1}{2} + \frac{1}{2}\sqrt{5}$ m, since it can't be negative.

CHAPTER 11 SIMULTANEOUS EQUATIONS

11.1 Simultaneous linear equations

Exercise 11A

1. a) $x = 5, y = 10$ b) $x = 18, y = 6$
   c) $x = 12, y = 48$ d) $x = 6, y = 18$
   e) $x = 12.5, y = 2.5$ f) $x = 0.5, y = 4.5$
2. a) $x = 13, y = 7$ b) $x = 9, y = 14$
   c) $x = 10, y = -4$ d) $x = 0.5, y = 4$
   e) $x = 5.5, y = 14.5$ f) $x = 2, y = 8$
3. a) $x = 4, y = 1$ b) $x = 1, y = 4$
   c) $x = 3, y = 1$ d) $x = 2.25, y = 6.5$
   e) $x = 4, y = 3$ f) $x = 5, y = 3$
4. a) $x = 2, y = -3$ b) $x = 7, y = 3$
   c) $x = 2.5, y = 1.5$ d) $x = 6, y = 3$
   e) $x = 0.5, y = -0.75$ f) $x = 1.5, y = 0.75$
5. a) $x = 5, y = 1$ b) $x = 3, y = 8$
   c) $x = 9, y = 1$ d) $x = 3, y = -2$
   e) $x = 2, y = \frac{1}{2}$ f) $x = -2, y = -3$
   g) $x = -\frac{1}{2}, y = -6\frac{1}{2}$ h) $x = 3\frac{1}{2}, y = 1\frac{1}{2}$
   i) $x = -2\frac{1}{2}, y = -3\frac{1}{2}$

6. (1, −2) is the solution to equations A and C; (−1, 3) is the solution to equations A and D; (2, 1) is the solution to B and C; (3, −3) is the solution to B and D.
7. Intersection points are (0, 6), (1, 3) and (2, 4). Area is 2 cm^2.
8. Intersection points are (0, 3), (6, 0) and (4, −1). Area is 6 cm^2.

11.2 Linear and non-linear equations

Exercise 11B

1. a) (5, −1) b) (4, 1) c) (8, −1)
2. a) (1, 2) and (−2, −1) b) (−4, 1) and (−2, 2)
3. a) (3, 4) and (4, 3) b) (0, 3) and (−3, 0)
   c) (3, 2) and (−2, 3)
4. a) (2, 5) and (−2, −3) b) (−1, −2) and (4, 3)
   c) (3, 3) and (1, −1)
5. a) (−3, −3), (1, 1) b) (3, −2), (−2, 3)
   c) (−2, −1), (1, 2) d) (2, −1), (3, 1)
   e) (−2, 1), (3, 6) f) (1, −4), (4, 2)
   g) (4, 5), (−5, −4)

11.3 Using graphs to solve simultaneous linear equations

Exercise 11C

1. a) $x = 2, y = 6$ b) $x = 6, y = 2$ c) $x = 4, y = 4$
   d) $x = 0, y = 4$
2. a) $x = 4, y = 8$ b) $x = 6, y = 4$ c) $x = 6, y = 0$
3. (4, 1)
4. (2, 3)
5. (3, 10)
6. 5, 5)
7. (1, 5)
8. (3, 16)
9. (−2, 6)
10. (−6, −9)
11. (1, −1)
12. (2, 6)
13. (2, 8)
14. $\left(7\frac{1}{2}, 3\frac{1}{2}\right)$
15. a) P and R b) R and S c) P and Q
   d) Q and S e) $\left(-\frac{2}{5}, -\frac{3}{5}\right)$
16. (0, 0), (−3, 3), (−3, −3), (−3, 2), (−2, 2), (2, 2)

11.4 Using graphs to solve simultaneous equations, one linear and one non-linear

Exercise 11D

1. a) (1, 1), (−3, −3) b) (6, 12), (−1, −2)
   c) (4, −3), (−3, 4) d) (0.8, 1.8), (−1.8, −0.8)

e) (4.6, 8.2), (0.5, 0) **f)** (3, 6), (−2, 1)

g) (4.8, 6.6), (0.2, −2.6) **h)** (2.6, 1.6), (−1.6, −2.6)

2. a) (1, 0)

b) Only one intersection point

c) $x^2 + x(3 - 5) + (-4 + 5) = 0 \Rightarrow x^2 - 2x + 1 = 0$

d) $(x - 1)^2 = 0 \Rightarrow x = 1$

e) Only one solution as line is a tangent to curve

3. a) There is no solution

b) The graphs do not intersect

c) $x^2 + x + 4 = 0$

d) $b^2 - 4ac = -15$

e) There is no solution as the discriminant is negative and a negative number has no square root.

11.5 Solving equations by the method of intersection

Exercise 11E

1. a) i −1.4, 4.4 **ii** −2, 5 **iii** −0.6, 3.6 **b)** 2.6, 0.4

2. a) −5, 1 **b) i** −5.3, 1.3 **ii** −4.8, 0.8 **iii** −3.4, −0.6

3. a) i 0, 6 **ii** 4.3, 0.7 **b) i** 4.8, 0.2 **ii** 5.4, −0.4

4. a) i −1.6, 2.6 **ii** 1.4, −1.4

b) i 2.3, −2.3 **ii** 2, −2

5. a) 0, 2 **b)** 2.5 **c)** −0.6, 1, 1.6

d) −2, 2.8 **e)** −0.8, 0.6, 2.2

6. a) −0.4, 4.4 **b)** −1, 5

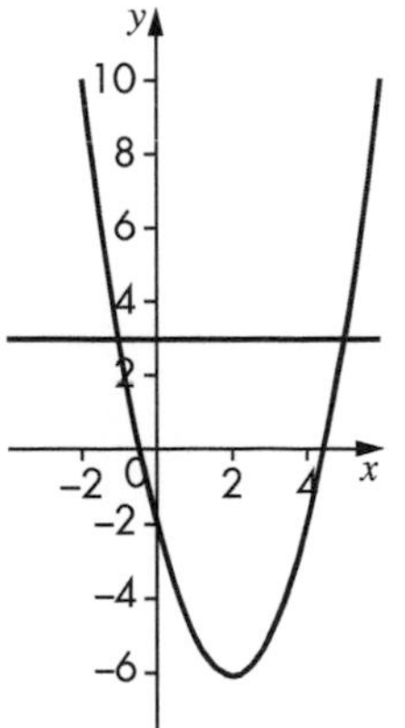

7. a) 1.6, −1.6 **b)** −1.2, 1.2

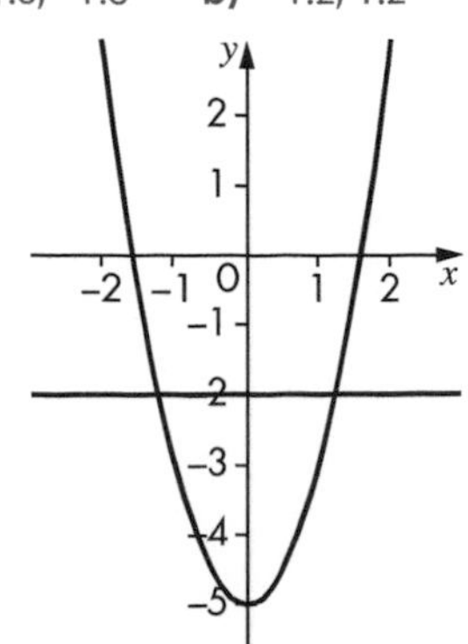

8. a) 2.2, −2.2 **b)** −1.8, 2.8

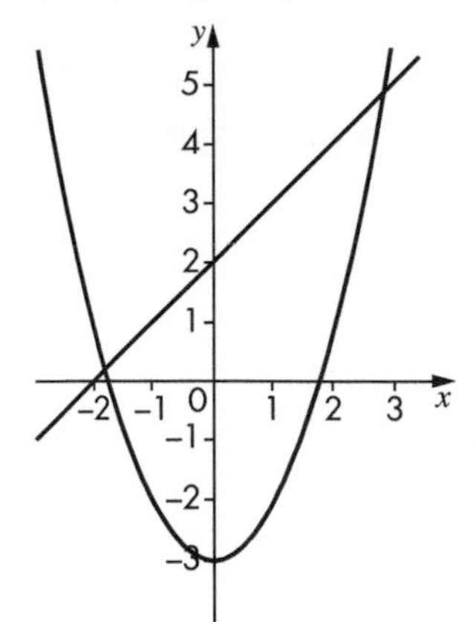

9. a) 3.3, −0.3 **b)** 4.8, 0.2

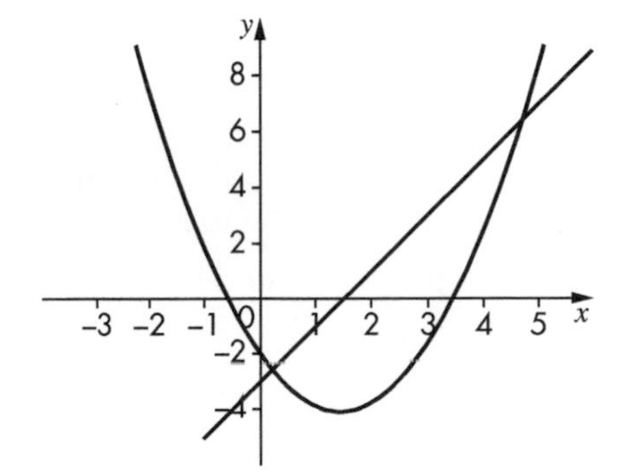

10. a) 2 **b)** 2.5

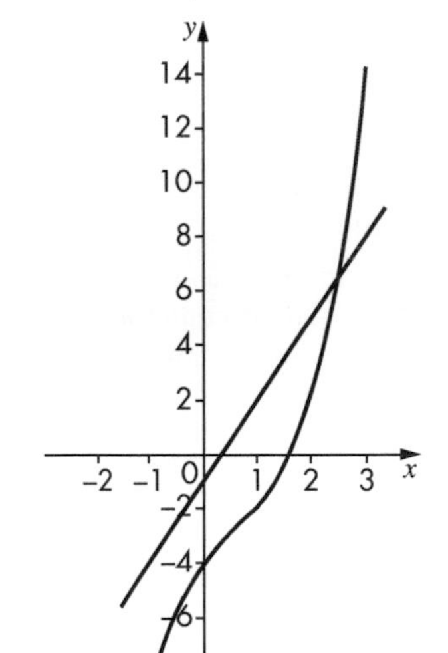

11. a) C and D **b)** A and D

c) $x^2 + 4x - 1 = 0$ **d)** (−1.5, −10.25)

12. a) a $(x + 2)(x - 1) = 0$ **b)** 5 − −2 = + 7, not − 7

c) $y = 2x + 7$

Exam-style questions

1. $x = 5, y = -2$
2. $x = 7, y = 1$
3. $x = 2, y = 5$
4. $x = 4, y = -3$
5. $x = 7, y = 3$
6. $x = 4, y = 2$
7. $x = 5$ and $y = 1$, or $x = -3$ and $y = -3$
8. (5, 0) and (–4, 3)
9. (4, 6) and (4, –6)
10. $x = 1.62$ and $y = 4.24$, or $x = -0.62$ and $y = -0.24$

CHAPTER 12 INEQUALITIES

12.1 Linear inequalities

Exercise 12A

1. **a)** $x < 3$ **b)** $t > 8$ **c)** $p \geqslant 10$ **d)** $x < 5$
   e) $y \leqslant 3$ **f)** $t > 5$ **g** $x < 6$ **h)** $y \leqslant 15$
   i) $t \geqslant 18$ **j)** $x < 7$ **k)** $x \leqslant 3$ **l)** $t \geqslant 5$
2. **a)** 8 **b)** 6 **c)** 16 **d)** 3 **e)** 7
3. **a)** 11 **b)** 16 **c)** 16
4. **a)** Because 3 + 4 = 7, which is less than the third side of length 8
   b $x + x + 2 > 10, 2x + 2 > 10, 2x > 8, x > 4$, so smallest value of x is 5
5. **a)** $x = 6$ and $x < 3$ scores –1 (nothing in common), $x < 3$ and $x > 0$ scores 1 (1 in common for example), $x > 0$ and $x = 2$ scores 1 (2 in common), $x = 2$ and $x \geqslant 4$ scores –1 (nothing in common), so the score is –1 + 1 + 1 – 1 = 0
   b) $x > 0$ and $x = 6$ scores +1 (6 in common), $x = 6$ and $x \geqslant 4$ scores +1 (6 in common), $x \geqslant 4$ and $x = 2$ scores –1 (nothing in common), $x = 2$ and $x < 3$ scores +1 (2 in common), +1 + 1 – 1 + 1 = 2
   c Any acceptable combination, e.g. $x = 2, x < 3, x > 0, x \geqslant 4, x = 6$
6. **a)** $x \geqslant -6$ **b)** $t \leqslant \frac{8}{3}$ **c)** $y \leqslant 4$ **d)** $x \geqslant -2$
   e) $w \leqslant 5.5$ **f)** $x \leqslant \frac{14}{5}$
7. **a)** $x \leqslant 2$ **b)** $x > 38$ **c)** $x < 6\frac{1}{2}$ **d)** $x \geqslant 4.5$
   e) $t > 15$ **f)** $y \leqslant \frac{7}{5}$
8. **a)** 4 **b)** 99 **c)** 11 **d)** 11 **e)** 6
9. **a)** 0, 10, –10 **b)** $x < 16$
10. **a)** $x < 9$ **b)** $x \geqslant 11$ **c)** $x \geqslant 3$
11. **a)** $x \geqslant 7.5$ **b)** $x \leqslant -2$ **c)** $x < 6$ **d)** $x > 1.5$
    e) $x \geqslant -5$ **f)** $x < 0.5$
12. **a)** $3 < x < 6$ **b)** $2 < x < 5$ **c)** $x < 6$ **d)** $1 \leqslant x < 4$
    e) $2 \leqslant x < 4$ **f)** $0 \leqslant x \leqslant 5$

12.2 Quadratic inequalities

Exercise 12B

1. **a)** $-4 \leqslant x \leqslant 4$

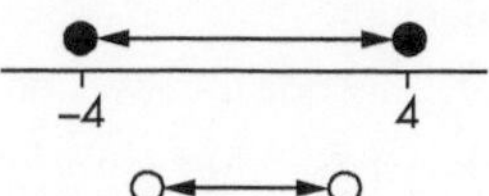

   b) $-2 < x < 2$

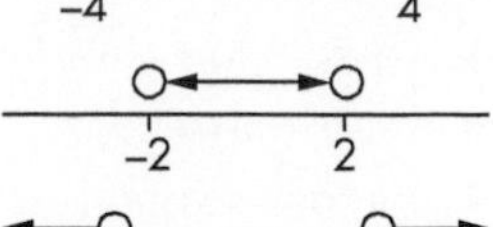

   c) $x < -2.5$ or $x > 2.5$

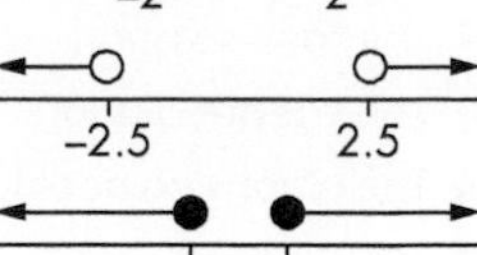

   d $x \leqslant -1$ or $x \geqslant 1$

2. **a)** $-3 < x < 3$
   b) $x < -5$ or $x > 5$
   c) $x \leqslant -1.5$ or $x \geqslant 1.5$
   d $-0.5 \leqslant x \leqslant 0.5$
3. **a)** $-2 \leqslant x \leqslant 2$ (not $-2 \leqslant x \leqslant 4$)
   b) $x < -3.5$ or $x > 3.5$
   c) $-2.5 < x < 2.5$
   d $-3 \leqslant x \leqslant 3$
4. **a)** $-3 \leqslant x \leqslant 2$ **b)** $x > \frac{2}{3}$ or $x < -3$
   c) $-1 \leqslant x \leqslant \frac{5}{9}$ **d** $x \geqslant \frac{3}{5}$ or $x \leqslant -5$ **e** $-\frac{1}{3} \leqslant x \leqslant 4$
5. 2, 3, 4
6. 1, 2

12.3 GRAPHICAL INEQUALITIES

Exercise 12C

1. **a) & b)**

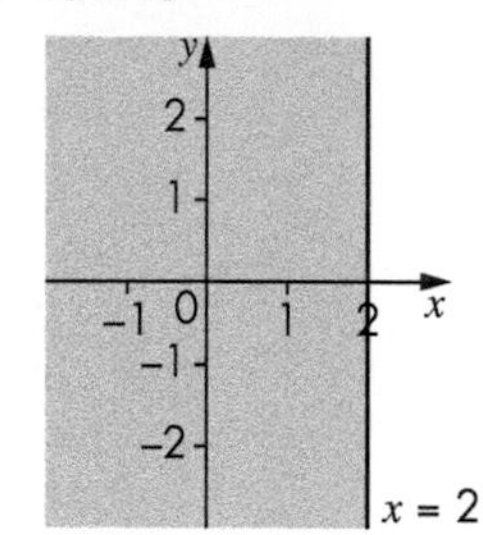

2. **a) & b)**

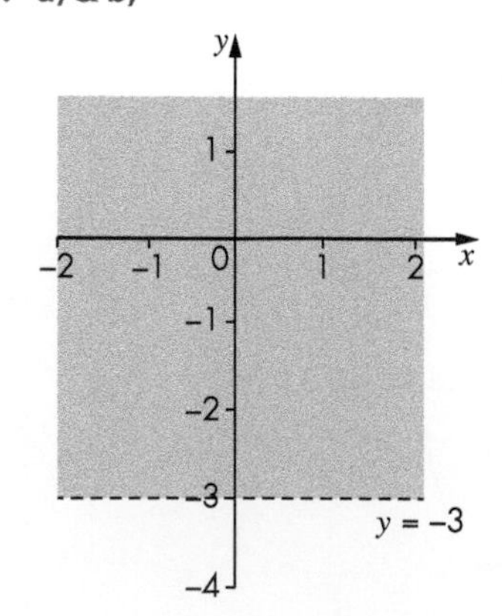

3. a)–c)

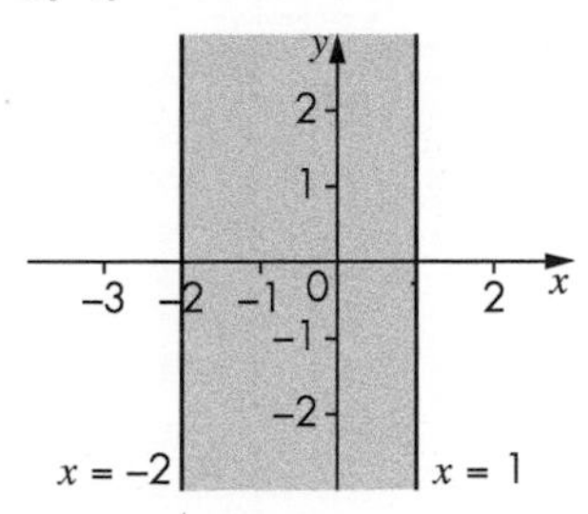

4. a)–c)

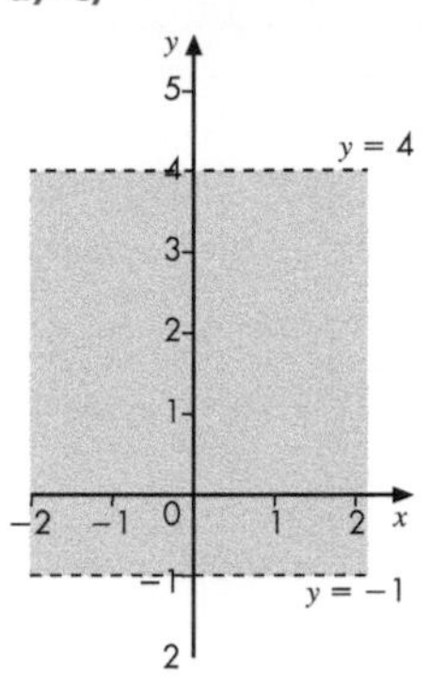

5.

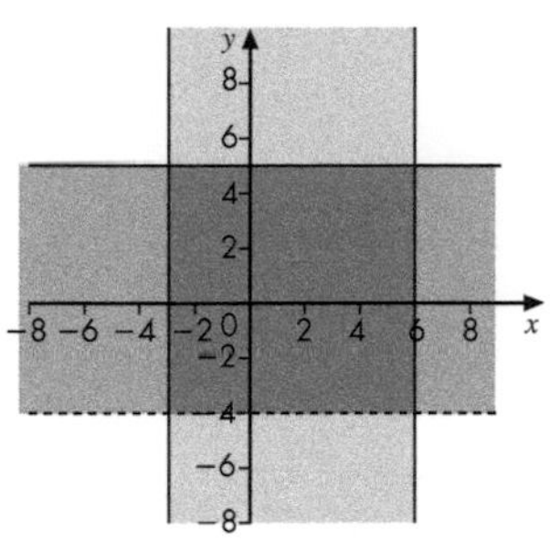

b) i Yes **ii** Yes **iii** No

6. a) & b)

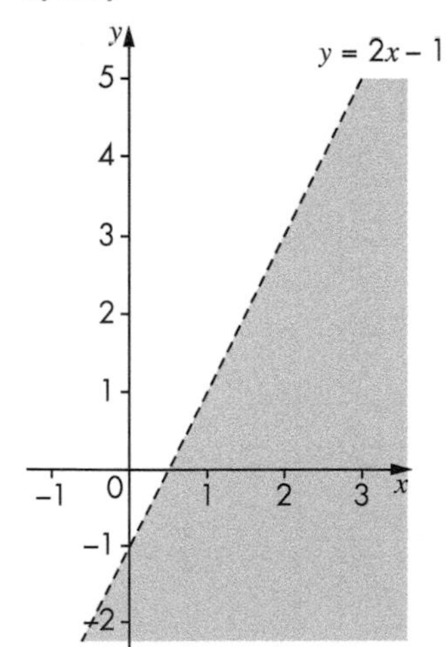

7. a) & b)

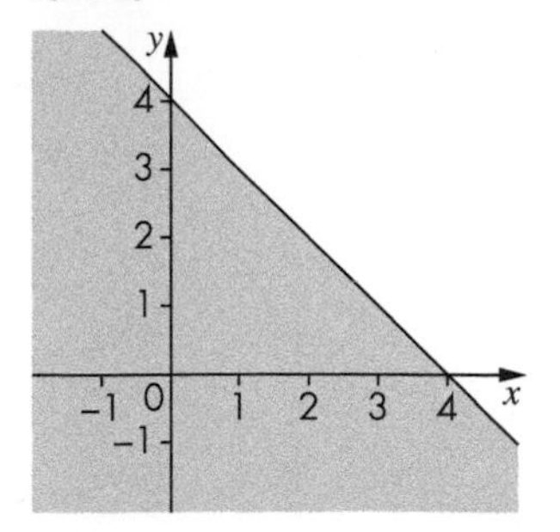

8. a) & b)

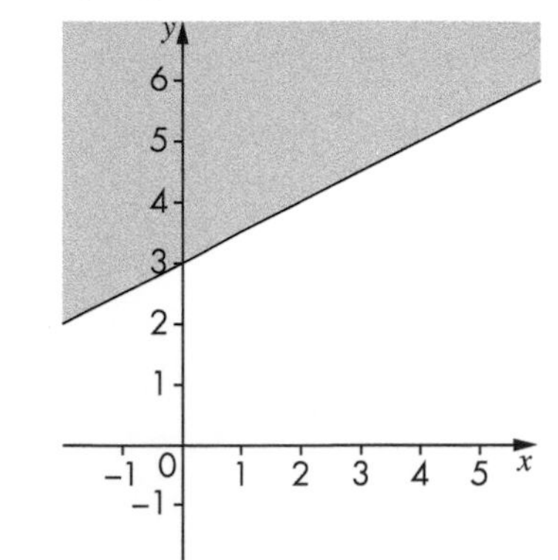

9.

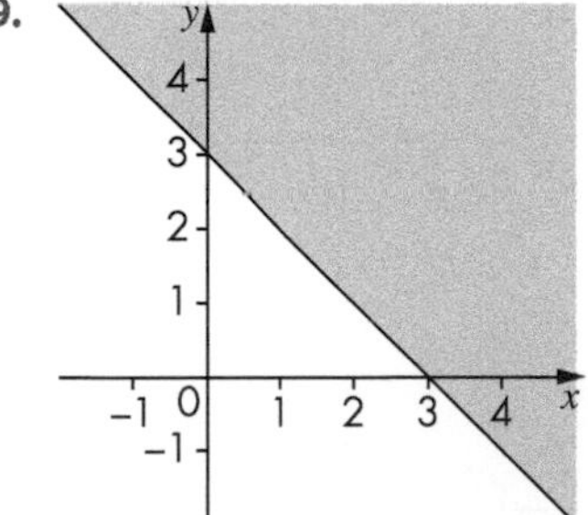

Exercise 12D

1.

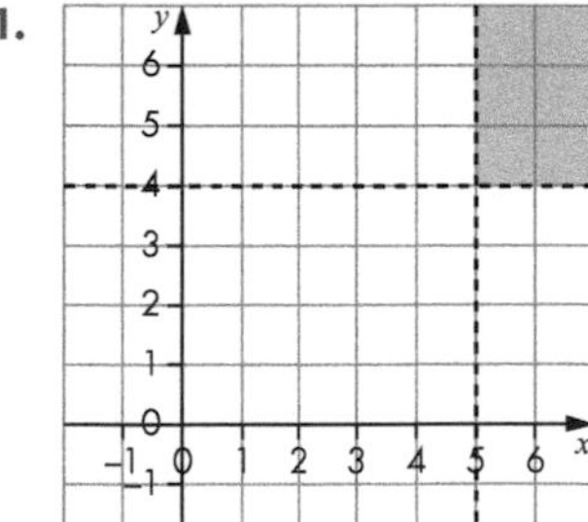

2.

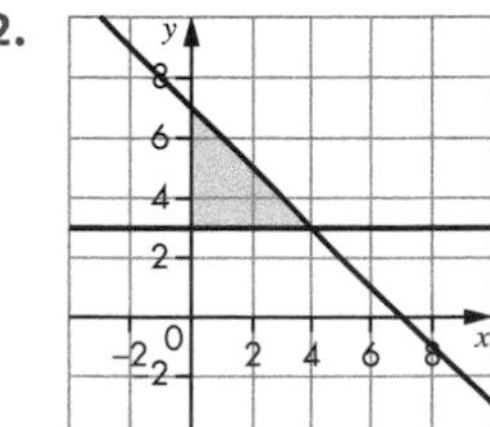

3.

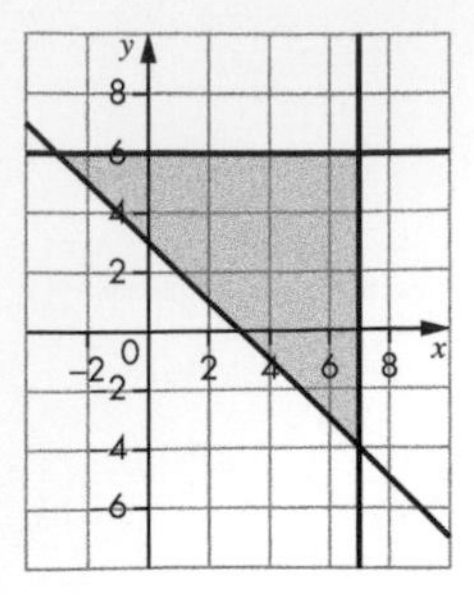

4.

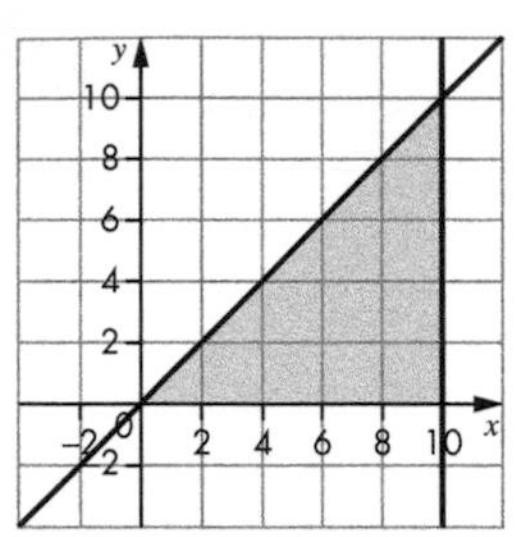

5.

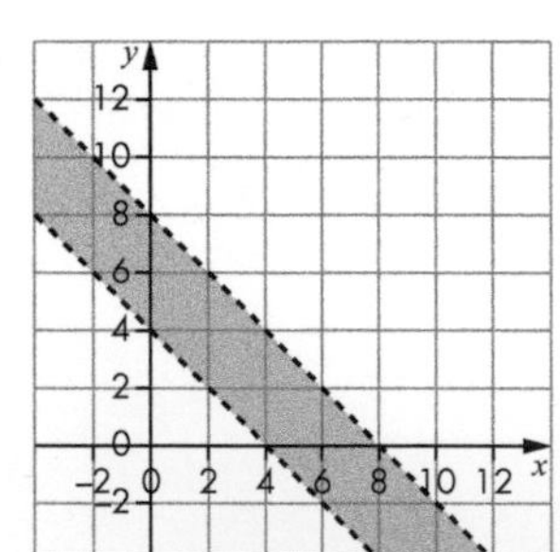

6.

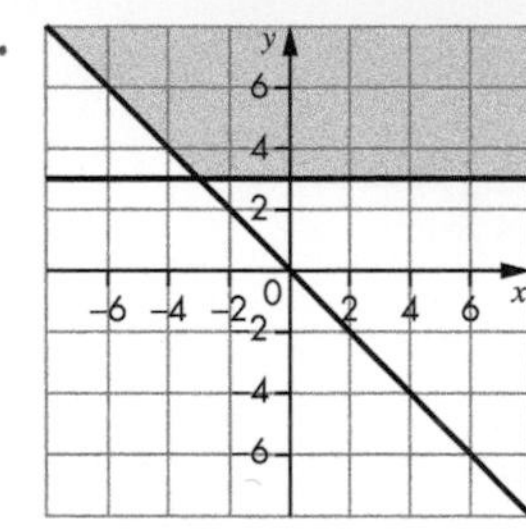

7. $x \geqslant -2$ and $y \geqslant 3$

8. $x + y \geqslant 5$ and $x \geqslant 2$ and $y \geqslant 1$

9.

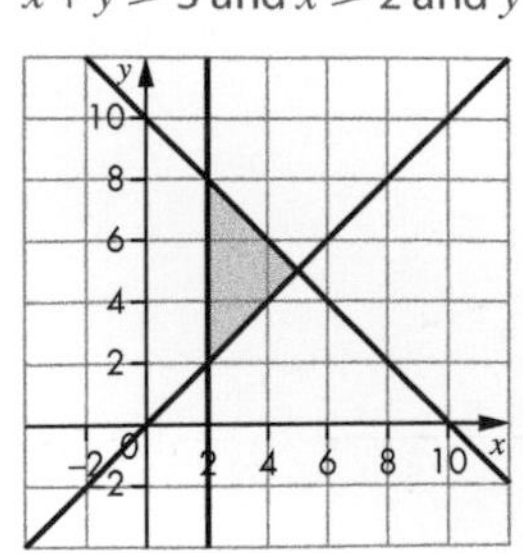

10.

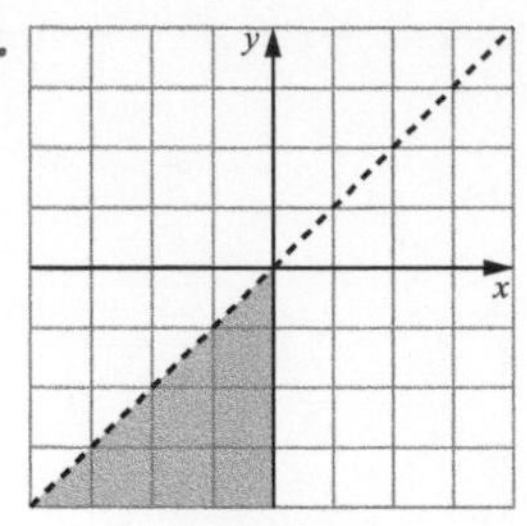

Exercise 12E

1. **a)-d)**

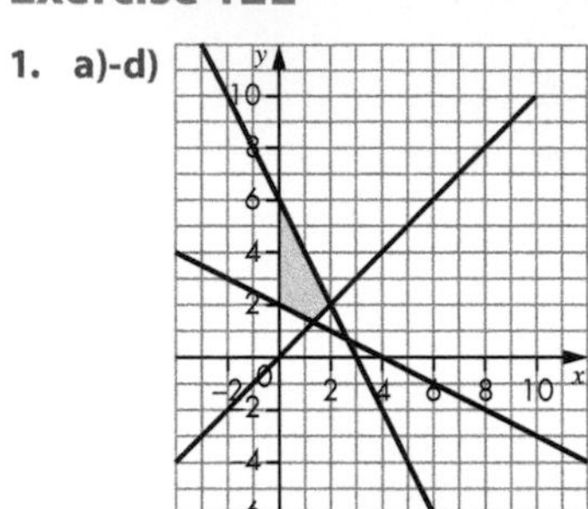

e) i No **ii** Yes **iii** Yes

2. **a) i-iii**

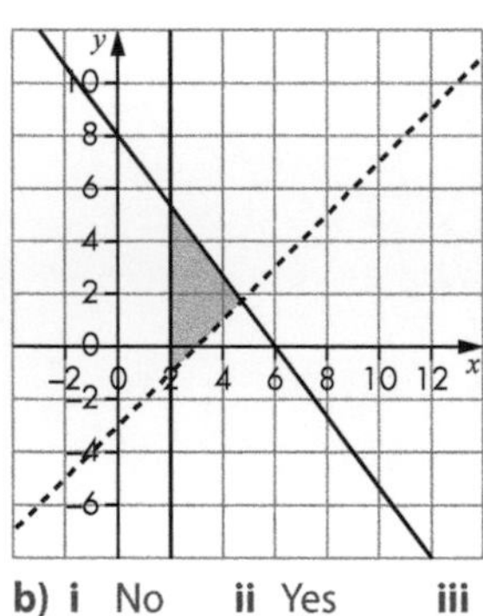

b) i No **ii** Yes **iii** Yes **iv** No

3. **a) & b)**

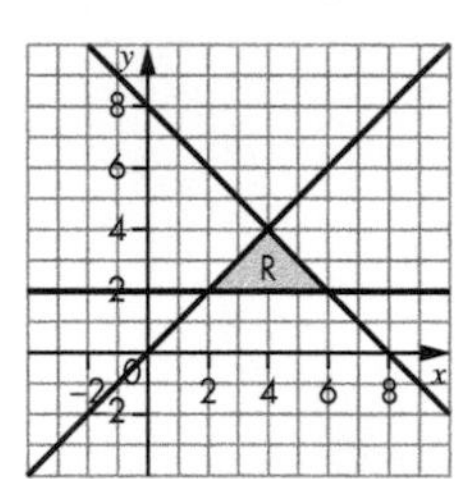

4. **a) & b)**

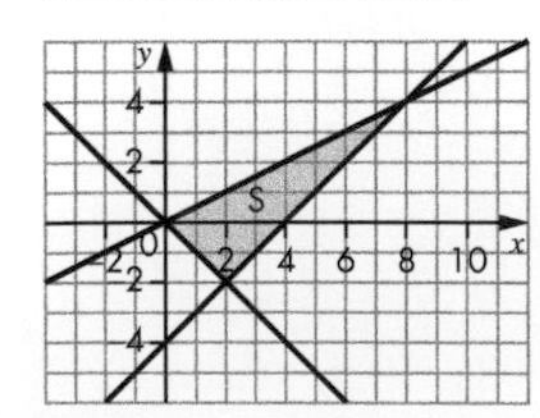

c) 4 **d)** −2 **e)** 12

5. Test a point such as the origin (0, 0), so $0 < 0 + 2$, which is true. So the side that includes the origin is the required side.

6. a) $x + y \geqslant 3$, $y \leqslant \frac{1}{2}x + 3$ and $y \geqslant 5x - 15$
   b) 9   c) 3 at (3,0)

7. a) & b)

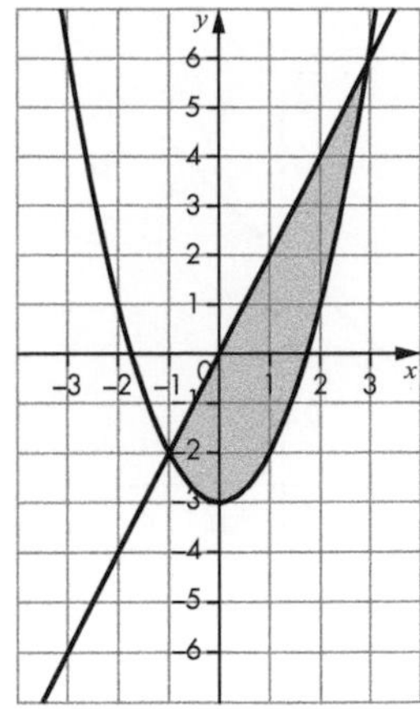

Exam-style questions

1. $x \geqslant 4.4$
2. $x < 2$
3. $x < -5$ or $x > -1$
4. −1, 0, 1, 2, 3
5. a)

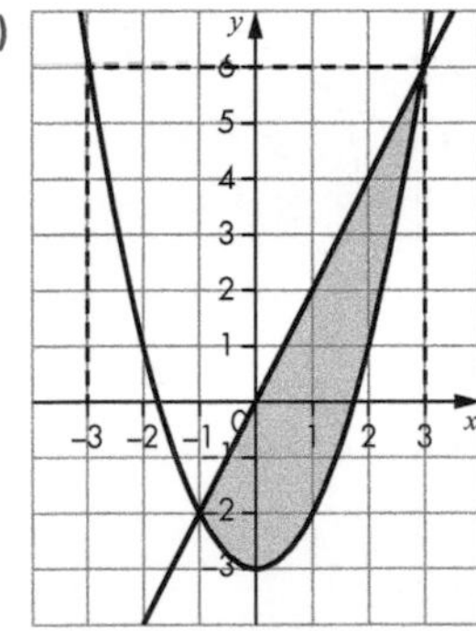

   b) $-1 \leqslant x \leqslant 3$
6. a) $x(3x + 4) \geqslant 160$
   b) $x \leqslant -8$ or $x \geqslant 6\frac{2}{3}$; AD is at least $6\frac{2}{3}$ cm
7. a) $\frac{1}{2}N^2 \sin 60° \geqslant 100$ b) 16

CHAPTER 13 COORDINATE GEOMETRY

13.1 Cartesian grids and straight-line graphs

Exercise 13A

1. a) i $y = 2$ ii $x = -2$ iii $y = -3$ b) B c) C
2. a) (1, 2) b) i (−2, −1) ii (0, −2) iii (3, 1) c) R
3. a) $\left(3\frac{1}{2}, -\frac{1}{2}\right)$ and $\left(-2\frac{1}{2}, -\frac{1}{2}\right)$ b) $y = -\frac{1}{2}$

Exercise 13B

1. Extreme points are (0, −3), (10, 2)
2. Extreme points are (−6, 2), (6, 6)
3. a) Extreme points are (0, −2), (5, 13) and (0, 1), (5, 11)
   b) (3, 7)
4. a) Extreme points are (0, −5), (5, 15) and (0, 3), (5, 13)
   b) (4, 11)
5. a) Extreme points are (0, −1), (12, 3) and (0, −2), (12, 4)
   b) (6, 1)
6. a) Extreme points are (0, 1), (4, 13) and (0, −2), (4, 10)
   b) Do not cross because they are parallel

Exercise 13C

1. a) At $x = 6$ b) At $y = 3$
   c)

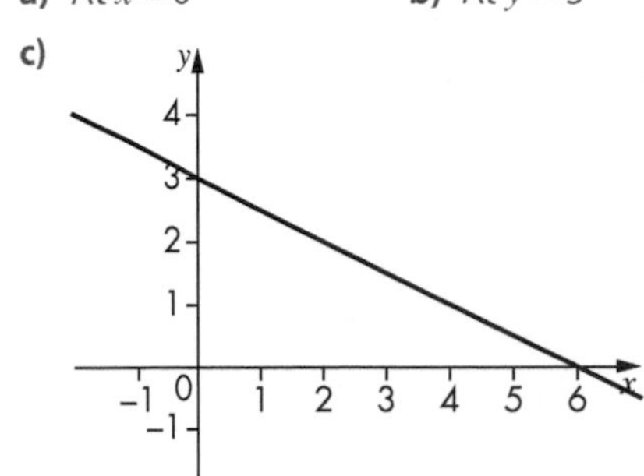

2.

3.

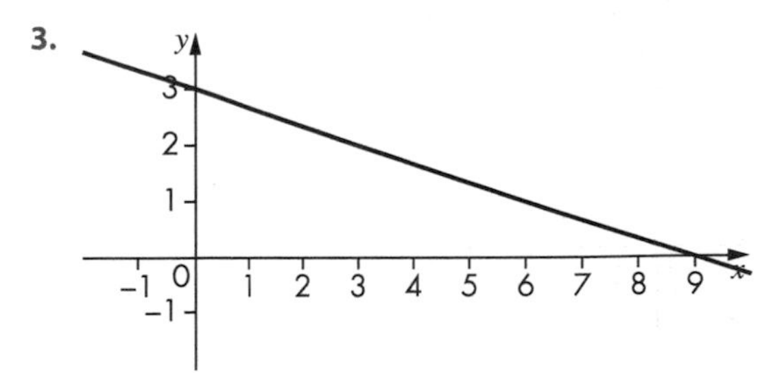

4.

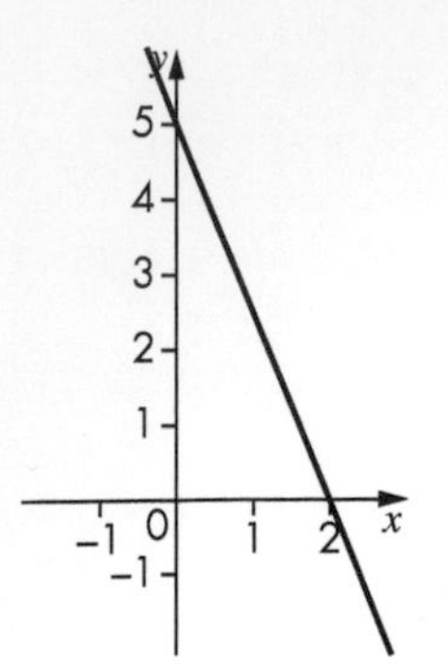

5.

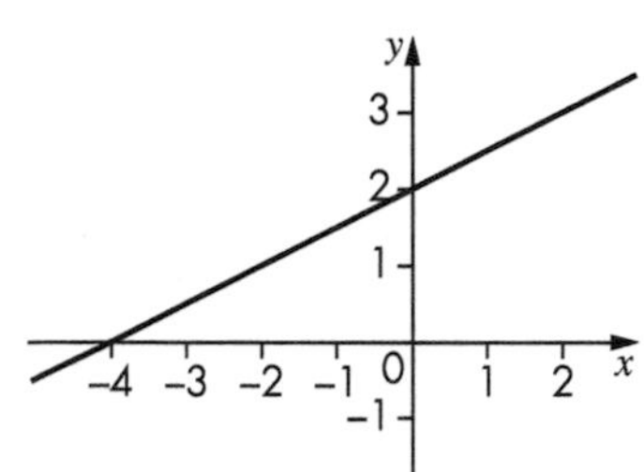

Exercise 13D

1. **a)** $\frac{1}{3}$ **b)** -3 **c)** $-\frac{1}{4}$

2.

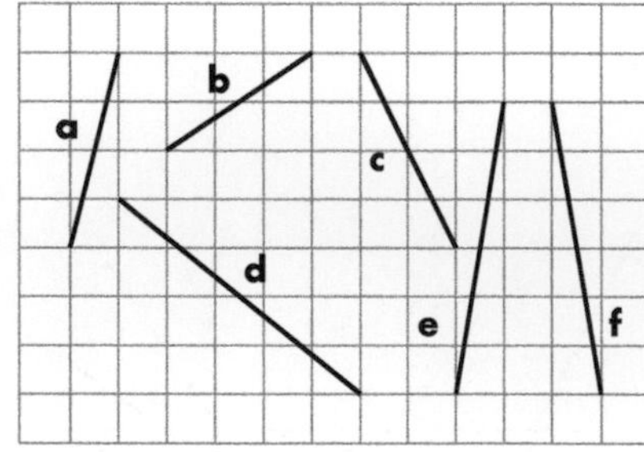

3. **a)** 1 **b)** -1
   They are symmetrical about the axes. One is the mirror image of the other.
   They are perpendicular to each other.

4. **a)** 0.5 **b)** 0.4 **c)** 0.2 **d)** 0.1 **e)** 0

5. **a)** $1\frac{2}{3}$ **b)** 2 **c)** $3\frac{1}{3}$ **d)** 10 **e)** $\propto$

6. **a)** $300 \div 3380 \approx 0.1$ (1 dp) equivalent to $300 \div 0.6$ feet per mile = 500 feet per mile
   b) i Because the gradient of the curve is steepest here.
   ii $-520 \div 3696 \approx -0.2$ (1 dp) equivalent to $-520 \div 0.7$ feet per mile = -743 feet per mile
   c) MB

7. **a)** 4 **b)** $\frac{1}{4}$ **c)** $2\frac{1}{2}$ **d)** 10 **e)** -2
   f) $-\frac{1}{5}$ **g)** 0 **h)** $-1\frac{1}{2}$

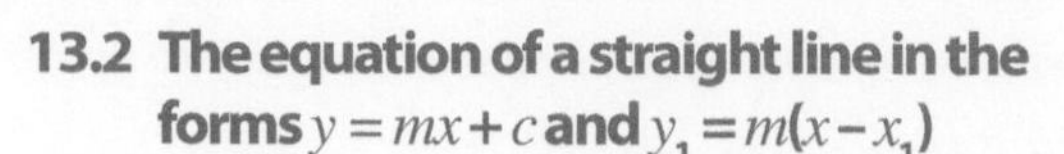

13.2 The equation of a straight line in the forms $y = mx + c$ and $y_1 = m(x - x_1)$

Exercise 13E

1.

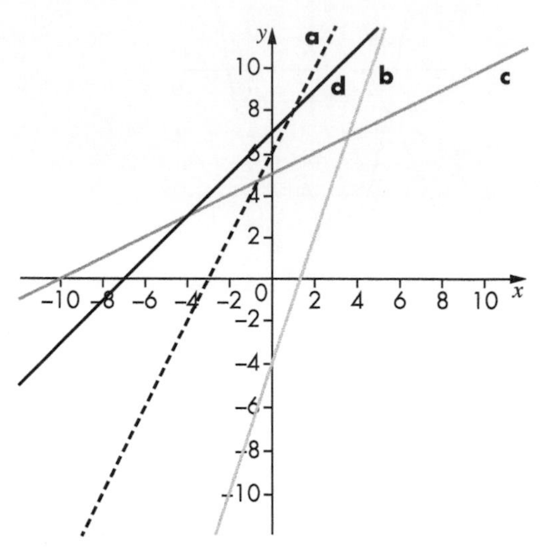

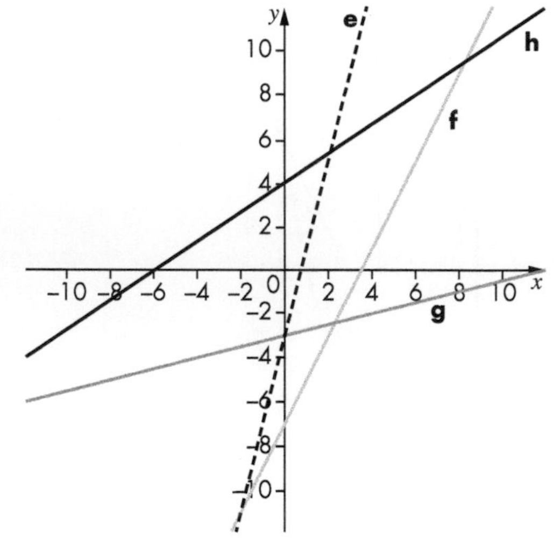

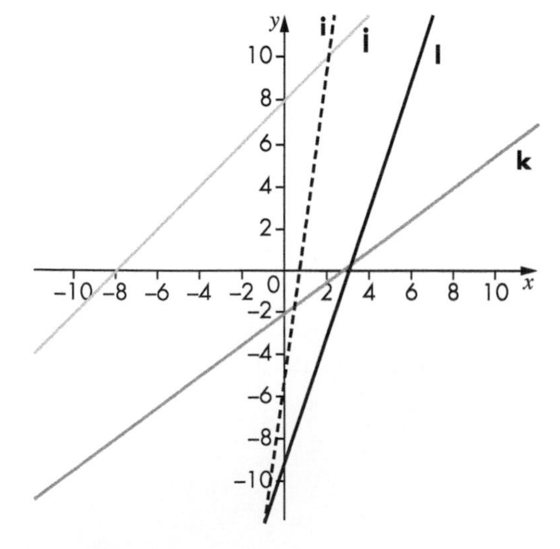

2. a)

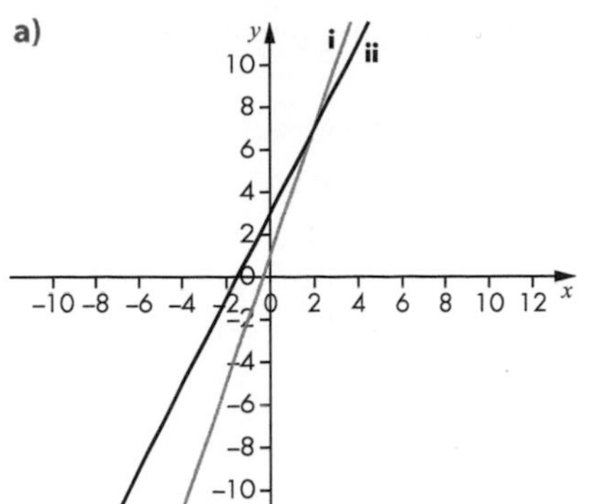

b) (2, 7)

3. a)

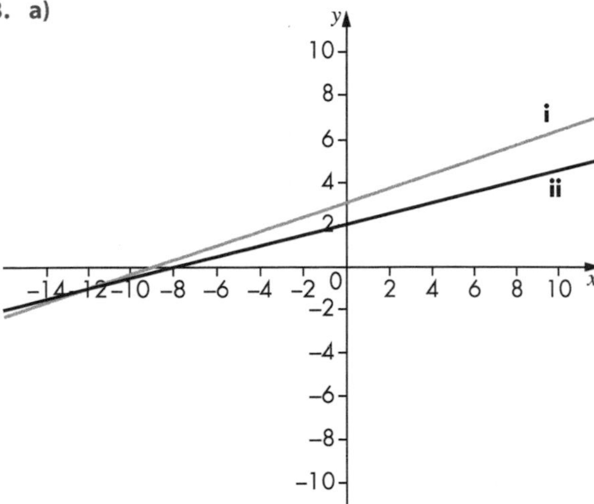

b) (−12, −1)

4. a)

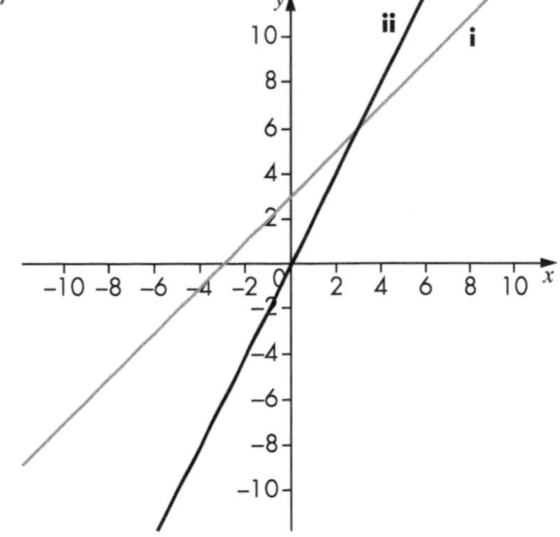

b) (3, 6)

5. a) They have the same gradient (3).

b) They intercept the y-axis at the same point (0, −2).

c) (−1, −4)

6. a) −2 **b)** $\frac{1}{2}$ **c)** 90° (They are perpendicular.)

d) Their product is −1, one is the negative reciprocal of the other.

e) $-\frac{1}{3}$

Exercise 13F

1.

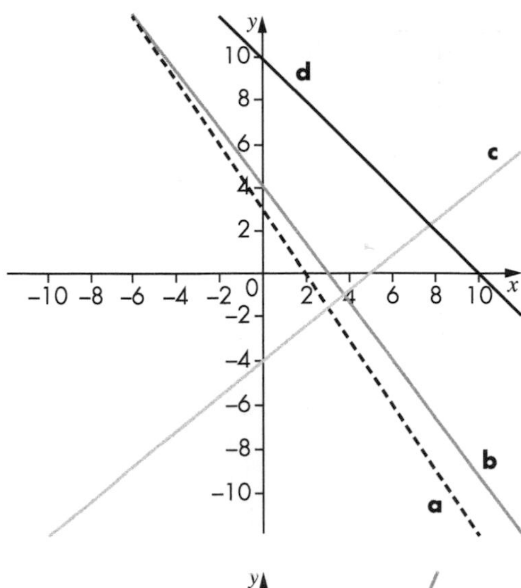

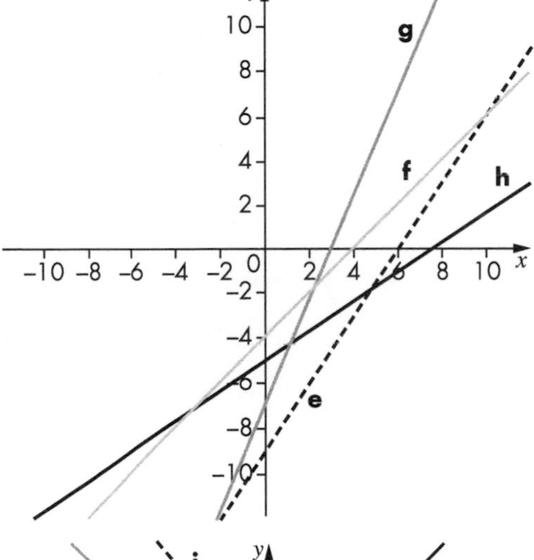

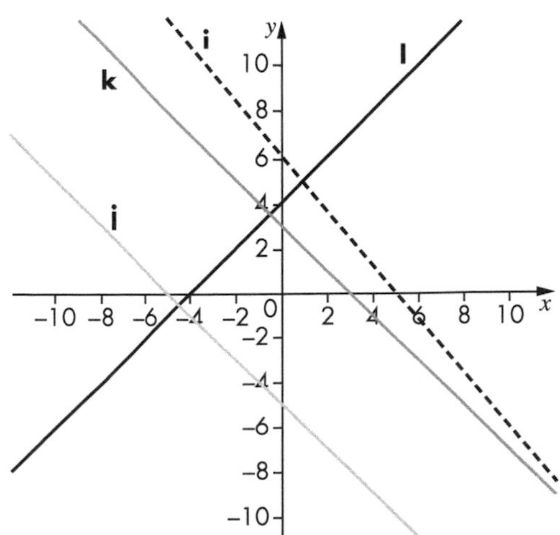

2. a)

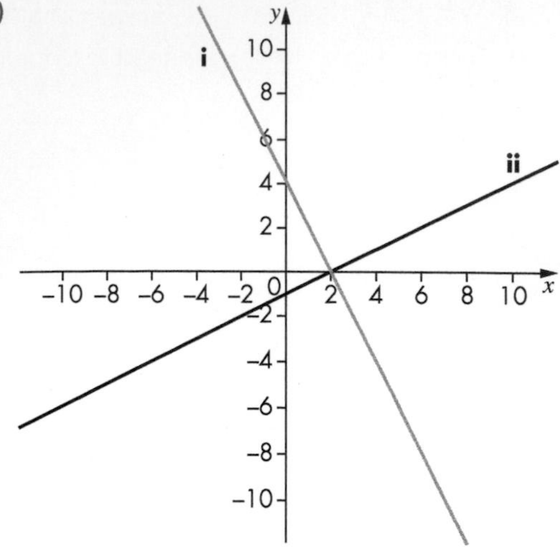

b) (2, 0)

3. a)

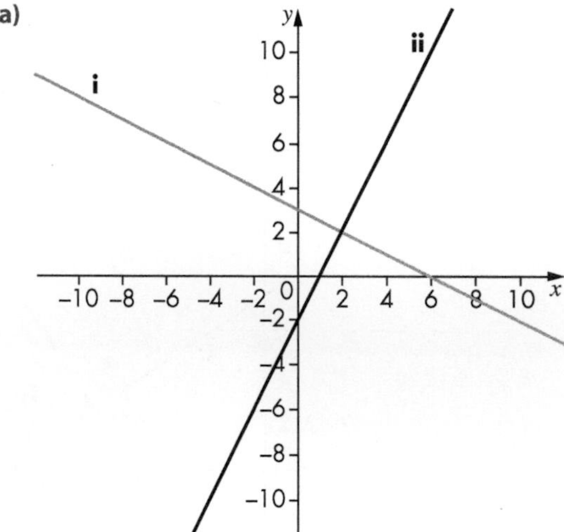

b) (2, 2)

4. a)

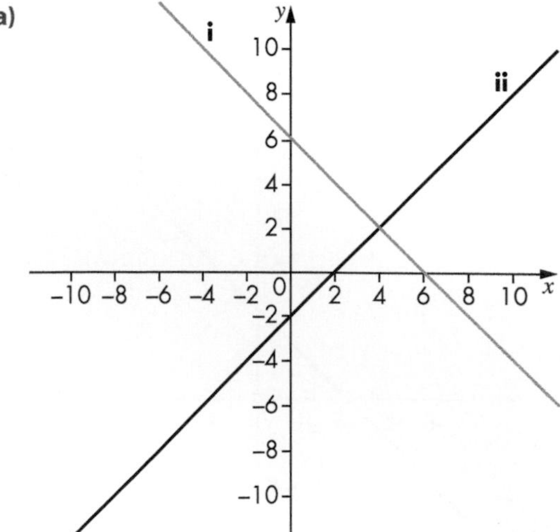

b) (4, 2)

5. a) They both have a y-intercept of 6; they intersect at (6, 0).

b) They both pass through (intersect at) the point (0, −3)

c) They have the same gradient, they are parallel

d) $-2x + 9y = 18$

6. a) **i** $x = 3$ **ii** $x - y = 4$ **iii** $y = -3$ **iv** $x + y = -4$ **v** $x = -3$ **vi** $y = x + 4$

b) **i** −3 **ii** $\frac{1}{3}$ **iii** $-\frac{1}{3}$

Exercise 13G

1. a) $y = x + 1$ **b)** $y = 2x - 3$ **c)** $y = \frac{4}{3}x - 2$ or $3y = 4x - 6$

2. a) i $y = 2x + 1, y = -2x + 1$

ii Reflection in y-axis and $y = 1$; rotation 180° about (0, 1)

iii Different signs

b) i $5y = 2x - 5, 5y = -2x - 5$

ii Reflection in y-axis and $y = -1$; rotation 180° about (0, −1)

iii Different signs

c) i $y = x + 1, y = -x + 1$

ii Reflection in y-axis and $y = 1$; rotation 180° about (0, 1)

iii Different signs

3. a) The x-coordinates go from $2 \rightarrow 1 \rightarrow 0$ and y-coordinates go from $5 \rightarrow 3 \rightarrow 1$.

b) The x-step between the points is 1 and the y-step is 2.

c) $y = 3x + 2$

4. a) $y = -2x + 1$ **b)** $2y = -x$ **c)** $y = -x + 1$ **d)** $5y = -2x$ **e)** $y = -\frac{3}{2}x - 3$ or $2y = -3x - 6$

5. a) i $2y = -x + 1, y = -2x + 1$

ii Reflection in $x = y$

iii Gradients are reciprocals of each other

b) i $2y = 5x + 5, 5y = 2x - 5$

ii Reflection in $x = y$

iii Gradients are reciprocals of each other

c) i $y = 2, x = 2$

ii Reflection in $x = y$

iii Gradients are reciprocals of each other (reciprocal of zero is infinity)

6. a) 3 **b)** $\frac{1}{2}$ **c)** 4 **d)** −1 **e)** $-\frac{1}{2}$ **f)** $\frac{2}{3}$

7. a) $y = 2x - 3$ **b)** $y = \frac{1}{2}x + 4$ **c)** $y = 4x - 2$ **d)** $y = -3x + 8$

8. a) (5, 3) **b)** (4, 5) **c)** (3, 2) **d)** (3, 3) **e)** (1, 3.5) **f)** (−0.5, 0)

9. a)

b) $y = 0.5x + 6.5$ c) (–1, 3)
d) $y = -x + 8$

10. a) $y = \frac{3}{4}x + \frac{1}{2}$ b) $y = \frac{1}{2}x + 3\frac{1}{2}$
c) $y = -2x + 7$ d) $y = x + 2$

13.4 Parallel and perpendicular lines

Exercise 13H

1. a) Line A does not pass through (0, 1).
b) Line C is perpendicular to the other two.
c) (i)

2. a) $-\frac{1}{2}$ b) $\frac{1}{3}$ c) $-\frac{1}{5}$ d) 1
e) –2 f) –4 g) 3 h) $\frac{3}{2}$
i) $-\frac{2}{3}$ j) $-\frac{1}{10}$ k) $\frac{1}{6}$ l) $-\frac{3}{4}$

3. a) $y = -\frac{1}{2}x - 1$ b) $y = \frac{1}{3}x + 1$
c) $y = -x + 2$ d) $y = x + 2$
e) $y = -2x + 3$ f) $y = -4x - 3$
g) $y = 3x$ h) $y = 1.5x - 5$

4. a) $y = 4x + 1$ b) $y = \frac{1}{2}x - 2$ c) $y = -x + 3$

5. a) $y = -\ x - 1$ b) $y = 3x + 5$ c) $y = -x + 1$

6. a) $y = -x + 14$ b) $y = x + 2$

7. $y = 2x + 6$

8. $y = -\frac{1}{4}x + 2$

9. a) (0, –20) b) $y = -\frac{1}{5}x + 6$ c) $y = 5x - 20$

10. $y = -\frac{1}{2}x + 5$

11. a) $y = 3x - 6$
b) Bisector of AB is $y = -2x + 9$, bisector of AC is $y = \frac{1}{2}x + \frac{3}{2}$, solving these equations shows the lines intersect at (3, 3).
c) (3, 3) lies on $y = 3x - 6$ because $(3\times3) - 6 = 3$

12. $y = 2x + \frac{9}{2}$

13.4 Applications of coordinate geometry

Exercise 13I

1. a) 5 units b) 13 units c) 10 units d) 15 units
e) 17 units f) 25 units

2. a) $3\sqrt{2}$ units b) $2\sqrt{5}$ units c) $2\sqrt{13}$ units
d) $5\sqrt{2}$ units e) $4\sqrt{5}$ units f) $2\sqrt{17}$ units

3. a) (2, 3) b) (3.5, 2) c) (–3, 3) d) (5, 2)
e) (0, 0) f) (2, 3)

4. C(7, –12)

5. F(–11, 11)

6. C(0, 15)

7. Z(15, –5)

8. Q(–8, 15)

9. Gradient of AB = gradient DC = $\frac{1}{3}$. Gradient BC = gradient AD = $\frac{3}{2}$.
So ABCD is a parallelogram as AB ≠ BC.

10. 19.5 square units

Exam-style questions

1. 0.08 or $\frac{2}{25}$

2. (0, –20)

3. $x + 2y = -1$

4. $3y = 20 - 2x$

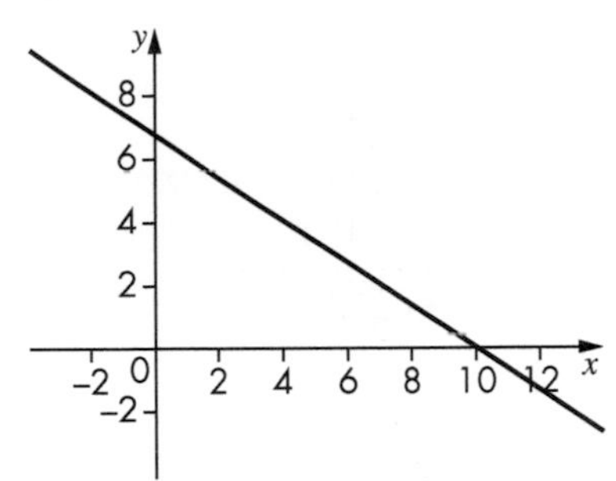

5. $3y = x - 10$

6. AB and DC both have gradient –4. DA and CB both have gradient $\frac{1}{4}$ so they are parallel and they are perpendicular to the other two sides because $-4 \times \frac{1}{4} = -1$.

7. (8.25, 5.5)

8. a) CA = CB = $\sqrt{40}$ b) $x + y = 3$
c) 16 square units

CHAPTER 14 THE EQUATION OF A CIRCLE

14.1 The equation of a circle centred on the origin

Exercise 14A

1. a) $x^2+y^2=49$ b) $x^2+y^2=81$ c) $x^2+y^2=36$
d) $x^2+y^2=2$ e) $x^2+y^2=5$ f) $x^2+y^2=12$

2. a) 10 b) 12 c) 15 d) $\sqrt{5}$
e) $\sqrt{28}$ or $2\sqrt{7}$ f) $3\sqrt{5}$

3. a) (4, 2) and (–2, –4) b) $\sqrt{72}$ or $6\sqrt{2}$

4. 10

5. a) (4, –3), (–3, 4) b) (0.8, 1.8), (–1.8, –0.8)
c) (2.6, 1.6), (–1.6, –2.6)

14.2 The equation of a circle centred on any point (a, b)

Exercise 14B

1. a) $(x-1)^2+(y-2)^2=9$ b) $(x+1)^2+(y-3)^2=16$
   c) $(x-4)^2+(y+2)^2=25$ d) $(x-3)^2+(y-3)^2=5$
   e) $(x-8)^2+(y+3)^2=18$ f) $(x-4)^2+(y+4)^2=12$
2. a) Centre (1, 4), radius 5 units
   b) Centre (–2, 3), radius 3 units
   c) Centre (5, –5), radius 6 units
   d) Centre (1, –3), radius $\sqrt{6}$ units
   e) Centre (2, –3), radius $\sqrt{20}$ units or $2\sqrt{5}$ units
   f) Centre (–8, 1), radius $\sqrt{24}$ units or $2\sqrt{6}$ units
3. $(x-3)^2+(y+1)^2=13$
4. a) (5, –1) and (4, 0) b) AB = $\sqrt{2}$ units
5. PQ = $9\sqrt{2}$ units
6. $\begin{pmatrix} 4 \\ -7 \end{pmatrix}$
7. $(x+1)^2+(y-2)^2=25$

Exam-style questions

1. a) (0, 0) b) 5 c) At (4, 3) and (–4, –3)
2. a) $x^2+y^2=52$
   b) It is inside. The distance from the origin to (5, 5) is $\sqrt{50}$ and this is less than the radius of $\sqrt{52}$ or $2\sqrt{13}$.
3. 9
4. a) $(x-3)^2+y^2=9$ b) $y=0.5x-3$
   c) (1.2, –2.4)
5. a) $(x-8)^2+(y+7)^2=50$
   b) The radius is $\sqrt{50}$ which is between 7 and 8. The centre is 8 units from the x-axis so the circle does not cross that axis. The centre is 7 units from the y-axis so the circle does cross that axis.
6. $(x-3)^2+(y-5)^2=34$

CHAPTER 15 INDICES

15.1 Using indices

Exercise 15A

1. a) 2^4 b) 3^5 c) 7^2 d) 5^3 e) 10^7
   f) 6^4 g) 4^1 h) 1^7 i) 0.5^4 j) 100^3
2. a) $3\times3\times3\times3$
   b) $9\times9\times9$
   c) 6×6
   d) $10\times10\times10\times10\times10$
   e) $2\times2\times2\times2\times2\times2\times2\times2\times2\times2$
   f) 8
   g) $0.1\times0.1\times0.1$
   h) 2.5×2.5
   i) $0.7\times0.7\times0.7$
   j) 1000×1000
3. a) 16 b) 243 c) 49 d) 125
   e) 10 000 000 f) 1296 g) 4 h) 1
   i) 0.0625 j) 1 000 000
4. a) 81 b) 729 c) 36 d) 100 000 e) 1024
   f) 8 g) 0.001 h) 6.25 i) 0.343 j) 1 000 000
5. 125 m^3
6. b) 10^2 c) 2^3 d) 5^2
7. a) 1 b) 4 c) 1 d) 1 e) 1
8. Any power of 1 is equal to 1.
9. 10^6
10. a) $2x$ b) xy c) $\frac{x}{y}$ d) x^5 e) x^2y^2 f) $\frac{x^2}{y^2}$

15.2 Rules of indices

Exercise 15B

1. a) 5^4 b) 5^3 c) 5^6 d) 5^9 e) 5^5
   f) 6^3 g) 6^3 h) 6^2 i) 6^3 j) 6^1
2. a) a^3 b) a^5 c) a^7 d) a^4 e) a^2 f) a^1
3. a) Any two values such that $x+y=10$
   b) Any two values such that $x-y=10$
4. a) 4^6 b) 4^{15} c) 4^6
   d) 4^6 e) 4^8 f) $4^0=1$
5. a) $\frac{1}{a}$ b) $\frac{1}{a^2}$ c) a^3 d) $\frac{1}{a^3}$ e a
6. a) 7 b) 4
7. a) a^4 b) a^8 c) a^3 d) a^5
8. a) $3a$ b) $4a^3$ c) $3a^4$
   d) $\frac{6}{a}$ e) $4a^3$ f) $\frac{5}{a^4}$
9. 12 ($a=2, b=1, c=3$)

15.3 Negative indices

Exercise 15C

1. a) $\frac{1}{5^3}$ b) $\frac{1}{6}$ c) $\frac{1}{10^5}$ d) $\frac{1}{3^2}$ e) $\frac{1}{8^2}$
   f) $\frac{1}{9}$ g) $\frac{1}{w^2}$ h) $\frac{1}{t}$ i) $\frac{1}{x^m}$ j) $\frac{4}{m^3}$
2. a) 3^{-2} b) 5^{-1} c) 10^{-3} d) m^{-1} e) t^{-n}
3. a) i 2^4 ii 2^{-1} iii 2^{-4} iv -2^3
   b) i 10^3 ii 10^{-1} iii 10^{-2} iv 10^6
   c) i 5^3 ii 5^{-1} iii 5^{-2} iv 5^0
   d) i 3^2 ii 3^{-3} iii 3^0 iv -3^5
4. a) $\frac{5}{x^3}$ b) $\frac{6}{t}$ c) $\frac{7}{m^2}$ d) $\frac{4}{q^4}$ e) $\frac{10}{y^5}$ f) $\frac{1}{2x^3}$
   g) $\frac{1}{2^m}$ h) $\frac{3}{4t^4}$ i) $\frac{4}{5y^3}$ j) $\frac{7}{8x^5}$
5. a) $7x^{-3}$ b) $10p^{-1}$ c) $5t^{-2}$ d) $8m^{-5}$ e) $3y^{-1}$

6. a) i 25 ii $\frac{1}{125}$ iii $\frac{4}{5}$
 b) i 64 ii $\frac{1}{16}$ iii $\frac{5}{256}$
 c) i 8 ii $\frac{1}{32}$ iii $\frac{9}{2}$ or $4\frac{1}{2}$
 d) i 1 000 000 ii $\frac{1}{1000}$ iii $\frac{1}{4}$

7. a) a^{-7} b) a^2 c) a^4 d) a^{-5} e) a^{-6} f) a^6

15.4 Fractional indices

Exercise 15D

1. a) 5 b) 10 c) 8 d) 9 e) 25
 f) 3 g) 4 h) 10 i) 5 j) 8
 k) 12 l) 20 m) 5 n) 3 o) 10
 p) 3 q) 2 r) 2 s) 6 t) 6
 u) $\frac{1}{4}$ v) $\frac{1}{2}$ w) $\frac{1}{3}$ x) $\frac{1}{5}$ y) $\frac{1}{10}$

2. a) $\frac{5}{6}$ b) $1\frac{2}{3}$ c) $\frac{8}{9}$ d) $1\frac{4}{5}$ e) $\frac{5}{8}$
 f) $\frac{3}{5}$ g) $\frac{1}{4}$ h) $2\frac{1}{2}$ i) $\frac{4}{5}$ j) $1\frac{1}{7}$

3. $\left(x^{\frac{1}{n}}\right)^n = x^{\frac{1}{n}\times n} = x^1 = x$, but $\left(\sqrt[n]{x}\right)^n = \sqrt[n]{x} \times \sqrt[n]{x} \ldots n$ times $= x$, so $x^{\frac{1}{n}} = \sqrt[n]{x}$

4. $64^{-\frac{1}{2}} = \frac{1}{8}$, others are both $\frac{1}{2}$

5. Possible answer: The negative power gives the reciprocal, so $27^{-\frac{1}{3}} = \frac{1}{27}^{\frac{1}{3}}$. The power one-third means cube root, so you need the cube root of 27 which is 3, so $27^{\frac{1}{3}} = 3$ and $\frac{1}{27}^{\frac{1}{3}} = \frac{1}{3}$

6. Possible answer: $x = 1$ and $y = -1$, $x = 8$ and $y = \frac{1}{64}$.

7. a) 3 b) $\frac{1}{3}$ c) 0 d) $\frac{1}{2}$ e) $\frac{1}{2}$ f) $\frac{1}{4}$
 g) $\frac{1}{4}$ h) $\frac{1}{3}$ i) $\frac{1}{3}$ j) $\frac{1}{2}$ k) $\frac{1}{3}$ l) $\frac{1}{7}$

Exercise 15E

1. a) 16 b) 25 c) 216 d) 81

2. a) $t^{\frac{2}{3}}$ b) $m^{\frac{3}{4}}$ c) $k^{\frac{2}{5}}$ d) $x^{\frac{3}{2}}$

3. a) $\frac{1}{5}$ b) $\frac{1}{6}$ c) $\frac{1}{2}$ d) $\frac{1}{3}$
 e) $\frac{1}{4}$ f) $\frac{1}{2}$ g) $\frac{1}{2}$ h) $\frac{1}{3}$

4. a) $\frac{1}{125}$ b) $\frac{1}{216}$ c) $\frac{1}{8}$ d) $\frac{1}{27}$
 e) $\frac{1}{256}$ f) $\frac{1}{4}$ g) $\frac{1}{4}$ h $\frac{1}{9}$

5. a) $\frac{1}{100000}$ b) $\frac{1}{12}$ c) $\frac{1}{25}$ d) $\frac{1}{27}$
 e) $\frac{1}{32}$ f) $\frac{1}{32}$ g) $\frac{1}{81}$ h) $\frac{1}{13}$

6. $8^{-\frac{2}{3}} = \frac{1}{4}$, others are both $\frac{1}{8}$

7. a) $\frac{27}{8}$ b) $\frac{9}{25}$ c) $\frac{1024}{243}$ d) $\frac{8}{343}$
 e) $\frac{16}{9}$ f) $\frac{8}{27}$ g) $\frac{625}{256}$ h) $\frac{32}{243}$

8. a) $\frac{25}{9}$ b) $\frac{27}{64}$ c) $\frac{125}{729}$ d) $\frac{243}{32}$
 e) $\frac{16}{25}$ f) $\frac{512}{125}$ g) $\frac{243}{32}$ h) $\frac{32}{243}$

9. a) x^4 b) x^{-1} c) $4y^2$ d) $10x^2$ e) $20x^{-1}$ f) $\frac{1}{3}y$

10. a) x b) d^{-1} c) $t^{\frac{3}{2}}$ d) x^2 e) $y^{\frac{1}{2}}$ f) a^4

11. a) $x^{\frac{1}{2}}$ b) y^{-1} c) $a^{\frac{5}{3}}$ d) t^{-2} e) d^2 f) 1

12. a) x^3 b) x^{-1} c) x d) x e) $x^{\frac{3}{2}}$ f) x

15.5 Solving equations with indices

Exercise 15F

1. a) $x = 64$ b) $x = 25$ c) $x = 8$
 d) $x = 1000$ e) $x = \frac{1}{8}$ f) $x = \frac{4}{5}$

2. a) $x = 2\sqrt{2}$ b) $x = 2\sqrt{5}$ c) $x = 3\sqrt{3}$
 d) $x = \sqrt{2}$ e) $x = 2\sqrt{3}$ f) $x = \frac{1}{\sqrt{2}}$ or $x = \frac{\sqrt{2}}{2}$

3. a) $x = 81$ b) $x = 49$ c) $x = 3$
 d) $x = 18$ e) $x = 20$ f) $x = 125$

4. a) $x = 8$ b) $x = 5\sqrt{5}$ c) $x = 81$
 d) $x = 375\sqrt{3}$ e) $x = 4\sqrt{2}$ f) $x = 64$

5. a) $x = 27$ b) $x = 8$ c) $x = 16$
 d) $x = 1$ e) $x = 32$ f) $x = 100\,000$

6. a) $x = \frac{1}{4}$ b) $x = \frac{1}{3}$ c) $x = \frac{1}{\sqrt{2}}$ or $x = \frac{\sqrt{2}}{2}$
 d) $x = \frac{1}{5}$ e) $x = \frac{1}{3}$ f) $x = 2$

7. a) $x = \frac{1}{49}$ b) $x = \frac{1}{64}$ c) $x = \frac{1}{81}$
 d) $x = \frac{1}{32}$ e) $x = \frac{1}{6}$ f) $x = \frac{1}{2\sqrt{2}}$ or $x = \frac{\sqrt{2}}{4}$

8. a) $x = \frac{1}{4}$ b) $x = \frac{1}{27}$ c) $x = \frac{1}{81}$
 d) $x = \frac{1}{3125}$ e) $x = 9$ f) $x = 3\sqrt{3}$

Exam-style questions

1. a) For example, $3xy \times 3xy^4$
 b) For example, $24x^3y^5 \div 2x$

2. a) $\frac{1}{216}$ b) 4

3. x

4. $x^{\frac{1}{6}}$

5. a) $x = 512$ b) $x = \frac{1}{9}$

6. a) z^{-2} b) $z^{-\frac{1}{2}}$ c) $z = \frac{1}{5}$

CHAPTER 16 CALCULUS

16.1 The gradient of a curve

Exercise 16A

1. **a)** The missing numbers are 0, −1, 0

b)

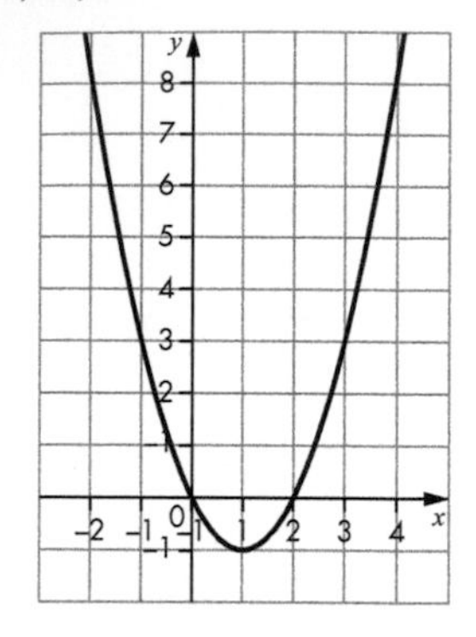

c) $2x - 2$ **d)** 4 **e)** 6 **f)** Student's choice

g) (1, −1) **h)** Student's check

2. **a)** $2x - 6$ **b)** −6 **c)** 4 **d)** (4, 7)

3. **a)** $4x$ **b)** 8 **c)** −4 **d)** (3, 8)

4. **a)** $4 - 2x$ **b)** 4 and −4 **c)** (1, 3) **d)** (1.5, 3.75)

5. **a)** $2x + 1$ **b)** $2x - 7$ **c)** $8x - 1$ **d)** $0.6x - 1.5$

e) $-2 + 2x$ **f)** $3 - 2x$ **g)** 2 **h)** 0

6. $2x + 2$

7. **a)** $4x + 2$ **b)** $2x + 7$ **c)** $2x$

8. **a)** (0, −5) **b)** 2

16.2 More complex curves

Exercise 16B

1. **a)** $6x^2$ **b)** 6 and 24

2. **a)** $3x^2 - 12x + 8$ **b)** If $x = 0$ or 2 or 4, $y = 0$

c) 8; −4; 8

3. **a)** $8x^3$ **b)** $15x^2 - 2$ **c)** $9x^2 + 5$

d) $-3x^2$ **e)** $4x^3 - 1$

4. 16 at (2, 0); −16 at (−2, 0); 0 at (0, 0)

5. $x^2 - 5 = 4$ has two solutions, $x = 3$ or −3. Points are (3, −2) and (−3, 10)

16.3 Stationary points and curve sketching

Exercise 16C

1. **a)** $2x - 4$ **b)** $2x - 4 = 0 \Rightarrow x = 2$; (2, −1)

c) Minimum **d)** $x \geqslant 2$

2. **a)** (−3, −12) **b)** Minimum

3. **a)** $5 - 2x$ **b)** (2.5, 7.25)

c) Maximum **d)** $x \leqslant 2.5$

4. **a)** $3x^2 - 6x$ **b)** $x = 0$ or 2

c) (0, 0) and (2, −4) **d)** $0 \leqslant x \leqslant 2$

5. **a)** If $x = -2$ or 5, $y = 0$ **b)** $2x - 3$

c) (1.5, −12.25); Minimum

d

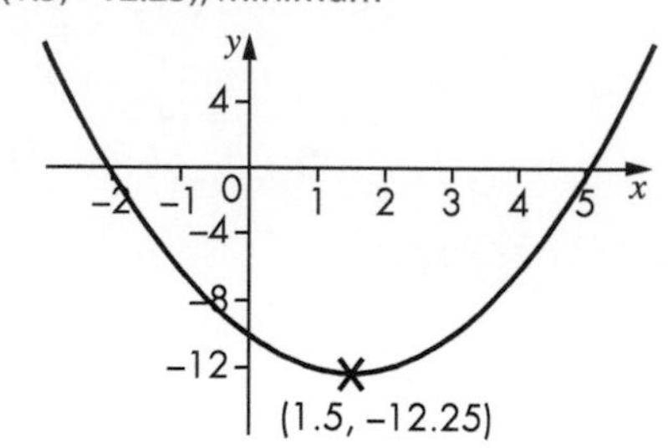

e) $x = 1.5$

6. **a)** $6x^2 - 6$ **b)** (1, 0) minimum, (−1, 8) maximum

c)

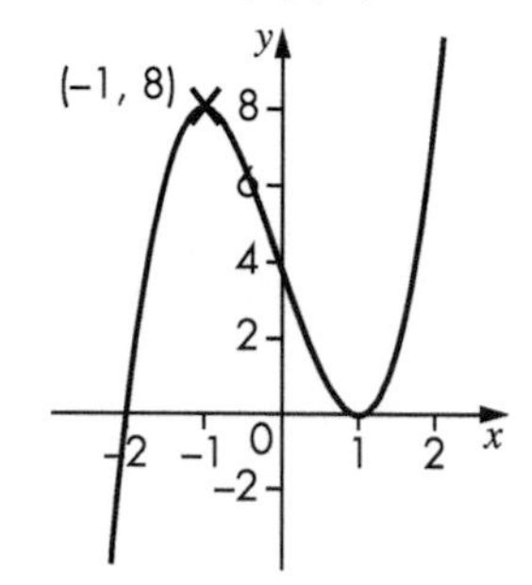

7. **a)** $\frac{dy}{dx} = 3x^2 + 12x + 12$

b) (−2, −8), a point of inflection

8. **a)** $\frac{dy}{dx} = 3x^2 - 30x + 75$, $\frac{dy}{dx} = 3(x - 5)^2$. The only stationary point is at $x = 5$, the point (5, 131).

b) When $x = 4.9$, $\frac{dy}{dx} = 0.03$; when $x = 5$, $\frac{dy}{dx} = 0$; when $x = 5.1$, $\frac{dy}{dx} = 0.03$, so it is a point of inflection.

16.4 The equation of a tangent and normal at any point on a curve

Exercise 16D

1. **a)** $\frac{dy}{dx} = 6x$ **b)** $y = 12x - 12$ **c)** $12y + x = 146$

2. **a)** $\frac{dy}{dx} = 2x - 4$ **b)** $y = -2x + 2$ **c)** $4y + x = 16$

3. **a)** $\frac{dy}{dx} = 3x^2 + 2x - 1$ **b)** $y = 4x + 2$

c) $x + 7y - 19 = 0$

4. $y = 5x - 15$ and $y = -5x - 10$

5. **a)** $y = -x + 5$ **b)** (0, 5)

6. **a)** $y = -4x + 9$ **b)** $2y + x = 7$

c) Solving the equations in **a** and **b** simultaneously for x gives $-4x + 9 = \frac{1}{2}(7 - x)$, from which $x = \frac{11}{7}$

Exam-style questions

1. **a)** $8x^3 - 12x$ **b)** $10x - 3x^2$ **c)** $2x - 7$

2. **a)** −9 **b)** 2 and −2

3. a) 10 **b)** (–1, 6) **c)** $\frac{1}{4}$

4. a) If $x = 2$ then $y = 8 - 8 + 5 = 5$ **b)** 8

c) (0, –11)

5. a) $\frac{dy}{dx} = 3x^2 - 6x$. If $x = 0$ then $y = 4$ and $\frac{dy}{dx} = 0$, so (0, 4) is a turning point.

b) (2, 0)

c)

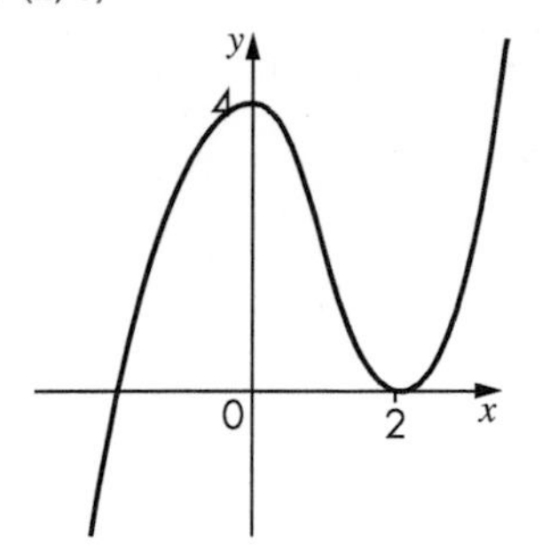

6. a) If $x = 2$ then $y = 16 - 16 + 5 = 5$ **b)** $y = 16x - 11$

c) $\frac{dy}{dx} = 0 \Rightarrow 4x^3 - 8x = 0 \Rightarrow x(x^2 - 2) = 0 \Rightarrow x = 0$ or $\pm\sqrt{2}$.

Three values give three turning points.

7. a) $y = x - 8$ **b)** (4, –4)

8. 10

CHAPTER 17 RATIOS OF ANGLES AND THEIR GRAPHS

17.1 Trigonometric ratios of angles between 90° and 360°

ACTIVITY 1

a)

x	$\sin x$	x	$\sin x$	x	$\sin x$	x	$\sin x$
0°	0	180°	0	180°	0	360°	0
15°	0.259	165°	0.259	195°	–0.259	345°	–0.259
30°	0.500	150°	0.500	210°	–0.500	330°	–0.500
45°	0.707	135°	0.707	225°	–0.707	315°	–0.707
60°	0.866	120°	0.866	240°	–0.866	300°	–0.866
75°	0.966	105°	0.966	255°	–0.966	285°	–0.966
90°	1	90°	1	270°	–1	270°	–1

b) They have the same value, or –(the same value)

c) The graph is shown in 17.1, ACTIVITY 1

d) For example, it has line symmetry about the line $x = 90°$, and rotational symmetry about (180°, 0)

Exercise 17A

1. a) 36.9°, 143.1° **b)** 53.1°, 126.9°

c) 48.6°, 131.4° **d)** 224.4°, 315.6°

e) 194.5°, 345.5° **f)** 198.7°, 341.3°

g) 190.1°, 349.9° **h)** 234.5°, 305.5°

i) 28.1°, 151.9° **j)** 185.6°, 354.4°

2. sin 234°, as the others all have the same numerical value

3. a) 438° or 78° + 360n° **b)** –282° or 78° –360n°

c) Line symmetry about ±90n° where n is an odd integer. Rotational symmetry about ±180n° where n is an integer

ACTIVITY 2

a)

x	$\cos x$	x	$\cos x$	x	$\cos x$	x	$\cos x$
0°	1	180°	–1	180°	–1	360°	1
15°	0.966	165°	–0.966	195°	–0.966	345°	0.966
30°	0.866	150°	–0.866	210°	–0.866	330°	0.866
45°	0.707	135°	–0.707	225°	–0.707	315°	0.707
60°	0.500	120°	–0.500	240°	–0.500	300°	0.500
75°	0.259	105°	–0.259	255°	–0.259	285°	0.259
90°	0	90°	0	270°	0	270°	0

b) They have the same value, or –(the same value)

c) The graph is shown in 17.1, ACTIVITY 2

d) For example, it has line symmetry about the $x = 180°$, and rotational symmetry about (90°, 0)

Exercise 17B

1. a) 53.1°, 306.9° b) 54.5°, 305.5°
   c) 62.7°, 297.3° d) 54.9°, 305.1°
   e) 79.3°, 280.7° f) 143.1°, 216.9°
   g) 104.5°, 255.5° h) 100.1°, 259.9°
   i) 111.2°, 248.8° j) 166.9°, 193.1°
2. cos 58°, as the others are negative
3. a) 492° or 132° + $360n$° b) –228° or 132° – $360n$°
   c) Line symmetry about $\pm 180n$° where n is an integer. Rotational symmetry about $\pm 90n$° where n is an odd integer

Exercise 17C

1. a) 0.707 b) –1 (–0.9998)
   c) –0.819 d) 0.731
2. a) –0.629 b) –0.875
   c) –0.087 d) 0.999
3. a) 21.2°, 158.8° b) 209.1°, 330.9° c) 50.1°, 309.9°
   d) 150.0°, 210.0° e) 60.9°, 119.1° f) 29.1°, 330.9°
4. 30°, 150°
5. –0.755
6. a) 1.41 b) –1.37 c) –0.0367
   d) –0.138 e) 1.41 f) –0.492
7. True
8. a) cos 65° b) cos 40°
9. a) 10°, 130° b) 12.7°, 59.3°
10. 38.2°, 141.8°

Exercise 17D

1. a) 14.5°, 194.5° b) 38.1°, 218.1°
   c) 50.0°, 230.0° d) 61.9°, 241.9°
   e) 68.6°, 248.6° f) 160.3°, 340.3°
   g) 147.6°, 327.6° h) 135.4°, 315.4°
   i) 120.9°, 300.9° j) 105.2°, 285.2°
2. tan 235°, as the others have a numerical value of 1
3. a) 425° or 65° + $180n$°, $n \geqslant 2$
   b) –115° or 65° – $180n$°
4. c) No line symmetry. Rotational symmetry about $\pm 180n$° where n is an integer

17.2 The circular function graphs

Exercise 17E

1. 115°
2. 327°
3. 324°
4. 195°
5. 216°
6. 331°
7. 210°, 330°
8. 135°, 225°
9. 120° and 300°
10. a) Say 32°, sin 32° = 0.53, cos 58° = 0.53
    b) Say 70°, sin 70° = 0.94, cos 20° = 0.94
    c) $\sin x = \cos (90 - x)°$
    d) $\cos x = \sin (90 - x)°$
11. a) 64° b) 206°, 334° c) 116°, 244°
12. 113°
13. 124.7°. The calculator gives the value of the acute angle but you are asked for the obtuse angle; the angle 124.7° has the same positive sign.
14. The calculator shows 1.1307 for sin A, which gives an error. If you tried to draw this triangle accurately then you would see that the line that is 12 units long does not intersect with the base.
15. a–f) All true g) False

17.3 Special right-angled triangles

Exercise 17F

1. a) 3 cm b) $4\sqrt{3}$ cm c) 1.5 cm
   d) $\frac{5}{\sqrt{3}}$ cm or $\frac{5\sqrt{3}}{3}$ cm e) $3\sqrt{2}$ cm f) 2 cm
2. a) $3\sqrt{2}$ cm b) $(5\sqrt{3} - 5)$ cm
   c) $\frac{3\sqrt{2}}{4}$ cm

17.4 Trigonometrical expressions and equations

Exercise 17G

1. a) 17.5° and 162.5° b) 78.5° and 281.5°
   c) 50.2° and 230.2° d) 36.9° and 143.1°
   e) 0° and 360° f) 35.0° and 215.0°
2. a) 197.5° and 322.5° b) 101.5° and 258.5°
   c) 129.8° and 309.8° d) 216.9° and 343.1°
   e) 180° f) 145.0° and 325.0°
3. a) 194.5° and 345.5° b) 113.6° and 246.4°
   c) 45° and 225° d) 45° and 135°
   e) 70.5° and 250.5° f) 0°, 180° and 360°
4. a) 63.4° b) 76.0°
   c) 161.6° d) 158.2°
   e) 135° f) 104.0°
5. a) $\sin^3 x + \sin x \cos^2 x \equiv \sin x (\sin^2 x + \cos^2 x) = \sin x$
   b) $\cos x \tan x \equiv \cos x \frac{\sin x}{\cos x} = \sin x$
   c) $\frac{1 - \cos^2 x}{\cos^2 x} = \frac{\sin^2 x}{\cos^2 x} = \tan^2 x$

d) $\frac{(\sin^2 x+\cos^2 x)\sin x}{\tan x} = \frac{\sin x\cos x}{\sin x} = \cos x$

e) $\frac{1}{1+\tan^2 x} \equiv \frac{1}{1+\frac{\sin^2 x}{\cos^2 x}} = \frac{\cos^2 x}{\cos^2 x+\sin^2 x} = \cos^2 x$

f) $\sin^4 x - \cos^4 x \equiv (\sin^2 x - \cos^2 x)(\sin^2 x + \cos^2 x)$
$= \sin^2 x - \cos^2 x = \sin^2 x - (1 - \sin^2 x)$
$= \sin^2 x - 1 + \sin^2 x = 2\sin^2 x - 1$

6. a) 0°, 180°, 360° b) 45°, 135°, 225°, 315°
c) 0°, 60°, 180°, 240°, 360° d) 120°, 180°, 240°
e) 63.4°, 135°, 243.4°, 315° f) 30°, 150°

Exam-style questions

1. a)

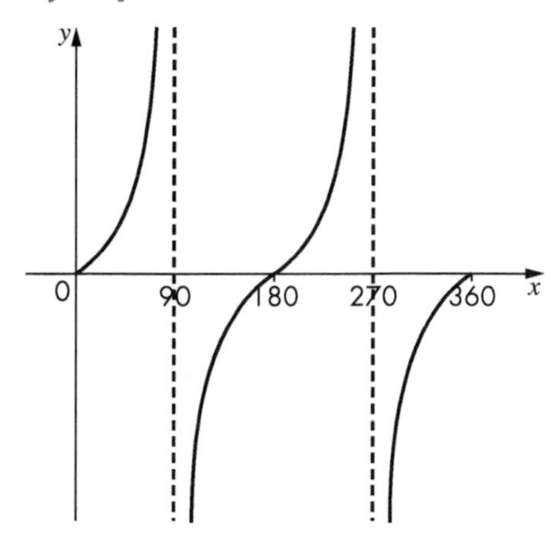

b) 256°

2. a) 153° b) 207° and 233°

3. a) −0.17 b) −0.94 c) 0.17

4. $6 + 2\sqrt{3}$ cm

5. The base is $\frac{1}{2}a$ so the area $= \frac{1}{2} \times \frac{1}{2}a \times a \sin 60°$
$= \frac{1}{4}a^2 \times \frac{\sqrt{3}}{2} = \frac{\sqrt{3}a^2}{8}$ cm²

6. a) $\frac{4}{5}$ b) $\frac{3}{4}$ c) $-\frac{3}{5}$

7. $(1 + \cos\theta)(1 - \cos\theta) = 1 - \cos^2\theta = \sin^2\theta$ and $\sin\theta$
$\cos\theta\tan\theta = \sin\theta\cos\theta \times \frac{\sin\theta}{\cos\theta} = \sin^2\theta$, which shows that the expressions are identical.

8. $x = 18.4°$

9. $x = 30°, 150°, 210°$ or $330°$

CHAPTER 18 PROOF

18.1 Algebraic proof

Exercise 18A

1. a) Odd, yes
b) $(2n + 1) + 2m = 2(n + m) + 1$, which is odd.

2. a) For example, $2m + 2n = 2(m + n)$, which is even
b) For example, $(2m)(2n) = 2(2mn)$, which is even
c) For example, $(2m + 1)(2n) = 2(2mn + n)$, which is even
d) For example, $(2m + 1)(2n + 1) = 2(2mn + m + n) + 1$, which is odd
e) For example, $n + (n + 1) + (n + 2) + (n + 3) = 2(2n + 3)$, which is even
f) For example, from **e**, $\frac{1}{2} \times 2(2n + 3) = 2(n + 1) + 1$, which is odd

3. a) 3, 5, 8, 13, 21, 34, 55
b) $3a + 5b, 5a + 8b, 8a + 13b, 13a + 21b, 21a + 34b$
c) $(8a + 13b) - (2a + 3b) = 6a + 10b = 2(3a + 5b)$

4. a) Substitute $n = 11$ and $n = 12$ in the formula and add them to get 144 which is a square number.
b) Substitute $(n+1)$ for n in the formula:
$\frac{1}{2}(n + 1)((n + 1) + 1) = \frac{1}{2}(n + 1)(n + 2)$
c) $\frac{1}{2}n(n + 1) + \frac{1}{2}(n + 1)(n + 2)$ simplifies to $(n + 1)^2$, a square number.

5. a) $\frac{1}{2}(6)(6 + 1) = 21 = 1 + 2 + 3 + 4 + 5 + 6$
b) The sum of each respective term of the two series is $(n + 1)$,
so the sum is $n(n + 1)$
c) From **b**, twice the sum of the first n integers is $n(n + 1)$.
Hence the sum is $\frac{1}{2}n(n + 1)$

6. a) $8T + 1 = 8 \times \frac{1}{2}n(n + 1) + 1$ which simplifies to $(2n + 1)^2$, a square number.
b) $9T + 1 = 9 \times \frac{1}{2}n(n + 1) + 1 = \frac{1}{2}(9n^2 + 9n + 2)$
$= \frac{1}{2}(3n + 1)(3n + 2)$
$= \frac{1}{2}(3n + 1)((3n + 1) + 1)$, which is the $(3n+1)$th triangular number

7. a) For example, let the numbers be $(a + b)$ and $(a - b)$
Then $\frac{1}{2}[(a + b)^2 + (a - b)^2]$ expands and simplifies to $a^2 + b^2$, the sum of two squares.
b) For example, let the numbers be $(a + b)$ and $(a - b)$
Then $2[(a + b)^2 + (a - b)^2]$ expands and simplifies to $(2a)^2 + (2b)^2$, the sum of two squares.
c) This can be shown to be true.

8. Set $(2n)^2 + (n^2 - 1)^2 = (n^2 + 1)^2$, then by expanding and simplifying show that the LHS = RHS.

9. Yes, this has been shown to be true.

10. $(2x + 1)^2 - (x - 1)^2 = 4x^2 + 4x + 1 - (x^2 - 2x + 1)$
$= 4x^2 + 4x + 1 - x^2 + 2x - 1 = 3x^2 + 6x = 3x(x + 2)$
$= (2x + 1 + x - 1)(2x + 1 - (x - 1))$

18.2 Geometric proof

Exercise 18B

1. $\angle DFE = 180° - (90° + \frac{x°}{2}) = 90° - \frac{x°}{2}$

   $\angle DEF = 180° - x° - (90° - \frac{x°}{2}) = 90° - \frac{x°}{2}$

   $\angle DFE = \angle DEF$ hence triangle DEF is isosceles.

2. The exterior angle of a triangle is equal to the sum of the opposite 2 interior angles.

   $x° = \frac{x°}{2} + \frac{x°}{2}$, hence the triangle is isosceles

3. $\angle AOC = 2x°$, hence $\angle ABC = x°$, reflex angle AOC $= 2y°$, hence $\angle ADC = y°$

   But $2x° + 2y° = 360°$ (angles around a point) hence $2(x° + y°) = 360°$ giving $x° + y° = 180°$

4. $\angle CED + \angle AEC = 180°$ (angles on a straight line)

   $\angle ABC + \angle AEC = 180°$ (cyclic quadrilateral)

   But $\angle ABC = \angle ACB$ (isosceles triangle)

   Hence $\angle ACB = \angle CED$

5. PS = QR, RS = PQ, both triangles share side QS hence by SSS triangles are congruent

6. **a)**

   For example, the diagram shows a circle, centre O, with tangent PT at T.

   Angle PTO is 90° (angle between tangent and radius)

   Therefore angle $x = 90° - y$

   Angle TBO = angle BTO = y (isosceles triangle BOT)

   Therefore angle BOT is $180° - 2y$ (angles in a triangle), and angle TAP = $90° - y$ (angle at circumference is half the angle at the centre subtended by arc PT)

   Hence angle PTB = angle TAB, or the angle between a tangent and a chord is equal to the angle in the opposite segment.

   b) By the alternate segment theorem $\angle TXA = \angle TYB$ hence AX is parallel to BY

7. **a)** $\overrightarrow{YW} = \overrightarrow{YZ} + \overrightarrow{ZW} = 2a + b + a + 2b = 3a + 3b = 3(a + b) = 3\overrightarrow{XY}$

   b) 3 : 1

   c) They lie on a straight line.

   d) Points are A(6, 2), B(1, 1) and C(2, –4). Using Pythagoras' theorem, $AB^2 = 26$, $BC^2 = 26$ and $AC^2 = 52$ so $AB^2 + BC^2 = AC^2$ hence $\angle ABC$ must be a right angle

Exam-style questions

1. Let the digits of the number be a and b. If you look at all the two-digit multiples of 9 (except 99), the sum of the digits is 9. So $a + b = 9$.

   The value of the number will be $10a + b$ and when reversed the value will be $10b + a$.

   Adding the two numbers gives $11a + 11b = 11(a + b)$.

   But $(a + b) = 9$, so the sum is always 99.

2. **a)** Substitute for a and b to show LHS = RHS

   b) Expand each of the LHS and RHS to show they are algebraically the same.

   c) $(a + b)^2 + (a - b)^2 = a^2 + 2ab + b^2 + a^2 - 2ab + b^2$

   $= 2a^2 + b^2$

   $= 2(a^2 + b^2)$

3. $(a + b)^2 - (a - b)^2 = a^2 + 2ab + b^2 - (a^2 - 2ab + b^2) = 4ab$

4. **a)** $7^2 - 5^2 = 24$, and $2(7 + 5) = 24$

   b) $a^2 - b^2 = (a - b)(a + b) = 2(a + b)$, because $(a - b) = 2$

   c) $a^2 - b^2 = (a - b)(a + b) = n(a + b)$, because $(a - b) = n$

5. **a)** $4 \times 5 \times 6 \times 7 + 1 = 841 = 29^2$

   b) Expand and simplify the LHS to the expression on the RHS.

   c) $n(n - 2)(n - 1)(n + 1) + 1 = n(n^2 - 1)(n - 2) + 1$

   $= n^4 - 2n^3 - n^2 + 2n + 1 = (n^2 - n - 1)2$ (from **b**).

6. Let the integers be n and $(n + 1)$. Then $n^2 + (n + 1)^2$ expands and simplifies to $2(n^2 + n) + 1$, which is odd.

7. **a)** $10^2 = 100$, $1 + 8 + 27 + 64 = 100$

   b) $(\frac{1}{2}n(n + 1))^2 = \frac{1}{4}n^2(n + 1)^2$

   c) $1 + 8 + 27 + 64 + 125 + 216 = 441$

   $\frac{1}{4} \times 36 \times 49 = 441$

8. $\angle QAT = \angle QTA$ (isosceles triangle)

   $\angle PTB = \angle QTA$ (vertically opposite angles)

   $\angle PTB = \angle PBT$ (isosceles triangle)

   Hence $\angle PBT = \angle QAT$ and PB is parallel to AQ